Handbook of Discrete-Valued Time Series

Chapman & Hall/CRC
Handbooks of Modern Statistical Methods

Series Editor

Garrett Fitzmaurice
Department of Biostatistics
Harvard School of Public Health
Boston, MA, U.S.A.

Aims and Scope

The objective of the series is to provide high-quality volumes covering the state-of-the-art in the theory and applications of statistical methodology. The books in the series are thoroughly edited and present comprehensive, coherent, and unified summaries of specific methodological topics from statistics. The chapters are written by the leading researchers in the field, and present a good balance of theory and application through a synthesis of the key methodological developments and examples and case studies using real data.

The scope of the series is wide, covering topics of statistical methodology that are well developed and find application in a range of scientific disciplines. The volumes are primarily of interest to researchers and graduate students from statistics and biostatistics, but also appeal to scientists from fields where the methodology is applied to real problems, including medical research, epidemiology and public health, engineering, biological science, environmental science, and the social sciences.

Published Titles

Handbook of Mixed Membership Models and Their Applications
Edited by Edoardo M. Airoldi, David M. Blei,
Elena A. Erosheva, and Stephen E. Fienberg

Handbook of Markov Chain Monte Carlo
Edited by Steve Brooks, Andrew Gelman,
Galin L. Jones, and Xiao-Li Meng

Handbook of Discrete-Valued Time Series
Edited by Richard A. Davis, Scott H. Holan,
Robert Lund, and Nalini Ravishanker

Handbook of Design and Analysis of Experiments
Edited by Angela Dean, Max Morris,
John Stufken, and Derek Bingham

Longitudinal Data Analysis
Edited by Garrett Fitzmaurice, Marie Davidian,
Geert Verbeke, and Geert Molenberghs

Handbook of Spatial Statistics
Edited by Alan E. Gelfand, Peter J. Diggle,
Montserrat Fuentes, and Peter Guttorp

Handbook of Cluster Analysis
Edited by Christian Hennig, Marina Meila,
Fionn Murtagh, and Roberto Rocci

Handbook of Survival Analysis
Edited by John P. Klein, Hans C. van Houwelingen,
Joseph G. Ibrahim, and Thomas H. Scheike

Handbook of Missing Data Methodology
Edited by Geert Molenberghs, Garrett Fitzmaurice,
Michael G. Kenward, Anastasios Tsiatis, and Geert Verbeke

Chapman & Hall/CRC
Handbooks of Modern
Statistical Methods

Handbook of Discrete-Valued Time Series

Edited by

Richard A. Davis
Columbia University, USA

Scott H. Holan
University of Missouri, USA

Robert Lund
Clemson University, USA

Nalini Ravishanker
University of Connecticut, USA

CRC Press
Taylor & Francis Group
Boca Raton London New York

CRC Press is an imprint of the
Taylor & Francis Group, an **informa** business
A CHAPMAN & HALL BOOK

CRC Press
Taylor & Francis Group
6000 Broken Sound Parkway NW, Suite 300
Boca Raton, FL 33487-2742

First issued in paperback 2020

Version Date: 20150817

ISBN 13: 978-0-367-57039-2 (pbk)
ISBN 13: 978-1-4665-7773-2 (hbk)

Visit the Taylor & Francis Web site at
http://www.taylorandfrancis.com

and the CRC Press Web site at
http://www.crcpress.com

Contents

Preface

Statisticians continually face new modeling challenges as data collection becomes more widespread and the type of data more varied. Modeling techniques are constantly expanding to accommodate a wider range of data structures. Traditionally, time series modeling has been applied to data that are continuously valued. Unfortunately, many continuous models do not adequately describe discrete data. The most common violation lies with the case of counts. Examples include cases when the time series might contain daily counts of individuals infected with a rare disease, the number of trades of a stock in a minute, the yearly number of hurricanes that make landfall in Florida, etc. For these examples, the counts tend to be small and may include zeros—discreteness of the observations cannot be adequately modeled with a continuous distribution. During the past 30+ years, much progress has been made for modeling a wide range of count time series. Having reached a reasonable level of maturity, this seems an ideal time to assemble a handbook on state-of-the-art methods for modeling time series of counts, including frequentist and Bayesian approaches and methodology for discrete-valued spatio-temporal data and multivariate data. While the focus of this handbook is on time series of counts, some of the techniques discussed can be extended to other types of discrete-valued time series, such as binary-valued or categorical time series.

The main thrust of classical time series methodology considers modeling stationary time series. This assumes that some form of *preprocessing* of the original time series, which might include the application of transformations and various filters, has been applied to reduce the series to a stationary one. Linear time series models, the mainstay of stationary time series modeling, are motivated by the celebrated Wold's decomposition. This decomposition says that every nondeterministic stationary time series admits a linear representation. The autoregressive moving-average (ARMA) models are a special family of parsimonious short memory models for linear stationary time series. Box and Jenkins' groundbreaking 1970 text provided a paradigm for identifying, fitting, and forecasting ARMA models that were accessible to practitioners.

For stationary Gaussian processes, which are completely determined by their covariance functions, ARMA models provide a useful description. For non-Gaussian series, ARMA models can only be assured to capture an approximation of the covariance structure of the process. Some dependence structures are invisible to the covariance function and hence are suboptimally described in the ARMA paradigm. Nonlinear models are available in this case and have received considerable attention since the 1980s.

Discrete-valued time series, and time series of counts in particular, present modeling challenges not encountered for continuous-valued responses. First, representing a count time series in an additive Wold-type form may not make sense. Second, the covariance function, which is a measure of linear dependence, is not always a useful measure of dependence for a time series of counts. As a result, one is led to consider nonlinear models. Perhaps the most common and flexible modeling formulation for count data is based on a state-space formulation. In this setup, the counts are assumed to be independent and, say, Poisson distributed, given a latent series $\lambda_t > 0$ at time t. Serial dependence and the effect of possible covariates are incorporated into λ_t. Borrowing from the generalized linear models (GLM) literature, the log-link function is used so that $\log(\lambda_t)$ is modeled as a linear regression model with correlated Gaussian errors. Zeger (1988) was an early proponent of such models and applied them to a time series of monthly polio incidences in the United States.

In the late 1970s, Jacobs and Lewis proposed mixing tactics now referred to as discrete ARMA (DARMA) methods. A DARMA process can have any prescribed marginal distribution while possessing the covariance structure of an ARMA series. Unfortunately, this class of models generated series that remained constant in time for long runs and were subsequently dismissed as unrealistic. In the 1980s, integer ARMA (INARMA) methods were introduced by McKenzie. These methods used probabilistic thinning techniques to keep the support set of the series integer valued and are still in use today. Covariance structures produced by DARMA and INARMA models are not as general as ARMA; for example, they are always nonnegative.

From a Bayesian perspective, discrete-valued time series have experienced tremendous growth since the emergence of the dynamic generalized linear model in West et al. in the mid-1980s. This work was followed by Fahrmeir et al. in the late 1980s and 1990s. In the late 1990s, Waller et al., Wikle, and others proposed hierarchical spatio-temporal models for discrete-valued data. In both time and space–time, this has become an area of significant research as data collection and sampling methods, including Markov chain Monte Carlo and approximation algorithms, improve with technological and computational advances.

This handbook addresses a plethora of diverse topics on modeling discrete-valued time series, and in particular time series of counts. The reader will find both frequentist and Bayesian methods that consider issues of model development, residual diagnostics, applications in business, changepoint analyses, binary series, etc. Theoretical, methodological and practical issues are pursued.

This handbook is arranged as follows: Section I contains eight chapters that essentially narrate history and some of the current methods for modeling and analysis of univariate count series. Section II is a short interlude into diagnostics and applications, containing only three chapters. Section III moves to binary and categorical time series. Section IV is our guide to modern methods for discrete-valued spatio-temporal data. Section V concludes with several chapters on multivariate and long-memory count series. Several topics are notably absent, for example, chapters devoted to copula and empirical likelihood methods. We hope that future editions of the handbook will remedy this shortcoming.

The list of contributors to this handbook is distinguished, and all chapters were solicited from topical experts. Considerable effort was undertaken to make the chapters flow together rather than be disjoint works. Taken as a whole, themes emerge. Many of the models used in this handbook can be categorized as either *parameter driven* or *observation driven*, a nomenclature that is originally due to Cox (1981). Essentially, a parameter-driven model is one that has a state-space formulation, where the *state variable* is viewed as the parameter. An observation-driven model expresses the current value of the time series as a function of past observations and an independent noise term. It turns out that likelihood methods are difficult to implement for the state-space models considered in this book since the evaluation of the likelihood requires the computation of a large-dimensional integral. Hence, for such models, one often turns to approximate likelihood methods, estimating equations, simulation methods, or a fully Bayesian implementation for model fitting. On the other hand, for observation-driven models, it is often straightforward to use likelihood methods and the theory has been well-developed in a large number of cases. There are several chapters devoted to these topics in this handbook, which describe the advantages and limitations of the various modeling approaches.

The technical level of the individual chapters is modest and should be accessible to masters level students with an elementary class in statistical time series. We have made an effort to make the chapters accessible and nontechnical, keeping probabilistic technicalities to a minimum.

Preparation of this handbook has been easeful toil on the editors. The timeliness of chapter authors is especially appreciated, as is the cohesiveness of the editorial team. We are also indebted to Rob Calver of Chapman & Hall for unwavering support during the preparation of this handbook. The comments and critiques of numerous unnamed reviewers supported our efforts.

References

Cox, D. R. (1981). Statistical analysis of time series: Some recent developments. *Scandinavian Journal of Statistics*, 8:93–115.

Zeger, S. L. (1988). A regression model for time series of counts. *Biometrika*, 75(4):621–629.

Editors

Richard A. Davis is chair and Howard Levene Professor of Statistics at Columbia University, New York. He is currently president (2015–16) of the Institute of Mathematical Statistics. He received his PhD in mathematics from the University of California at San Diego in 1979 and has held academic positions at MIT, Colorado State University, and visiting appointments at numerous other universities. He is coauthor (with Peter Brockwell) of the bestselling books, *Time Series: Theory and Methods* and *Introduction to Time Series and Forecasting*. Together with Torben Andersen, Jens-Peter Kreiss, and Thomas Mikosch, he coedited the *Handbook in Financial Time Series*. In 1998, he won (with collaborator W.T.M. Dunsmuir) the Koopmans Prize for Econometric Theory. He has served on the editorial boards of major journals in probability and statistics and most recently was editor in chief of the *Bernoulli Journal* (2010–2012). His research interests include time series, applied probability, extreme value theory, and spatial-temporal modeling.

Scott H. Holan is a professor in the Department of Statistics at the University of Missouri, Columbia, Missouri. He received his PhD in statistics from Texas A&M University in 2004. His research is primarily focused on time-series analysis, spatio-temporal methodology, Bayesian methods, and hierarchical models and is largely motivated by problems in federal statistics, econometrics, ecology, and environmental science. He is a fellow of the American Statistical Association and an elected member of the International Statistics Institute.

Robert Lund received his PhD in statistics from the University of North Carolina in 1993. He is currently professor in the Department of Mathematical Sciences at Clemson University, after an 11-year stint in the Statistics Department at the University of Georgia. He is a 2007 Fellow of the American Statistical Association and was the 2005–2007 chief editor of the *Journal of the American Statistical Association*, Reviews section. He has published over 80 refereed papers and has graduated 17 doctoral students. His interests are in time series, applied probability, and statistical climatology.

Nalini Ravishanker is a professor in the Department of Statistics at the University of Connecticut, Storrs. Her primary research interests include time series and times-to-events analysis, Bayesian dynamic modeling, and signal processing. Her primary interdisciplinary research areas are ecology, finance, marketing, and transportation engineering. She has an undergraduate degree in statistics from Presidency College, Chennai, India, and a PhD in statistics and operations research from the Stern School of Business, New York University. She is coauthor of the textbook *A First Course in Linear Model Theory*. She is a fellow of the American Statistical Association and elected member of the International Statistical Institute, the theory and methods editor of *Applied Stochastic Models in Business and Industry*, and an associate editor for the *Journal of Forecasting*.

Contributors

Carlos A. Abanto-Valle
Institute of Mathematics
Federal University of Rio de Janeiro
Rio de Janeiro, Brazil

Tevfik Aktekin
College of Business and Economics
University of New Hampshire
Durham, New Hampshire

Ana Corberán-Vallet
Department of Statistics and Operations Research
University of Valencia
Burjassot, Spain

Richard A. Davis
Department of Statistics
Columbia University
New York, New York

William T.M. Dunsmuir
Department of Statistics
School of Mathematics and Statistics
University of New South Wales
Sydney, New South Wales, Australia

Konstantinos Fokianos
Department of Mathematics & Statistics
University of Cyprus
Nicosia, Cyprus

Dani Gamerman
Institute of Mathematics
Federal University of Rio de Janeiro
Rio de Janeiro, Brazil

Scott H. Holan
Department of Statistics
University of Missouri
Columbia, Missouri

Mevin B. Hooten
U.S. Geological Survey
Colorado Cooperative Fish and Wildlife Research Unit
Department of Fish, Wildlife, and Conservation Biology
Department of Statistics
Colorado State University
Fort Collins, Colorado

Shan Hu
Predictive Modeler
Plymouth Rock Assurance
Boston, Massachusetts

Harry Joe
Department of Statistics
University of British Columbia
Vancouver, British Columbia, Canada

Robert C. Jung
Institute of Economics
University of Hohenheim
Stuttgart, Germany

Dimitris Karlis
Department of Statistics
Athens University of Economics and Business
Athens, Greece

Benjamin Kedem
Department of Mathematics
University of Maryland
College Park, Maryland

Bumsoo Kim
Sogang Business School
Sogang University
Seoul, South Korea

Claudia Kirch
Department of Mathematics
Institute for Mathematical Stochastics
Otto-von-Guericke University Magdeburg
Magdeburg, Germany

Andrew B. Lawson
Department of Public Health Sciences
Medical University of South Carolina
Charleston, South Carolina

James Livsey
Center for Statistical Research
and Methodology
U.S. Census Bureau
Washington, DC

Robert Lund
Department of Mathematical Sciences
Clemson University
Clemson, South Carolina

Iain L. MacDonald
Actuarial Science
University of Cape Town
Rondebosch, Cape Town, South Africa

Thiago G. Martins
Yahoo!
Trondheim, Norway

Brendan P.M. McCabe
Management School
University of Liverpool
Liverpool, United Kingdom

Nalini Ravishanker
Department of Statistics
University of Connecticut
Storrs, Connecticut

Ralph S. Silva
Institute of Mathematics
Federal University of Rio de Janeiro
Rio de Janeiro, Brazil

Refik Soyer
Department of Decision Sciences
The George Washington University
Washington, DC

David Stoffer
Department of Statistics
University of Pittsburgh
Pittsburgh, Pennsylvania

Joseph Tadjuidje Kamgaing
Department of Mathematics
University of Kaiserslautern
Kaiserslautern, Germany

Aerambamoorthy Thavaneswaran
Department of Statistics
University of Manitoba
Winnipeg, Manitoba, Canada

Dag Tjøstheim
Department of Mathematics
University of Bergen
Bergen, Norway

A.R. Tremayne
Management School
University of Liverpool
Liverpool, United Kingdom

Rajkumar Venkatesan
Darden School of Business
University of Virginia
Charlottesville, Virginia

Christopher K. Wikle
Department of Statistics
University of Missouri
Columbia, Missouri

Yanbing Zheng
Department of Statistics
University of Kentucky
Lexington, Kentucky

Jun Zhu
Department of Statistics
and
Department of Entomology
University of Wisconsin–Madison
Madison, Wisconsin

Walter Zucchini
Zentrum für Statistik
Georg-August-Universität Göttingen
Göttingen, Germany

Section I

Methods for Univariate Count Processes

1

Statistical Analysis of Count Time Series Models: A GLM Perspective

Konstantinos Fokianos

CONTENTS

1.1 Introduction

In this chapter we discuss statistical models and methods that have been developed over the past few years for the analysis of count time series which occur in diverse application areas, like finance, biostatistics, environmentrics, and others. The analysis of count time series data has attracted considerable attention, see Kedem and Fokianos (2002, Secs. 4 & 5) for several references and Fokianos (2012) for a more recent review of this research area. In what follows, I will present the generalized linear methodology (GLM) advanced by Nelder and Wedderburn (1972) and McCullagh and Nelder (1989). This framework naturally generalizes the traditional ARMA methodology and includes several complicated data generating processes besides count data such as binary and categorical data. In addition, the fitting of such models can be carried out by likelihood methods; therefore testing, diagnostics, and all type of likelihood arguments are available to the data analyst. The Bayesian point of view towards the analysis of such models will be discussed later in this handbook; see chapters by Gamerman et al. (2015; Chapter 8 in this volume) and Soyer et al. (2015; Chapter 11 in this volume).

However, a theoretical study of GLM-based models poses challenging problems some of which are still open for further research; see the review article by Tjøstheim (2012) and Section 1.4. The main problem posed by count time series in theoretical analysis is that the observed response is discrete valued and therefore it might not be strong mixing (see Andrews, 1984). The theoretical analysis of count time series has been based so far on a perturbation argument (see Fokianos et al. 2009), weak dependence (Doukhan et al. 2012, 2013), and Markov chain theory without irreducibility assumptions (see Woodard et al. 2011, Douc et al. 2013). On the other hand, Neumann (2011) has proved a number of mixing results by employing coupling techniques. Since most of the theoretical results will be presented in other chapters of this handbook, we will confine our attention to data analysis tools which are useful for estimation, testing goodness of fit, and prediction.

In Section 1.2, we present several statistical models that mimic the usual AR models for count time series analysis. In Section 1.3, we discuss quasi-maximum likelihood estimation (QMLE) of the unknown parameters. In Sections 1.4 and 1.5, we develop goodness-of-fit methodology and prediction for count time series, respectively. Finally, we conclude the chapter with a discussion of possible research topics. The present contribution can be thought of as a companion paper to Fokianos (2012), where detailed properties for GLM-based count time series models were discussed.

As a final introductory remark, we note that there are other alternative classes of regression models for count time series; the most prominent being the integer autoregressive models. These models are based on the notion of thinning. Accordingly, integer autoregressive models imitate the structure of the usual autoregressive processes, in the sense that thinning is applied instead of scalar multiplication. These models will be discussed in other chapters of this handbook, and therefore, they will not be included in this presentation.

1.2 Modeling

To motivate our discussion, consider the transactions data used by Fokianos et al. (2009). This data set consists of the number of transactions per minute for the stock Ericsson B over a single day. To economize on space, we do not show plots of the data or their autocorrelation since these can be easily obtained from the reference cited earlier. However, we report that the mean of the data is 9.909 and the variance is 32.837; this is a clear case of overdispersion where the mean is less than the variance. The minimum value of this series is 0 and the maximum is 37. The sample autocorrelation function is positive; at lag 1 (respectively, lag 2) it is equal to 0.405 (respectively, 0.340) and decays slowly toward zero; for related plots see Fokianos et al. (2009). These data are counts and therefore a reasonable model for their analysis can be based on a counting process. This is the point of view taken by Rydberg and Shephard (2000) who argue that the number of transactions influence the price of an asset. Hence they study models of the form (1.5)—and more generally of the form (1.7)—to analyze transactions within a small time interval. These models are natural analogues of real-valued time series models and their development can be based on their similarities with other econometric models; see Bollerslev (1986) for the GARCH family of models. We will be more specific in what follows.

Let $\{Y_t, t \geq 1\}$ denote a count time series, and let $\{\lambda_t, t \geq 1\}$ be a latent mean process. Let $\mathcal{F}_t^{Y,\lambda} = \sigma(Y_s, s \leq t, \lambda_0)$, where λ_0 denotes a starting value, denote the past of the process up to and including time t. We will assume that the observed count time series conditionally on its past is Poisson distributed with mean $Z_t\lambda_t$, where $\{Z_t\}$ denotes an i.i.d. sequence of positive random variables with mean 1, which is independent of Y_s, for all $s < t$. That is

$$Y_t \mid \mathcal{F}_t^{Y,\lambda} \sim \text{Po}(Z_t\lambda_t). \tag{1.1}$$

In the context of transactions data, (1.1) implies that the total number of transactions at time t is a realization of a Poisson random variable with a time varying stochastic intensity. The random variables $\{Z_t\}$ correspond to random fluctuations (different trading strategies, delays, and other). Model (1.1) is called a mixed Poisson model (Mikosch, 2009) and its usefulness is not confined to the analysis of econometric data, but can also be employed for modeling medical, environmental, and other data. In fact, (1.1) can be viewed as a stochastic intensity model, similar to the stochastic volatility model.

To study (1.1) in detail consider the following representation for the process $\{Y_t, t \geq 1\}$,

$$Y_t = N_t(0, Z_t\lambda_t], \quad t \geq 1. \tag{1.2}$$

In (1.2), $\{N_t, t \geq 1\}$ is an i.i.d. sequence from a standard homogeneous Poisson process (that is, a Poisson process with rate equal to 1) and $\{Z_t\}$ denotes an i.i.d. sequence of positive random variables with mean 1, which is independent of N_t, for all t. In addition, we assume that $\{Z_t\}$ is independent of Y_s, for all $s < t$. This class of models is large and includes two important distributions routinely employed for the analysis of count time series, viz. the Poisson distribution

$$P[Y_t = y \mid \mathcal{F}_{t-1}^{Y,\lambda}] = \frac{\exp(-\lambda_t)\lambda_t^y}{y!}, \quad y = 0, 1, 2, \ldots \tag{1.3}$$

and the negative binomial distribution

$$P[Y_t = y \mid \mathcal{F}_{t-1}^{Y,\lambda}] = \frac{\Gamma(\nu + y)}{\Gamma(y+1)\Gamma(\nu)} \left(\frac{\nu}{\nu + \lambda_t}\right)^\nu \left(\frac{\lambda_t}{\nu + \lambda_t}\right)^y, \quad y = 0, 1, 2, \ldots, \tag{1.4}$$

where the dispersion parameter ν is positive. It is obvious that (1.3) is a special case of (1.2) when $\{Z_t\}$ is a sequence of degenerate random variables with mean 1, and (1.4) is a special case of (1.2) when $\{Z_t\}$ are i.i.d. Gamma variables with mean 1 and variance $1/\nu$. Regardless of the choice of $Z's$ in (1.2), the conditional mean of Y_t is always equal to λ_t. Furthermore, the conditional variance of Y_t is given by $\lambda_t + \sigma_Z^2\lambda_t^2$, with $\sigma_Z^2 = \text{Var}(Z_t)$. The conditional variance of the Poisson distribution is equal to λ_t, whereas the conditional variance of the negative binomial variable is equal to $\lambda_t + \lambda_t^2/\nu$. Hence, although a Poisson-based conditional model is capable of accommodating overdispersion (i.e., the mean of the data is less than the variance), we anticipate that the negative binomial distribution will be a more suitable model for accommodating this fact.

In what follows, we review some standard models that have been suggested over the past few decades for modeling count time series data. Since most of them have been already discussed in the recent survey by Fokianos (2012) and Tjøstheim (2012), we only give their definitions and some motivation. The interested reader is referred to the earlier two articles for further details.

1.2.1 The Linear Model

The linear model for the analysis of count time series is specified by

$$\lambda_t = d + a_1\lambda_{t-1} + b_1 Y_{t-1}, \quad t \geq 1, \tag{1.5}$$

where the parameters d, a_1, b_1 are assumed to be nonnegative and to satisfy $0 < a_1 + b_1 < 1$. If $a_1 = b_1 = 0$, then we obtain a model for i.i.d data with mean d. When $a_1 = 0$, (1.5) reduces to an ordinary AR(1) model; in this case the mean of the process is influenced by its past. In the context of transactions data, (1.5), with $a_1 = 0$, implies that the mean number of transactions at time t depends on the number of transactions at time $t - 1$. This is a sensible way to analyze the data and generalizes naturally the framework of AR models. When $a_1 > 0$, (1.5) can be viewed as a parsimonious way to analyze count time series whose sample autocorrelation function decays slowly towards zero; this is the case of the transactions data. In fact, when both a_1 and b_1 are positive, (1.5) has the same second-order properties as those of an ARMA(1,1) model. For more details about (1.5), see Rydberg and Shephard (2000), Streett (2000), Heinen (2003), Ferland et al. (2006), and more recently Fokianos et al. (2009) for the case of Poisson distributed data. The case of negative binomial distributed data has been studied by Zhu (2011), Davis and Liu (2015), and Christou and Fokianos (2014).

The mean parametrization based on (1.4) yields a stationary region which is independent of the additional dispersion parameter ν. This implies that fitting count time series models, like (1.5), to data is implemented by constrained optimization of the log-likelihood function with respect to the regression parameters (a_1, b_1 in the case of (1.5)). The parameter ν (equivalently σ_Z^2 in the general framework) is estimated separately. We emphasize this point further by comparing the proposed parametrization (1.4) with that of Zhu (2011) given by

$$P[Y_t = y \mid \mathcal{F}_{t-1}^{Y,\lambda}] = \binom{y - r - 1}{r - 1} p_t^r (1 - p_t)^y, \quad y = 0, 1, 2 \ldots. \tag{1.6}$$

In (1.6), r is a positive integer and p_t denotes the conditional probability of success. Zhu (2011) models the odds ratio, denoted by $\lambda_t = p_t/(1 - p_t)$, as a function of past values of itself and past values of the response. Stationarity conditions for the model studied by Zhu (2011) depend upon the parameter r in (1.6). Therefore, it is challenging to maximize the log-likelihood function since such an optimization problem imposes restrictions on the regression parameters and r. Furthermore, it is well known that the estimation of the odds ratio might be problematic when probabilities are either very small or very large. For related work on negative binomial–based models, see also Davis and Liu (2015).

Ferland et al. (2006) enlarge the class of models (1.5) to the general model of order (p, q):

$$Y_t \mid \mathcal{F}_{t-1}^{Y,\lambda} \sim \text{Poisson}(\lambda_t), \quad \lambda_t = d + \sum_{i=1}^{p} a_i \lambda_{t-i} + \sum_{j=1}^{q} b_j Y_{t-j}, \quad t \geq \max(p, q), \tag{1.7}$$

and show that it is second-order stationary provided that $0 < \sum_{i=1}^{p} a_i + \sum_{j=1}^{q} b_j < 1$ under the Poisson assumption. These conditions are true for the mixed Poisson process (1.2) provided that $E(Z_t) = 1$ (see Christou and Fokianos 2014).

The properties of (1.5) have been studied in detail by Fokianos (2012). Here, we only mention that by repeated substitution in (1.5), the mean process λ_t is given by

$$\lambda_t = d\frac{1 - a_1^t}{1 - a_1} + a_1^t \lambda_0 + b_1 \sum_{i=0}^{t-1} a_1^i Y_{t-i-1}.$$

In other words, the hidden process $\{\lambda_t\}$ is determined by past functions of lagged responses and the initial value λ_0. Therefore (1.5) belongs to the class of observation-driven models in the sense of Cox (1981).

1.2.2 Log-Linear Models for Count Time Series

As empirical experience has shown, (1.5) and its variants can successfully accommodate dependent count data, especially when there exists positive autocorrelation and there are no covariates. However, in applications, we observe time series data that might be negatively correlated, and possibly with covariates. In this case, the logarithmic function is the most popular link function for modeling count data. In fact, this choice corresponds to the canonical link of generalized linear models. Log-linear models for dependent count data have been considered by Zeger and Qaqish (1988), Li (1994), MacDonald and Zucchini (1997), Brumback et al. (2000), Kedem and Fokianos (2002), Benjamin et al. (2003), Davis et al. (2003), Fokianos and Kedem (2004), Jung et al. (2006), Creal et al. (2008), Fokianos and Tjøstheim (2011), and Douc et al. (2013), among others.

Recall that $\{Y_t\}$ denotes a count time series and following the notation introduced in (1.2), let $\nu_t \equiv \log \lambda_t$. A log-linear model with feedback for the analysis of count time series (Fokianos and Tjøstheim 2011) is defined as

$$\nu_t = d + a_1 \nu_{t-1} + b_1 \log(Y_{t-1} + 1), \quad t \geq 1. \tag{1.8}$$

In general, the parameters d, a_1, b_1 can be positive or negative but they need to satisfy certain conditions so that we obtain a stationary time series. Note that the lagged observations of the response Y_t are fed into the autoregressive equation for ν_t via the term $\log(Y_{t-1}+1)$. This is a one-to-one transformation of Y_{t-1} which avoids taking logarithm of zero data values. Moreover, both λ_t and Y_t are transformed into the same scale. Covariates can be easily accommodated by model (1.8). When $a_1 = 0$, we obtain an AR(1) type model in terms of $\log(Y_{t-1} + 1)$. In addition, the log-intensity process of (1.8) can be rewritten as

$$\nu_t = d\frac{1 - a_1^t}{1 - a_1} + a_1^t \nu_0 + b_1 \sum_{i=0}^{t-1} a_1^i \log(1 + Y_{t-i-1}),$$

after repeated substitution. Hence, we obtain again that the hidden process $\{\nu_t\}$ is determined by past functions of lagged responses. Equivalently, the log-linear model (1.8) belongs to the class of observation-driven models and possesses similar properties to the linear model (1.5). For more details, see Fokianos (2012).

1.2.3 Nonlinear Models for Count Time Series

A large class of models for the analysis of count time series is given by the following nonlinear mean specification

$$\lambda_t = f(\lambda_{t-1}, Y_{t-1}), \quad t \geq 1, \tag{1.9}$$

where $f(\cdot)$ is known up to an unknown finite dimensional parameter vector such that $f : (0,\infty) \times \mathbb{N} \rightarrow (0,\infty)$. The function $f(\cdot)$ is assumed to satisfy the contraction condition

$$|f(\lambda, y) - f(\lambda^{'}, y^{'})| \leq \alpha_1 |\lambda - \lambda^{'}| + \gamma_1 |y - y^{'}|, \tag{1.10}$$

for (λ, y) and $(\lambda^{'}, y^{'})$ in $(0,\infty) \times \mathbb{N}$, where $\alpha_1 + \gamma_1 < 1$; see Fokianos et al. (2009), Neumann (2011), Doukhan et al. (2012, 2013), and Fokianos and Tjøstheim (2012).

An interesting example of a nonlinear regression model for count time series is given by

$$f(\lambda, y) = d + (a_1 + c_1 \exp(-\gamma \lambda^2))\lambda + b_1 y, \tag{1.11}$$

where d, a_1, c_1, b_1, γ are positive parameters, which is similar to the exponential autoregressive model (Haggan and Ozaki 1981). In Fokianos et al. (2009), (1.11) was studied for the case $d = 0$. The parameter γ introduces a perturbation of the linear model (1.5), in the sense that when γ tends either to 0 or infinity, then (1.11) approaches two distinct linear models. It turns out that the regression coefficients of model (1.11) must satisfy the condition $a_1 + b_1 + c_1 < 1$ to guarantee ergodicity and stationarity of the joint process (Y_t, λ_t); see Doukhan et al. (2012) for more precise conditions. Model (1.11) shows that there is a smooth transition between two linear models for count time series in terms of the unobserved process. This transition might be difficult to estimate because of the nonlinear parameter γ and the fact that λ_t is not directly observed. An alternative method for introducing the smooth transition is by employing the observed data. In other words, instead of (1.11), we can consider

$$f(\lambda, y) = d + a_1 \lambda + \left(b_1 + c_1 \exp\left(-\gamma y^2\right)\right) y.$$

The previous model is interpreted in a similar manner to (1.11) and must satisfy the condition $a_1 + b_1 + c_1 < 1$ to obtain ergodicity and stationarity of the joint process (Y_t, λ_t).

Another nonlinear model studied by Fokianos and Tjøstheim (2012) is given by

$$f(\lambda, y) = \frac{d}{(1 + \lambda_{t-1})^{\gamma}} + a_1 \lambda_{t-1} + b_1 Y_{t-1}, \tag{1.12}$$

where all regression parameters are assumed positive. Here, the inclusion of γ introduces a nonlinear perturbation, in the sense that small values of γ cause (1.12) to approach (1.5).

Moderate values of γ introduce a stronger perturbation. Models of the form (1.9) have also been studied in the context of the negative binomial distribution by Christou and Fokianos (2014). The condition $\max\{a_1, d\gamma - a_1\} + b_1 < 1$ guarantees ergodicity and stationarity of the joint process (Y_t, λ_t). Following the arguments made earlier in connection with model (1.11), we can alternatively consider the following modification of (1.12):

$$f(\lambda, y) = \frac{d}{(1 + Y_{t-1})^{\gamma}} + a_1 \lambda_{t-1} + b_1 Y_{t-1},$$

with the required stationarity condition $\max\{b_1, d\gamma - b_1\} + a_1 < 1$.

An obvious generalization of model (1.9) is given by the following specification of the mean process (see Franke 2010 and Liu 2012):

$$\lambda_t = f(\lambda_{t-1}, \ldots, \lambda_{t-p}, Y_{t-1}, \ldots, Y_{t-q}), \tag{1.13}$$

where $f(.)$ is a function such that $f : (0, \infty)^p \times \mathbb{N}^q \to (0, \infty)$. It should be clear that models (1.11) and (1.12) can be extended according to (1.13). Such examples are provided by the class of smooth transition autoregressive models of which the exponential autoregressive model is a special case (cf. Teräsvirta 1994, Teräsvirta et al. 2010). Further examples of nonlinear time series models can be found in Tong (1990) and Fan and Yao (2003). These models have not been considered earlier in the literature in the context of generalized linear models for count time series, and they provide a flexible framework for studying dependent count data. For instance, nonlinear models can be quite useful when testing departures from linearity; this topic is partially addressed in Section 1.6.1. A more general approach would have been to estimate the function f of (1.13) by employing nonparametric methods. However, such an approach is missing from the literature.

1.3 Inference

Maximum likelihood inference for the Poisson model (1.3) and the negative binomial model (1.4) has been developed by Fokianos et al. (2009), Fokianos and Tjøstheim (2012), and Christou and Fokianos (2014). They develop estimation procedures based on the Poisson likelihood function, which for the Poisson model (1.3) is obviously the true likelihood. However, for the negative binomial model (1.4), and more generally for mixed Poisson models, this method resembles the QMLE method for GARCH models which employs the Gaussian likelihood function irrespective of the assumed error distribution. The QMLE method, in the context of GARCH models, has been studied in detail by Berkes et al. (2003), Francq and Zakoïan (2004), Mikosch and Straumann (2006), Bardet and Wintenberger (2010), and Meitz and Saikkonen (2011), among others. This approach yields consistent estimators of regression parameters under a correct mean specification and it bypasses complicated likelihood functions (Godambe and Heyde 1987, Zeger and Qaqish 1988, Heyde 1997).

In the case of mixed Poisson models (1.2), it is impossible, in general, to have a readily available likelihood function since the distribution of Z's is generally unknown. Hence, we resort to QMLE methodology and, for defining properly the QMLE, we consider the

Poisson log-likelihood function, as in Fokianos et al. (2009), where θ denotes the unknown parameter vector,

$$l(Y;\theta) = \sum_{t=1}^{n} l_t(\theta) = \sum_{t=1}^{n} \left(Y_t \log \lambda_t(\theta) - \lambda_t(\theta)\right). \tag{1.14}$$

The quasi-score function is defined by

$$S_n(\theta) = \frac{\partial l(Y;\theta)}{\partial \theta} = \sum_{t=1}^{n} \frac{\partial l_t(\theta)}{\partial \theta} = \sum_{t=1}^{n} \left(\frac{Y_t}{\lambda_t(\theta)} - 1 \right) \frac{\partial \lambda_t(\theta)}{\partial \theta}, \tag{1.15}$$

where the vector $\partial \lambda_t(\theta)/\partial \theta$ is defined recursively depending upon the mean specification employed. The solution of the system of nonlinear equations $S_n(\theta) = 0$, if it exists, yields the QMLE of θ which we denote by $\hat{\theta}$. The conditional information matrix is defined by

$$G_n(\theta) = \sum_{t=1}^{n} \mathrm{Var}\left[\frac{\partial l_t(\theta)}{\partial \theta} \middle| \mathcal{F}_{t-1}^{Y,\lambda} \right] = \sum_{t=1}^{n} \left(\frac{1}{\lambda_t(\theta)} + \sigma_Z^2 \right) \left(\frac{\partial \lambda_t(\theta)}{\partial \theta} \right) \left(\frac{\partial \lambda_t(\theta)}{\partial \theta} \right)'.$$

The asymptotic properties of $\hat{\theta}$ have been studied in detail by Fokianos et al. (2009) and Fokianos and Tjøstheim (2012) for the case of the Poisson model (1.3). For the case of the negative binomial distribution, see Christou and Fokianos (2014). In both cases, consistency and asymptotic normality of $\hat{\theta}$ is achieved. In fact, it can be shown that the QMLE is asymptotically normally distributed, i.e.,

$$\sqrt{n}(\hat{\theta} - \theta_0) \xrightarrow{D} \mathcal{N}(0, G^{-1}(\theta_0) G_1(\theta_0) G^{-1}(\theta_0)),$$

where the matrices G and G_1 are given by the following:

$$G(\theta) = E\left(\frac{1}{\lambda_t(\theta)} \left(\frac{\partial \lambda_t}{\partial \theta} \right) \left(\frac{\partial \lambda_t}{\partial \theta} \right)' \right), \quad G_1(\theta) = E\left(\left(\frac{1}{\lambda_t(\theta)} + \sigma_Z^2 \right) \left(\frac{\partial \lambda_t}{\partial \theta} \right) \left(\frac{\partial \lambda_t}{\partial \theta} \right)' \right). \tag{1.16}$$

For the case of Poisson distribution (1.3), we obtain that

$$\sqrt{n}(\hat{\theta} - \theta_0) \xrightarrow{D} \mathcal{N}(0, G^{-1}(\theta_0)),$$

since $\sigma_Z^2 = 0$. If $\sigma_Z^2 > 0$, then we need to estimate this parameter for maximum likelihood inference. We propose to estimate σ_Z^2 by solving the equation

$$\sum_{t=1}^{n} \frac{(Y_t - \hat{\lambda}_t)^2}{\hat{\lambda}_t (1 + \hat{\lambda}_t \sigma_Z^2)} = n - m, \tag{1.17}$$

where m denotes the dimension of θ and $\hat{\lambda}_t = \lambda_t(\hat{\theta})$; see Lawless 1987. In particular, we recognize that for the negative binomial case, $\sigma_Z^2 = 1/\nu$. Therefore, the previous estimation method is standard in applications; see Cameron and Trivedi (1998, Ch. 3) among others.

Although the earlier mentioned formulas are given for the linear model (1.5), they can be modified suitably for the log-linear model (1.8) and the nonlinear model (1.9).

1.3.1 Transactions Data

Recall the transactions data discussed in Section 1.2. To fit model (1.5) to those data we proceed as follows: we set $\lambda_0 = 0$ and $\partial\lambda_0/\partial\theta = 0$ for initialization of the recursions; recall (1.15). Starting values for the parameter vector θ are obtained after observing that (1.5) can be expressed as an ARMA(1,1) model of the form

$$\left(Y_t - \frac{d}{1-(a_1+b_1)}\right) = (a_1+b_1)\left(Y_{t-1} - \frac{d}{1-(a_1+b_1)}\right) + \epsilon_t - a_1\epsilon_{t-1} \tag{1.18}$$

where $\epsilon_t = Y_t - \lambda_t$ is a white noise process.

The first three lines of Table 1.1 report estimation results after fitting model (1.5) to these data. The first line reports estimates of the regression parameters (previously reported in Fokianos 2012). Regardless of the assumed distribution, the estimators are identical because we maximize the log-likelihood function (1.14). If the Poisson distribution is assumed to be the true data generating process, then these estimators are MLE; otherwise they are QMLE. If the true data generating process is assumed to be the negative binomial distribution, then we can also estimate the dispersion parameter ν by means of (1.17) and the fact that $\sigma_Z^2 = 1/\nu$. The next two rows report the standard errors of the estimators. For the regression parameters, these are calculated either by using the Poisson assumption and the matrix G from (1.16) (second row of Table 1.1) or by using the sandwich matrix obtained by G and G_1 from (1.16) (third row of Table 1.1). We note that the standard errors obtained from the sandwich matrix are somewhat larger than those obtained from G. These are robust standard errors in the sense that we are using a working likelihood—namely the Poisson likelihood—to carry out inference (see White 1982). The standard error of $\hat{\nu}$ has been computed by parametric bootstrap. In other words, given $\hat{\theta}$ and $\hat{\nu}$, generate a large number

TABLE 1.1

QMLE and their standards errors (in parentheses) for the Linear Model (1.5), the Nonlinear Model (1.12) with $\gamma = 0.5$ and the Loglinear Model (1.8), for the total number of transactions per minute for the Stock Ericsson B for the time period between July 2 and July 22, 2002

	Maximum Likelihood Estimates			Estimates of ν
	$\hat{d}$	$\hat{a}_1$	$\hat{b}_1$	$\hat{\nu}$
Linear model (1.5)	0.581	0.745	0.199	7.158
	(0.148)	(0.030)	(0.022)	(0.907)
	(0.236)	(0.047)	(0.035)	
Nonlinear model (1.12)	1.327	0.774	0.186	7.229
	(0.324)	(0.026)	(0.021)	(1.028)
	(0.506)	(0.041)	(0.034)	
Log-linear model (1.8)	0.105	0.746	0.207	6.860
	(0.032)	(0.028)	(0.022)	(1.067)
	(0.125)	(0.081)	(0.070)	

Note: The total number of observations is 460.

of count time series models by means of (1.5) and using (1.4). For each of the simulated count time series, carry out QML estimation and get an estimator of ν by using (1.17). The standard error of these replicates is reported in Table 1.1, underneath $\hat{\nu}$.

We now fit model (1.12) to these data with $\gamma = 0.5$. Although we have assumed for simplicity that γ is known, it can be estimated along with the regression parameters as outlined in Fokianos and Tjøstheim (2012). In principle, additional nonlinear parameters can be estimated by QMLE, but the sample size should be sufficiently large. For fitting this model, we set again $\lambda_0 = 0$ and $\partial\lambda_0/\partial\theta = 0$. Starting values for the parameter vector θ are obtained by initially fitting a linear model (1.5). Table 1.1 shows the results of this exercise. The estimators of a_1 and b_1 from model (1.12) are close to those obtained from fitting model (1.5). The same observation holds for $\hat{\nu}$ and the standard errors of these coefficients which are computed in an analogous manner to that of the linear model. It is worth pointing out that the sum of $\hat{a}_1$ and $\hat{b}_1$ is close to unity. This fact provides some evidence of nonstationarity when we fit these types of models to transactions data. These observations are repeated when the log-linear model (1.8) is fitted to those data; see the last three lines of Table 1.1. Some empirical experience with these models has shown that when there is positive correlation among the data, then all models will produce similar output. In general, the log-linear model will be more useful when some covariate information is available.

1.4 Diagnostics

A detailed discussion concerning diagnostic methods for count time series models has been given by Kedem and Fokianos (2002, Sec. 1.6); also see the chapter by Jung et al. (2015; Chapter 9 in this volume). Various quantities, like Pearson and deviance residuals, for example, have been suggested and it was shown that they can be easily calculated using standard software. In what follows we focus our attention on the so-called Pearson residuals and on a new test statistic proposed recently by Fokianos and Neumann (2013) for testing the goodness of fit of the model under a conditional Poisson model. However, properties of different types of residuals (raw and deviance residuals for instance) have not been investigated in the literature. A slightly different definition of the residuals than the one which we will be using has been given recently by Zhu and Wang (2010). They also study the large sample behavior of the autocorrelation function of the residuals that they propose, but for a model of the form (1.5) without feedback.

1.4.1 Residuals

To examine the adequacy of the fit, we consider the so-called Pearson residuals. Set

$$e_t \equiv \frac{Y_t - E\left(Y_t \mid \mathcal{F}_{t-1}^{Y,\lambda}\right)}{\sqrt{\mathrm{Var}\left(Y_t \mid \mathcal{F}_{t-1}^{Y,\lambda}\right)}} = \frac{Y_t - \lambda_t}{\sqrt{\lambda_t + \lambda_t^2 \sigma_Z^2}}, \quad t \geq 1, \tag{1.19}$$

where the first equality is the general definition of the Pearson residuals and the second equality follows from (1.2). Under the true model, the sequence $\{e_t, t \geq 1\}$ is a white noise sequence with $\mathrm{Var}(e_t) = 1$. It is straightforward to see that under the Poisson assumption, (1.19) becomes

$$e_t = \frac{Y_t - \lambda_t}{\sqrt{\lambda_t}}, \quad t \geq 1,$$

while for the case (1.4), we obtain

$$e_t = \frac{Y_t - \lambda_t}{\sqrt{\lambda_t + \lambda_t^2/\nu}}, \quad t \geq 1.$$

To compute the Pearson residuals in either case, substitute λ_t by $\hat{\lambda}_t \equiv \lambda_t(\hat{\theta})$ and σ_Z^2 by $\hat{\sigma}_Z^2$. Construction of their autocorrelation function and cumulative periodogram plots (see Brockwell and Davis 1991, Sec. 10.2) give some clue about the whiteness of the sequence $\{e_t, t \geq 1\}$.

Figure 1.1 shows the plots of the autocorrelation function and the cumulative periodogram of the Pearson residuals obtained after fitting (1.5) to the transactions data. The upper panel corresponds to the case of Poisson distribution and the lower panel is constructed using the negative binomial assumption. We observe that both models fit the data quite adequately. We have also computed the Pearson residuals for models (1.8) and (1.11). The corresponding plots are not shown because the results are quite analogous to the case of the simple linear model.

1.4.2 Goodness-of-Fit Test

A goodness-of-fit test for model (1.5), and more generally of (1.9), was recently proposed by Fokianos and Neumann (2013), by considering two forms of hypotheses. The first of these refers to the simple hypothesis

$$H_0^{(s)}: \ f = f_0 \quad \text{against} \quad H_1^{(s)}: \ f \neq f_0,$$

for some completely specified function f_0 which satisfies (1.10). However, in applications, the most interesting testing problem is given by the following composite hypotheses

$$H_0: \ f \in \{f_\theta : \theta \in \Theta\} \quad \text{against} \quad H_1: \ f \notin \{f_\theta : \theta \in \Theta\}, \tag{1.20}$$

where $\Theta \subseteq \mathbb{R}^m$ and the function f_θ is known up to a parameter θ and again satisfies (1.10).

The methodology for testing (1.20) is quite general and can be applied to all models considered so far. Recall that $\hat{\theta}$ denotes the QMLE and $\hat{\lambda}_t = \lambda_t(\hat{\theta})$. If $\hat{e}_t$ are the Pearson residuals (1.19), the statistic for testing (1.20) is given by

$$\hat{T}_n = \sup_{x \in \Pi} |\hat{G}_n(x)|, \quad \hat{G}_n(x) = \frac{1}{\sqrt{n}} \sum_{t=1}^{n} \hat{e}_t \, w(x - \hat{I}_{t-1}), \tag{1.21}$$

where $x \in \Pi := [0, \infty)^2$, $\hat{I}_t = (\hat{\lambda}_t, Y_t)'$, and $w(\cdot)$ is some suitably defined weight function. In the applications, we can consider the weight function to be of the form $w(x) = w(x_1, x_2) = K(x_1)K(x_2)$ where $K(\cdot)$ is a univariate kernel and $x = (x_1, x_2) \in \Pi$. We can employ the

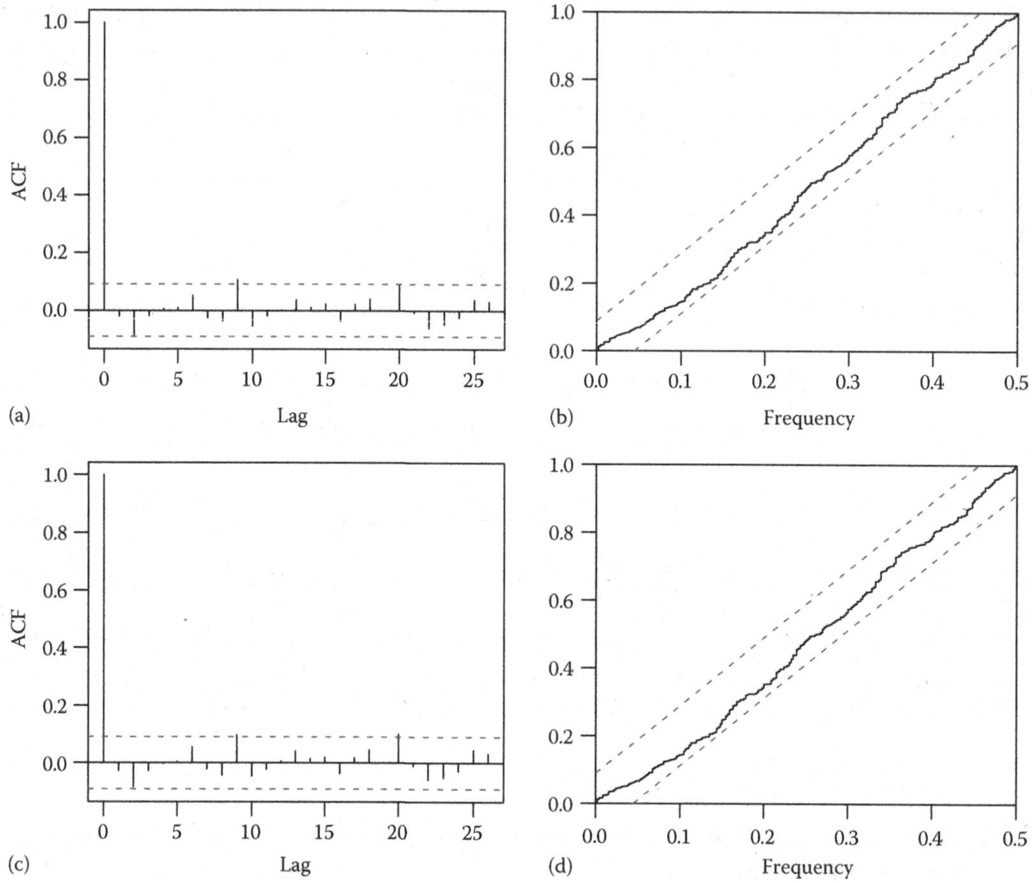

FIGURE 1.1
Diagnostic plots for the Pearson residuals (1.19) after fitting model (1.5) to the transactions data. (a) Autocorrelation function using the Poisson assumption. (b) Cumulative periodogram plot using the Poisson assumption. (c) Autocorrelation function using the negative binomial assumption. (d) Cumulative periodogram plot using the negative binomial assumption.

uniform, Gaussian and the Epanechnikov kernels. For instance, when the uniform kernel is employed, compute the test statistic (1.21) by using the weights

$$w(x - I_{t-1}) = K(x_1 - \lambda_{t-1})K(x_2 - Y_{t-1}) = \frac{1}{4}\mathbb{1}(|x_1 - \lambda_{t-1}| \le 1)\mathbb{1}(|x_2 - Y_{t-1}| \le 1),$$

where $\mathbb{1}(A)$ is the indicator function of a set A. Then, the test statistic (1.21) becomes

$$\hat{T}_n = \sup_{x \in \Pi} |\hat{G}_n(x)|,$$

where

$$\hat{G}_n(x_1, x_2) = \frac{1}{4\sqrt{n}} \sum_{t=1}^{n} \hat{e}_t \mathbb{1}(|x_1 - \hat{\lambda}_{t-1}| \le 1)\mathbb{1}(|x_2 - Y_{t-1}| \le 1).$$

Obvious formulas hold for other kernel functions. It turns out that (1.21) yields a consistent procedure when testing against Pitman's local alternatives. It converges weakly with the usual parametric rate under some regularity conditions on the kernel function; see Fokianos and Neumann (2013) for more details.

An alternative test statistic for testing goodness of fit for count time series can be based on supremum-type tests of the following form (Koul and Stute 1999):

$$\hat{H}_n = \sup_{x \in \Pi} |H_n(x)|, \quad H_n(x) = n^{-1/2} \sum_{t=1}^{n} \hat{e}_t \mathbb{1}(\widehat{I}_{t-1} \leq x), \tag{1.22}$$

using the same notations as before. Although the asymptotic behavior of supremum-type test statistics based on (1.22) has not been studied in the literature, it is possible to develop a theory following the arguments of Koul and Stute (1999) and utilizing the recent results on weak dependence properties obtained by Doukhan et al. (2012), at least for some classes of models.

Regardless of the chosen statistic and the distributional assumption, we can calculate critical values by using parametric bootstrap; see Fokianos and Neumann (2013) for details under the Poisson assumption. More specifically, to compute the p-value of the test statistic, (1.21) or (1.22) is recalculated for B parametric bootstrap replications of the data set. Then, if $\hat{T}_n$ denotes the observed value of the test statistic and $\widehat{T}^{\star}_{i;n}$ denotes the value of the test statistic in the ith bootstrap run, the corresponding p-value used to determine acceptance/rejection is given by

$$p\text{-value} = \frac{\#\left\{i : \widehat{T}^{\star}_{i;n} \geq \widehat{T}_n\right\}}{B+1}.$$

A similar result holds for $\widehat{H}_n$.

Test statistics (1.21) (with the uniform and Epanechnikov kernels) and (1.22) were computed for the transactions data for testing the goodness of fit of the linear model (1.5). Under the Poisson assumption, the observed values of these test statistics were calculated to be 0.212, 0.234, and 1.390, respectively. Under the negative binomial assumption, the observed values were equal to 0.146, 0.164, and 0.818, respectively. Table 1.2 shows the bootstrap p-value of the test statistics which have been obtained by parametric bootstrap as explained earlier. We note that the test statistics formed by (1.21) yield identical conclusions; that is, the linear model can be used for fitting the transactions data regardless of the assumed distribution. However, the test statistic (1.22) raises some doubt about the linearity, under the Poisson assumption.

TABLE 1.2

p-values for the transactions data when testing for the Linear Model (1.5)

	Test statistic		
Distribution	**(1.22)**	**(1.21) with Uniform Kernel**	**(1.21) with Epanechnikov Kernel**
Poisson	0.024	0.350	0.279
Negative binomial	0.659	0.611	0.585

Note: Results are based on $B = 999$ bootstrap replications.

1.5 Prediction

Following Gneiting et al. (2007), we take the point of view that predictions should be probabilistic in nature. In addition, they should strive to maximize the sharpness of the predictive distribution subject to calibration. Calibration refers to the statistical consistency between the predictive distribution and the observations. The notion of sharpness refers to the concentration of the predictive distribution and is a property of the forecasts only. It follows that if the predictive distribution is more concentrated, then the forecasts are sharper. In this section, we provide diagnostic tools to evaluate the predictive performance. Note that calculation of all these measures requires an assumption on the conditional distribution of the process; hence general processes of the form (1.2) cannot be fitted without simulating from the mixing variables Z_t. Predictive performance based on the following diagnostic tools has been examined recently by Jung and Tremayne (2011) and Christou and Fokianos (2015).

1.5.1 Assessment of Probabilistic Calibration

To ascertain whether or not the negative binomial distribution is a better choice than the Poisson distribution, we use the diagnostic tool of the Probability Integral Transformation (PIT) histogram, as explained below. This tool is used for checking the statistical consistency between the predictive distribution and the distribution of the observations. If the observation is drawn from the predictive distribution, then the PIT has a standard uniform distribution. In the case of count data, the predictive distribution is discrete and therefore the PIT is no longer uniform. To remedy this, several authors have suggested a randomized PIT. However, Czado et al. (2009) recently proposed a nonrandomized uniform version of the PIT. We explain their approach in the context of count time series models. Note that the approach is quite general and can accommodate various data generating processes.

In our context, we fit any model discussed earlier to the data by using the quasi-likelihood function (1.14). After obtaining consistent estimators for the regression parameters, we estimate the mean process λ_t by $\hat{\lambda}_t = \lambda_t(\hat{\theta})$ and the parameter ν by $\hat{\nu}$. Then, the PIT is based on the conditional cumulative distribution

$$F(u|Y_t = y) = \begin{cases} 0 & u \le P_{y-1}, \\ (u - P_{y-1})/(P_y - P_{y-1}) & P_{y-1} \le u \le P_y, \\ 1 & u \ge P_y, \end{cases}$$

where P_y is equal to the conditional c.d.f. either of the Poisson distribution (1.3) evaluated at $\hat{\lambda}_t$, or of the negative binomial p.m.f. (1.4) evaluated at $\hat{\lambda}_t$ and $\hat{\nu}$. Subsequently, we form the mean PIT by

$$\bar{F}(u) = \frac{1}{n}\sum_{t=1}^{n} F^{(t)}(u|y_t), \quad 0 \le u \le 1.$$

The mean PIT is compared to the c.d.f. of the standard uniform distribution. The comparison is performed by plotting a nonrandomized PIT histogram, which can be used as a diagnostic tool. After selecting the number of bins, say J, we compute

$$f_j = \bar{F}\left\{\frac{j}{J}\right\} - \bar{F}\left\{\frac{j-1}{J}\right\}$$

for equally spaced bins $j = 1, \ldots, J$. Then we plot the histogram with height f_j for bin j and check for uniformity. Deviations from uniformity hint at reasons for forecasting failures and model deficiencies. U-shaped histograms point at underdispersed predictive distributions, while hump or inverse–U shaped histograms indicate overdispersion.

Figure 1.2 shows nonrandomized PIT histograms with 10 equally spaced bins for two different situations for the transactions data. The left plots show the PIT histograms when

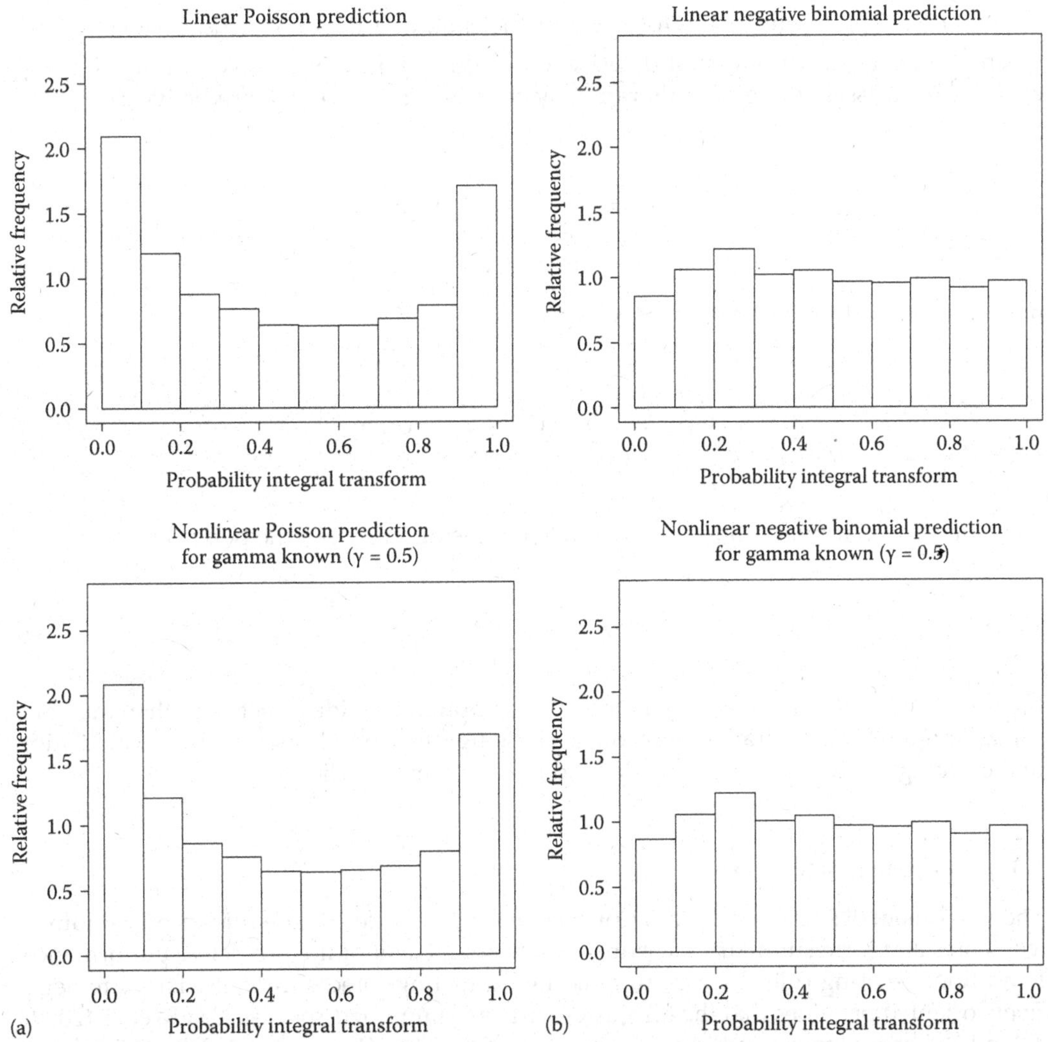

FIGURE 1.2
PIT histograms applied to the number of the transactions per minute for the stock Ericsson B during July 2, 2002. From top to bottom: PIT histograms for model (1.5) and model (1.12) for $\gamma = 0.5$. (a) The conditional distribution is Poisson. (b) The conditional distribution is negative binomial.

the fit is based on the Poisson distribution, for the linear and nonlinear models. Clearly, the plots show deviations from the Poisson distribution, indicating underdispersed predictive distributions. The right plots indicate no apparent deviations from uniformity; these plots are based on the negative binomial distribution (1.4). Similar findings were obtained after fitting the log-linear model (1.8) to the transactions data.

1.5.2 Assessment of Marginal Calibration

We now turn to the question of examining marginal calibration. We suppose that the observed time series $\{Y_t,\ t \geq 1\}$ is stationary with marginal c.d.f. $G(\cdot)$. In addition, we assume that we pick a probabilistic forecast in the form of a predictive c.d.f. $P_t(x) = P(Y_t \leq x \mid \mathcal{F}_{t-1}^{Y,\lambda})$. In our case, $P_t(\cdot)$ is either the c.d.f. of a Poisson random variable with mean $\hat{\lambda}_t$, or a negative binomial distribution evaluated at $\hat{\lambda}_t$ and $\hat{\nu}$. We follow Gneiting et al. (2007) to assess marginal calibration by comparing the average predictive c.d.f.

$$\bar{P}(x) = \frac{1}{n}\sum_{t=1}^{n} P_t(x), \quad x \in \mathbb{R},$$

to the empirical c.d.f. of the observations given by

$$\hat{G}(x) = \frac{1}{n}\sum_{t=1}^{n} \mathbb{1}(Y_t \leq x), \quad x \in \mathbb{R}.$$

To display the marginal calibration plot, we plot the difference of the two c.d.f.,

$$\bar{P}(x) - \hat{G}(x), \quad x \in \mathbb{R}. \tag{1.23}$$

Figure 1.3 shows that the negative binomial assumption provides a better fit than the Poisson assumption for the transactions data. These figures were drawn by direct calculation of the average c.d.f. $\bar{P}$ and the empirical c.d.f, as explained earlier.

1.5.3 Assessment of Sharpness

The assessment of sharpness is accomplished via scoring rules. These rules provide numerical scores and form summary measures for the assessment of the predictive performance. In addition, scoring rules help us to rank the competing forecast models. They are negatively oriented penalties that the forecaster wishes to minimize, see also Czado et al. (2009). Table 1.3 shows a few examples of scoring rules, following Czado et al. (2009). The calculation of all these scores requires an assumption on the conditional distribution of the process. The squared error score is identical for both the Poisson and the negative binomial distributions, since the conditional means are equal. Note that the normalized square error score is formed by the Pearson residuals defined in (1.19).

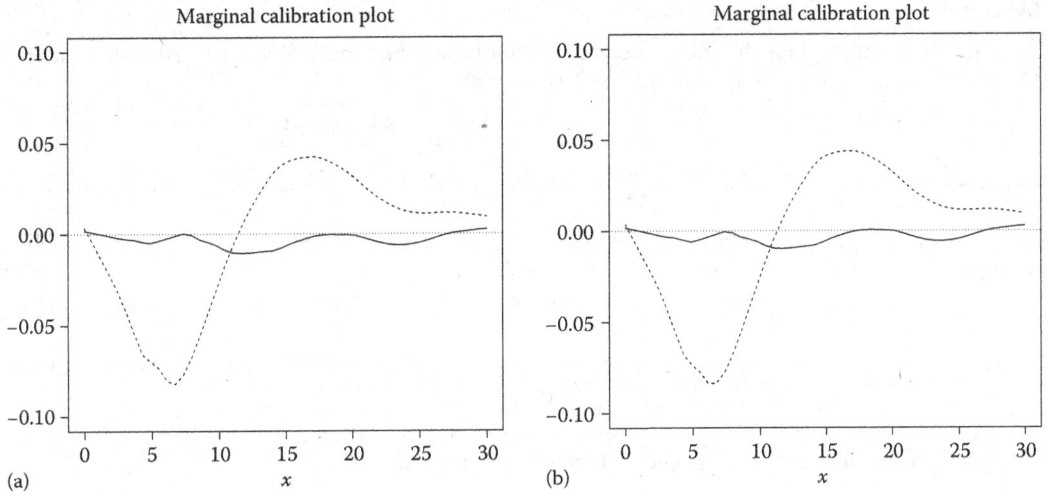

FIGURE 1.3
Marginal calibration plot for the transactions data. (a) corresponds to model (1.5) and (b) corresponds to model (1.12). Solid line corresponds to the negative binomial prediction, while dashed line is for the Poisson forecast. A similar plot is obtained for the case of log-linear model (1.8) but it is not shown.

TABLE 1.3
Definition of scoring rules

Scoring Rule	Notation	Definition
Logarithmic score	*logs*	$-\log p_y$
Quadratic or Brier score	*qs*	$2p_y + \|p\|^2$
Spherical score	*sphs*	$-p_y/\|p\|$
Ranked probability score	*rps*	$\sum_{x=0}^{\infty}(P_t(x) - \mathbb{1}(Y_t \leq x))^2$
Dawid–Sebastiani score	*dss*	$\left((Y_t - \mu_{P_t})/\sigma_{P_t}\right)^2 + 2\log \sigma_{P_t}$
Normalized squared error score	*nses*	$\left((Y_t - \mu_{P_t})/\sigma_{P_t}\right)^2$
Squared error score	*ses*	$(Y_t - \mu_{P_t})^2$

Note: For notational purposes, set $p_y = P(Y_t = y \mid \mathcal{F}_{t-1}^{Y,\lambda})$, $\|p\|^2 = \sum_{y=0}^{\infty} p_y^2$, and μ_{P_t} and σ_{P_t} are the mean and the standard deviation of the predictive distribution P_t, respectively.

Table 1.4 shows all scoring rules applied to transactions data. It is clear that the negative binomial model fits these data considerably better than the Poisson model regardless of the assumed model. We note again that all models yield almost identical scoring rules. This further supports our point that when there exists positive persistent correlation in the data, then all models will produce similar output.

For the transactions data, a simple linear model of the form (1.5) under the negative binomial assumption seems to describe the data adequately. This conclusion is a direct consequence of the earlier findings with additional evidence provided by the values of goodness-of-fit test (1.21) reported in Table 1.2.

TABLE 1.4

Scoring rules calculated for the transactions data after fitting the Linear Model (1.5), the Nonlinear Model (1.12) for $\gamma = 0.5$, and the Log-Linear Model (1.8)

		Scoring Rules						
	Forecaster	*logs*	*qs*	*sphs*	*rps*	*dss*	*nses*	*ses*
Linear model (1.5)	Poisson	3.126	−0.076	−0.276	3.633	4.585	2.326	23.477
	NegBin	**2.902**	**−0.080**	**−0.292**	**3.284**	**4.112**	**0.993**	**23.477**
Nonlinear model (1.12)	Poisson	3.123	−0.075	−0.274	3.605	4.579	2.318	23.435
	NegBin	**2.901**	**−0.080**	**−0.289**	**3.267**	**4.107**	**0.985**	**23.435**
Log-linear model (1.8)	Poisson	3.144	−0.081	−0.286	3.764	4.633	2.376	23.894
	NegBin	**2.910**	**−0.083**	**−0.300**	**3.334**	**4.132**	**0.993**	**23.894**

Note: The two forecasters are compared by the mean logarithmic, quadratic, spherical, ranked probability, Dawid–Sebastiani, normalized squared error and squared error scores. Bold face numbers in each column indicate the minimum value obtained between the two forecasters.

1.6 Other Topics

In this section, we will discuss other interesting research topics in the context of count time series analysis. This list is not exhaustive, and several other interesting topics will be covered in the following chapters. The list below reflects our personal research interests in the framework of generalized linear models.

1.6.1 Testing for Linearity

Consider the nonlinear model (1.11) and suppose that we are interested in testing the hypothesis $H_0 : c_1 = 0$ which is equivalent to testing linearity of the model. This testing problem is not standard because under the hypothesis, the nonlinear parameter γ is not identifiable. Furthermore, $c_1 = 0$ implies that the parameter is on the boundary of the parameter space. Similar comments can be made for model (1.12) when testing the hypothesis $H_0 : \gamma = 0$, but without the additional challenge implied by the nonidentifiability issue. These type of testing problems have been recently discussed by Christou and Fokianos (2013).

Suppose that, in general, the vector of unknown parameters can be decomposed as $\theta = \left(\theta^{(1)}, \theta^{(2)}\right)$ and let $S_n = \left(\mathbf{S}_n^{(1)}, \mathbf{S}_n^{(2)}\right)$ be the corresponding partition of the score function. Consider testing $H_0 : \theta^{(2)} = \mathbf{0}$ vs. $H_1 : \theta^{(2)} > \mathbf{0}$, componentwise. This problem is attacked by using the score test statistic which is given by

$$LM_n = \mathbf{S}_n^{(2)'}(\tilde{\theta}_n)\tilde{\mathbf{\Sigma}}^{-1}(\tilde{\theta}_n)\mathbf{S}_n^{(2)}(\tilde{\theta}_n),$$

where $\tilde{\theta}_n = \left(\tilde{\theta}_n^{(1)}, \mathbf{0}\right)$ is the QMLE of θ under the hypothesis and $\tilde{\mathbf{\Sigma}}$ is an appropriate estimator for the covariance matrix $\mathbf{\Sigma} = \text{Var}\left(\frac{1}{\sqrt{n}}\mathbf{S}_n^{(2)}\left(\tilde{\theta}_n\right)\right)$. If all the parameters are identified under the null hypothesis, the score statistic (1.11) follows asymptotically a $\mathcal{X}^2_{m_2}$ distribution under the null, where $m_2 = \dim(\theta^{(2)})$ (Francq and Zakoïan 2010, Ch. 8). Model (1.12) belongs to this class.

When the parameters are not identified under the null, a supremum type test statistic resolves this problem; see for instance Davies (1987). Consider model (1.11), for example, and let Γ be a grid of values for the nuisance parameter, denoted by γ. Then the sup-score test statistic is given by

$$LM_n = \sup_{\gamma \in \Gamma} LM_n(\gamma).$$

Critical values of the test statistics can be either based on the asymptotic chi-square approximation or by employing parametric bootstrap as in the case of the test statistic (1.21).

1.6.2 Intervention Analysis

Occasionally, some time series data may show that both variation and the level of the data change during some specific time interval. Additionally, there might exist outlying values (unusual values) at some time points. This is the case for the campylobacterosis infections data reported from January 1990 to the end of October 2000 in the north of the Province of Québec, Canada; see Fokianos and Fried (2010, Fig. 1). It is natural to ask whether these fluctuations can be explained by (1.5) or whether the inclusion of some interventions will yield better results; see Box and Tiao (1975), Tsay (1986) and Chen and Liu (1993) among others.

Generally speaking, types of intervention effects on time series data are classified according to whether their impact is concentrated on a single or a few data points, or whether they affect the whole process from some specific time $t = \tau$ on. In classical linear time series methodology, an intervention effect is included in the observation equation by employing a sequence of deterministic covariates $\{X_t\}$ of the form

$$X_t = \xi(\mathcal{B}) I_t(\tau), \tag{1.24}$$

where $\xi(\mathcal{B})$ is a polynomial operator, $\mathcal{B}$ is the shift operator such that $\mathcal{B}^i X_t = X_{t-i}$, and $I_t(\tau)$ is an indicator function, with $I_t(\tau) = 1$ if $t = \tau$, and $I_t(\tau) = 0$ if $t \neq \tau$. The choice of the operator $\xi(\mathcal{B})$ determines the kind of intervention effect: additive outlier (AO), transient shift (TS), level shift (LS), or innovational outlier (IO). Since models of the form (1.5) are not defined in terms of innovations, we focus on the first three types of interventions (but see Fried et al. 2015 for a Bayesian point of view).

However, a model like (1.5) is determined by a latent process. Therefore, a formal linear structure, as in the case of the Gaussian linear time series model, does not hold any more and interpretation of the interventions is a more complicated issue. Hence, a method which allows the detection of interventions and estimation of their size is needed so that structural changes can be identified successfully. Important steps to achieve this goal are the following; see Chen and Liu (1993):

1. A suitable model for accommodating interventions in count time series data.
2. Derivation of test procedures for their successful detection.

3. Implementation of joint maximum likelihood estimation of model parameters and outlier sizes.
4. Correction of the observed series for the detected interventions.

All these issues and possible directions for further developments of the methodology have been addressed by Fokianos and Fried (2010, 2012) for the linear model (1.5) and the log-linear model (1.8), under the Poisson assumption.

1.6.3 Robust Estimation

The previous work on intervention analysis is complemented by developing robust estimation procedures for count time series models. The works by El Saied (2012) and El Saied and Fried (2014) address this research topic in the context of the linear model (1.5) when $a_1 = 0$. In the context of log-linear model (1.8), the work of Kitromilidou and Fokianos (2015) develops robust estimation for count time series by adopting the methods suggested by Künsch et al. (1989) and Cantoni and Ronchetti (2001). In particular, Cantoni and Ronchetti (2001) robustified the quasi-likelihood approach for estimating the regression coefficient of generalized linear models by considering robust deviances which are natural generalizations of the quasi-likelihood functions. The robustification proposed by Cantoni and Ronchetti (2001) is performed by bounding and centering the quasi-score function.

1.6.4 Multivariate Count Time Series Models

Another interesting topic of research is the analysis of multivariate count time series models; see Liu (2012), Pedeli and Karlis (2013), and Section V of this volume which contains many interesting results. The main issue for attacking the problem of multivariate count time series is that multivariate count distributions are quite complex to be analyzed by maximum likelihood methods.

Assume that $\{\mathbf{Y}_t = (Y_{i,t}), t = 1, 2, \ldots, n\}$ denotes a p-dimensional count time series and suppose further that $\{\boldsymbol{\lambda}_t = (\lambda_{i,t}), t = 1, 2, \ldots, n\}$ is a corresponding p-dimensional intensity process. Here the notation p denotes dimension but not order as in (1.13). Then, a natural generalization of (1.5) is given by

$$Y_{i,t} = N_{i,t}(0, \lambda_{i,t}], \quad i = 1, 2, \ldots, p, \quad \boldsymbol{\lambda}_t = \mathbf{d} + \mathbf{A}\boldsymbol{\lambda}_{t-1} + \mathbf{B}\mathbf{Y}_{t-1}, \tag{1.25}$$

where $\mathbf{d}$ is a p-dimensional vector and $\mathbf{A}$, $\mathbf{B}$ are $p \times p$ matrices, all of them unknowns to be estimated. Model (1.25) is a direct extension of the linear autoregressive model (1.5) and assumes that marginally the count process is Poisson distributed. However, the statistical problem of dealing with the joint distribution of the vector process $\{\mathbf{Y}_t\}$ requires further research; some preliminary results about ergodicity and stationarity of (1.25) have been obtained by Liu (2012). More on multivariate models for count time series is given in the chapter by Karlis (2015; Chapter 19 in this volume), and an application is discussed by Ravishanker et al. (2015; Chapter 20 in this volume).

1.6.5 Parameter-Driven Models

So far we have discussed models that fall under the framework of observation-driven models. This implies that even though the mean process $\{\lambda_t\}$ is not observed directly, it can still be recovered explicitly as a function of the past responses. However, a different point of view has been taken by Zeger (1988), who introduced regression models for time series of counts by assuming that the observed process is driven by a latent (unobserved) process. To be more specific, suppose that, conditional on an unobserved process $\{\xi_t, t \geq 1\}$, $\{Y_t, t \geq 1\}$, is a sequence of independent counts such that

$$E[Y_t \mid \xi_t] = \mathrm{Var}[Y_t \mid \xi_t] = \xi_t \exp(d + a_1 y_{t-1}). \tag{1.26}$$

In (1.26) we consider a simple model for illustration, but more complex models that include higher-order lagged values of the response and any covariates can be assumed. It can be proved that the earlier formulation, although similar to a Poisson log-linear model, reveals that the observed data are overdispersed. Estimation of all unknown parameters is discussed by Zeger (1988). A further detailed study of model (1.26) can be found in Davis et al. (2000), where the authors address the problem of existence of the latent stochastic process $\{\xi_t\}$ and derive the asymptotic distribution of the regression coefficients when the latter exist. In the context of negative binomial regression, the latent process model (1.26) has been extended by Davis and Wu (2009). See also Harvey and Fernandes (1989) for a state-space approach with conjugate priors for the analysis of count time series and Jørgensen et al. (1999) for multivariate longitudinal count data. More generally, state space models for count time series are discussed in West and Harrison (1997), Durbin and Koopman (2001), and Cappé et al. (2005), among others.

1.7 Other Extensions

There are several other possible directions for extending the theory and methods discussed in this chapter. Threshold models have been considered recently by Woodard et al. (2011), Douc et al. (2013), and Wang et al. (2014). Several questions are posed by such models such as estimation of regression and threshold or/and delay parameters. The concept of mixture models for the analysis of count time series data is a topic closely related to that of threshold models; for the real-valued case, see Wong and Li (2001) among others. Finally, I would like to bring forward the possibility of introducing local stationarity to count time series in the sense of Dahlhaus (1997, 2000). Such models pose several problems; for instance, the question of existence of a stationary approximation and estimation of the time-varying parameters by nonparametric likelihood inference.

Acknowledgments

Many thanks to the editors for all their efforts and for inviting me to contribute to this volume. Special thanks are due to V. Christou, N. Papamichael, D. Tjøstheim, and a referee for several useful comments and suggestions. Support was provided by the Cyprus Research Promotion Foundation grant TEXNOLOGIA/THEPIS/0609(BE)/02.

References

Andrews, D. (1984). Non-strong mixing autoregressive processes. *Journal of Applied Probability*, 21:930–934.

Bardet, J.-M. and Wintenberger, O. (2010). Asymptotic normality of the quasi-maximum likelihood estimator for multidimensional causal processes. *The Annals of Statistics*, 37:2730–2759.

Benjamin, M. A., Rigby, R. A., and Stasinopoulos, D. M. (2003). Generalized autoregressive moving average models. *Journal of the American Statistical Association*, 98:214–223.

Berkes, I., Horváth, L., and Kokoszka, P. (2003). GARCH processes: Structure and estimation. *Bernoulli*, 9:201–227.

Bollerslev, T. (1986). Generalized autoregressive conditional heteroskedasticity. *Journal of Econometrics*, 31:307–327.

Box, G. E. P. and Tiao, G. C. (1975). Intervention analysis with applications to economics and environmental problems. *Journal of the American Statistical Association*, 70:70–79.

Brockwell, P. J. and Davis, R. A. (1991). *Time Series: Data Analysis and Theory*, 2nd edn. Springer, New York.

Brumback, B. A., Ryan, L. M., Schwartz, J. D., Neas, L. M., Stark, P. C., and Burge, H. A. (2000). Transitional regression models with application to environmental time series. *Journal of the American Statistical Association*, 85:16–27.

Cameron, A. C. and Trivedi, P. K. (1998). *Regression Analysis of Count Data*. Cambridge University Press, Cambridge, U.K.

Cantoni, E. and Ronchetti, E. (2001). Robust inference for generalized linear models. *Journal of the American Statistical Association*, 96:1022–1030.

Cappé, O., Moulines, E., and Rydén, T. (2005). *Inference in Hidden Markov Models*. Springer, New York.

Chen, C. and Liu, L.-M. (1993). Joint estimation of model parameters and outlier effects in time series. *Journal of the American Statistical Association*, 88:284–297.

Christou, V. and Fokianos, K. (2013). Estimation and testing linearity for mixed Poisson autoregression. (Submitted).

Christou, V. and Fokianos, K. (2014). Quasi-likelihood inference for negative binomial time series models. *Journal of Time Series Analysis*, 35:55–78.

Christou, V. and Fokianos, K. (2015). On count time series prediction. *Journal of Statistical Computation and Simulation*, 2:357–373.

Cox, D. R. (1981). Statistical analysis of time series: Some recent developments. *Scandinavian Journal of Statistics*, 8:93–115.

Creal, D., Koopman, S. J., and Lucas, A. (2008). A general framework for observation driven time–varying parameter models. Technical Report TI 2008–108/4, Tinbergen Institute, Amsterdam, the Netherlands.

Czado, C., Gneiting, T., and Held, L. (2009). Predictive model assessment for count data. *Biometrics*, 65:1254–1261.

Dahlhaus, R. (1997). Fitting time series models to nonstationary processes. *The Annals of Statistics*, 25:1–37.

Dahlhaus, R. (2000). A likelihood approximation for locally stationary processes. *The Annals of Statistics*, 28:1762–1794.

Davies, R. B. (1987). Hypothesis testing when a nuisance parameter is present only under the alternative. *Biometrika*, 74:33–43.

Davis, R. and Wu, R. (2009). A negative binomial model for time series of counts. *Biometrika*, 96:735–749.

Davis, R. A., Dunsmuir, W. T. M., and Streett, S. B. (2003). Observation-driven models for Poisson counts. *Biometrika*, 90:777–790.

Davis, R. A., Dunsmuir, W. T. M., and Wang, Y. (2000). On autocorrelation in a Poisson regression model. *Biometrika*, 87:491–505.

Davis, R. A. and Liu, H. (2015). Theory and inference for a class of observation-driven models with application to time series of counts. *Statistica Sinica*. doi:10.5705/ss.2014.145t (to appear).

Douc, R., Doukhan, P., and Moulines, E. (2013). Ergodicity of observation-driven time series models and consistency of the maximum likelihood estimator. *Stochastic Processes and their Applications*, 123:2620–2647.

Doukhan, P., Fokianos, K., and Tjøstheim, D. (2012). On weak dependence conditions for Poisson autoregressions. *Statistics & Probability Letters*, 82:942–948.

Doukhan, P., Fokianos, K., and Tjøstheim, D. (2013). Correction to "on weak dependence conditions for Poisson autoregressions" [Statist. Probab. Lett. 82 (2012) 942—948]. *Statistics & Probability Letters*, 83:1926–1927.

Durbin, J. and Koopman, S. J. (2001). *Time Series Analysis by State Space Methods*. Oxford University Press, Oxford, U.K.

El Saied, H. (2012). *Robust Modelling of Count Time Series: Applications in Medicine*. PhD thesis, TU Dortmund University, Dortmund, Germany.

El Saied, H. and Fried, R. (2014). Robust fitting of INARCH models. *Journal of Time Series Analysis*, 35:517–535.

Fan, J. and Yao, Q. (2003). *Nonlinear Time Series*. Springer-Verlag, New York.

Ferland, R., Latour, A., and Oraichi, D. (2006). Integer-valued GARCH processes. *Journal of Time Series Analysis*, 27:923–942.

Fokianos, K. (2012). Count time series. In Rao, T. S., Rao, S. S., and Rao, C. R., eds., *Handbook of Statistics: Time Series Analysis–Methods and Applications*, vol. 30, pp. 315–347. Elsevier B. V., Amsterdam, the Netherlands.

Fokianos, K. and Fried, R. (2010). Interventions in INGARCH processess. *Journal of Time Series Analysis*, 31:210–225.

Fokianos, K. and Fried, R. (2012). Interventions in log-linear Poisson autoregression. *Statistical Modelling*, 12:299–322.

Fokianos, K. and Kedem, B. (2004). Partial likelihood inference for time series following generalized linear models. *Journal of Time Series Analysis*, 25:173–197.

Fokianos, K. and Neumann, M. H. (2013). A goodness-of-fit test for Poisson count processes. *Electronic Journal of Statistics*, 7:793–819.

Fokianos, K., Rahbek, A., and Tjøstheim, D. (2009). Poisson autoregression. *Journal of the American Statistical Association*, 104:1430–1439.

Fokianos, K. and Tjøstheim, D. (2011). Log-linear Poisson autoregression. *Journal of Multivariate Analysis*, 102:563–578.

Fokianos, K. and Tjøstheim, D. (2012). Non-linear Poisson autoregression. *Annals of the Institute of Statistical Mathematics*, 64:1205–1225.

Francq, C. and Zakoïan, J.-M. (2004). Maximum likelihood estimation of pure GARCH and ARMA-GARCH processes. *Bernoulli*, 10:605–637.

Francq, C. and Zakoïan, J.-M. (2010). *GARCH Models: Structure, Statistical Inference and Financial Applications*. Wiley, London, U.K.

Franke, J. (2010). Weak dependence of functional INGARCH processes. Report in Wirtschaftsmathematik 126, University of Kaiserslautern, Kaiserslautern, Germany.

Fried, R., Agueusop, I., Bornkamp, B., Fokianos, K., Fruth, J., and Ickstat, K. (2015). Retrospective Bayesian outlier detection in INGARH series. *Statistics & Computing*, 25:365–374.

Gamerman, D., Abanto-Valle, C. A., Silva, R. S., and Martins, T. G. (2015). Dynamic Bayesian models for discrete-valued time series. In R. A. Davis, S. H. Holan, R. Lund and N. Ravishanker, eds., *Handbook of Discrete-Valued Time Series*, pp. 165–188. Chapman & Hall, Boca Raton, FL.

Gneiting, T., Balabdaoui, F., and Raftery, A. E. (2007). Probabilistic forecasts, calibration and sharpness. *Journal of the Royal Statistical Society: Series B (Statistical Methodology)*, 69:243–268.

Godambe, V. P. and Heyde, C. C. (1987). Quasi-likelihood and optimal estimation. *International Statistical Review*, 55:231–244.

Haggan, V. and Ozaki, T. (1981). Modelling non-linear random vibrations using an amplitude-dependent autoregressive time series model. *Biometrika*, 68:189–196.

Harvey, A. C. and Fernandes, C. (1989). Time series models for count or qualitative observations. *Journal of Business & Economic Statistics*, 7:407–422. with discussion.

Heinen, A. (2003). Modelling time series count data: An autoregressive conditional Poisson model. Technical Report MPRA Paper 8113, University Library of Munich, Munich, Germany. Available at http://mpra.ub.uni-muenchen.de/8113/.

Heyde, C. C. (1997). *Quasi-Likelihood and its Applications: A General Approach to Optimal Parameter Estimation*. Springer, New York.

Jørgensen, B., Lundbye-Christensen, S., Song, P. X.-K., and Sun, L. (1999). A state space model for multivariate longitudinal count data. *Biometrika*, 86:169–181.

Jung, R. C., McCabe, B. P. M., and Tremayne, A.R. (2015). Model validation and diagnostics. In R. A. Davis, S. H. Holan, R. Lund and N. Ravishanker, eds., *Handbook of Discrete-Valued Time Series*, pp. 189–218. Chapman & Hall, Boca Raton, FL.

Jung, R. and Tremayne, A. (2011). Useful models for time series of counts or simply wrong ones? *AStA Advances in Statistical Analysis*, 95:59–91.

Jung, R. C., Kukuk, M., and Liesenfeld, R. (2006). Time series of count data: Modeling, estimation and diagnostics. *Computational Statistics & Data Analysis*, 51:2350–2364.

Karlis, D. (2015). Models for multivariate count time series. In R. A. Davis, S. H. Holan, R. Lund and N. Ravishanker, eds., *Handbook of Discrete-Valued Time Series*, pp. 407–424. Chapman & Hall, Boca Raton, FL.

Kedem, B. and Fokianos, K. (2002). *Regression Models for Time Series Analysis*. Wiley, Hoboken, NJ.

Kitromilidou, S. and Fokianos, K. (2015). Robust estimation methods for a class of count time series log-linear models. *Journal of Statistical Computation and Simulation* (to appear).

Koul, H. L. and Stute, W. (1999). Nonparametric model checks for time series. *The Annals of Statistics*, 27:204–236.

Künsch, H. R., Stefanski, L. A., and Carroll, R. J. (1989). Conditionally unbiased bounded-influence estimation in general regression models, with applications to generalized linear models. *Journal of the American Statistical Association*, 84:460–466.

Lawless, J. F. (1987). Negative binomial and mixed Poisson regression. *The Canadian Journal of Statistics*, 15:209–225.

Li, W. K. (1994). Time series models based on generalized linear models: Some further results. *Biometrics*, 50:506–511.

Liu, H. (2012). Some models for time series of counts. PhD thesis, Columbia University, New York.

MacDonald, I. L. and Zucchini, W. (1997). *Hidden Markov and Other Models for Discrete–valued Time Series*. Chapman & Hall, London, U.K.

McCullagh, P. and Nelder, J. A. (1989). *Generalized Linear Models* 2nd edn. Chapman & Hall, London, U.K.

Meitz, M. and Saikkonen, P. (2011). Parameter estimation in non-linear AR—GARCH models. *Econometric Theory*, 27:1236–1278.

Mikosch, T. (2009). *Non-Life Insurance Mathematics–An Introduction with the Poisson Process*, 2nd edn. Springer-Verlag, Berlin, Germany.

Mikosch, T. and Straumann, D. (2006). Stable limits of martingale transforms with application to the estimation of GARCH parameters. *The Annals of Statistics*, 34:493–522.

Nelder, J. A. and Wedderburn, R. W. M. (1972). Generalized linear models. *Journal of the Royal Statistical Society, Series A*, 135:370–384.

Neumann, M. (2011). Absolute regularity and ergodicity of Poisson count processes. *Bernoulli*, 17:1268–1284.

Pedeli, X. and Karlis, D. (2013). Some properties of multivariate INAR(1) processes. *Computational Statistics & Data Analysis*, 67:213–225.

Ravishanker, N., Venkatesan, R., and Hu, S. (2015). Dynamic models for time series of counts with a marketing application. In R. A. Davis, S. H. Holan, R. Lund and N. Ravishanker, eds., *Handbook of Discrete-Valued Time Series*, pp. 425–446. Chapman & Hall, Boca Raton, FL.

Rydberg, T. H. and Shephard, N. (2000). A modeling framework for the prices and times of trades on the New York stock exchange. In Fitzgerlad, W. J., Smith, R. L., Walden, A. T., and Young, P. C., editors, *Nonlinear and Nonstationary Signal Processing*, pp. 217–246. Isaac Newton Institute and Cambridge University Press, Cambridge, U.K.

Soyer, R., Aktekin, T., and Kim, B. (2015). Bayesian modeling of time series of counts with business applications. In R. A. Davis, S. H. Holan, R. Lund and N. Ravishanker, eds., *Handbook of Discrete-Valued Time Series*, pp. 245–266. Chapman & Hall, Boca Raton, FL.

Streett, S. (2000). Some observation driven models for time series of counts. PhD thesis, Colorado State University, Department of Statistics, Fort Collins, CO.

Teräsvirta, T. (1994). Specification, estimation, and evaluation of smooth transition autoregressive models. *Journal of the American Statistical Association*, 89:208–218.

Teräsvirta, T., Tjøstheim, D., and Granger, C. W. J. (2010). *Modelling Nonlinear Economic Time Series*. Oxford University Press, Oxford, U.K.

Tjøstheim, D. (2012). Some recent theory for autoregressive count time series. *TEST*, 21:413–438.

Tong, H. (1990). *Nonlinear Time Series:A Dynamical System Approach*. Oxford University Press, New York.

Tsay, R. S. (1986). Time series model specification in the presence of outliers. *Journal of the American Statistical Association*, 81:132–141.

Wang, C., Liu, H., Yao, J.-F., Davis, R. A., and Li, W. K. (2014). Self-excited Threshold Poisson Autoregression. *Journal of the American Statistical Association*, 109:777–787.

West, M. and Harrison, P. (1997). *Bayesian Forecasting and Dynamic Models*, 2nd edn. Springer, New York.

White, H. (1982). Maximum likelihood estimation of misspecified models. *Econometrica*, 50:1–25.

Wong, C. S. and Li, W. K. (2001). On a mixture autoregressive conditional heteroscedastic model. *Journal of the American Statistical Association*, 96:982–995.

Woodard, D. W., Matteson, D. S., and Henderson, S. G. (2011). Stationarity of count-valued and nonlinear time series models. *Electronic Journal of Statistics*, 5:800–828.

Zeger, S. L. (1988). A regression model for time series of counts. *Biometrika*, 75:621–629.

Zeger, S. L. and Qaqish, B. (1988). Markov regression models for time series: A quasi-likelihood approach. *Biometrics*, 44:1019–1031.

Zhu, F. (2011). A negative binomial integer-valued GARCH model. *Journal of Time Series Analysis*, 32:54–67.

Zhu, F. and Wang, D. (2010). Diagnostic checking integer-valued ARCH(p) models using conditional residual autocorrelations. *Computational Statistics and Data Analysis*, 54:496–508.

2

Markov Models for Count Time Series

Harry Joe

CONTENTS

2.1 Introduction

The focus of this chapter is on the construction of count time series models based on thinning operators or a joint distribution on consecutive observations, and comparison of the properties of the resulting models.

The models for count time series considered here are mainly intended for low counts with the possibility of 0. If all counts are large and "far" from 0, the models considered here can be used as well as models that treat the count response as continuous.

Count data are often overdispersed relative to Poisson. There are many count regression models with covariates, examples are regression models with negative binomial (NB), generalized Poisson (GP), zero-inflated Poisson, zero-inflated NB, etc.

If the count data are observed as a time series sequence, then the count regression model can be adapted in two ways: (1) add previous observations as covariates and (2) make use of some models for stationary count time series. Methodology for case (1) is covered in Davis et al. (2000) and Fokianos (2011), and here, we discuss the quite different methodology for case (2). The advantage of a time series regression model with univariate margins corresponding to a count regression model is that predictions as a function of covariates can be made with or without preceding observations. That is, this is useful if one is primarily interested in regression but with time-dependent observations.

Common parametric models for count regression are NB and GP, and these include Poisson regression at the boundary. In this chapter, we use these two count regression models for concreteness, but some approaches, such as copula-based models, can accommodate other count distributions.

The remainder of the chapter is organized as follows. Section 2.2 summarizes some count regression models and contrasts some properties of count time series models constructed under different approaches. Sections 2.3 through 2.5 provide some details for count time series models based, respectively, on thinning operators, multivariate distributions with random variables in a convolution-closed infinitely divisible class, and copulas for consecutive observations. Section 2.6 compares the fits of different models for one data set.

Some conventions and notation that are subsequently used are as follows: f is used for probability mass functions (pmf) and F is used for cumulative distribution functions (cdfs) with the subscript used to indicate the margin or random vector; $\sum_{i=1}^{y} k_i = 0$ when $y = 0$; $\mathcal{N}_0$ is the set of nonnegative integers; ϵ_t is used for the innovation at time t (that is, ϵ_t is independent of random variables at time $t-1, t-2, \ldots$ in the stochastic representation).

2.2 Models for Count Data

In this section, we show how NB and GP distributions have been used for count regression and count time series.

NB and GP regression models are nonunique in how regression coefficient βs are introduced into the univariate parameters. If the mean is assumed to be loglinear in covariates, there does not exist a unique model because the mean involves the convolution parameter and a second parameter that links to overdispersion.

Brief details are summarized as follows, with F_{NB} and F_{GP} denoting the cdfs and f_{NB} and f_{GP} denoting the pmfs:

1. (NB): θ convolution parameter, π probability parameter, $\xi = \pi^{-1} - 1 \geq 0$; mean $\mu = \theta\xi = \theta(1-\pi)/\pi$, variance $\sigma^2 = \mu(1+\xi) = \theta(1-\pi)/\pi^2$, and

$$f_{NB}(y;\theta,\xi) = \frac{\Gamma(\theta+y)}{\Gamma(\theta)\, y!}\frac{\xi^y}{(1+\xi)^{\theta+y}}, \quad y = 0,1,2,\ldots, \quad \theta > 0,\ \xi > 0.$$

   If $\theta \to \infty$, $\xi \to 0$ with $\theta\xi$ fixed, the Poisson distribution is obtained.

2. (GP): θ convolution parameter, second parameter $0 \leq \eta < 1$; mean $\mu = \theta/(1-\eta)$, variance $\sigma^2 = \theta/(1-\eta)^3$, and

$$f_{GP}(y;\theta,\eta) = \frac{\theta(\theta+\eta y)^{y-1}}{y!} e^{-\theta-\eta y}, \quad y = 0,1,2,\ldots, \quad \theta > 0,\ 0 \leq \eta < 1.$$

   If $\eta = 0$, the Poisson distribution is obtained.

Cameron and Trivedi (1998) present the NB$k(\mu, \gamma)$ parametrization where $\theta = \mu^{2-k}\gamma^{-1}$ and $\xi = \mu^{k-1}\gamma$, $1 \leq k \leq 2$. For the NBk model, $\log\mu = z^T\beta$ depends on the covariate vector z, and either or both θ and ξ are covariate dependent. For the NB1 parametrization:

$k = 1, \theta = \mu\gamma^{-1}, \xi = \gamma$; that is, θ depends on covariates, and the dispersion index $\xi = \gamma$ is constant. For the NB2 parametrization: $k = 2, \theta = \gamma^{-1}, \xi = \mu\gamma$ and this is the same as in Lawless (1987); that is, θ is constant and ξ is a function of the covariates, and the dispersion index varies with the covariates. For $1 < k < 2$, one could interpolate between these two models using the NBk parametrization. Similarly, GP1 and GP2 regression models can be defined.

Next, we consider stationary time series $\{Y_t : t = 1, 2, \ldots\}$, where the stationary distribution is NB, GP, or general F_Y.

A Markov order 1 time series can be constructed based on a common joint distribution F_{12} for (Y_{t-1}, Y_t) for all t with marginal cdfs $F_1 = F_2 = F_Y = F_{NB}(\cdot; \theta, \xi)$ or $F_{GP}(\cdot; \theta, \eta)$ (or another parametric univariate margin). Let f_{12} and f_Y be the corresponding bivariate and univariate pmfs. The Markov order 1 transition probability is

$$\Pr(Y_t = y_{new} | Y_{t-1} = y_{prev}) = \frac{f_{12}(y_{prev}, y_{new})}{f_Y(y_{prev})}$$

A Markov order 2 time series can be constructed based on a common joint distribution F_{123} for (Y_{t-2}, Y_{t-1}, Y_t) for all t with univariate marginal cdfs $F_1 = F_2 = F_3$ and bivariate margins $F_{12} = F_{23}$. The ideas extend to higher-order Markov. However for count time series with small counts, simpler models are generally adequate for forecasting.

There are two general approaches to obtain the transition probabilities; the main ideas can be seen with Markov order 1.

1. Thinning operator for Markov order 1 dependence: $Y_t = R_t(Y_{t-1}; \alpha) + \epsilon_t(\alpha)$, $0 \le \alpha \le 1$, where R_t are independent realizations of a stochastic operator, the ϵ_t are appropriate innovation random variables, and typically $\mathrm{E}[R_t(y; \alpha) | Y_{t-1} = y] = \alpha y$ for $y = 0, 1, \ldots$.
2. Copula-based transition probability from $F_{12} = C(F_Y, F_Y; \delta)$ for a copula family C with dependence parameter δ.

The review paper McKenzie (2003) has a section entitled "Markov chains" but copula-based transition models were not included. Copulas are multivariate distributions with $U(0, 1)$ margins and they lead to flexible modeling of multivariate data with the dependence structure separated from the univariate margins. References for use of copula models are Joe (1997) and McNeil et al. (2005).

Some properties and contrasts are summarized below, with details given in subsequent sections. Weiß (2008) has a survey of many thinning operators for count time series models, and Fokianos (2012) has a survey of models based on thinning operators and conditional Poisson. Some references where copulas are used for transition probabilities are Joe (1997) (Chapter 8), Escarela et al. (2006), Biller (2009), and Beare (2010).

For thinning operators, the following hold:

- The stationary margin is infinitely divisible (such as NB, GP).
- The serial correlations are positive.
- The operator is generally interpretable and the conditional expectation is linear.
- For extension to include covariates (and/or time trends), the ease depends on the operator; covariates can enter into a parameter for the innovation distribution, but in this way, the marginal distribution does not necessarily stay in the same family.

- For extension to higher Markov orders, there are "integer autoregressive" models, such as INAR(p) in Du and Li (1991) or GINAR(p) in Gauthier and Latour (1994), and constructions as in Lawrance and Lewis (1980), Alzaid and Al-Osh (1990), and Zhu and Joe (2006) to keep margins in a given family. Without negative serial correlations, the range of autocorrelation functions is not as flexible as the Gaussian counterpart.
- Numerical likelihood inference is simple if the transition probability has closed form; for some operators, only the conditional probability generating function (pgf) has a simple form and then the approach of Davies (1973) can be used to invert the pgf. Conditional least squares (CLS) and moment methods can estimate mean parameters but are not reliable for estimating the overdispersion parameter.
- There are several different thinning operators for NB or GP and these can be differentiated based on the conditional heteroscedasticity $\mathrm{Var}(Y_t|Y_{t-1} = y)$.
- Although thinning operators can be used in models that are analogues of Gaussian AR(p), MA(q), and ARMA models, not as much is known about probabilistic properties such as stationary distributions.

For copula-based transition, the following hold:

- The stationary margin can be anything, and positive or negative serial dependence can be attained by choosing appropriate copula families.
- The conditional expectation is generally nonlinear and different patterns are possible. The tail behavior of the copula family affects the conditional expectation and variance for large values.
- It is easier to combine the time series model with covariates in a univariate count regression model.
- The extension from Markov order 1 to higher-order Markov is straightforward.
- Theoretically, the class of autocorrelation functions is much wider than those based on thinning operators.
- Likelihood inference is easy if the copula family has a simple form.
- The Gaussian copula is a special case; for example, autoregressive-to-anything (ARTA) in Biller and Nelson (2005).
- As a slight negative compared with thinning operators, for NB/GP, the copula approach does not use any special univariate property.

2.3 Thinning Operators

This section has more detail on thinning operators. Operators are initially presented and discussed without regard to stationarity and distributional issues. Notation similar to that in Jung and Tremayne (2011) is used.

A general stochastic model is

$$Y_t = R_t(Y_{t-1}, Y_{t-2}, \ldots) + \epsilon_t,$$

where ϵ_t is the innovation random variable at time t and R_t is a random variable that depends on the previous observations. In order to get a stationary model with margin F_Y, the choice of distribution for $\{\epsilon_t\}$ depends on the distribution of $R_t(Y_{t-1}, Y_{t-2}, \ldots)$. If one is not aiming for a specific stationary distribution, there is no constraint on the distribution of $\{\epsilon_t\}$.

If $R_t(Y_{t-1}, Y_{t-2}, \ldots) = R_t(Y_{t-1})$, then a Markov model of order 1 is obtained and if $R_t(Y_{t-1}, Y_{t-2}, \ldots) = R_t(Y_{t-1}, Y_{t-2})$, then a Markov model of order 2 is obtained, etc.

For Markov order 1 models, the conditional pmf of $[R(Y)|Y = y]$ is the same as the pmf of $R(y)$ and the unconditional pmf of $R(Y)$ is

$$f_{R(Y)}(x) = \sum_{y=0}^{\infty} f_{R(y)}(x) f_Y(y).$$

The following classes of operators are included in the review in Weiß (2008):

- Binomial thinning (Steutel and Van Harn 1979): $R(y) = \alpha \circ y \stackrel{d}{=} \sum_{i=1}^{y} I_i(\alpha)$ has a $\text{Bin}(y, \alpha)$ distribution, where $I_1(\alpha), I_2(\alpha), \ldots$ are independent Bernoulli random variables with mean $\alpha \in (0, 1)$. Hence $\text{E}[R(y)] = \alpha y$ and $\text{Var}[R(y)] = \alpha(1 - \alpha)y$.
- Expectation or generalized thinning (Latour, 1998; Zhu and Joe, 2010a): $R(y) = K(\alpha) \circledast y \stackrel{d}{=} \sum_{i=1}^{y} K_i(\alpha)$ where $K_1(\alpha), K_2(\alpha), \ldots$ are independent random variables that are replicates of $K(\alpha)$ which has support on $\mathcal{N}_0$ and satisfies $\text{E}[K(\alpha)] = \alpha \in [0, 1]$. For the boundary case $K(0) \equiv 0$ and $K(1) \equiv 1$. Hence $\text{E}[R(y)] = \alpha y$ and $\text{Var}[R(y)] = y \,\text{Var}[K(\alpha)]$.
- Random coefficient thinning (Zheng et al., 2006, 2007): $R(y) = A \circ y \stackrel{d}{=} \sum_{i=1}^{y} I_i(A)$ where A has support on $(0, 1)$ and given $A = a$, $I_i(a)$ are independent Bernoulli(a) random variables. Hence $\Pr(A \circ y = j) = \int_0^1 \binom{y}{j} a^j (1 - a)^{y-j} dF_A(a)$. If A has mean α, then $\text{E}[R(y)] = \text{E}[Ay] = \alpha y$ and $\text{Var}[R(y)] = \text{E}[A(1 - A)y] + \text{Var}(Ay) = \alpha(1 - \alpha)y + y(y - 1)\,\text{Var}(A)$.

Interpretations are provided later, with a subscript on the thinning operator to emphasize that thinnings are performed at time t.

- Time series based on binomial thinning:

$$Y_t = \alpha \circ_t Y_{t-1} + \epsilon_t = \sum_{i=1}^{Y_{t-1}} I_{ti}(\alpha) + \epsilon_t, \tag{2.1}$$

where the $I_{ti}(\alpha)$ are independent over t and i. It can be considered that $\alpha \circ_t Y_{t-1}$ consists of the "survivors" (continuing members) from time $t - 1$ to time t (with each individual having a probability α of continuing), and ϵ_t consists of the "newcomers" (innovations) at time t.

- Time series based on generalized thinning:

$$Y_t = K(\alpha) \circledast_t Y_{t-1} + \epsilon_t = \sum_{i=1}^{Y_{t-1}} K_{ti}(\alpha) + \epsilon_t \tag{2.2}$$

where the $K_{ti}(\alpha)$ are independent over t and i. This can be viewed as a dynamic system so that $K(\alpha) \circledast_t Y_{t-1}$ is a sum where each countable unit at time $t-1$ may be absent, present, or split into more than one new unit at time t, and ϵ_t consists of the new units at time t. Also $K(\alpha) \circledast y$ can be considered as a compounding or branching operator, and the time series model can be considered as a branching process model with immigration.

- Time series based on random coefficient thinning: this is random binomial thinning, where the chance of survival to the next time is a random variable that depends on t.

$$Y_t = A_t \circ Y_{t-1} + \epsilon_t = \sum_{i=1}^{Y_{t-1}} I_{ti}(A_t) + \epsilon_t, \tag{2.3}$$

A beta-binomial thinning operator based on the construction in Section 2.4 fits within this class.

Because all of the above models have a conditional expectation that is linear in the previous observation, they have been called integer-autoregressive models of order 1, abbreviated INAR(1). The models are not truly autoregressive in the sense of linear in the previous observations (because such an operation would not preserve the integer domain).

2.3.1 Analogues of Gaussian AR(p)

An extension of (2.3) to a higher-order Markov time series model is given in Section 2.4 for one special case. Otherwise, binomial thinning is a special case of generalized thinning. We next extend (2.2) to higher-order Markov:

$$Y_t = \sum_{j=1}^{p} K(\alpha_j) \circledast_t Y_{t-j} + \epsilon_t = \sum_{j=1}^{p} \sum_{i=1}^{Y_{t-j}} K_{tji}(\alpha_j) + \epsilon_t, \tag{2.4}$$

where $0 \le \alpha_j \le 1$ for $j = 1, \ldots, p$ and the $K_{tji}(\alpha_j)$ are independent over t, j and i, and ϵ_t is the innovation at time t. This is called GINAR(p) in (Gauthier and Latour 1994; Latour 1997, 1998). It can also be interpreted as a branching process model with immigration, where a unit at time t has independent branching at times $t+1, \ldots, t+p$. The most common form of INAR(p) in the statistical literature involves the binomial thinning operator; see Du and Li (1991). For the binomial thinning operator, Alzaid and Al-Osh (1990) define INAR(p) in a different way from the above with a conditional multinomial distribution for $(\alpha_1 \circ Y_t, \ldots, \alpha_p \circ Y_t)$. Because the survival/continuation interpretation for (2.1) does not extend to second and higher orders, it is better to consider (2.4) with more general thinning operators; if the K_{tji} are Bernoulli random variables, this can still be interpreted as a branching process model with immigration (with limited branching).

More specifically, for a GINAR(2) model based on compounding, unit i at time t' contributes $K_{t'+1,i}(\alpha_1)$ units to the next time and $K_{t'+2,i}(\alpha_2)$ units in two time steps. That is, at time t, the total count comes from branching of units at times $t-1$ and $t-2$ plus the innovation count.

It will be shown below that the GINAR(p) model has an overdispersion property if $\{K(\alpha)\}$ satisfies $\mathrm{Var}[K(\alpha)] = \sigma^2_{K(\alpha)} = a_K\alpha(1-\alpha)$ where $a_K \geq 1$. A sufficient condition for this is the self-generalizability of $\{K(\alpha)\}$ (defined below in Section 2.3.3).

Let $R_t(y_{t-1},\ldots,y_{t-p}) = \sum_{j=1}^{p}\sum_{i=1}^{y_{t-j}} K_{tji}(\alpha_j)$. Then its conditional mean and variance are $\sum_{j=1}^{p}\alpha_j y_{t-j}$ and $\sum_{j=1}^{p} y_{t-j}\,\mathrm{Var}[K(\alpha_j)]$, respectively. That is, this GINAR(p) model has linear conditional expectation and variance, given previous observations. The mean is the same for all $\{K(\alpha)\}$ that have $\mathrm{E}[K(\alpha)] = \alpha$, but the conditional variance depends on the family of $\{K(\alpha)\}$. With a self-generalized family, the conditional variance is $a_K\sum_{j=1}^{p} y_{t-j}\alpha_j(1-\alpha_j)$ so that different families of $\{K(\alpha)\}$ lead to differing amounts of conditional heteroscedasticity, and a larger value of a_K leads to more heteroscedasticity.

The condition for stationarity of (2.4) is $\sum_{j=1}^{p}\alpha_j < 1$. In this case, in stationary state, the equations for the mean and variance lead to

$$\mu_Y = \mu_Y\sum_{j=1}^{p}\alpha_j + \mu_\epsilon, \quad \mu_Y = \frac{\mu_\epsilon}{(1-\alpha_1-\cdots-\alpha_p)}, \tag{2.5}$$

and

$$\sigma_Y^2 = \sigma_Y^2\left[\sum_{j=1}^{p}\alpha_j^2 + 2\sum_{1\le j<k\le p}\rho_{k-j}\alpha_j\alpha_k\right] + \mu_Y\sum_{j=1}^{p}\sigma^2_{K(\alpha_j)} + \sigma^2_\epsilon, \tag{2.6}$$

where ρ_ℓ is the autocorrelation at lag ℓ. If the innovation is overdispersed relative to Poisson (that is, $\sigma^2_\epsilon/\mu_\epsilon \geq 1$), then we show that the stationary distribution of Y is also overdispersed. From (2.5) and (2.6), and assuming $\sigma^2_{K(\alpha)} = a_K\alpha(1-\alpha)$,

$$\begin{aligned}\frac{\sigma_Y^2}{\mu_Y} &= \frac{a_K\sum_{j=1}^{p}\alpha_j(1-\alpha_j) + \sigma^2_\epsilon\left[1-\sum_{j=1}^{p}\alpha_j\right]/\mu_\epsilon}{1-\sum_{j=1}^{p}\alpha_j^2 - 2\sum_{1\le j<k\le p}\rho_{k-j}\alpha_j\alpha_k}\\ &\geq \frac{\sum_{j=1}^{p}\alpha_j(1-\alpha_j) + \left[1-\sum_{j=1}^{p}\alpha_j\right]}{1-\sum_{j=1}^{p}\alpha_j^2 - 2\sum_{1\le j<k\le p}\rho_{k-j}\alpha_j\alpha_k}\\ &= \frac{1-\sum_{j=1}^{p}\alpha_j^2}{1-\sum_{j=1}^{p}\alpha_j^2 - 2\sum_{1\le j<k\le p}\rho_{k-j}\alpha_j\alpha_k} \geq 1,\end{aligned}$$

because $\rho_\ell \geq 0$ and the denominator is positive. The inequality is strict for $p > 1$ with $\rho_1 > 0$. If the innovation is Poisson with $\sigma^2_\epsilon/\mu_\epsilon = 1$ and $a_K = 1$ for binomial thinning, then one still has $\sigma_Y^2/\mu_Y > 1$ for $p > 1$, so that the stationary distribution cannot be Poisson. A Markov model of order p with stationary Poisson marginal distributions and Poisson innovations is developed in Section 2.4.

With $p=1$ and $\alpha_1=\alpha$, the above becomes $D=\sigma_Y^2/\mu_Y=(1+\alpha)^{-1}[a_K\alpha + \sigma^2_\epsilon/\mu_\epsilon]$. For a GINAR($p$) stationary time series without a self-generalized family $\{K(\alpha)\}$, no general overdispersion property can be proved.

2.3.2 Analogues of Gaussian Autoregressive Moving Average

To define analogues of Gaussian moving-average (MA) and ARMA-like models, let $\{\epsilon_t : \ldots, -1, 0, 1, \ldots\}$ be a sequence of independent and identically distributed random variables with support on $\mathcal{N}_0$; ϵ_t is an innovation random variable at time t. In a general context, the extension of moving average of order q becomes q-dependent where observations more than q apart are independent.

The model, denoted as INMA(1), is

$$Y_t = \epsilon_t + K(\alpha') \circledast \epsilon_{t-1} = \epsilon_t + \sum_{i=1}^{\epsilon_{t-1}} K_{ti}(\alpha'), \tag{2.7}$$

with independent $K_{ti}(\alpha')$ over t, i, and the model denoted as INMA(q) is

$$Y_t = \epsilon_t + \sum_{j=1}^{q} \sum_{i=1}^{\epsilon_{t-j}} K_{tji}(\alpha'_j), \tag{2.8}$$

with independent $K_{tji}(\alpha'_j)$ over t, j, i. The model denoted as INARMA($1, q$), with a construction analogous to the Poisson ARMA($1, q$) in McKenzie (1986), is the following:

$$\begin{aligned} Y_t &= W_{t-q} + \sum_{j=0}^{q-1} K(\alpha'_j) \circledast \epsilon_{t-j} = W_{t-q} + \sum_{j=0}^{q-1} \sum_{i=1}^{\epsilon_{t-j}} K_{tji}(\alpha'_j), \\ W_s &= K(\alpha) \circledast W_{s-1} + \epsilon_s = \sum_{\ell=1}^{W_{s-1}} K_{s\ell}(\alpha) + \epsilon_s, \end{aligned} \tag{2.9}$$

with independent $K_{tji}(\alpha'_j), K_{s\ell}(\alpha)$ over t, j, i, s, ℓ. If $\alpha = 0$, then $W_s = \epsilon_s$ and $Y_t = \epsilon_{t-q} + \sum_{j=0}^{q-1} K(\alpha'_j) \circledast \epsilon_{t-j}$ is q-dependent (but not exactly the same as (2.8)).

2.3.3 Classes of Generalized Thinning Operators

In this subsection, some classes of generalized thinning operators and known results about the stationary distribution for GINAR series with $p = 1$ are summarized. The following definitions are needed:

Definition 2.1 (Generalized discrete self-decomposability and innovation).

(a) A nonnegative integer-valued random variable Y is *generalized discrete self-decomposable* (GDSD) with respect to $\{K(\alpha)\}$ if and only if (iff)

$$Y \stackrel{d}{=} K(\alpha) \circledast Y + \epsilon(\alpha) \quad \text{for each } \alpha \in [0, 1].$$

In this case, $\epsilon(\alpha)$ has pgf $G_Y(s)/G_Y(G_K(s; \alpha))$.

(b) Under expectation thinning compounding and a GDSD marginal (with pgf $G_Y(s)$), the stationary time series model is (2.2), where the innovation ϵ_t has pgf $G_Y(s)/G_Y(G_K(s;\alpha))$.

Definition 2.2 (Self-generalized $\{K(\alpha)\}$). Consider a family of $K(\alpha) \sim F_K(\cdot;\alpha)$ with $\mathrm{E}[K(\alpha)] = \alpha$ and pgf $G_K(s;\alpha) = \mathrm{E}[s^{K(\alpha)}]$, $\alpha \in [0,1]$. Then $\{F_K(\cdot;\alpha)\}$ is *self-generalized* iff

$$G_K(G_K(s;\alpha);\alpha') = G_K(s;\alpha\alpha'), \quad \forall\, \alpha,\alpha' \in (0,1).$$

For binomial thinning, the class of possible margins is called the discrete self-decomposable (DSD) class. Note that unless Y is Poisson and $\{K(\alpha)\}$ corresponds to binomial thinning, the distribution of the innovation is in a different parametric family than F_Y.

The terminology of self-generalizability is used in Zhu and Joe (2010b), and the concept is called a semigroup operator in Van Harn and Steutel (1993). Zhu and Joe (2010a) show that (1) $\mathrm{Var}[K(\alpha)] = \sigma^2_{K(\alpha)} = a_K\alpha(1-\alpha)$, where $a_K \geq 1$ for a self-generalized family $\{K(\alpha)\}$ and (2) that generalized thinning operators without self-generalizability lack some closure properties. Also self-generalizability is a nice property for embedding into a continuous-time process.

For NB, Zhu and Joe (2010b) show that $\mathrm{NB}(\theta,\xi)$ is GDSD for three self-generalizable thinning operators that are given below. For $\mathrm{NB}(\theta,\xi)$, with parametrization as given in Section 2.2, the pgf is $G_{NB}(s;\theta,\xi) = [\pi/\{1-(1-\pi)s\}]^\theta$, for $s > 0$, $\theta > 0$ and $\xi > 0$.

Three types of thinning operators based on $\{K(\alpha)\}$ are given below in terms of the pgf, together with $\mathrm{Var}[K(\alpha)]$; the second operator (**I2**) has been used by various authors in several different parametrizations; the specification is simplest via pgfs. The different $\{K(\alpha)\}$ families allow different degrees of conditional heteroscedasticity.

(**I1**) (binomial thinning) $G_K(s;\alpha) = (1-\alpha) + \alpha s$, with $\mathrm{Var}[K(\alpha)] = \alpha(1-\alpha)$.

(**I2**) $G_K(s;\alpha;\gamma) = \frac{(1-\alpha)+(\alpha-\gamma)s}{(1-\alpha\gamma)-(1-\alpha)\gamma s}$, $0 \leq \gamma \leq 1$, with $\mathrm{Var}[K(\alpha)] = \alpha(1-\alpha)(1+\gamma)/(1-\gamma)$. Note that $\gamma = 0$ implies $G_K(z;\alpha) = (1-\alpha) + \alpha s$.

(**I3**) $G_K(s;\alpha;\gamma) = \gamma^{-1}[1+\gamma-(1+\gamma-\gamma s)^\alpha]$, $0 \leq \gamma$, with $\mathrm{Var}[K(\alpha)] = \alpha(1-\alpha)(1+\gamma)$. Note that $\gamma \to 0$ implies $G_K(s;\alpha) = (1-\alpha) + \alpha s$.

For $\mathrm{NB}(\theta,\xi)$, GDSD with respect to $\mathbf{I2}(\gamma)$ holds for $0 \leq \gamma \leq 1-\pi = \xi/(1+\xi)$, and GDSD with respect to $\mathbf{I3}(\gamma)$ holds for $0 \leq \gamma \leq (1-\pi)/\pi = \xi$. For $\mathrm{GP}(\theta,\eta)$, the property of DSD is shown in Zhu and Joe (2003), and it can be shown that $\mathrm{GP}(\theta,\eta)$ is GDSD with respect to $\mathbf{I2}(\gamma(\eta))$, where $\gamma(\eta)$ increases as the overdispersion η increases. Note that the GP distribution does not have a closed-form pgf.

2.3.4 Estimation

For parameter estimation in count time series models, a common estimation approach is CLS. This involves the minimization of $\sum_{i=2}^n (y_i - \mathrm{E}[Y_i|y_{i-1},y_{i-2},\ldots])^2$ for a time series of length n. For a stationary model, it is straightforward to get point estimators of μ_Y and some autocorrelation parameters. One problem with conditional least squares (CLS) is that it cannot distinguish overdispersed Poisson models for ϵ_t and Y_t. For example, if a NB or GP time series is assumed with one of the above generalized thinning operators, then the overdispersion cannot be reliably estimated with an extra moment equation after CLS.

We next mention what can be done for computations of pmfs and the likelihood for binomial thinning and generalized thinning.

1. Zhu and Joe (2006) have an iterative method for computing pmfs with binomial thinning and a DSD stationary margin.
2. The pgf of $K \circledast y$ has closed form if the pgf of $K(\alpha)$ has closed form and the pgf of the innovation has closed form if the pgf of Y has closed form. In this case, Zhu and Joe (2010b) invert a characteristic function for the pgf of $G_{K(\alpha)\circledast y}G_{\epsilon(\alpha)}$ using an algorithm of Davies (1973) to compute the conditional pmf of Y_t given $Y_{t-1} = y_{t-1}$. Let $\varphi_W(s) = \mathrm{E}\left(e^{isW}\right) = G_W\left(e^{is}\right)$ for a nonnegative integer random variable W and define $a(w) := \frac{1}{2} - (2\pi)^{-1}\int_{-\pi}^{\pi}\mathrm{Re}\left(\frac{\varphi_W(u)e^{-iuw}}{1-e^{-iu}}\right)du$. Then $\Pr(W < w) = a(w)$. The pmf of W is

$$f_W(0) = \Pr(W < 1) = a(1), \quad f_W(w) = a(w+1) - a(w), \quad w = 1, 2, \ldots.$$

   This works for NB but not GP because the latter does not have a closed-form pgf.

2.3.5 Incorporation of Covariates

For a NB(θ, ξ) stationary INAR(1) model, the pdf of the innovation is

$$\frac{G_{NB}(s;\theta,\xi)}{G_{NB}(G_K(s;\alpha);\theta,\xi)}.$$

For a time-varying θ_t that depends on covariates with fixed ξ (fixed overdispersion index), suppose the innovation ϵ_t has pgf

$$\frac{G_{NB}(s;\theta_t,\xi)}{G_{NB}(G_K(s;\alpha);\theta_t,\xi)}. \tag{2.10}$$

An advantage of this assumption is that a NB stationary margin results when θ_t is constant.

More generally, for GINAR(p) series where the stationary distribution does not have a simple form, the simplest extension to accommodate covariates is to assume an overdispersed distribution for ϵ_t and absorb a function of covariates into the mean of ϵ_t. Alternatively, other parameters can be made into functions of the covariates.

2.4 Operators in Convolution-Closed Class

The viewpoint in this section is to construct a stationary time series of order p based on a joint pmf $f_{1\cdots(p+1)}$ for $(Y_{t-p}, \ldots, Y_t)$, where marginal pmfs satisfy $f_{1:m} = f_{(1+i):(m+i)}$ for $i = 1, \ldots, p+1-m$ and $m = 2, \ldots, p$. Suppose the univariate marginal pmfs of $f_{1\cdots(p+1)}$ are all $f_Y = f_{Y_t}$. From this, one has a transition probability $f_{p+1|1\cdots p} = f_{Y_t|Y_{t-p},\ldots,Y_{t-1}}$. For $p = 1$, $f_{2|1}$ leads to a stationary Markov time series of order 1. For $p = 2$, $f_{3|12}$ leads to a stationary Markov time series of order 2 if f_{123} has bivariate marginal pmfs $f_{12} = f_{23}$.

There is some theory that covers several count time series models when f_Y is convolution-closed and infinitely divisible (CCID), because there is a way to construct a joint multivariate distribution based on these properties and they lead to thinning operators. This theory provides a bridge between the thinning operator approach of Section 2.3 and the general Markov approach with copulas in Section 2.5.

The operators have been studied in specific discrete cases by McKenzie (1985, 1986, 1988), Al-Osh and Alzaid (1987), and Alzaid and Al-Osh (1993), and in a more general framework in Joe (1996) and Jørgensen and Song (1998).

The general operator is presented first for the Markov order 1 case and then it is mentioned how it can be extended to higher-order Markov or q-dependent, etc. series. Also it will be mentioned how covariates can be accommodated. For this construction, Markov order 1 implies linear conditional expectation but not Markov orders of 2 or higher.

Let $\{F(\cdot;\theta) : \theta > 0\}$ be a CCID parametric family such that $F(\cdot;\theta_1) * F(\cdot;\theta_2) = F(\cdot;\theta_1+\theta_2)$, where $*$ is the convolution operator; $F(\cdot;0)$ corresponds to the degenerate distribution at 0. For $X_j \sim F(\cdot;\theta_j), j = 1,2$, with X_1, X_2 independent, let $H(\cdot;\theta_1,\theta_2,y)$ be the distribution of X_1 given that $X_1 + X_2 = y$. Let $R(\cdot) = R(\cdot;\alpha,\theta)$ $(0 < \alpha \leq 1)$ be a random operator such that $R(Y)$ given $Y = y$ has distribution $H(\cdot;\alpha\theta,(1-\alpha)\theta,y)$, and $R(Y) \sim F(\cdot;\alpha\theta)$ when $Y \sim F(\cdot;\theta)$.

A stationary time series with margin $F(\cdot;\theta)$ and autocorrelation $0 < \alpha < 1$ (at lag 1) can be constructed as

$$Y_t = R_t(Y_{t-1}) + \epsilon_t, \quad R_t(y_{t-1}) \sim H(\cdot;\alpha\theta,(1-\alpha)\theta,y_{t-1}), \tag{2.11}$$

since $F(\cdot;\theta) = F(\cdot;\theta\alpha) * F(\cdot;\theta(1-\alpha))$, when the innovations ϵ_t are independent and identically distributed with distribution $F(\cdot;(1-\alpha)\theta)$. Note that $\{R_t : t \geq 1\}$ are independent replications of the operator R.

The intuitive reasoning is as follows. A consecutive pair (Y_{t-1}, Y_t) has a common latent or unobserved component X_{12} through the stochastic representation:

$$Y_{t-1} = X_{12} + X_1, \quad Y_t = X_{12} + X_2,$$

where X_{12}, X_1, X_2 are independent random variables with distributions $F(\cdot;\alpha\theta)$, $F(\cdot;(1-\alpha)\theta)$, $F(\cdot;(1-\alpha)\theta)$, respectively. The operator $R_t(Y_{t-1})$ "recovers" the unobserved common component X_{12}; hence the distribution of $R_t(y)$ given $Y_{t-1} = y$ must be the same as the distribution of X_{12} given $X_{12} + X_1 = y$.

Examples of CCID operations for the infinite divisible distributions of Poisson, NB and GP are given below.

1. If $F(\cdot;\theta)$ is Po(θ), then $H(\cdot;\alpha\theta,(1-\alpha)\theta,y)$ is Bin(y,α). The resulting operator is binomial thinning.
2. If $F(\cdot;\theta) = F_{NB}(\cdot;\theta,\xi)$ with fixed $\xi > 0$, then $H(\cdot;\alpha\theta,(1-\alpha)\theta,y)$ or $\Pr(X_1 = x \mid X_1 + X_2 = y)$ with X_j independently NB(θ_j,ξ), is Beta-binomial($y,\alpha\theta,(1-\alpha)\theta$) independent of ξ. The pmf of H is

$$h(x;\theta_1,\theta_2,y) = \binom{y}{x}\frac{B(\theta_1+x,\theta_2+y-x)}{B(\theta_1,\theta_2)}, \quad x = 0,1,\ldots,y,$$

The operator matches the random coefficient thinning in Section 2.3, but not binomial thinning or generalized thinning. This first appeared in McKenzie (1986). For (2.11) based on this operator $E[Y_t|Y_{t-1}=y]=\alpha y+(1-\alpha)\theta\xi$, and

$$\text{Var}(Y_t|Y_{t-1}=y)=\frac{(1-\alpha)\theta\xi(1+\xi)+y(\theta+y)\alpha(1-\alpha)}{(\theta+1)}.$$

The conditional variance is quadratically increasing in y for large y, and hence this process has more conditional heteroscedasticity than those based on compounding operators in Section 2.3.

3. If $F(\cdot;\theta)=F_{GP}(\cdot;\theta,\eta)$ with $0<\eta<1$ fixed, then $H(\cdot;\alpha\theta,(1-\alpha)\theta,y)$ or $\Pr(X_1=x\mid X_1+X_2=y)$ with X_j independently $\text{GP}(\theta_j,\eta)$ is a quasi-binomial distribution with parameters $\pi=\theta_1/(\theta_1+\theta_2)$, $\zeta=\eta/(\theta_1+\theta_2)$. The quasi-binomial pmf is:

$$h(x;\pi,\zeta,y)=\binom{y}{x}\frac{\pi(1-\pi)}{1+\zeta y}\left[\frac{\pi+\zeta x}{1+\zeta y}\right]^{x-1}\left[\frac{1-\pi+\zeta(y-x)}{1+\zeta y}\right]^{y-x-1},$$

for $x=0,1,\ldots,y$. For (2.11) with this operator, $E[Y_t|Y_{t-1}=y]=\alpha y+(1-\alpha)\theta/(1-\eta)$,

$$\text{Var}[Y_t|Y_{t-1}=y]=\alpha(1-\alpha)\left[y^2-\sum_{j=0}^{y-2}\frac{y!\zeta^j}{(y-j-2)!(1+y\zeta)^{j+1}}\right]+\frac{(1-\alpha)\theta}{(1-\eta)^3},\quad \zeta=\frac{\eta}{\theta};$$

see Alzaid and Al-Osh (1993). Numerically this is superlinear and asymptotically $O(y^2)$ as $y\to\infty$.

These operators can be used for INMA(q) and INARMA($1,q$) models in an analogous manner to the models in (2.8) and (2.9).

Next, we present the Markov order 2 extension of Joe (1996) and Jung and Tremayne (2011). Consider the following model for three consecutive observations

$$\begin{aligned}Y_{t-2}&=X_{123}+X_{12}+X_{13}+X_1\\Y_{t-1}&=X_{123}+X_{12}+X_{23}+X_2,\\Y_t&=X_{123}+X_{23}+X_{13}+X_3,\end{aligned}\tag{2.12}$$

where $X_1,X_2,X_3,X_{12},X_{13},X_{23},X_{123}$ have distributions in the family $F(\cdot;\theta)$ with respective parameters $\theta_1'=\theta-\theta_0-\theta_1-\theta_2$, $\theta_2'=\theta-\theta_0-2\theta_1$, θ_1', θ_1, θ_2, θ_1, θ_0 (θ is defined so that θ_1',θ_2' are nonnegative). The conditional probability $\Pr(Y_t=y_{new}|Y_{t-1}=y_{prev1},Y_{t-2}=y_{prev2})$ does not lead to a simple operator for Markov order 1, so that computationally one can just use

$$\Pr(Y_t=w_3|Y_{t-1}=w_2,Y_{t-2}=w_1)=\frac{\Pr(Y_{t-2}=w_1,Y_{t-1}=w_2,Y_t=w_3)}{\Pr(Y_{t-2}=w_1,Y_{t-1}=w_2)}.$$

The numerator involves a quadruple sum:

$$\sum_{x_{123}=0}^{w_1\wedge w_2\wedge w_3} \sum_{x_{12}=0}^{(w_1-x_{123})\wedge(w_2-x_{123})} \sum_{x_{23}=0}^{(w_3-x_{123})\wedge(w_2-x_{123}-x_{12})} \sum_{x_{13}=0}^{(w_1-x_{123}-x_{12})\wedge(w_3-x_{123}-x_{23})}$$

$$f(x_{123},\theta_0)f(x_{12};\theta_1)f(x_{23};\theta_1)f(x_{13};\theta_2)f(w_1-x_{123}-x_{12}-x_{13};\theta_1')$$
$$\cdot f(w_2-x_{123}-x_{12}-x_{23};\theta_2')f(w_3-x_{123}-x_{23}-x_{13};\theta_1').$$

For a model simplication, let $\theta_{13}=0$ so that $X_{13}=0$; then the numerator becomes a triple sum. Letting $X_{13}=0$ is sufficient to get a one-parameter extension of Markov order 1; this Markov order 2 model becomes the Markov order 1 model when $\theta_0=\alpha^2\theta, \theta_1=\alpha(1-\alpha)\theta$, $\theta_1'=(1-\alpha)\theta, \theta_2'=(1-\alpha)^2\theta$.

When $\theta_{13}=0$, and $\alpha_1\geq\alpha_2\geq 0$ are the autocorrelations at lags 1 and 2, the time series model has a stochastic representation:

$$Y_t = R_t(Y_{t-1}, Y_{t-2}) + \epsilon_t, \quad \epsilon_t \sim F(\cdot;(1-\alpha_1)\theta), \tag{2.13}$$

where via (2.12), $R_t(y_{t-1}, y_{t-2})$ has the the conditional distribution of $X_{123}+X_{12}+X_{23}$ given $Y_{t-1}=y_{t-1}, Y_{t-2}=y_{t-2}$ and the convolution parameters of $X_{123}, X_{12}, X_1, X_2$ are, respectively, $\alpha_2\theta, (\alpha_1-\alpha_2)\theta, (1-\alpha_1)\theta, (1-2\alpha_1+\alpha_2)\theta$ with $\alpha_2\geq 2\alpha_1-1$. If $\theta_{13}>0$, then the convolution parameters of X_{123}, X_{12}, X_{13}, X_1, X_2 are, respectively, $\theta_0,\theta_1,\theta_2,\theta_1',\theta_2'$ with $\alpha_1=(\theta_1+\theta_0)/\theta, \alpha_2=(\theta_2+\theta_0)/\theta$.

The pattern extends to higher-order Markov but numerically the transition probability becomes too cumbersome because the most general p-dimensional distribution of this type involves 2^p-1 independent X_S for S being a nonempty subset of $\{1,\ldots,p\}$. As mentioned by Jung and Tremayne (2011), the autocorrelation structure of this Markov model for $p\geq 2$ with Poisson, NB, or GP margins does not mimic the Gaussian counterpart, because of a nonlinear conditional mean function.

Because the distribution of the innovation is in the same family as the stationary marginal distribution, the models can be extended easily so that the convolution parameter of Y_t is θ_t, which depends on time-varying covariates. For example, for the Markov order 1 model, with $Y_t\sim F(\cdot;\theta_t)$, $R_t(Y_{t-1})\sim F(\cdot;\alpha\theta_{t-1})$ and $\epsilon_t\sim F(\cdot;\zeta_t)$ with $\zeta_t=\theta_t-\alpha\theta_{t-1}\geq 0$ (Joe 1997, Section 8.4.4). For NB and GP, this means the univariate regression models are NB1 and GP1, respectively.

2.5 Copula-Based Transition

The copula modeling approach is a way to get a joint distribution for $(Y_{t-p},\ldots,Y_t)$ without an assumption of infinite divisibility. Hence univariate margins can be any distribution in the stationary case. However, the property of linear conditional expectation for the Markov order 1 process will be lost. For a $(p+1)$-variate copula $C_{1:(p+1)}$, then $F_{1:(p+1)} = C_{1:(p+1)}(F_Y,\ldots,F_Y)$ is a model for the multivariate discrete distribution of $(Y_{t-p},\ldots,Y_t)$. For stationarity, marginal copulas satisfy $C_{1:m}=C_{(1+i):(m+i)}$ for $i=1,\ldots,p+1-m$ and $m=2,\ldots,p$. The resulting transition probability $\Pr(Y_t=y_t|Y_{t-p}=y_{t-p},\ldots,Y_{t-1}=y_{t-1})$ can be computed from $F_{1:(p+1)}$. If Y were a continuous random variable, there is a simple stochastic representation for the copula-based Markov model in terms of $U(0,1)$ random variables, but this is not the case for Y discrete.

If there are time-varying covariates z_t so that $F_{Y_t} = F(\cdot;\beta, z_t)$, then one can use $F_{1:(p+1)} = C_{1:(p+1)}(F_{Y_{t-p}},\ldots,F_{Y_t})$ for the distribution of $(Y_{t-p},\ldots,Y_t)$ with Markov dependence and a time-varying parameter in the univariate margin.

For q-dependence, one can get a time series model $\{F_Y^{-1}(U_t)\}$ with stationary margin F_Y if $\{U_t\}$ is a q-dependent sequence of $U(0,1)$ random variables. For mixed Markov/q-dependent, a copula model that combines features of Markov and q-dependence can be defined. Chapter 8 of Joe (1997) has the copula time series models for Markov dependence and 1-dependence.

More specific details of parametric models are given for Markov order 1, followed by brief mention of higher-order Markov, q-dependent and mixed Markov/q-dependent.

For a stationary time series model, with stationary univariate distribution F_Y, let $F_{12} = C(F_Y, F_Y;\delta)$ be the distribution of (Y_{t-1}, Y_t) where C is a bivariate copula family with dependence parameter δ. Then the transition probability $\Pr(Y_t = y_t|Y_{t-1} = y_{t-1})$ is

$$f_{2|1}(y_t|y_{t-1}) = \frac{F_{12}(y_{t-1}, y_t) - F_{12}(y_{t-1}^-, y_t) - F_{12}(y_{t-1}, y_t^-) + F_{12}(y_{t-1}^-, y_t^-)}{f_Y(y_{t-1})},$$

where y_i^- is shorthand for $y_i - 1$ for $i = t-1$ and t.

Below are a few examples of one-parameter copula models that include independence, perfect positive dependence, and possibly an extension to negative dependence. Different tail behavior of the copula leads to different asymptotic tail behavior of the conditional expectation and variance, but the conditional expectation is roughly linear in the middle. If a copula C is the distribution of a bivariate uniform vector (U_1, U_2), then the distribution of the reflection $(1-U_1, 1-U_2)$ is $\widehat{C}(u_1,u_2) := u_1 + u_2 - 1 + C(1-u_1, 1-u_2)$. The copula C is reflection symmetric if $C = \widehat{C}$. Otherwise for a reflection asymmetric bivariate copula C, one can also consider $\widehat{C}$ as a model with the opposite direction of tail asymmetry.

The bivariate Gaussian copula can be considered as a baseline model from which other copula families deviate from in tail behavior. Based on Jeffreys' and Kullback–Leibler divergences of Y_1, Y_2 that are NB or GP, the bivariate distribution F_{12} from the binomial thinning operator or the beta/quasi-binomial operators are very similar, with typically a sample size of over 500 needed to distinguish the models when the (lag 1) correlation is moderate (0.4–0.7).

Below is a summary of bivariate copula families with different tail properties and hence different tail behavior of the conditional mean $E(Y_t|Y_{t-1}=y)$ and variance $\mathrm{Var}(Y_t|Y_{t-1}=y)$ as $y\to\infty$, when $F_Y = F_{NB}$ or F_{GP}.

1. *Bivariate Gaussian:* reflection symmetric, with Φ, Φ_2 being the univariate and bivariate Gaussian cdf with mean 0 and variance 1, $C(u_1,u_2;\rho) = \Phi_2(\Phi^{-1}(u_1), \Phi^{-1}(u_2);\rho)$, $-1 < \rho < 1$. The conditional mean is asymptotically slightly sublinear and the conditional variance is asymptotically close to linear.
2. *Bivariate Frank:* reflection symmetric, $C(u_1,u_2;\delta) = -\delta^{-1}\log[1-(1-e^{-\delta u_1})(1-e^{-\delta u_2})/(1-e^{-\delta})]$, $-\infty < \delta < \infty$. Because the upper tail behaves like $1 - u_1 - u_2 + C(u_1,u_2) \sim \zeta(1-u_1)(1-u_2)$ for some $\zeta > 0$ as $u_1, u_2 \to 1^-$, the conditional mean and variance are asymptotically flat.

3. *Bivariate Gumbel:* reflection asymmetric with stronger dependence in the joint upper tail. $C(u_1, u_2; \delta) = \exp\{-[(-\log u_1)^\delta + (-\log u_2)^\delta]^{1/\delta}\}$ for $\delta \geq 1$. The conditional mean is asymptotically linear and the conditional variance is asymptotically sublinear.
4. *Reflected or survival Gumbel:* reflection asymmetric with stronger dependence in the joint lower tail. $C(u_1, u_2; \delta) = u_1 + u_2 - 1 + \exp\{-[(-\log\{1-u_1\})^\delta + (-\log\{1-u_2\})^\delta]^{1/\delta}\}$ for $\delta \geq 1$. The conditional mean and variances are asymptotically sublinear.

The Gumbel or reflected Gumbel copula can be recommended when there is some tail asymmetry relative to Gaussian. The Gumbel copula can be recommended when it is expected that there is some clustering of large values (exceeding a large threshold). The Frank copula is the simple copula that is reflection symmetric and can allow negative dependence. However its bivariate joint upper and lower tails are lighter than the Gaussian's copula, and this has implication that the conditional expectation $E(Y_t|Y_{t-1}=y)$ converges to a constant for large y. For Gaussian, Gumbel, and reflected Gumbel, $E(Y_t|Y_{t-1} = y)$ is asymptotically linear or sublinear for large y for $\{Y_t\}$ with a stationary distribution that is exponentially decreasing (like NB and GP). Some of these results can be proved with the techniques in Hua and Joe (2013).

For second order Markov chains, one just needs a trivariate copula that satisfies a good choice is the trivariate Gaussian copula with lag 1 and lag 2 latent correlations being ρ_1, ρ_2 respectively. If closed-form copula functions are desired, a class to consider has form

$$C_{\psi,H}(u_1, u_2, u_3) = \psi\left(\sum_{j\in\{1,3\}}\left[-\log H(e^{-0.5\psi^{-1}(u_j)}, e^{-0.5\psi^{-1}(u_2)}) + \tfrac{1}{2}\psi^{-1}(u_j)\right]\right),$$

where ψ is the Laplace transform of a positive random variable, H is a bivariate permutation symmetric max-infinite divisible copula; it has bivariate margins:

$$C_{j2} = \psi\left(-\log H\left(e^{-0.5\psi^{-1}(u_j)}, e^{-0.5\psi^{-1}(u_2)}\right) + \tfrac{1}{2}\psi^{-1}(u_j) + \tfrac{1}{2}\psi^{-1}(u_2)\right), \quad j = 1, 3,$$

and $C_{13}(u_1, u_3) = \psi(\psi^{-1}(u_1) + \psi^{-1}(u_3))$. This $C_{\psi,H}$ is a suitable copula, with closed-form cdf, for the transition of a stationary time series of Markov order 2, when there is more dependence for measurements at nearer time points. If a model with clustering of large values is desired, then one can take H to be the bivariate Gumbel copula and ψ to be the positive stable Laplace transform, and then $C_{\psi,H}$ is a trivariate extreme value copula. Other simple choices used for the data set in Section 2.6 are the Frank copula for H together with the positive stable or logarithmic series Laplace transform for ψ.

Both the Gaussian copula and $C_{\psi,H}$ can be extended to AR(p). Other alternatives for copulas for Markov order $p \geq 2$ are based on discrete D-vines (see Panagiotelis et al. 2012).

For copula versions of Gaussian MA(q) series, analogues of (2.8) can be constructed for dependent $U(0,1)$ sequences which can then be converted with the inverse probability transform F_Y^{-1}. For a q-dependent sequence, a $(q+1)$-variate copula $K_{1:(q+1)}$ is needed with margin $K_{1:q}$ being an independence copula. Then $K_{1:(q+1)}$ is the copula of $(\epsilon_{t-q}, \ldots, \epsilon_{t-1}, U_t)$, where $\{\epsilon_t\}$ is a sequence of independent $U(0,1)$ innovation random variables, and $U_t = K_{q+1|1:q}^{-1}(\epsilon_t|\epsilon_{t-q}, \ldots, \epsilon_{t-1})$. Here $K_{q+1|1:q}$ is the conditional distribution of variable $q+1$ given variables $1, \ldots, q$.

For count time series, q-dependence for some positive integer q is generally not expected based on context, as there may be no reason for independence to occur with longer lags. Mixed Markov/q-dependent models with small p, q are better models to consider if the dependence is not as simple as Markov dependence.

The copula version with $p = q = 1$ is defined below (it does extend to $p, q \geq 1$ by combining the above constructions for Markov order p and q-dependence). Let C_{12} and K_{12} be two different bivariate copulas. Define

$$W_s = C_{2|1}^{-1}(\epsilon_s | W_{s-1}), \quad U_t = K_{2|1}^{-1}(\epsilon_t | W_{t-1}), \quad Y_t = F_Y^{-1}(U_t),$$

where $\{\epsilon_t\}$ is a sequence of innovation random variables and $\{W_s\}$ is a sequence of unobserved $U(0,1)$ random variables.

2.6 Statistical Inference and Model Comparisons

For NB margins, numerical maximum likelihood is possible for all of the Markov models in the preceding three sections. For GP margins, all are possible except the ones based on the generalized thinning operators denoted as **I2** and **I3**. With numerical maximum likelihood for Markov models, the asymptotic likelihood inference theory is similar to that for independent observations, using the results in Billingsley (1961). The CLS method can estimate the mean of a stationary distribution but not additional parameters such as overdispersion.

For applications to count time series with small counts, generally low-order Markov models are adequate. The q-dependent models which are the analogies of Gaussian MA(q) are usually less appealing within the context of count data. The analogue of Gaussian ARMA(1,1) has been used in the literature on count time series, but there is less experience with such models. It would be desirable to have the ARMA(1,1) equivalent of the above models in the three categories. The joint likelihood of $Y_1, \ldots, Y_n$ (time series of length n) involves either a high-dimensional sum or integral so that maximum likelihood estimation is not feasible. However for the ARMA(1,1) analogue in the CCID class or copula approach, the joint pmf of three consecutive observations is, respectively, a triple sum, and a triple sum plus an integral. Therefore, likelihood inference could proceed with composite likelihood based on sum of log-likelihoods of subsets of three consecutive observations, as in Davis and Yau (2011) and Ng et al. (2011). Because of space constraints, the data analysis below will only involve Markov models.

In the remainder of this section, the three classes of Markov count time series models are fitted and discussed for one data set, and some results are briefly summarized for another data set that had previously been analyzed.

For some count time data sets with low counts, serial dependence and explanation of trends from covariates, we consider monthly downloads to specialized software. A source for such series is `http://econpapers.repec.org/software/`. Such data are considered here because the next month's total download is plausibly based on the current month's total download through a generalized thinning operator. A similar data set with daily downloads for the program TeXpert is used in Weiß (2008).

The specific series consist of the monthly downloads of `Signalpak` (a package for `octave`, which is a Matlab clone) from September 2000 to March 2013. See Figure 2.1.

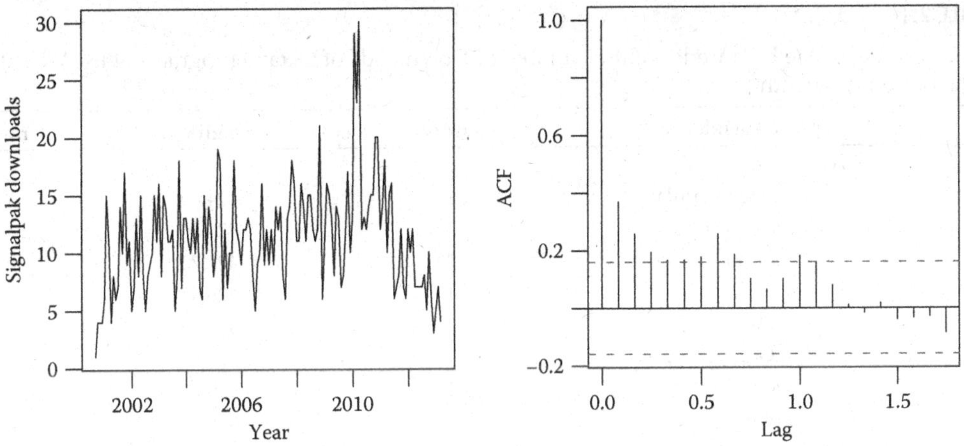

FIGURE 2.1
Time series plot of monthly download of `signalpak`; and the autocorrelation function.

This series looks close to stationary. For longer periods of time, there is no reason to expect the series to be stationary because software can go through periods or local trends of more or less demand. We also consider a covariate for forecasting that is a surrogate for the popularity of the `octave` software; the surrogate is the total monthly abstract views of `octave` codes that are at the website referred to earlier. A summary of model fits is given in Table 2.1.

The use of the covariate marginally improves the log-likelihood and root mean square prediction error. Otherwise the Markov order 1 models fit about the same, and Markov 2 models did not add explanatory power. As might be expected, the generalized thinning operators fit a little better than binomial thinning since those operators are more intuitive for the context of these data. For thinning operators **I2** and **I3**, the γ parameter was set to be the largest possible so that they lead to more conditional heteroscedasticity than the model with binomial thinning. Experience with **I2** and **I3** is that for short time series, if γ is estimated, it is usually at one of the boundaries.

Table 2.1 lists maximum likelihood estimates and corresponding standard errors for one of the better fitting models without/with the covariate. The estimates of the univariate parameters are similar for the different models as well as the strength of serial dependence at lag 1. Note that the addition of the covariate decreases the estimation of overdispersion and lag 1 serial dependence.

For another count time series data set, we also comment on comparisons of fits of models for the data set in Zhu and Joe (2006) on monthly number of claims of short-term disability benefits, made by workers in the logging industry with cut injuries, to the B.C. Workers' Compensation Board for the period of 10 years from 1985 to 1994. This data set of claims is one of few data sets in the literature where "survivor" interpretion of binomial thinning is plausible. There is clear seasonality from the time series plot and autocorrelation function plot, so we use the covariates $(\sin(2\pi t/12), \cos(2\pi t/12))$. This led to root mean square prediction errors (RMSPE) that were from 2.78 to 2.86 with no seasonal covariates and 2.64 to 2.73 with seasonal covariates. The Markov model with a Gaussian copula fitted a little better than those based on thinning operators in terms of

TABLE 2.1

The Covariate Is One Hundredth of the Number of Downloads of Octave Programs at the Website in the Preceding Month

regr.	Dependence	Order	−Loglik	rmspe
	No covariate			
NB	Gaussian copula	1	424.49	4.17
GP	Gaussian copula	1	424.36	4.17
NB	Gaussian copula	2	422.72	4.13
NB	Frank copula	1	428.65	4.24
NB	Gumbel copula	1	427.35	4.24
NB	refl.Gumbel copula	1	424.07	4.15
NB	conv-closed oper.	1	425.24	4.18
NB	binomial thinning	1	426.49	4.18
NB	**I2** thinning	1	425.36	4.18
NB	**I3** thinning	1	425.83	4.18
	One covariate			
NB1	Gaussian copula	1	416.07	4.02
NB2	Gaussian copula	1	416.03	4.02
GP1	Gaussian copula	1	415.99	4.02
NB1	Gaussian copula	2	415.79	4.01
NB2	Gaussian copula	2	415.76	4.02
NB1	Frank copula	1	416.98	4.05
NB2	Frank copula	1	416.90	4.05
NB1	Gumbel copula	1	416.00	4.01
NB2	Gumbel copula	1	416.25	4.01
NB1	refl.Gumbel copula	1	417.14	4.02
NB2	refl.Gumbel copula	1	417.04	4.02
NB1	conv-closed oper.	1	415.99	4.02
NB	binomial thinning	1	416.10	4.02
NB	**I2** thinning	1	415.63	4.02
NB	**I3** thinning	1	415.86	4.02
	Model (Markov order 1)		MLE	SE
NB with Gaussian copula and no covariates			$\hat{\beta}_0 = 2.40$	0.05
			$\hat{\xi} = 0.89$	0.27
			$\hat{\rho} = 0.42$	0.08
NB1 with Gaussian copula			$\hat{\beta}_0 = 1.93$	0.11
			$\hat{\beta}_1 = 0.09$	0.02
			$\hat{\xi} = 0.48$	0.18
			$\hat{\rho} = 0.22$	0.09

Note: For the models with covariates based on the **I2** and **I3** thinning operators, the formulation of Section 2.3.5 is used. The root mean square prediction error (rmspe) is defined as $\{(n-p)^{-1}\sum_{t=1+p}^{n}(y_t - \hat{\text{E}}(Y_t|Y_{t-1} = y_{t-1}, \ldots, Y_{t-p} = y_{t-p})\}^{1/2}$, where n is the length of the series and p is the Markov order and $\hat{\text{E}}$ means the conditional expectation with parameter equal to the maximum likelihood estimate (MLE) of the model. In the bottom, MLEs and corresponding standard errors are given for one of the better fitting models without and with the covariate.

log-likelihood, but the Frank copula model with NB2 regression margins was the best in terms of log-likelihood and RMSPE.

Similar to Table 2.1, there is a bit more variation among copula model fits than those based on thinning operators (the latter models tend to be close to the Gaussian copula model). An explanation is that different copula families have a wide variety of shapes of conditional expectation and variance functions.

Finally, we indicate an advantage of count time series models with known univariate marginal distributions. In this case, univariate and conditional probabilities can be easily obtained and also predictive intervals with and without the previous observation(s). A predictive interval without the previous time point is a regression inference and a predictive interval with the previous time point is an inference that combines regression and forecasting. The first type of predictive interval would not be easy to do with other classes of count time series models that are based on conditional specifications.

Acknowledgments

This work was supported by an NSERC Discovery grant.

References

Al-Osh, M. A. and Alzaid, A. A. (1987). First-order integer-valued autoregressive INAR(1) process. *Journal of Time Series Analysis*, 8:261–275.

Alzaid, A. A. and Al-Osh, M. (1990). An integer-valued pth-order autoregressive structure INAR(p) process. *Journal of Applied Probability*, 27(2):314–324.

Alzaid, A. A. and Al-Osh, M. A. (1993). Some autoregressive moving average processes with generalized Poisson marginal distributions. *Annals of the Institute of Statistical Mathematics*, 45:223–232.

Beare, B. K. (2010). Copulas and temporal dependence. *Econometrica*, 78(1):395–410.

Biller, B. (2009). Copula-based multivariate input models for stochastic simulation. *Operations Research*, 57(4):878–892.

Biller, B. and Nelson, B. L. (2005). Fitting time-series input processes for simulation. *Operations Research*, 53(3):549–559.

Billingsley, P. (1961). *Statistical Inference for Markov Processes*. University of Chicago Press, Chicago, IL.

Cameron, A. C. and Trivedi, P. K. (1998). *Regression Analysis of Count Data*. Cambridge University Press, Cambridge, U.K.

Davies, R. B. (1973). Numerical inversion of a characteristic function. *Biometrika*, 60:415–417.

Davis, R. A., Dunsmuir, W. T. M., and Wang, Y. (2000). On autocorrelation in a Poisson regression model. *Biometrika*, 87:491–505.

Davis, R. A. and Yau, C. Y. (2011). Comments on pairwise likelihood in time series models. *Statistica Sinica*, 21(1):255–277.

Du, J.-G. and Li, Y. (1991). The integer-valued autoregressive (INAR(p)) model. *Journal of Time Series Analysis*, 12(2):129–142.

Escarela, G., Mena, R. H., and Castillo-Morales, A. (2006). A flexible class of parametric transition regression models based on copulas: Application to poliomyelitis incidence. *Statistical Methods in Medical Research*, 15:593–609.

Fokianos, K. (2011). Some recent progress in count time series. *Statistics*, 45(1):49–58.

Fokianos, K. (2012). Count time series models. In Rao, T. S., Rao, S. S., and Rao, C. R., eds., *Time Series Analysis: Methods and Applications*, vol. 30, Chapter 12, pp. 315–347. Elsevier B.V., Amsterdam, the Netherlands.

Gauthier, G. and Latour, A. (1994). Convergence forte des estimateurs des paramètres d'un processus GENAR(p). *Annales des Sciences Mathematique du Québec*, 18:49–71.

Hua, L. and Joe, H. (2013). Strength of tail dependence based on conditional tail expectation. *Journal of Multivariate Analysis*, 123:143–159.

Joe, H. (1996). Time series models with univariate margins in the convolution-closed infinitely divisible class. *Journal of Applied Probability*, 33(3):664–677.

Joe, H. (1997). *Multivariate Models and Dependence Concepts*. Chapman & Hall, London, U.K.

Jørgensen, B. and Song, P. X. K. (1998). Stationary time series models with exponential dispersion model margins. *Journal of Applied Probability*, 35(1):78–92.

Jung, R. C. and Tremayne, A. R. (2011). Convolution-closed models for count time series with applications. *Journal of Time Series Analysis*, 32(3):268–280.

Latour, A. (1997). The multivariate GINAR(p) process. *Advances in Applied Probability*, 29(1): 228–248.

Latour, A. (1998). Existence and stochastic structure of a non-negative integer-valued autoregressive process. *Journal of Time Series Analysis*, 19:439–455.

Lawless, J. F. (1987). Negative binomial and mixed Poisson regression. *Canadian Journal of Statistics*, 15:209–225.

Lawrance, A. J. and Lewis, P. A. W. (1980). The exponential autoregressive-moving average earma(p, q) process. *Journal of the Royal Statistical Society B*, 42:150–161.

McKenzie, E. (1985). Some simple-models for discrete variate time-series. *Water Resources Bulletin*, 21(4):645–650.

McKenzie, E. (1986). Autoregressive moving-average processes with negative-binomial and geometric marginal distributions. *Advances in Applied Probability*, 18(3):679–705.

McKenzie, E. (1988). Some ARMA models for dependent sequences of Poisson counts. *Advances in Applied Probability*, 20:822–835.

McKenzie, E. (2003). Discrete variate time series. In Shanbag, D. N. and Rao, C. R., eds., *Stochastic Processes: Modelling and Simulation, Handbook of Statistics*, vol. 21, pp. 573–606. Elsevier, Amsterdam, the Netherlands.

McNeil, A. J., Frey, R., and Embrechts, P. (2005). *Quantitative Risk Management*. Princeton University Press, Princeton, NJ.

Ng, C. T., Joe, H., Karlis, D., and Liu, J. (2011). Composite likelihood for time series models with a latent autoregressive process. *Statistica Sinica*, 21(1):279–305.

Panagiotelis, A., Czado, C., and Joe, H. (2012). Pair copula constructions for multivariate discrete data. *Journal of the American Statistical Association*, 107:1063–1072.

Steutel, F. W. and Van Harn, K. (1979). Discrete analogs of self-decomposability and stability. *Annals of Probability*, 7(5):893–899.

Van Harn, K. and Steutel, F. W. (1993). Stability equations for processes with stationary independent increments using branching-processes and Poisson mixtures. *Stochastic Processes and Their Applications*, 45(2):209–230.

Weiß, C. H. (2008). Thinning operations for modeling time series of counts—A survey. *ASTA-Advances in Statistical Analysis*, 92(3):319–341.

Zheng, H., Basawa, I. V., and Datta, S. (2006). Inference for pth-order random coefficient integer-valued autoregressive processes. *Journal of Time Series Analysis*, 27(3):411–440.

Zheng, H., Basawa, I. V., and Datta, S. (2007). First-order random coefficient integer-valued autoregressive processes. *Journal of Statistical Planning and Inference*, 137(1):212–229.

Zhu, R. and Joe, H. (2003). A new type of discrete self-decomposability and its application to continuous-time Markov processes for modeling count data time series. *Stochastic Models*, 19(2):235–254.

Zhu, R. and Joe, H. (2006). Modelling count data time series with markov processes based on binomial thinning. *Journal of Time Series Analysis*, 27(5):725–738.

Zhu, R. and Joe, H. (2010a). Count data time series models based on expectation thinning. *Stochastic Models*, 26(3):431–462.

Zhu, R. and Joe, H. (2010b). Negative binomial time series models based on expectation thinning operators. *Journal of Statistical Planning and Inference*, 140(7):1874–1888.

3

Generalized Linear Autoregressive Moving Average Models

William T.M. Dunsmuir

CONTENTS

3.1 Introduction

Generalized linear autoregressive moving average (GLARMA) models are a class of observation-driven non-Gaussian nonlinear state space models in which the state process depends linearly on covariates and nonlinearly on past values of the observed process. Conditional on the state process, the observations are independent and have a distribution from the exponential family. This could include continuous responses but, in this chapter, we focus entirely on discrete responses such as binary, binomial, Poisson, or negative binomial distributions.

The main advantage of GLARMA models over other observation or parameter-driven models is that they can be fit relatively easily to long time series or to many individual time series. Examples of the latter will be given herein. They provide a natural extension of the GLM modeling framework to include serial dependence terms and provide rapid assessment of the presence and form of serial dependence. This model-based assessment of serial dependence is particularly useful for discrete response data for which standard sample autocorrelation methods can be misleading—see Davis et al. (2000) for example.

Since their genesis in the unpublished paper by Shephard (1995), GLARMA models have found application in many disciplines including financial modeling (Rydberg and Shephard, 2003; Liesenfeld et al., 2006), epidemiological assessments (Davis et al., 2003; Turner et al., 2011), clinical management (Buckley and Bulger, 2012), analysis of crime statistics (Dunsmuir et al., 2008), and primate behavior (Etting and Isbell, 2014).

Benjamin et al. (2003) provide a review of generalized autoregressive models of which the GLARMA models are a subclass. Davis et al. (1999) review various approaches to modeling count time series, whereas Davis et al. (2003, 2005) consider GLARMA models for Poisson response series. Kedem and Fokianos (2002) provide a comprehensive coverage of observation-driven models for discrete response time series with covariates.

In this chapter, we first review the properties of GLARMA models for single time series and describe how they can be estimated using the `glarma` package (Dunsmuir et al., 2014) for the R language (R Core Team, 2014)—see also Dunsmuir and Scott (2015). Additionally, we describe how univariate GLARMA fitting software can be adapted to model multiple independent series either by a fixed effects analysis or by a random effects analysis. Multiple independent time series share many common features with longitudinal data or panel data collections, but differ in two key aspects. First, in a longitudinal data setting there are typically a large number of independent short temporal observation trajectories, whereas in the applications we discuss, there are a small to moderate number of time series each of which is long. Second, longitudinal data analyses typically restrict the modeling of serial dependence in each trajectory to be the same across the trajectories, whereas for the applications we have in mind, the serial dependence can vary across the series. Indeed, some series may not require serial dependence terms at all, whereas others can have substantial serial dependence. A major focus of this chapter is on describing how to combine GLARMA models for single time series with random effects on regressors between series.

3.2 GLARMA Models for Single Time Series

In this section, we describe the GLARMA model class for a single time series. Let $y_1, \ldots, y_n$ be the available observations on the discrete response series. Associated with these are vectors, $\mathbf{x}_t$, of K regressors observed for $t = 1, \ldots, n$. Let $\mathcal{F}_t = \{Y_s : s < t, \mathbf{x}_s : s \leq t\}$ denote the past information available on the response series and the past and present information on the regressors. The distribution of Y_t conditional on $\mathcal{F}_t$ is assumed to be of exponential family form

$$f(y_t|W_t) = \exp\left\{y_t W_t - a_t b(W_t) + c_t\right\}, \tag{3.1}$$

where $\{a_t\}$ and $\{c_t\}$ are sequences of constants, with c_t often depending on the observations y_t. The response distribution could also be the negative binomial for which an additional parameter is required. The information in $\mathcal{F}_t$ is summarized in the state variable W_t; that is, W_t is a function of the elements in $\mathcal{F}_t$. We denote the conditional means and variances of the responses as $\mu_t := E(Y_t|W_t)$ and $\sigma_t^2 := \text{var}(Y_t|W_t)$. Throughout, we will use the canonical link connecting μ_t and W_t, in which case, $\mu_t = a_t\dot{b}(W_t)$ and $\sigma_t^2 = a_t\ddot{b}(W_t)$, where $\dot{b}(u)$ and $\ddot{b}(u)$ are the first and second derivatives, respectively, of $b(u)$ with respect to u.

While (3.1) is not the fully general form of the exponential family (see McCullagh and Nelder, 1989), it covers several popular and useful distributions. The basic theory and computational methodology presented here can be readily modified to include other response distributions or more general specifications of link functions. An example of such an extension is the use of the negative binomial response distribution. Let $\mu_t = \exp(W_t)$. The `glarma` package uses the negative binomial density in the form

$$f(y_t|W_t,\alpha) = \frac{\Gamma(\alpha + y_t)}{\Gamma(\alpha)\Gamma(y_t+1)}\left[\frac{\alpha}{\alpha+\mu_t}\right]^{\alpha}\left[\frac{\mu_t}{\alpha+\mu_t}\right]^{y_t}, \quad y_t = 0,1,2,\ldots. \tag{3.2}$$

Note that $\sigma_t^2 = \mu_t + \mu_t^2/\alpha$. As $\alpha \to \infty$, the negative binomial density converges to the Poisson density. Also note that if α is known, this density belongs to the one-parameter exponential family with appropriate definitions of θ_t, $b(\theta_t)$, a_t, c_t. If α is not known, (3.2) is not a member of the one-parameter exponential family.

Specification of W_t for observation-driven models takes various forms; see Benjamin et al. (2003) for a general discussion or Kedem and Fokianos (2002) for a comprehensive treatment of various models. Following our previous work (Davis et al., 1999, 2003), we consider the case where the state vector in (3.1) is linear in the covariates

$$W_t = \mathbf{x}_t^T\beta + Z_t. \tag{3.3}$$

Here the "noise" process $\{Z_t\}$, which induces serial dependence in the states and hence in the observations, is of the form

$$Z_t = \sum_{j=1}^{\infty}\gamma_j(\psi)e_{t-j}, \tag{3.4}$$

with the parameters ψ specified separately from the regression coefficient β.

The predictive residuals in (3.4) are defined as

$$e_t = \frac{Y_t - \mu_t}{\nu_t} \tag{3.5}$$

where the scaling sequence $\{\nu_t\}$ is to be selected. We consider three choices (currently as supported in the `glarma` package): $\nu_{P,t} = \sigma_t$, giving classical Pearson residuals; $\nu_{S,t} = \sigma_t^2$, giving the "score-type" residuals suggested by Creal et al. (2008); and, $\nu_{I,t} = 1$ (referred to as "identity" scaling), mainly used for binary response GLARMA models. Note that $\{e_t\}$ are martingale differences, and hence are zero mean and uncorrelated. The Pearson type $\{e_t\}$ also have unit variance, and hence are weakly stationary white noise. Often, when the identity scaling is used in the Poisson and negative binomial cases, the resulting state equation can be explosive leading to infinite means or zero means, both forms of degeneracy.

This may not be very crucial for the binomial response case but little is currently known about such stability issues. We discuss this topic in more detail later in this chapter.

Following Davis et al. (1999), for a GLARMA model, the general form of $\{Z_t\}$ in (3.4) is specified via an ARMA-type recursion

$$Z_t = \sum_{i=1}^{p} \phi_i(Z_{t-i} + e_{t-i}) + \sum_{i=1}^{q} \theta_i e_{t-i}. \tag{3.6}$$

The $\{Z_t\}$ defined in this way can be thought of as the best linear predictor of a stationary invertible ARMA process with driving noise $\{e_t\}$ of scaled predictive residuals.

The recursions to construct the Z_t component of the state and its derivatives require initialization. In the `glarma` package we set $e_t = 0$ and $Z_t = 0$ for $t \leq 0$ ensuring that the conditional and unconditional expected values of e_t are zero for all t.

3.2.1 The GLARMA Likelihood

Given n successive observations y_t $(t = 1, \ldots, n)$ on the response series, and fixed initial conditions (such as using zeros) for the recursions in (3.6), the likelihood is constructed as the product of conditional densities of Y_t given $\mathcal{F}_t$ (or equivalently the state W_t) giving the log-likelihood corresponding to the distribution (3.1) as

$$l(\boldsymbol{\delta}) = \sum_{t=1}^{n} \{y_t W_t(\boldsymbol{\delta}) - a_t b(W_t(\boldsymbol{\delta})) + c_t\}, \tag{3.7}$$

where $\boldsymbol{\delta} = (\boldsymbol{\beta}, \boldsymbol{\phi}, \boldsymbol{\theta})$. For the negative binomial response distribution, the log-likelihood is more complicated because the shape parameter α also has to be estimated along with $\boldsymbol{\beta}$, $\boldsymbol{\phi}$, and $\boldsymbol{\theta}$. For this case, we expand the parameters to $\boldsymbol{\delta} = (\boldsymbol{\beta}, \boldsymbol{\phi}, \boldsymbol{\theta}, \alpha)$. Because the e_t in (3.5), the Z_t in (3.6), and the W_t in (3.3) are functions of the unknown parameter $\boldsymbol{\delta}$, they must be recomputed at each iteration to maximize the likelihood. For $t = 1, \ldots, n$, recursive expressions for calculating e_t, Z_t, and W_t as well as their first and second partial derivatives with respect to $\boldsymbol{\delta}$ are available in Davis et al. (2005) for the Poisson case. Corresponding formulae for the binomial and negative binomial cases are easily derived in a similar way. The essential computational cost lies with the recursions for Z_t and W_t $(t = 1, \ldots, n)$ and their first and second derivatives. These recursions are common to all response distributions. It is only the computation of e_t and the $\log f_{Y_t|W_t}(y_t|W_t; \boldsymbol{\delta})$ that depend on the response distribution being used and choice of scaling ν_t.

It may be tempting to consider maximization of (3.7) by using readily available existing GLM (generalized linear modeling) software. Without modification, this is not possible because, for $t = 1, \ldots, n$, the e_t and hence Z_t appearing in the state equation for W_t are functions of unknown parameters. One suggestion that is sometimes made is, for an initial value of $\boldsymbol{\delta}$, calculate $e_t(\boldsymbol{\delta})$ and hence via (3.6), $Z_t(\boldsymbol{\delta})$. Using these, create an augmented set of covariates $\tilde{\mathbf{x}}_t^T = \left(\mathbf{x}_t^T, Z_{t-1}, \ldots, Z_{t-p}, e_{t-1}, \ldots, e_{t-\max(p,q)}\right)$ and update the parameters estimates using standard GLM software for the regression term $\tilde{\mathbf{x}}_t^T \boldsymbol{\delta}$ to update the estimate of $\boldsymbol{\delta}$. Recursion of this method does converge but unfortunately not to the maximum likelihood estimate and hence is biased. In any case, there is not much additional computational cost using the GLARMA likelihood calculations directly.

The log-likelihood is maximized from a suitable starting value $\boldsymbol{\delta}$ using Newton–Raphson iteration or a Fisher scoring approximation. Define the vector of first and second derivatives of the log-likelihood by $d(\boldsymbol{\delta}) = \partial l(\boldsymbol{\delta})/\partial \boldsymbol{\delta}$ and $D_{\text{NR}}(\boldsymbol{\delta}) = \partial^2 l(\boldsymbol{\delta})/\partial \boldsymbol{\delta} \partial \boldsymbol{\delta}^T$, where the matrix of second derivatives of the log-likelihood for (3.7) is given by

$$D_{\text{NR}}(\boldsymbol{\delta}) = \sum_{t=1}^{n} \left[y_t - a_t \dot{b}(W_t) \right] \frac{\partial^2 W_t}{\partial \boldsymbol{\delta} \partial \boldsymbol{\delta}^T} - \sum_{t=1}^{n} a_t \ddot{b}(W_t) \frac{\partial W_t}{\partial \boldsymbol{\delta}} \frac{\partial W_t}{\partial \boldsymbol{\delta}^T}. \tag{3.8}$$

At the true parameter $\boldsymbol{\delta}$, $E[y_t - a_t \dot{b}(W_t) | \mathcal{F}_t] = 0$. Hence the expected value of the first summation in (3.8) is zero which motivates the Fisher scoring–type approximation based only on first derivatives given by

$$D_{\text{FS}}(\boldsymbol{\delta}) = -\sum_{t=1}^{n} a_t \ddot{b}(W_t) \frac{\partial W_t}{\partial \boldsymbol{\delta}} \frac{\partial W_t}{\partial \boldsymbol{\delta}^T}. \tag{3.9}$$

Note that $E[D_{\text{FS}}(\boldsymbol{\delta})] = E[D_{\text{NR}}(\boldsymbol{\delta})]$. Also, using the fact that $d(\boldsymbol{\delta})$ is a sum of martingale differences, the usual identity $E[D_{\text{NR}}(\boldsymbol{\delta})] = -E\left[d(\boldsymbol{\delta})d(\boldsymbol{\delta})^T\right]$ holds. These expectations cannot be computed in closed form. From an initial value for $\boldsymbol{\delta}$, Newton–Raphson (using D_{NR}) or approximate Fisher scoring (using D_{FS}) methods are used to find the maximum likelihood estimate $\hat{\boldsymbol{\delta}}$.

3.2.2 Parameter Identifiability

The GLARMA component Z_t of the state variable given in (3.6) can be rewritten as

$$Z_t = \sum_{i=1}^{p} \phi_i Z_{t-i} + \sum_{i=1}^{\tilde{q}} \tilde{\theta}_i e_{t-i}, \tag{3.10}$$

where $\tilde{q} = \max(p, q)$ and if $p \le q$, $\tilde{\theta}_j = \theta_j + \phi_j$ for $j = 1, \ldots, p$ and $\tilde{\theta}_j = \theta_j$ for $j = p + 1, \ldots, q$, while if $p > q$, $\tilde{\theta}_j = \theta_j + \phi_j$ for $j = 1, \ldots, q$ and $\tilde{\theta}_j = \phi_j$ for $j = q + 1, \ldots, p$. Hence, if $Z_t = 0$ for $t \le 0$ and $e_t = 0$ for $t \le 0$) and if $\tilde{\theta}_j = 0$ for $j = 1, \ldots, \tilde{q}$, the recursion (3.10) renders $Z_t = 0$ for all t, there is no serial dependence in the GLARMA model, and it reduces to a standard GLM model. This is equivalent to $\phi_j = -\theta_j$ for $j = 1, \ldots, p$ and $\theta_j = 0$ for $j = p + 1, \ldots, \tilde{q}$. Consequently, the null hypothesis of no serial dependence requires only these constraints on $\boldsymbol{\theta}$ and $\boldsymbol{\phi}$ and there can be nuisance parameters that cannot be estimated. This has implications for (1) convergence of the iterations required to optimize the likelihood and (2) on testing that there is no serial dependence in the observations, beyond that induced by the regression component $\mathbf{x}_t^T \boldsymbol{\beta}$.

In cases where $p > 0$ and $q = 0$ (equivalent to an ARMA(p, p) specification with constraint $\theta_j = \phi_j$) or where $p = 0$ and $q > 0$ (a pure MA(q)) case, identification issues do not arise and there are no nuisance parameters to take into account when using a score, Wald or likelihood ratio test for the GLARMA parameters. However, in other "mixed" model cases where both $p > 0$ and $q > 0$, some care is needed when fitting models and testing for serial dependence. To illustrate, consider the case where no serial dependence exists but $p = q > 0$ is specified. Then the likelihood iterations are unlikely to converge because the likelihood surface will be "ridge-like" on the manifold where $\phi_j = -\theta_j$, an issue that is encountered for standard ARMA model fitting. Corresponding to this, the second derivative matrix $D_{\text{NR}}(\boldsymbol{\delta})$ will be singular or the state variable W_t can degenerate or diverge. Because of this possibility, it is prudent to start with low orders for p and q and avoid specifying them as equal. Once stability of estimation is reached for a lower-order specification, increasing the values of p or q could be attempted.

The likelihood ratio test that there is no serial dependence versus the alternative that there is GLARMA-like serial dependence with $p = q > 0$ will not have a standard chi-squared distribution because the parameters ϕ_j, for $j = 1, \ldots, p$, are nuisance parameters which cannot be estimated under the null hypothesis. Testing methods such as those proposed by Hansen (1996) or Davies (1987) need to be developed for this situation. Further details on these points can be found in Dunsmuir and Scott (2015).

3.2.3 Distribution Theory for Likelihood Estimation

The consistency and asymptotic distribution for the maximum likelihood estimate $\hat{\boldsymbol{\delta}}$ is rigorously established only in a limited number of special cases. In the stationary Poisson response case in Davis et al. (2003) where $x_t^T \equiv 1$ (intercept only) and $p = 0$ and $q = 1$, these results have been proved rigorously. Similarly, for simple models in the Bernoulli stationary case in Streett (2000) these results hold. Simulation results are also reported in Davis et al. (1999, 2003) for nonstationary Poisson models. Other simulations not reported in the literature support the supposition that $\hat{\boldsymbol{\delta}}$ has a multivariate normal distribution for large samples for a range of regression designs and for the various response distributions considered here.

For inference in the GLARMA model, it is assumed that the central limit theorem holds so that

$$\hat{\boldsymbol{\delta}} \overset{d}{\approx} N(\boldsymbol{\delta}, \hat{\Omega}), \tag{3.11}$$

where the approximate covariance matrix is estimated by $\hat{\Omega} = -D_{\text{NR}}(\hat{\boldsymbol{\delta}})^{-1}$ or $\hat{\Omega} = -D_{\text{FS}}(\hat{\boldsymbol{\delta}})^{-1}$. In the `glarma` package, this distribution is used to obtain standard errors and to construct Wald tests of the hypotheses that subsets of $\boldsymbol{\delta}$ are zero. It is also assumed that Wald tests and equivalent likelihood ratio tests will be asympotically chi-squared with the correct degrees of freedom, results which would follow straightforwardly from (3.11) and its proof when available. Regardless of the technical issues involved in establishing a general central limit theorem, the earlier approximate result seems plausible since, for these models, the log-likelihood is a sum of elements in a triangular array of martingale differences. Conditions under which this result would likely hold include identifiability conditions as discussed earlier, conditions on the regressors similar to those used in Davis et al. (2000, 2003) and Davis and Wu (2009), where the covariates $\mathbf{x}_t$ are assumed to be a realization of a stationary time series or is defined as $\mathbf{x}_t = \mathbf{x}_{nt} = \mathbf{f}(t/n)$ where $\mathbf{f}(u)$ is a piecewise continuous function from $u \in [0, 1]$ to $\mathbb{R}^{\mathbb{K}}$. Additional conditions on the coefficients are also needed to ensure that Z_t and hence W_t do not degenerate or grow without bound. Indeed, little is known so far about suitable conditions to ensure this.

3.2.4 Convergence of GLARMA Recursions: Ergodicity and Stationarity

To date, the stationarity and ergodicity properties of the GLARMA model are only partially understood. These properties are important to ensure that the process is capable of generating sample paths that do not degenerate to zero or do not explode as time progresses, as well as for establishing the large sample distributional properties of parameter estimates. Davis et al. (2003) provide results for the simplest of all models: Poisson responses specified with $p = 0$, $q = 1$, and $\mathbf{x}_t^T \beta = \beta$. Results for simple examples of the stationary Bernoulli case are given in Streett (2000).

For the Poisson response distribution GLARMA model, failure to scale by the variance or standard deviation will lead to unstable Poisson means (that diverge to infinity or collapse to zero as an absorbing state for instance) and existence of stationary and ergodic solutions to the recursive state equation is not assured—see Davis et al. (1999, 2003, 2005) for details. For the binomial situation, this lack of scaling should not necessarily lead to instability in the success probability as time evolves since the success probabilities, p_t, and observed responses, Y_t, are both bounded between 0 and 1. Thus, degeneracy can only arise if the regressors $\mathbf{x}_t$ become unbounded. As recommended in Davis et al. (1999), temporal trend regressors should be scaled using a factor relating to the sample size n.

Asymptotic results for various types of observation-driven models without covariates are increasingly becoming available. Tjøstheim (2012) (also see Tjøstheim [2015; Chapter 4 in this volume]) has provided a review of the ergodic and stationarity properties for various observation-driven models—primarily in the Poisson response context—as well as presenting large sample theory for likelihood estimation. Wang and Li (2011) discuss the binary (BARMA) model and present some asymptotic results for that case. However, the state equation for the BARMA model differs from that for the binary GLARMA model in that the latter involved scaled residuals while the former uses identity residuals and is also structurally different in its use of past observation of $\{Y_t\}$. Davis and Liu (2015) present general results for the one-parameter exponential family response distributions and a semi-parametric observation-driven specification of the state equation. Woodard et al. (2011) present some general results on stationarity and ergodicity for the GARMA models similar to those available in Benjamin et al. (2003). However, because the state equation recursions involve applying the link function to both the responses and the mean responses, none of these results apply to the specific form of GLARMA models presented here. Also, none of these recent results consider the case of covariates; hence they are not, as yet, applicable to likelihood estimation for regression models for discrete outcome time series.

3.3 Application of Univariate GLARMA Models

We now illustrate the fitting of GLARMA models to binomial and binary time series arising in the study of listener responses to a segment of electroacoustic music. This example also motivates studying multiple independent time series as an ensemble (as will be discussed in the next section). The background to the analysis presented here is in Dean et al. (2014b). Members of a panel comprising three musical expertise groups (8 electroacoustic musicians, 8 musicians, and 16 nonmusicians) provided real-time responses to a segment of electroacoustic music. Aims of these types of experiments are to determine the way in which features of the music (in this case the sound intensity) impact listener response measured in various ways (we concentrate on the "arousal" response). Dean et al. (2014b) present a variety of standard time series methods for modeling the arousal responsiveness in terms of lags of the musical intensity. Questions such as: "Do the musical expertise groups display differences between them with respect to the impact of intensity on their group average responses?" and "Are there substantial differences between individuals within the panel or within each musical group?" were addressed. For example, in Dean et al. (2014b), the transfer function coefficients for the impact of *changes*, at lag 1, of musical intensity on the *changes* in arousal were modeled using 11 lag transfer function models of the form

$$\nabla Y_{jt} = \omega_0 + \sum_{k=1}^{11} \omega_{jk} \nabla X_{t-k} + \alpha_{jt}, \tag{3.12}$$

where $\nabla A_t = A_t - A_{t-1}$ and α_{jt} was modeled by an autoregression of at most order 3. There was evidence that variation between individuals in their responsiveness was substantial and suggested use of a cross-sectional time series analysis as in Dean et al. (2014a). However, analysis of individual responses suggested varying levels of "stasis"; that is, frequent and sometime prolonged periods during which their arousal response did not change. In some cases, this led to a high level of zeros in their differenced responses, which has the potential to impact the validity of model estimates based on traditional Gaussian linear time series analysis. In fact, listeners varied substantially in the amount of time that their responses are in stasis, for example, from 15% to 90%. Response distributions with such large numbers of zeros constitute a challenge for conventional time series analysis.

To examine how robust their findings were based on standard Gaussian linear times series methods, Dean et al. (2014b) also considered an approach similar to that employed in Rydberg and Shephard (2003) for decomposing transaction level stock price data into components of change and size of change. For each listener, binary responses were defined as $D_{jt} = 1$ if the change in their response from time $t-1$ to t was positive, otherwise $D_{jt} = 0$. This is one of the two components of a potential trinomial model for the change process for which the approach of Liesenfeld et al. (2006) could be used; currently, the `glarma` package does not handle trinomial responses.

In order to answer "Are there variations between the three musical groups with respect to their group average responses to the same musical excerpt?" the binary time series were aggregated at each time into binomial counts of $8, 8, and\, 16$ respondents at each time in the EA, M, and NM musical groups. The multiple GLARMA fixed effects modeling (described in the next section) was used to examine the differences between group average responses. For this, it is assumed that the aggregated counts obey a binomial distribution. The independence of trials assumption is not in doubt since the individuals responded independently to the musical excerpt. However, the assumption that each individual in the group shares the same probability of response at each time appears in doubt as we now explain.

We consider the nonmusician group of 16 listeners. Let $S_t = \sum_{j=1}^{16} D_{jt}$ count the number of respondents whose arousal change was positive and assume that $S_t \sim \text{Bin}(\pi_t, 16)$, where

$$\text{logit}(\pi_t) = \omega_0 + \sum_{k=1}^{11} \omega_k \nabla X_{t-k} + \alpha_t \tag{3.13}$$

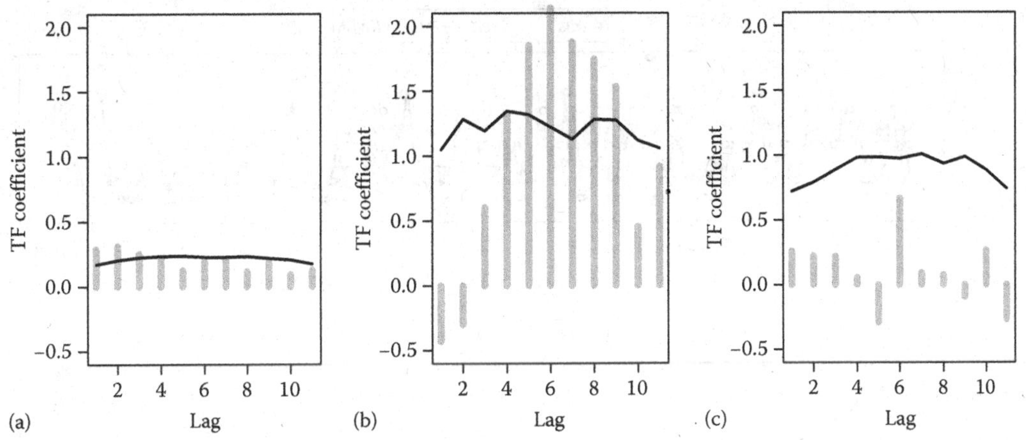

FIGURE 3.1
Transfer function coefficients for aggregated binomial series. (a) 16 nonmusicians (b) Listener 22 (c) Listener 23.

and α_t follows a suitably specified GLARMA process. For this group, we found that $(p, q) = (3, 0)$ was adequate (based on minimum AIC). The fitted transfer function coefficients with individual 95% confidence limits under the assumption of normality are shown in Figure 3.1a. Most transfer functions coefficients are individually significant. The observed binary responses (as probabilities) along with the fixed effects fit and GLARMA model fit is shown in Figure 3.2a. The nonrandomized probability integral transform (PIT) residual plot for this fit is shown in Figure 3.3. Clearly the binomial assumption is not correct as the PIT plot suggests that the binomial distribution is not providing a good prediction of the probability of small or large counts. This is not surprising since the 16 individuals in this group show substantial variability in their "stasis" levels, which of course will impact the average probability of a positive change in arousal. The PIT analysis suggests that aggregation of individual binary responses in this way is not appropriate.

Consequently, we turn to analysis of two individual binary responses to illustrate the application of GLARMA model for binary data using the individual models

$$\text{logit}(\pi_{jt}) = \omega_{j,0} + \sum_{k=1}^{11} \omega_{j,k} \nabla X_{t-k} + \alpha_{jt}, \tag{3.14}$$

where $\{\alpha_{jt}\}$ is a GLARMA process for the jth series. For most of the series, $p = 1$ and $q = 0$ seemed appropriate; hence,we settle on this for all 32 series. We illustrate the results of such fits on two listeners: listener 22, who had 88% stasis, and listener 23, who had 52% stasis. With such a high proportion of zero changes in arousal responses, it is not clear that application of standard Gaussian time series transfer function modeling would be reasonable for these two cases. We modeled the binary response series $\{D_{jt}\}$ using a binary GLARMA model with probabilities specified as in (3.14). Figure 3.1b and c shows the estimated values of the transfer function coefficients $\omega_{j,k}$ along with 95% significance levels. There are clear differences between the two individual responses to the same musical excerpt. The fitted values for these cases are shown in Figure 3.2b and c. There are also substantial differences between overall level as measured by the intercept terms ($\hat{\omega}_{22,0} = -3.24 \pm 0.39$, $\hat{\omega}_{23,0} = -1.21 \pm 0.54$), which is consistent with the relative levels of stasis observed for

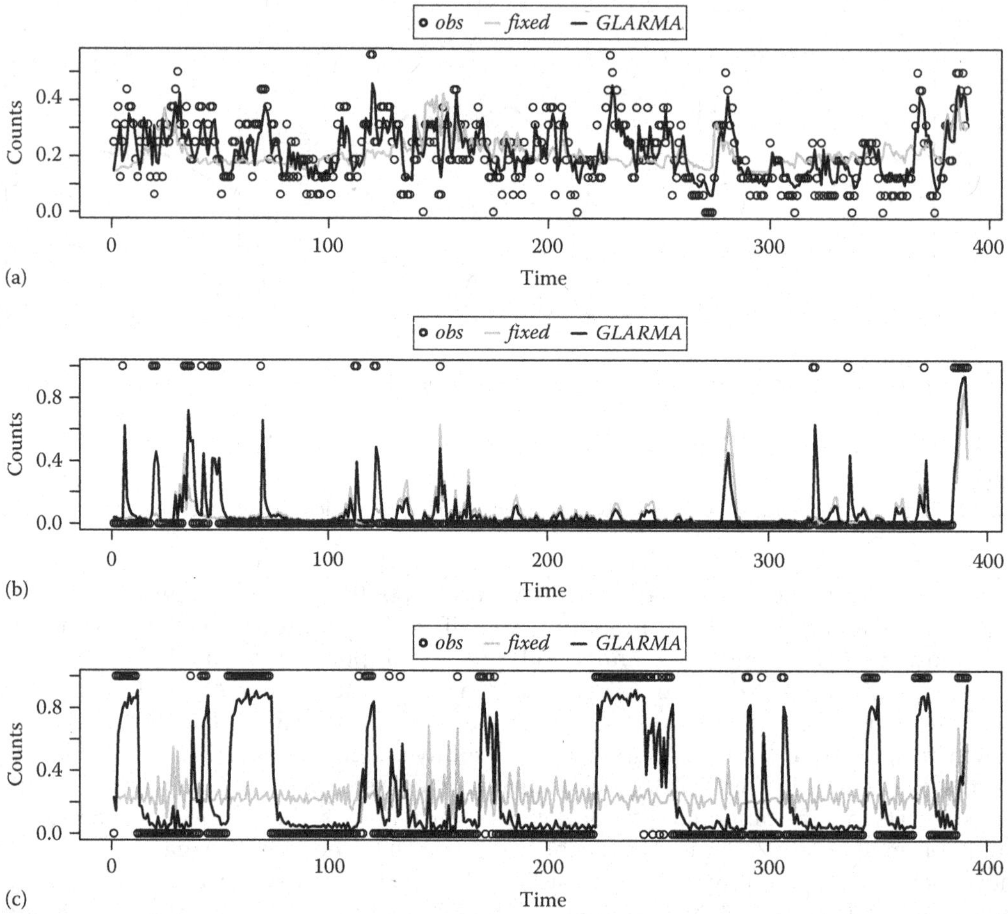

FIGURE 3.2
Binomial time series for 16 NM groups, response with GLARMA fits of response to changes in musical intensity and two individual binary time series with fits. (a) Binomial Fit: 16 Nonmusician Listeners (b) Binary Fit: Listener 22 (c) Binary Fit: Listener 23.

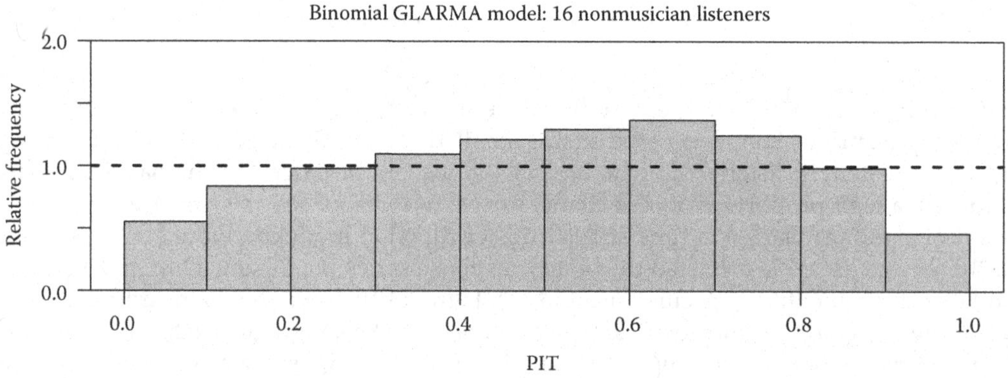

FIGURE 3.3
PIT residual plot for GLARMA model for binomial counts of positive responses in 16 nonmusician group listeners.

these two individuals as reported earlier. The GLARMA model autoregressive coefficients also substantially differed ($\hat{\phi}_{22} = 0.687 \pm 0.079$ and $\hat{\phi}_{23} = 0.894 \pm 0.024$). This also holds when all 32 individual responses are modeled in this way.

We have also applied GLARMA modeling with the `glarma` package on binary and binomial time series to study responsiveness to musical features in the complete panel of 32 listeners. A clear conclusion from this extended analysis is that there is a strong need to consider modeling of individual responses in order to address questions such as: "Are there differences between musical expertise groups?". Also clear is the need to accommodate differing amounts of serial dependence for each series in the ensemble, something that even current longitudinal data analysis for mixed models does not readily allow. In the next section we explain two approaches to allowing for variation between individual time series.

3.4 GLARMA Models for Multiple Independent Time Series

3.4.1 Examples

The musicology example of Section 3.3 is an example where ensembles of individual independent responses need to be modeled together. In a public health policy setting, Bernat et al. (2004) assessed, using a pooled cross-sectional random effects analysis, the impact of lowering the legal allowable blood alcohol concentration (BAC) in motor vehicle drivers from 0.10 to 0.08 on monthly counts of single vehicle night time fatalities in 17 U.S. states. In that study, serial dependence was detected in some series but could not be modeled with software and methods available at that time. The purpose of the remainder of this chapter is to present methods that overcome these gaps.

Two approaches are developed here. The first is referred to as the fixed effects plus GLARMA specification. Here, regression effects and serial dependence parameters may be constrained across series to test various hypotheses of interest primarily about regression effects, but also, about variation in serial dependence between series. For long longitudinal data each series can be estimated and modeled individually using the GLARMA models previoulsy discussed and combined likelihoods for constrained parameterizations constructed on which to base inference. Here, the length of the individual time series allows considerable flexibility in serial dependence structures between individual series. In traditional longitudinal data analysis where the number of repeated measures is low, it may not be possible to allow for individuality of this type.

The second approach is based on a random effect specification of the regression component while allowing individual series to have different serial dependence structures and strengths. This is not possible for traditional longitudinal data and the methods proposed here is a substantial extension of existing methodology, which is enabled by the length of the individual series.

Before defining models that combine fixed or random effects with GLARMA serial dependence, we mention some other recent examples along these lines. Xu et al. (2007) present a parameter-driven specification for serial dependence with random effects on covariates. Their approach is limited by the requirement that only autoregressive serial dependence is covered and all series must share the same structure and parameter values. Additionally, the method has not been demonstrated, neither in simulation nor in application, on series longer than 20 time points. Zhang et al. (2012) specify an exponential decay autocorrelation on a latent process and use marginal estimation methods for their combined serial dependence and random effects specification. They also assume that the structure and strength must be the same for all series. These approaches, which force the same dependence structure on all series, may be a legacy from the modeling of traditional longitudinal data in which the individual series are very short and for which the ability to detect different serial dependence properties between series is limited. It is our view that for long longitudinal data, this restriction is artificial for two reasons. First, in all examples we have encountered, there is strong evidence that the serial dependence structure and strength varies between series; some series will have strong serial dependence, while others will have none at all. One explanation for this is that the covariates used for the ensemble for fixed and random effect terms may not be inclusive of covariates that are unavailable and impact individual series—in such circumstances serial dependence may be stronger simply because it acts as a proxy for unobserved covariates. The second reason relates to the fact that the use of random effects on regression variables should logically extend to their use for serial dependence parameters. The models and methods presented here provide considerably more flexibility than those proposed by Xu et al. (2007) and Zhang et al. (2012). However, being based on GLARMA models, which require regular time spacing of observed outcomes, they cannot handle irregular spacings. For this irregularly spaces observations, parameter-driven models for the individual series would be more appropriate.

It is our view that, more frequently, increasingly long longitudinal data will become available and methods such as those proposed here will be become needed. For example, studies with panels of subjects equipped with automatic data loggers measuring their physical condition and activity levels are now quite feasible.

3.4.2 General Model for Multiple Independent GLARMA Models with Random Effects

Let Y_{jt} be the observation at time $t = 1, \ldots, n_j$ on the jth series of counts, where $j = 1, \ldots, J$ and let $\mathbf{x}_{j,t}$ be the covariates for the jth series. We also let $\mathbf{r}_{jt}$ denote d random effect covariates that apply to all series through coefficients represented as vectors of normally distributed random effects $\mathbf{U}_j \sim$ i.i.d $N(\mathbf{0}, \Sigma(\lambda))$, where Σ is a $d \times d$ covariance matrix determined by parameters λ.

In addition to assuming that the series $\{Y_{jt}\}$ are independent across the J cases, for each j, we also assume that, given the state process $W_{j,t}$, Y_{jt} are independent with exponential family distribution (3.1), where

$$W_{j,t} = \mathbf{x}_{j,t}^T \boldsymbol{\beta}^{(j)} + \mathbf{r}_{jt}^T \mathbf{U}_j + Z_{jt} \tag{3.15}$$

is the linear state process for the jth case series. Typically, some of the covariates appear in both $\mathbf{x}_{j,t}$ and $\mathbf{r}_{jt}$ in which case corresponding components of the $\boldsymbol{\beta}^{(j)}$ will not depend on j and some or all of the covariates can be the same across all series.

Serial dependence is modeled with the Z_{jt} assumed to satisfy (3.6) with degrees $(p^{(j)}, q^{(j)})$

$$Z_{jt} = \sum_{l=1}^{p^{(j)}} \phi_l^{(j)} \left(Z_{j,t-l} + e_{j,t-l}\right) + \sum_{l=1}^{q^{(j)}} \theta_l^{(j)} e_{j,t-l}. \tag{3.16}$$

Let $\tau^{(j)} = \left(\phi_1^{(j)}, \ldots, \phi_{p^{(j)}}^{(j)}, \theta_1^{(j)}, \ldots, \theta_{q^{(j)}}^{(j)}\right)$. We distinguish three special cases for the general state equation (3.15).

1. *Generalized linear mixed model*: In this specification, the serial dependence term Z_{jt} is absent, giving the standard generalized linear mixed model for longitudinal data discussed in Diggle et al. (2002) and Fitzmaurice et al. (2012)

$$W_{j,t} = \mathbf{x}_{j,t}^T \boldsymbol{\beta}^{(j)} + \mathbf{r}_{jt}^T \mathbf{U}_j. \tag{3.17}$$

   The only source of within series correlation is in the random effects component. Various packages (such as SAS and R) are available for model fitting.

2. *Fixed effects multiple GLARMA model*: Random effects are not included but there is a serial dependence in the form of an observation-driven model:

$$W_{j,t} = \mathbf{x}_{j,t}^T \boldsymbol{\beta}^{(j)} + Z_{jt}, \tag{3.18}$$

   where Z_{jt} is given by (3.16).

3. *Random effects multiple GLARMA model*: Both random effects and serial dependence are present. This is the general specification for W_{jt} given by (3.15), with Z_{jt} specified by the GLARMA model (3.16).

Model Class 1 was used in Bernat et al. (2004). A previous example of Model Class 2 type of modeling was considered in Dunsmuir et al. (2004), who investigated the commonality of regression impacts on three series of daily asthma presentation counts. Extension to datasets with substantially more time series and more complex hypotheses on the fixed effects parameters required considerable development of the previous software. The random effects multiple GLARMA model is developed in detail in Dunsmuir et al. (2014). The BAC dataset will be used to illustrate fitting of both models (3.18) and (3.15).

3.5 Fixed Effects Multiple GLARMA Model

3.5.1 Maximum Likelihood Estimation

In this section, we develop maximum likelihood estimation for the model with individual series state equations specified as in (3.18). Let the log-likelihood for the jth series be as in (3.7) and denoted by $l_j\left(\theta^{(j)}\right)$ where $\theta^{(j)T} = \left(\boldsymbol{\beta}^{(j)T}, \tau^{(j)T}\right)$. The log-likelihood across all series is

$$l(\theta) = \sum_{j=1}^{J} l_j(\theta^{(j)}) \tag{3.19}$$

where $\boldsymbol{\beta}^T = \left(\boldsymbol{\beta}^{(1)T}, \ldots, \boldsymbol{\beta}^{(J)T}\right)$ and $\tau^T = (\tau^{(1)T}, \ldots, \tau^{(J)T})$. Let $\theta = \left(\boldsymbol{\beta}^T, \tau^T\right)^T$ denote all parameters.

A primary focus is on testing if the parameterization across series can be simplified so that regression coefficients or serial dependence parameters can be constrained to be the same. We consider only linear constraints of the form $\theta = A\psi$ where A has fewer columns than rows and ψ denotes the lower-dimensional vector of parameters in the constrained model. Typically $\psi^T = \left(\psi_\beta^T, \psi_\tau^T\right)$ since we will generally not be interested in relating the regression coefficients to the serial dependence parameters. In that case A will be block diagonal appropriately partitioned. Denote the log-likelihood with respect to the constrained parameters as $l(\psi) = l(A\theta)$.

Maximization of (3.19) with respect to the constrained parameters can be over a high-dimensional parameter space. Initial estimates of $\hat{\theta}$ are obtained by maximizing (3.19) without constraints which is the same as maximizing all individual likelihoods separately and combining the resulting $l_j\left(\hat{\beta}^{(j)}, \hat{\tau}^{(j)}\right)$. These unconstrained estimates are used to initialize the constrained parameters via $\hat{\psi}_{(0)} = (A^TA)^{-1}A^T\hat{\theta}$. Next, using the appropriate components of $\hat{\theta}_{(0)} = A\hat{\psi}_{(0)}$, each component log-likelihood $l_j\left(\hat{\theta}_{(0)}^{(j)}\right)$ and its derivatives are calculated using the standard GLARMA software. Finally, these are combined to get the overall $l\left(\hat{\psi}_{(0)}\right)$. Derivatives with respect to ψ can be obtained using the identities $\partial l(\psi)/\partial\psi = A^T\partial l(A\psi)/\partial\theta$ and $\partial^2 l(\psi)/\partial\psi\partial\psi^T = A^T\partial^2 l(A\psi)/\partial\theta\partial\theta^T A$. This procedure is then iterated to convergence using the Newton–Raphson or Fisher scoring algorithm. Fisher scoring was found to be more stable in the initial stages. Once the derivative $\partial l(\psi)/\partial\psi$ settles down, the iterative search for the optimum can be switched to the Newton–Raphson updates, which typically gives speedier convergence.

Similar to the single series case, the asymptotic properties of the MLEs $\hat{\psi}$ have not been established for the general model previously described. Asymptotic results for longitudinal data typically let the number of cases J tend to infinity and the lengths of individual series n_j are typically held fixed. For this scenario, asymptotic theory is typically straightforward since it relies on large numbers of independent trajectories. For our applications, J is often bounded and we perceive of all n_j as tending to infinity in which case asymptotic results rely on those for individual time series which, as previously noted, are rather underdeveloped. We assume, however, that asymptotic results hold and perform inference in the usual way. For example, the matrices of second derivatives $\ddot{l}(\hat{\psi})$, computed in the course of the Newton–Raphson or Fisher scoring maximization procedures, are used to estimate standard errors for individual parameters. Also, to test the null hypothesis of common regression effects, we use the likelihood ratio statistic $G^2 = -2\left[l(\hat{\psi}_0) - l(\hat{\psi}_1)\right]$, where $\hat{\psi}_0$ is the estimate obtained under the null hypothesis and $\hat{\psi}_1$ the estimate obtained under the alternative. Degrees of freedom, for the chi-squared approximate reference distribution for G^2, are calculated in the usual way. These ideas were first illustrated in Dunsmuir et al. (2004) for three series of daily asthma counts which are assessed for common seasonal patterns, day of the week, weather, and pollution effects.

3.5.2 Application of Multiple Fixed Effects GLARMA to Road Deaths

Bernat et al. (2004) assessed the impact of lowering the legal allowable BAC in motor vehicle drivers from 0.10 to 0.08 on monthly counts of single vehicle night time fatalities in 17 U.S. states for which at least 12 months of post intervention data were available. The study design selected 72 consecutive months of data with 36 months prior to the decrease in allowable BAC and up to 36 months after for each of the 17 states used in the analysis. The mixed effects model presented in Bernat et al. (2004) assumed that the observed counts of single vehicle night time deaths, $Y_{j,t}$ in month t for state j, had, conditional on the random effects $\mathbf{U}_j$, a Poisson distribution with log mean given by

$$W_{jt} = \beta_0 + \beta_1 I_1(t) + \beta_2 I_{2,j}(t) + \beta_3 x_{3,j,t} + \beta_4 x_{4,j,t} + O_{j,t} + \mathbf{r}_{jt}^T \mathbf{U}_j, \tag{3.20}$$

where $I_1(t)$ is the indicator variable for the change in BAC from 0.1 to 0.08 coded as 0 for $t \leq 36$ and 1 for $t > 36$ for all states in the study, and $I_{2,j}(t)$ the indicator variable taking the value 0 for $t < T_j$ and 1 for $t \geq T_j$, where T_j is the month in which an administrative license revocation law was enacted. This potential confounder was enacted in seven of the states during the period of data used there. The other two regression variables are $x_{3,j,t}$, the number of Friday and Saturday nights, and a control series $x_{4,j,t}$, the log of other motor vehicle deaths (adjusted for population and seasonal factors), in month t for state j. An offset term, $O_{j,t}$, was used to adjust for unique state population trends and seasonal factors. Finally, $\mathbf{U}_j \sim \mathrm{N}(\mathbf{0}, G)$ is the multivariate normal distribution with covariance matrix G for the random effects on the selected covariates in $\mathbf{r}_{jt}$. Further details concerning the rationale and definition of the model components mentioned earlier can be found in Bernat et al. (2004). Because of the differences between the time periods over which the data were collected from each of the 17 states and because of the control series used for each state, the assumption that the 17 series are independent was considered reasonable. We also confirmed this assumption using cross-correlation analysis of Pearson residuals from individual GLARMA model fits to the 17 series. Bernat et al. (2004) discussed the likely impact that any serial dependence might have on their key conclusion that there is a statistically significant lowering of overall average single vehicle night time fatalities associated with the lowering of the legal BAC level but suitable software was not available at that time to assess this statistically.

We now illustrate the fixed effects plus GLARMA model defined in (3.18) on these data. The first step was to use the `glarma` package to fit the model (3.18) to each of the individual series using regression and offset terms as specified in (3.20). Examination of the PIT residual plots for each series suggested overfitting relative to the Poisson distribution. We identified that the main contributor to overfitting was the seasonal offset term used in (3.20); this had been determined using standard seasonal adjustment methods available in the PROC X-11 package in SAS for each series separately and based on only $n = 72$ months of data. As an alternative, we used parametric harmonic seasonal terms to all 17 response series with the other regressors and the population offset only. In almost all cases, these seasonal harmonics were not significant suggesting that use of the control series (without seasonal adjustment) and lag 12 autoregressive terms in the GLARMA model are sufficient for modeling seasonality. In view of this, we performed our reanalysis of these data using the regression specification in (3.20), but dropping the seasonal adjustment in both the offset $O_{j,t}$ and logOMVD (so that the control series is now log other motor vehicle deaths per 100,000 population without seasonal adjustment), together with a GLARMA process for serial dependence. We compared various combinations of lags for the autoregressive and moving average terms in the GLARMA model to allow for both serial dependence at low orders and at seasonal lags. Based on AIC the overall best specification was an autoregression of order $p = 12$ with zero coefficients at lags 1 through 11. Hence we initially adopted this seasonal model for all series.

TABLE 3.1

Results for testing various fixed effect multiple GLARMA models for the Road Deaths Series in 17 U.S. States

Model	−2log L	S	G^2	d.f.	p-val
FE-I: Unrestricted	5345.69	92	—	—	—
FE-II: ϕ's in 6 groups	5347.87	81	$G^2_{\text{II v I}} = 2.18$	11	0.998
FE-III: BAC, ALR, FS same	5391.76	43	$G^2_{\text{III v II}} = 43.89$	38	0.236
FE-IV: BAC, ALR, FS, lnOVD same	5436.57	27	$G^2_{\text{IV v III}} = 44.81$	16	0.00015
			$G^2_{\text{IV v II}} = 88.61$	54	0.0021

We next check whether the regression coefficients in W_{jt} vary significantly between individual states. We begin with the overall unrestricted fit to all 17 states. We refer to this as Model FE-I, which has $-2\log L = 5345.69$ with $S = 92$ parameters. Examination of the individual estimates $\hat{\phi}_{12}$ suggested that they could be simplified as follows: Group 1 (State 11, $\hat{\phi}_{12} = -0.081 \pm 0.045$), Group 2 (States 1, 6, 7, 9, 10, 12:17, $\hat{\phi}_{12} = 0.005 \pm 0.013$), Group 3 (States 2, 4, $\hat{\phi}_{12} = 0.066 \pm 0.015$), Group 4 (State 5, $\hat{\phi}_{12} = 0.212 \pm 0.087$), Group 5 (State 8, $\hat{\phi}_{12} = 0.401 \pm 0.096$), Group 6 (State 5, $\hat{\phi}_{12} = 0.545 \pm 0.219$) in which, at most, 6 ϕ_{12} coefficients are significant.

The model with the ϕ_{12} restricted to these groups is referred to as Model FE-II in Table 3.1. Using the likelihood ratio test we obtain $G^2_{\text{II v I}} = 2.18$ on 11 d.f.; hence, restriction of the ϕ_{12} would not be rejected. From this model, we then examined whether or not some or all of the regression coefficients (other than the intercept which does vary substantially between states) take common values across all 17 states. Model FE-III restricts the coefficients for BAC, ALR, Friday–Saturday to be the same and (see Table 3.1) $G^2_{\text{III v II}} = 43.89$ on 38 d.f. and associated p-value of 0.24, which is not sufficiently strong evidence to suggest that the impact of these variables differs between individual states in a statistically significant way. Next, in Model FE-IV, log OMVD was allowed to differ between states. Compared with Model FE-III or Model FE-II, this risk control variable is strongly statistically significant between states with $G^2_{\text{IV v III}} = 44.81$ on 16 d.f. and associated p-value of 0.00015 and $G^2_{\text{IV v II}} = 88.61$ on 54 d.f. and associated p-value of 0.0021.

Hence, Model FE-III provides a useful summary of the commonality or otherwise of regression variable impacts on single vehicle night time road deaths across the 17 states. The fitted parameters and associated standard errors are reported in Table 3.2. The six groups for ϕ_{12} could be reduced to four by removing the nonsignificant cases of Groups 1 and 2. We did not pursue this here, preferring to move onto the use of a random effects analysis. The impact of lowering the legal BAC level is estimated to be $\hat{\beta}_2 = -0.072 \pm 0.022$ confirming the statistical significance of this association found in Bernat et al. (2004).

The fixed effects GLARMA model analysis provides a good starting point for the random effects GLARMA modeling that we turn to in the next section. In particular, it seems plausible from the results of Table 3.1 that random effects will be needed for the intercept term and the log OMVD term, but not for BAC, ALR, or Friday–Saturday effects. The parameter values reported for Model FE-III in Table 3.2 can provide useful starting values for the random effects model fitting. For fixed effects, we use the point estimates of coefficients for predictors that are common to all series, while for predictors that vary between series, we use the mean values of point estimates of the coefficients.

TABLE 3.2
Parameter estimates for the random effects model for the Road Deaths Series in 17 U.S. States

	RE-IV No GLARMA Random Effects		FE-III Multiple GLARMA Fixed Effects		RE-III Multiple GLARMA Random Effects	
	Estimate	s.e.	Estimate	s.e.	Estimate	s.e.
β_0 (intercept)	−1.649	0.116	−1.801	—	−1.705	0.119
β_1 (BAC change)	−0.054	0.022	−0.072	0.022	−0.060	0.022
β_2 (ALR term)	−0.063	0.035	−0.011	0.039	−0.047	0.037
β_3 (Frid-Sat)	0.032	0.011	0.037	0.011	0.037	0.011
β_4 (logOMVD)	0.395	0.063	0.314	—	0.367	0.061
Intercept RE s.d.	0.241	0.053	0.209	—	0.242	0.054
logOMVD RE s.d	0.160	0.066	0.140	—	0.145	0.072
ϕ_{Gp1}	—	—	−0.066	0.045	—	—
ϕ_{Gp2}	—	—	0.010	0.012	—	—
ϕ_{Gp3}	—	—	0.067	0.015	0.071	0.015
ϕ_{Gp4}	—	—	0.216	0.080	0.201	0.074
ϕ_{Gp5}	—	—	0.404	0.091	0.408	0.090
ϕ_{Gp6}	—	—	0.531	0.213	—	—
−2loglikelihood	5552.173		5391.1		5505.082	

The results labeled RE-IV is that reported in Bernat et al. (2004) using SAS PROC NLMIXED. The results labeled FE-III is the final fixed effects multiple GLARMA model discussed in Section 3.5.2.

Note: Values reported against the intercept β_0 and the logOMVD term β_4 are averages of the 17 individual values obtained while the values in the rows labeled "Intercept RE" and "logOMVD RE" are the standard deviations of these individual estimates, respectively. The results labeled RE-III is the final random effects multiple GLARMA model discussed in Section 3.6.3.

3.6 Random Effects Multiple GLARMA Model

3.6.1 Maximum Likelihood Estimation

Let W_{jt} be defined as in (3.15), where U_j are multivariate normal. Let $\theta = \left(\beta^{(1)}, \ldots, \beta^{(J)}, \tau^{(1)}, \ldots, \tau^{(J)}, \lambda\right)$ now be the collection of parameters in the GLARMA models and the random effects parameters. The joint log-likelihood is now

$$l(\theta) = \sum_{j=1}^{J} l_j(\beta^{(j)}, \tau^{(j)}, \lambda), \tag{3.21}$$

where

$$l_j(\beta^{(j)}, \tau^{(j)}, \lambda) = \log \int_{R^d} \exp(l_j(\beta^{(j)}, \tau^{(j)} | u) g_{\mathbf{U}}(\mathbf{u}; \Sigma(\lambda)) du, \tag{3.22}$$

and $g_U(\mathbf{u}; \Sigma(\lambda))$ is the multivariate normal density. To proceed further, we parameterize the covariance matrix as $\Sigma = LL^T$ where L is lower triangular and let $U_j = L\zeta_j$ where ζ_j are independent $N(0, I_d)$. Let $\boldsymbol{\lambda}$ =vech(L) be the half-vectorisation. With this parameterization, rewrite W_{jt} in (3.5) linearly in terms of λ as

$$W_{jt} = \mathbf{x}_{j,t}^T \boldsymbol{\beta}^{(j)} + \text{vech}\left(\zeta_j \mathbf{r}_{j,t}^T\right)^T \boldsymbol{\lambda} + Z_{jt}. \tag{3.23}$$

The log-likelihood (3.22) becomes

$$l_j(\boldsymbol{\beta}^{(j)}, \boldsymbol{\tau}^{(j)}, \boldsymbol{\lambda}) = \log \int_{\mathbb{R}^d} \exp\left(l_j\left(\boldsymbol{\beta}^{(j)}, \boldsymbol{\tau}^{(j)}, \boldsymbol{\lambda}|\zeta\right) g(\zeta) d\zeta\right. \tag{3.24}$$

where $g(\zeta)$ is the d-fold product of the standard normal density and

$$l_j\left(\boldsymbol{\beta}^{(j)}, \boldsymbol{\tau}^{(j)}, \boldsymbol{\lambda}|\zeta\right) = \sum_{t=1}^{n} \left[y_{jt} W_{jt} - a_t b(W_{jt})\right] + \sum_{t=1}^{n} c(y_{jt}).$$

Note that (3.23) is in the same form as (3.3) but the parameters $\boldsymbol{\lambda}$ are treated as regression parameters for any fixed value of the vector ζ and the random effects covariates $\mathbf{r}_{j,t}^T$.

The representation of the random effects covariance matrix as $\Sigma = LL^T$ allows the parameter $\boldsymbol{\lambda}$ to enter into the conditional log-likelihood linearly and without bounding constraints. Both properties enable existing GLARMA software to calculate the log-likelihood and derivatives with respect to the parameters. When some elements of Σ, and hence L, are specified as zero to reflect zero covariance between some of the random effects, $\boldsymbol{\lambda}$ is the half vectorization of L with the structural zeros removed. Covariance matrices in which certain combinations of random effects are specified to be zero cannot be represented in this form. However, these can often be accommodated by reordering the random effect variables and setting the appropriate elements of L to zero.

3.6.2 Laplace Approximation and Adaptive Gaussian Quadrature

For any fixed $\boldsymbol{\theta}$, computation of the log-likelihood $l(\boldsymbol{\theta})$ requires calculation of the J integrals defined in (3.24). We now outline an approximate method based on the Laplace approximation and adaptive Gaussian quadrature (AGQ). The integral in (3.24) can be rewritten as

$$L_j(\boldsymbol{\theta}) = \frac{1}{(2\pi)^{d/2}} \int_{R^d} \exp\left(F_j(\zeta|\boldsymbol{\theta})\right) d\zeta$$

where the exponent is considered as a function of ζ for fixed parameters $\boldsymbol{\theta}$ and is defined as

$$F_j(\zeta|\boldsymbol{\theta}) = \sum_{t=1}^{n_j} \left[y_{jt} W_{jt}(\zeta; \mathbf{x}_{jt}, \boldsymbol{\theta})) - a_{jt} b(W_{jt}(\zeta; \mathbf{x}_{jt}, \boldsymbol{\theta})) + c(y_{jt})\right] - \frac{\zeta^{\top}\zeta}{2}, \tag{3.25}$$

where

$$W_{jt}(\zeta; \mathbf{x}_{jt}, \theta) = \left(\mathbf{r}_{j,t}^T L\right)\zeta + \mathbf{x}_{j,t}^T \beta^{(j)} + Z_{jt} \tag{3.26}$$

is treated as a function of ζ for $\mathbf{x}_{j,t}^T \beta^{(j)}$ fixed. To find the Laplace approximation, we expand the exponent $F(\zeta)$ around its modal value in a second-order Taylor series, and ignore the remainder. The resulting integral can be obtained in closed form. Note that Z_{jt} in (3.26) is a function of ζ. Hence, the contribution to the first and second derivatives from the summation term in (3.25) required for the Taylor series expansion of F_j need to be calculated using the GLARMA software with ζ treated as a regression parameter for covariates $\left(\mathbf{r}_{j,t}^T L\right)$ and fixing $\mathbf{x}_{j,t}^T \beta^{(j)}$ as the offset term. To find the modal value, we need to find ζ_j^* which solves

$$\frac{\partial}{\partial \zeta_j} F_j(\zeta_j^*) = 0.$$

The Newton–Raphson method is used to find ζ_j^* and, at convergence, we set

$$\Sigma_j^* = -\left\{\frac{\partial^2}{\partial \zeta \partial \zeta^T} F_j(\zeta_j^*)\right\}^{-1}.$$

Since $\frac{\partial^2}{\partial \zeta \partial \zeta^T} F_j(\zeta_j^*)$ is almost surely positive definite for the canonical link exponential family, the Newton–Raphson method will converge to the modal solution from any starting point; we use $\zeta_j^{(0)} = 0$ to intitiate the recursions.

The Laplace approximation gives the approximate log-likelihood for the jth state as $\tilde{l}_j^{(1)}(\theta) = \log\det(\Sigma_j^*(\theta))^{1/2} + F_j(\zeta_j^*(\theta))$, which can be combined to give the overall approximate log-likelihood as

$$\tilde{l}^{(1)}(\theta) = \sum_{j=1}^{J} \tilde{l}_j^{(1)}(\theta) \tag{3.27}$$

AGQ methods can be used to improve the approximation as has been successfully done for likelihoods in other statistical models such as nonlinear and non-Gaussian mixed effects modeling. This approach is implemented in a number of widely used software systems as the default method—see Pinheiro and Bates (1995) and Pinheiro and Chao (2006) for examples. Our implementation of AGQ follows that of Pinheiro and Chao (2006). It relies on the mode, ζ_j^*, and Σ_j^* used in the Laplace approximation to center and scale Q quadrature points in each of d coordinates resulting in integrands evaluated at d^Q points. When $Q = 1$, the Laplace approximation is obtained.

The AGQ approximation to the jth integral is denoted by $\tilde{L}_j^{(Q)}(\theta)$, with corresponding approximation to the overall likelihood as

$$\tilde{l}^{(Q)}(\theta) = \sum_{j=1}^{J} \log \tilde{L}_j^{(Q)}(\theta). \tag{3.28}$$

Since $z_j^*(\theta)$ and $\Sigma_j^*(\theta)$ are functions of the unknown parameters θ, it is necessary to recompute the Laplace approximation at each iterate of θ to maximize (3.28).

Maximizing (3.28) using the optimizer `optim` in R proved to be very slow and unreliable for our applications. An alternative was to use Fisher Scoring or Newton–Raphson updates based on numerical derivatives obtained using the R package `numDeriv`. This also proved to be very slow. Analytical derivatives require implicit differentiation of $\zeta_j^*(\theta)$ and $\Sigma_j^*(\theta)$ which results in complex expressions requiring substantial modification to the current GLARMA software. We next describe an alternative approach that avoids all of these issues.

First derivatives of the log-likelihood (3.19) with respect to unknown parameters are

$$\dot{l}_j(\theta) = \frac{\partial}{\partial\theta} l(\theta) = \sum_{j=1}^{J} \frac{1}{L_j(\theta)} \int_{R^d} \frac{\partial}{\partial\theta}\left[l_j(\theta|\zeta\right] \exp(l_j(\theta|z)g(\zeta)d\zeta. \tag{3.29}$$

Second derivatives, $\ddot{l}_j(\theta)$, are also easy to derive and involve more integrals to be approximated. For any fixed ζ, the integrands in these derivative expressions can be calculated recursively using the unpackaged form of single series GLARMA software. If S denotes the number of parameters in θ, then there are $J \times (1+S+2S(S+1)/2) = J(1+S)^2$, d-dimensional integrals to calculate in order to implement the Newton–Raphson method. For instance, for the final model for the BAC example (Model RE-III) with two uncorrelated random effects, we have $J = 17$, $S = 10$ requiring calculation of 2057 $d = 2$ dimensional integrals at each step of the Newton–Raphson iterations. Fisher scoring is not available here because the summation to compute the whole likelihood is over J; hence, insufficient outer products of first derivative vectors would result in an ill-conditioned approximation to the second derivative matrix unless J is quite large.

In our experience, for long longitudinal data applications, the Laplace approximation can provide quite accurate single-point approximations to the integrals required for the likelihood itself. However, the first and second derivatives have integrands that are certainly not positive, nor are they unimodal, and so a single-point integral approximation is inadequate. However, AGQ can provide multipoint approximations for the integrals required for derivatives. In our experience, surprisingly few quadrature points are required to get approximations to the likelihood and the first and second derivatives which are sufficiently accurate for convergence to the optimum of the likelihood and which provide accurate standard errors for inferential purposes. We denote the estimates of $\dot{l}(\theta)$ and $\ddot{l}(\theta)$ obtained by applying AGQ with Q nodes by $\tilde{\dot{l}}^{(Q)}(\theta)$ and $\tilde{\ddot{l}}^{(Q)}(\theta)$, respectively. The same quadrature points and weights that are used for $\tilde{l}^{(Q)}(\theta)$ are also used to obtain $\dot{l}^{(Q)}(\theta)$ and $\ddot{l}^{(Q)}(\theta)$ using one pass of the GLARMA software.

Summary of algorithm to approximate likelihood and derivatives:

1. Initialize parameter value $\theta^{(k)}$
2. For each $j = 1, \ldots, J$
   a. Use GLARMA software treating ζ as a parameter to find the derivatives needed to find the J Laplace approximations ζ_j^*, Σ_j^*.
   b. Select Q quadrature points in each of d directions, relocate, and scale these using ζ_j^* and Σ_j^*
   c. Apply GLARMA software to calculate the integrands at the d^Q integrating points in order to estimate the likelihood and first and second derivatives at $\theta^{(k)}$.
3. Assemble the complete likelihood and derivatives over the J cases.
4. Use Newton–Raphson iteration to update $\theta^{(k)} \rightarrow \theta^{(k+1)}$. Repeat at Step 2 until convergence.

3.6.3 Application to Road Deaths Data

We used Model FE-III (central columns of Table 3.2) as the starting point for the random effects analysis. The correlation between the 17 pairs of intercepts and coefficients of the logOMVD variables was $\text{Corr}\left(\hat{\beta}_{0,j}, \hat{\beta}_{4,j}\right) = 0.531$ which is significantly different from zero. As a result, we used a correlated bivariate random effect for the intercept and this regressor. The results of fitting this model (Model RE-I) are given in Table 3.3. The point estimate of the parameter controlling the correlation between these two terms was not significant ($\hat{L}_{12} = 0.101 \pm 0.107$) and so we removed the correlation between the random effects to arrive at model RE-II. The likelihood ratio test (see Table 3.3) confirmed that these were not correlated. The random effect variance for the log OVMD regressor is marginally significant compared to its standard error. Use of a likelihood ratio test does not reject this simplification, even after adjusting for the fact that this is a test on the boundary using the standard methodology for mixed effects variance testing as described in Fitzmaurice et al. (2012). Similar to the fixed effects analysis, Groups 1 and 2 showed no significant autocorrelation; however, the large autocorrelation for the sixth group was no longer significant. Since there is clearly potential in these models for serially correlated effects to interact or trade-off with regression random effects, our next step was to refit the model after removing autocorrelation terms for Groups 1, 2, and 6. This resulted in Model RE-III and the likelihood ratio test confirms that autoregressive terms are not required for these three Groups. The final states for which significant autocorrelation is required are

TABLE 3.3

Results for testing various random effects GLARMA models for the Road Deaths Series

Model	ϕ_{12} Groups	Random Effects	$-2\log L$	S	G^2	d.f.	p-value
RE-I	6 levels	2 correlated	5500.85	14	—	—	—
RE-II	6 levels	2 uncorrelated	5501.87	13	$G^2_{\text{II v I}} = 0.98$	1	0.32
RE-III	3 levels	2 uncorrelated	5505.08	10	$G^2_{\text{III v II}} = 3.21$	3	0.36
RE-IV	No ϕ_{12}	2 uncorrelated	5552.17	7	$G^2_{\text{IV v III}} = 47.09$	3	3×10^{-10}

states 2, 4, 5, and 8 (corresponding to California, Florida, Idaho, and Maine). We next compared this model with the purely random effects model (the analogue of what was fit in Bernat et al., 2004), which is labeled Model RE-IV. The likelihood ratio test overwhelmingly rejects the hypothesis that autocorrelation terms can be removed from the model for these four states.

Did our inclusion of autoregressive terms impact the original conclusions of Bernat et al. (2004) concerning the strength and significance of the BAC intervention on single vehicle night time road deaths? They obtained $\hat{\beta}_1 = -0.052 \pm 0.021$ (p-value $= 0.013$). In the analysis presented here, we have removed the seasonal offset terms from the response distribution and the logOMVD control variable for reasons discussed earler. The analogous result is Model RE-IV and for that model $\hat{\beta} = -0.054 \pm 0.022$ (p-value $= 0.013$), a very similar finding to that in Bernat et al. (2004).

We report estimates for all parameters of Model RE-III as the final column pair in Table 3.2. Inclusion of significant serial dependence terms where needed has actually increased the size of the point estimate of the BAC effect relative to Model RE-IV (no serial dependence model) and left the standard error unchanged, resulting in a reduced p-value of 0.006 for this term and hence suggesting that the original finding of the significance of the BAC association may have been conservative.

3.6.4 Computational Speed and Accuracy

3.6.4.1 Impact of Increasing the Number of Quadrature Points

In several applications of the earlier method, we have found that usually $Q = 3, 5$, or 7 quadrature points in each of the d random effects coordinates are sufficient for use in optimizing the likelihood and for obtaining accurate standard errors and likelihood values and likelihood ratio statistics.

3.6.4.2 Comparison with Numerical Derivatives

For one iteration of the Newton–Raphson procedure, near convergence to the maximum likelihood, using first and second numerical derivatives calculated using the `numDeriv` package is of the order of 300 times longer than using the AGQ method for calculating derivatives that we propose. Use of `optim` for convergence and evaluation of the Hessian for standard error calculations were similarly slow. There was no substantial loss of accuracy either for convergence of the maximum likelihood updates or in the standard errors.

In summary, the AGQ method proposed here to calculate derivatives of the log-likelihood is two orders of magnitude faster than using `optim` without derivative or numerical derivatives based on `numDeriv` with no substantial loss of accuracy even for $Q = 3$. Use of the AGQ method makes it feasible to fit combined random GLARMA random effects models for long longitudinal data. We have experienced similar comparisons of speed and accuracy in more complex settings, such as the analysis of 32 binary time series of length 393 arising in the musicology study discussed earlier in which up to four random effects were included in the models and another study of 49 times series of length 336 of suicide counts for which a Poisson response was appropriate and up to three random effects were included in the model.

3.7 Interpretation of the Serially Dependent Random Effects Models

For single series GLARMA models, means, variances, and serial covariances for the state process $\{W_t\}$ can be readily derived using the definition of Z_t in (3.4). For the Poisson or negative binomial response GLARMA plus random effects model, the marginal interpretation of the fixed effects coefficients is approximately equal to the conditional interpretation since

$$E(Y_{jt}) \approx \exp\left(x_{0,t}^T \beta^{(0)} + \mathbf{x}_{j,t}^T \beta^{(j)}\right) \exp\left(\frac{\sigma_U^2}{2} + \frac{\gamma^2}{2}\right)$$

by simple extension of the argument in Davis et al. (2003).

For binomial and Bernoulli responses, calculation of means, variances, autocovariances for the response series and interpretation of regression coefficients are not straightforward. This is a typical issue for interpretation of random effects models and transition models in the binomial or Bernoulli case—see Diggle et al. (2002) for example.

3.8 Conclusions and Future Directions

This chapter has reviewed the fitting of GLARMA models for single time series of exponential family response distributions and illustrated this on some binary and binomial series arising in a study of listener responses to music features. Extensions and utilization of single series GLARMA modeling ideas and software to the long longitudinal data setting were explained. Two approaches to providing a combined analysis of all series in a panel of responses were considered. The first approach was based on a constrained fit across the panel using single series GLARMA software and allowed testing of parameter similarity in each series across the panel of series. The second approach modeled between series parameter variation using random effects. Again, single series GLARMA software can be used to compute a modal approximation to the integrals that constitute the likelihood when there are random effects. This modal approximation is based on the Laplace approximation and this can also be calculated using the GLARMA single series software. The use of AGQ to compute the very large number of integrals required to compute the first and second derivatives of the combined likelihood is explained and, somewhat surprisingly, these can be accurately and speedily computed using a small number of quadrature points, hence making the optimization of the likelihood based on Newton–Raphson iteration feasible in practice. The speed of this method compared to those based on numerical derivatives of standard optimizers is several hundred times faster per iteration. We illustrated the multiple independent time series approaches on a set of long longitudinal data arising in the study of the association between lowering the legal blood alcohol level in drivers on road deaths in 17 U.S. States.

The approach taken here allows considerable flexibility on the specification of the serial dependence in the individual time series. The examples presented here clearly require that flexibility. We know of no other current methods that have this flexibility; those that we have reviewed appear to continue assuming short longitudinal trajectories in which case it is difficult to allow flexibility in serial dependence specifications. For long longitudinal data, this restriction can be avoided as we have demonstrated in the examples presented herein.

We have also extended the random effects approach to parameter-driven models for individual time series serial dependence, again based on the use of a Laplace approximation as in Davis and Rodriguez-Yam (2005) and AGQ for the random effect integration. We aim to extend GLARMA random effects models covered in this chapter and in the parameter-driven version to allow for random effects on serial dependence parameters. This seems to us a natural extension within the modeling perspective of random effects for between series parameter variation.

There is an obvious need for rigorous asymptotic theory to properly justify the statistical inference that we have presented based on these new models. However, this will rely on similar theory being developed for individual observation-driven time series models, something that remains underdeveloped.

Finally, we have not discussed forecasting for GLARMA models in this review nor the somewhat related issue of missing data. GLARMA models, requiring recursive calculation of the state equation, are computationally intensive to forecast beyond one or two time points. This has implications for the computation of the likelihood when there are missing data, as it requires a conditional distribution of the response after gaps.

Acknowledgments

The Music Listener data used in Section 3.3 was provided by Roger Dean of the MARCS Institute at the University of Western Sydney and the U.S. Road Deaths data used in Sections 3.5.2 and 3.6.3 was that used in Bernat et al. (2004) and its source is acknowledged there. I would also like to thank Chris McKendry, my previous Honour's degree student, for his assistance in developing the adaptive Gaussian quadrature multiple independent random effects approach discussed in this chapter. Finally, helpful comments from a referee improved the clarity of the presentation.

References

Benjamin, M. A., Rigby, R. A., and Stasinopoulos, D. M. (2003). Generalized autoregressive moving average models. *Journal of the American Statistical Association*, 98(461):214–223.

Bernat, D. H., Dunsmuir, W., Wagenaar, A. C. et al. (2004). Effects of lowering the legal bac to 0.08 on single-vehicle-nighttime fatal traffic crashes in 19 jurisdictions. *Accident Analysis and Prevention*, 36(6):1089.

Buckley, D. and Bulger, D. (2012). Trends and weekly and seasonal cycles in the rate of errors in the clinical management of hospitalized patients. *Chronobiology International*, 29(7):947–954.

Creal, D., Koopman, S. J., and Lucas, A. (2008). A general framework for observation driven time-varying parameter models. Technical report, Tinbergen Institute Discussion Paper, Amsterdam, the Netherlands.

Davies, R. B. (1987). Hypothesis testing when a nuisance parameter is present only under the alternative. *Biometrika*, 74(1):33–43.

Davis, R. A., Dunsmuir, W., and Wang, Y. (1999). Modeling time series of count data. *Statistics TextBooks and Monographs*, 158:63–114.

Davis, R. A., Dunsmuir, W. T., and Streett, S. B. (2003). Observation-driven models for Poisson counts. *Biometrika*, 90(4):777–790.

Davis, R. A., Dunsmuir, W. T., and Streett, S. B. (2005). Maximum likelihood estimation for an observation driven model for Poisson counts. *Methodology and Computing in Applied Probability*, 7(2):149–159.

Davis, R. A., Dunsmuir, W. T., and Wang, Y. (2000). On autocorrelation in a Poisson regression model. *Biometrika*, 87(3):491–505.

Davis, R. A. and Liu, H. (2015). Theory and inference for a class of observation-driven models with application to time series of counts. *Statistica Sinica*. doi:10.5705/ss.2014.145t (to appear).

Davis, R. A. and Rodriguez-Yam, G. (2005). Estimation for state-space models based on a likelihood approximation. *Statistica Sinica*, 15(2):381–406.

Davis, R. A. and Wu, R. (2009). A negative binomial model for time series of counts. *Biometrika*, 96(3):735–749.

Dean, R. T., Bailes, F., and Dunsmuir, W. T. (2014a). Shared and distinct mechanisms of individual and expertise-group perception of expressed arousal in four works. *Journal of Mathematics and Music*, 8(3):207–223.

Dean, R. T., Bailes, F., and Dunsmuir, W. T. (2014b). Time series analysis of real-time music perception: Approaches to the assessment of individual and expertise differences in perception of expressed affect. *Journal of Mathematics and Music*, 8(3):183–205.

Diggle, P., Heagerty, P., Liang, K.-Y., and Zeger, S. (2002). *Analysis of Longitudinal Data*, 2nd edn. Oxford University Press, Oxford, U.K.

Dunsmuir, W. T., Li, C., and Scott, D. J. (2014). *Glarma: Generalized Linear Autoregressive Moving Average Models*. R package version 1.3-0. http://CRAN.Rproject.org/package=glarma

Dunsmuir, W. T. and Scott, D. J. (2015). The glarma package for observation driven time series regression of counts. *Journal of Statistical Software* (to appear).

Dunsmuir, W. T. M., Leung, J., and Liu, X. (2004). Extensions of observation driven models for time series of counts. In *Proceedings of the International Sri Lankan Statistical Conference: Visions of Futuristic Methodologies*, eds B. M. de Silva and N. Mukhopadhyay, RMIT University and University of Peradeniy, Peradeniy, Sri Lanka.

Dunsmuir, W. T. M., Tran, C., and Weatherburn, D. (2008). *Assessing the Impact of Mandatory DNA Testing of Prison Inmates in NSW on Clearance, Charge and Conviction Rates for Selected Crime Categories*. NSW Bureau of Crime Statistics and Research. http://www.bocsar.nsw.gov.au/lawlink/bocsar/ll_bocsar.nsf/pages/bocsar_pub_legislative.

Etting, S. F. and Isbell, L. A. (2014). Rhesus macaques (macaca mulatta) use posture to assess level of threat from snakes. *Ethology*, 120(12):1177–1184.

Fitzmaurice, G. M., Laird, N. M., and Ware, J. H. (2012). *Applied Longitudinal Analysis*, vol. 998. John Wiley & Sons. Hoboken, New Jersy.

Hansen, B. E. (1996). Inference when a nuisance parameter is not identified under the null hypothesis. *Econometrica: Journal of the Econometric Society*, 64(2):413–430.

Kedem, B. and Fokianos, K. (2002). *Regression Models for Time Series Analysis*. John Wiley & Sons. Hoboken, New Jersy.

Liesenfeld, R., Nolte, I., and Pohlmeier, W. (2006). Modelling financial transaction price movements: a dynamic integer count data model. *Empirical Economics*, 30(4):795–825.

McCullagh, P. and Nelder, J. A. (1989). *Generalized Linear Models (Monographs on Statistics and Applied Probability 37)*. Chapman & Hall, London, U.K.

Pinheiro, J. C. and Bates, D. M. (1995). Approximations to the log-likelihood function in the nonlinear mixed-effects model. *Journal of Computational and Graphical Statistics*, 4(1): 12–35.

Pinheiro, J. C. and Chao, E. C. (2006). Efficient laplacian and adaptive gaussian quadrature algorithms for multilevel generalized linear mixed models. *Journal of Computational and Graphical Statistics*, 15(1):58–81.

R Core Team (2014). *R: A Language and Environment for Statistical Computing*. R Foundation for Statistical Computing, Vienna, Austria.

Rydberg, T. H. and Shephard, N. (2003). Dynamics of trade-by-trade price movements: Decomposition and models. *Journal of Financial Econometrics*, 1(1):2–25.

Streett, S. (2000). Some observation driven models for time series of counts. PhD thesis, Colorado State University, Department of Statistics, Fort Collins, CO.

Tjøstheim, D. (2012). Some recent theory for autoregressive count time series. *Test*, 21(3):413–438.

Tjøstheim, D. (2015). Count time series with observation-driven autoregressive parameter dynamics. In R. A. Davis, S. H. Holan, R. Lund and N. Ravishanker, eds., *Handbook of Discrete-Valued Time Series*, pp. 77–100. Chapman & Hall, Boca Raton, FL.

Turner, R., Hayen, A., Dunsmuir, W., and Finch, C. F. (2011). Air temperature and the incidence of fall-related hip fracture hospitalisations in older people. *Osteoporosis International*, 22(4):1183–1189.

Wang, C. and Li, W. K. (2011). On the autopersistence functions and the autopersistence graphs of binary autoregressive time series. *Journal of Time Series Analysis*, 32(6):639–646.

Woodard, D. B., Matteson, D. S., and Henderson, S. G. (2011). Stationarity of generalized autoregressive moving average models. *Electronic Journal of Statistics*, 5:800–828.

Xu, S., Jones, R. H., and Grunwald, G. K. (2007). Analysis of longitudinal count data with serial correlation. *Biometrical Journal*, 49(3):416–428.

Zhang, Z., Albert, P. S., and Simons-Morton, B. (2012). Marginal analysis of longitudinal count data in long sequences: Methods and applications to a driving study. *The Annals of Applied Statistics*, 6(1):27–54.

4

Count Time Series with Observation-Driven Autoregressive Parameter Dynamics

Dag Tjøstheim

CONTENTS

4.1 Introduction

A count time series $\{Y_t\}$ is a time series that takes its values on a subset $\mathcal{N}_0$ of the nonnegative integers $\mathcal{N}$. Most often this subset will be all of $\mathcal{N}$, but it can also be the case that, for example, e.g. $\mathcal{N}_0 = \{0, 1\}$ or $\mathcal{N}_0 = \{0, 1, 2, \ldots, k\}$, for a binary and a binomial time series, respectively.

In this chapter, we will look at count time series with dynamics driven by an autoregressive mechanism. This means that the distribution of $\{Y_t\}$ is modeled by a parametric distribution, for example, a Poisson distribution, whose parameters are assumed to be stochastic processes. The dynamics of the $\{Y_t\}$ process is created through a recursive autoregressive scheme for the parameter process. More precisely, in the first order case,

$$P(Y_t = n|X_t) = p(n, X_t), \tag{4.1}$$

where $\sum_{n\in\mathcal{N}_0} p(n, X_t) = 1$, X_t is a parameter-driven process given by a possibly vector and possibly nonlinear AR(p)-type process

$$X_t = f(X_{t-1}, \ldots, X_{t-p}, \varepsilon_{t-1})$$

and $\{\varepsilon_t\}$ is a series of innovations or random shocks driving the process $\{X_t\}$. The parameter process is a genuine nonlinear autoregressive process if $\{\varepsilon_t\}$ consists of iid (independent identically distributed) random variables such that ε_{t-1} is independent of $\mathcal{F}^X_{t-1}$, the σ-algebra generated by $\{X_s, s \le t-1\}$. In the terminology of Cox (1981), the process $\{Y_t\}$ is then a parameter-driven process; see, for example, Davis and Dunsmuir (2015; Chapter 6 in this volume). However, this is not the case for the processes we will be mainly concerned with. We will rather look at the class of processes obtained by replacing $\{\varepsilon_t\}$ by $\{Y_t\}$ so that

$$X_t = f(X_{t-1}, \ldots, X_{t-p}, Y_{t-1}) \tag{4.2}$$

and the more general

$$X_t = f(X_{t-1}, \ldots, X_{t-p}, Y_{t-1}, \ldots, Y_{t-q}) \tag{4.3}$$

with appropriate initial conditions. Clearly these are not genuine AR or ARMA processes because of the presence of lagged values of $\{Y_t\}$, which themselves depend on lagged values of $\{X_t\}$. In the terminology of Cox (1981), the resulting $\{Y_t\}$ processes are examples of observation-driven processes. We will concentrate our analysis on (4.1) and (4.2) because it yields a Markov structure more or less directly, whereas (4.1) and (4.3) need a redefinition of the state space to obtain a Markov structure. Such models have been widely used in applications recently. For specific applications and many references the reader is referred to Fokianos (2015; Chapter 1 in this volume). In the current chapter, the emphasis will be on theory.

The main mathematical tool that has been used to handle the theory of these models is Markov chain theory. To see why, it is advantageous to rewrite the model slightly and at the same time make it more precise. To this end let $\{N_t\}$ be a sequence of nonnegative integer-valued random variables that are independent given $\{X_t\}$ and have the probability distribution function $p(\cdot, X_t)$. If $\{X_t\}$ is nonrandom and equal to a constant, then $\{N_t\}$ is an iid sequence. In the general case we can write

$$Y_t = N_t(X_t).$$

This means that as we move from $t-1$ to t, then first we obtain a value of X_t from (4.2) again with appropriate initial conditions. Then, given X_t, there is a separate and independent random mechanism where N_t, and as a result Y_t, is drawn from $p(\cdot, X_t)$. This makes $\{X_t\}$ into a pth order Markov chain with respect to the σ-field $\{\mathcal{F}^X_t\} = \{\sigma(X_s, s \le t)\}$ and $\{(X_t, Y_t)\}$ is a Markov chain on $\left\{\mathcal{F}^{X,Y}_t\right\} = \{\sigma(X_s, s \le t; Y_u, u \le t)\}$.

The perhaps simplest example of such a process is a first-order Poisson autoregression with $X_t = \lambda_t$, where λ_t is a scalar Poisson intensity parameter, and where

$$Y_t = N_t(\lambda_t). \tag{4.4}$$

Here $\{N_t(\cdot)\}$ could be looked at as a sequence of independent Poisson processes of unit intensity, and where

$$\lambda_t = d + a\lambda_{t-1} + bY_{t-1} \tag{4.5}$$

with a, b, d being nonnegative unknown scalars. This kind of model was treated in Fokianos et al. (2009) and other papers referred to in that paper. To start the recursion an initial value of λ_0 and Y_0 is needed. The process defined by (4.4) and (4.5) is often compared to a GARCH process, see, for example Francq and Zakoïan (2011),

$$Y_t = h_t^{1/2}\varepsilon_t \tag{4.6}$$

with h_t being the conditional variance of Y_t given its past, and where h_t is given by a recursive equation

$$h_t = d + ah_{t-1} + bY_{t-1}^2. \tag{4.7}$$

Here of course h_t corresponds to λ_t, and the series of iid random variables $\{\varepsilon_t\}$ with mean zero and variance 1 corresponds to the series of Poisson processes $\{N_t(\cdot)\}$ of unit intensity in (4.4). There are two problems which make the analysis of (4.4), (4.5) more difficult than the analysis of the GARCH system (4.6), (4.7): (1) $\{\lambda_t\}$ is driven by integer-valued innovations $\{Y_t\}$, whereas $\{\lambda_t\}$ itself, as an intensity parameter, is continuous valued. In the GARCH situation both X_t and h_t are usually taken to be continuous valued, although there are exceptions (Francq and Zakoïan 2004), and (2) the quite innocent looking relationship (4.4) in fact represents a complex nonlinear structure compared to the multiplicative structure of (4.6). These two problems will be discussed throughout the paper, as they are at the core of more or less everything that concerns these processes.

The analogy with the GARCH structure has led to the acronym INGARCH (integer generalized autoregressive conditional heteroscedastic) for these processes, but this is an acronym that I find to be unfortunate, since (4.4) is not in general a variance property. It may be difficult to change the terminology at the present point in time. I would rather prefer INGAR (integer generalized autoregressive processes). In an early version of the model it was called Bin(1) by Rydberg and Shephard (2001).

The Poisson distribution is of course just one example of a distribution $p(\cdot, X_t)$. Other examples are a binary probability where the probability of success $\{p_t\}$ would serve as a parameter process, Wang and Li (2011), or it could be a binomial distribution or a negative binomial. The case of the negative binomial has been treated by Christou and Fokianos (2014), Davis and Wu (2009), and Davis and Liu (2014), and we will return to processes governed by this distribution later.

When it comes to a specification of (4.1 through 4.3), there are two main categories of choices. The first and most obvious one has to do with the choice of the function f in (4.2) and (4.3). One special case is the first-order linear model (4.5). A nonlinear additive model with higher-order lags is the specification

$$f(X_{t-1}, \ldots, X_{t-p}, Y_{t-1}, \ldots, Y_{t-q}) = \sum_{i=1}^{p} g_i(X_{t-i}) + \sum_{i=1}^{q} h_i(Y_{t-i}) \tag{4.8}$$

for some (usually nonnegative) functions $\{g_i\}$ and $\{h_i\}$. So far, I do not know about any systematic attempts to analyse models such as (4.8) except in the case where g_i's and h_i's are linear functions and some special nonlinear models treated by Davis and Liu (2014). Clearly there are many other possibilities, and in the course of this paper we will put various restrictions on f.

The other category of specification has to do with the choice of distribution p in (4.1) and the choice of parametrization of this distribution. The parametrization is not unique and could have to do with the characterization of the exponential family of distributions, where there are canonical choices of parameters. For example, instead of the intensity λ_t as a choice of parameter, one could choose the canonical parameter $\nu_t = \ln \lambda_t$ in the Poisson case. For a binary parameter process with probability parameter p_t, one could choose the parameter $\alpha_t = \ln p_t/1 - p_t$. One obvious advantage of using ν_t instead of λ_t in the Poisson linear case is that in an equation

$$\nu_t = d + a\nu_{t-1} + b\ln(Y_{t-1} + 1)$$

corresponding to (4.5), it is not required any more that the parameters d, a, b be nonnegative, since ν_t itself can take negative values, and in a sense it is easier to implement explanatory variables. We have treated such processes in Fokianos and Tjøstheim (2011).

We will be concerned with both the probabilistic structure of the system (4.1), (4.2) and the asymptotic inference theory of parameter estimates of the parameters characterizing the autoregressive process $\{X_t\}$. Examples of parameters that have to be estimated are the parameters a, b, d in the linear case (4.5).

Somewhat different techniques have been used to characterize the probabilistic structure, but common for most of them is the Markov chain theory. This is of course an integral part of (nonlinear) AR processes (see Tjøstheim 1990), but it is made more difficult and nonstandard in the present case due to the incompatibility problems of values of $\{X_t\}$ or $\{\lambda_t\}$ as compared to $\{Y_t\}$.

One technique for obtaining asymptotic results for the parameter estimates uses standard Markov chain theory by perturbing the original recursive relationship (4.2) or (4.5) with a continuously distributed perturbation ε_t, so that in the linear case one obtains the perturbed equation

$$\lambda_t = d + a\lambda_{t-1} + bY_{t-1} + \varepsilon_t \tag{4.9}$$

and then letting this perturbation tend to zero. This is perhaps not as direct approach as one could wish for (cf. Doukhan 2012 for a critical view), but it leads relatively efficiently to results (Fokianos et al. 2009). A disadvantage of this approach is that one is not concerned so much with the probabilistic structure of the processes $(\{X_t\}, \{Y_t\})$ themselves but rather of the perturbed versions. To look at the probabilistic structure of (4.1), (4.2) and more specifically (4.4), (4.5), again there are different approaches. We will highlight all of this as we proceed.

As mentioned, we will cover both the probabilistic structure and the theory of inference. But the emphasis will be on the former because there are recent review papers, Fokianos (2012), Tjøstheim (2012), and Fokianos (2015; Chapter 1 in this volume), with focus on statistical inference and applications. We will start by a discussion of the existence of a stationary, that is invariant, probability measure for $(\{X_t\}, \{Y_t\})$. This problem is fundamental for most of what follows, such as ergodicity, irreducibility, and recurrence and henceforth for statistical inference. These topics are treated in Section 4.2. We draw the connection to consistency and asymptotic theory of parameter estimates in Section 4.3 and mention some extensions in Section 4.4.

4.2 Probabilistic Structure

4.2.1 Random Iteration Approach

The recursive system (4.1), (4.2) can be looked at as a random iteration scheme. General random iteration schemes have been studied among others by Diaconis and Freedman (1999). They look at an iterative scheme which in our notation can most conveniently be written as

$$X_0 = x, \quad X_1 = f_{N_0}(x),$$

or generally as

$$X_{t+1} = f_{N_t}(X_t). \tag{4.10}$$

In the context of our system (4.1), (4.2), $N_0, N_1, \ldots$ can be thought of as iterative and independent drawings from the distribution p, or in the linear Poisson case as independent drawings from the Poisson processes of unit intensity; that is, from the Poisson distribution with intensity parameter 1. The $\{X_t\}$ process of (4.10) can be directly identified with the $\{X_t\}$ process of (4.2) or λ_t of (4.5). The $\{Y_t\}$ process of (4.1) and of (4.4) is implicitly a part of (4.10) through the drawings from the distribution function p.

In the setup of Diaconis and Freedman (1999) $\{X_t\}$ has as its state space S, a complete metric space with a metric ρ. In the bulk of their paper, and in particular in their Theorems 1.1 and 5.1, p is not allowed to depend on $x \in S$, but they state (p. 49) that "Theorem 1.1 can be extended to cover p that depends on x, but further conditions are needed."

In our setup most of the time $\{X_t\}$ (or $\{\lambda_t\}$) would have $\mathcal{R}_+^1$ as its state space. Note that in order to use the results of Diaconis and Freedman the random mechanism N_t should not depend on X_t. For the Poisson setup in (4.4), (4.5) this is obtained by letting N_t be the realizations of Poisson processes of unit intensity. In, for example, Davis and Liu (2014), it is obtained by setting $X_t = E(Y_t|\mathcal{F}_{t-1})$ with $\mathcal{F}_{t-1} = \sigma(X_0, Y_0, \ldots, Y_{t-1})$, and by considering the inverse of the cumulative distribution function of an exponential family distribution of Y.

Theorems 1.1 and 5.1 of Diaconis and Freedman (1999) both give sufficient conditions for the existence of a unique stationary measure for the Markov chain $\{X_t\}$. This in turn is an essential condition for establishing limit results and a theory of inference for parameter estimates. The conditions of their Theorem 1.1 are somewhat more restrictive than those of Theorem 5.1, but they are easier to formulate and understand intuitively:

First, the functions $f_N(\cdot)$ are supposed to be Lipschitz such that

$$\rho(f_N(x), f_N(y)) \leq C_N \rho(x, y) \tag{4.11}$$

for some C_N and all x and y in S. In fact, f is assumed to be contracting in average, since it is assumed that $\int \ln C_N p(dN) < 0$ (which is a sum in our case since p is discrete). This makes $C_N < 1$ for a typical N. The statement in (4.11) is the statement of the stationarity condition of Bougerol and Picard (1992b), which has found use among other things in GARCH theory, Bougerol and Picard (1992a), Francq and Zakoïan (2004). Under the additional condition that $\int C_N p(dN) < \infty$ and $\int \rho[f_N(x_0), x_0]p(dN) < \infty$ for some $x_0 \in S$, there exists (Theorem 1.1 of Diaconis and Freedman (1999)) (1) a unique stationary distribution π for $\{X_t\}$ and such that (2) $\rho'[P^t(x,\cdot),\pi] \leq A_x r^t$ where ρ' is the Prokhorov metric induced by ρ, $P^t(x,dy)$ is the law of X_t given $X_0 = x$ such that $0 < A_x < \infty$, and $0 < r < 1$ not depending on t or x.

The conditions of Diaconis and Freedman (1999), Theorem 5.1 are somewhat more relaxed than those of Theorem 1.1. This is achieved by securing an appropriate tail behavior. This is the so-called algebraic tail behavior condition where a random variable U has an algebraic tail behavior if $P(U > u) < \alpha/u^{\beta}$ for all $u > 0$, $\alpha, \beta > 0$.

For the proof of these results we refer to Diaconis and Freedman (1999), but we briefly mention the role of the backward iterated process, since this concept also plays an important role in other proofs in the literature that will be mentioned in the sequel.

The forward iterated process is $X_t(x)$ defined by (writing f_t for f_{N_t})

$$X_t(x) = f_{t-1}(X_{t-1}) = (f_{t-1} \circ f_{t-2} \circ \cdots \circ f_0)(x), \quad X_0 = x,$$

and this in general does not converge (almost surely). However, the backward iterated process

$$B_t(x) = (f_0 \circ f_1 \circ \cdots \circ f_{t-1})(x)$$

does converge almost surely to a limit independent of x. Since X_t and B_t have the same distribution, this can be used as a key element in proving the existence of a stationary measure. To understand intuitively why the backward process converges and the forward not, consider the example of an ordinary autoregressive process

$$X_t = aX_{t-1} + \varepsilon_{t-1} \quad X_0 = x, \quad |a| < 1.$$

Here ε_t plays the role of N_t. Forward iteration yields

$$X_t = \varepsilon_{t-1} + a\varepsilon_{t-2} \cdots + a^{t-1}\varepsilon_0 + a^t X_0,$$

and this does not converge almost surely since there is always new random variation introduced by the term ε_{t-1}, and we just have convergence in distribution. On the other hand, the backward iterated process is given by

$$B_t = \varepsilon_0 + a\varepsilon_1 + \cdots + a^{t-1}\varepsilon_{t-1} + a^t X_0,$$

and this converges almost surely to $B_\infty = \sum_{s=0}^{\infty} a^s \varepsilon_s$ since here the new randomness ε_{t-1} is damped down by a^{t-1}. Neither of the two processes X_t or B_t are stationary due to a fixed initial condition, but a stationary process can be introduced as follows in the general case. We let

$$\ldots f_{-2}, f_{-1}, f_0, f_1, f_2, \ldots$$

be independent with common distribution p and

$$W_t = \lim_{s \to \infty} (f_{t-1} \circ f_{t-2} \circ \cdots \circ f_{t-s})(x).$$

Then $\{W_t\}$ is stationary with the transition probability of $\{X_t\}$, and B_∞ is distributed like any of the W_t. For the AR process, W_t is defined as

$$W_t = \sum_{s=0}^{\infty} a^s \varepsilon_{t-1-s}.$$

(Note that $\{W_t\}$ is not the backward process and does not converge almost surely.)

The log condition $\int \ln C_N p(dN) < 0$ is difficult to check for a given model. Wu and Shao (2004) replace this condition by conditions that are easier to verify. Using the same notation as in Diaconis and Freedman (1999), they introduce the following two conditions:

(i) There exists an $x_0 \in S$ and $\alpha > 0$ such that

$$E[\rho(x_0, f_N(x_0))^\alpha] = \int (\rho(x_0, f_N(x_0))^\alpha p(dN) < \infty$$

(ii) There exists an $y_0 \in S$, $\alpha > 0$, $r(\alpha) \in (0,1)$ and $C(\alpha) > 0$ such that

$$E[\rho(X_t(x), X_t(y_0))^\alpha] \leq C(\alpha)(r(\alpha))^t \rho(x, y_0)^\alpha$$

for all $x \in S$ and $t \in \mathcal{N}$.

Condition (i) corresponds to the condition $\int \rho(f_{N_0}(x_0), x_0)p(dN) < \infty$ in Diaconis and Freedman (1999), but is weaker. Condition (ii) replaces $\int \ln C_N p(dN) < 0$ and is also a contraction, "geometric moment contracting," condition in the terminology of Wu and Shao (2004). Under these conditions there exists a stationary measure π (unique). Moreover, Wu and Shao are able to say something about the convergence of the backward iterated process to B_∞, namely

$$E[\rho(B_t(x), B_\infty)^\alpha] \leq Cr^t(\alpha)$$

where $C > 0$ depends solely on x, x_0, b_0 and α, and where $0 < r(\alpha) < 1$. This mode of convergence can subsequently be exploited to obtain central limit theorems and asymptotic theory. The extension of a process from $t = 0, 1, 2, \ldots$ to $t = 0, \pm 1, \pm 2, \ldots$ as outlined earlier plays a role in this. Application to autoregressive count processes are considered by Davis and Liu (2014). We will return to this later in the survey.

4.2.2 The General Markov Chain Approach

In the preceding subsection, we briefly surveyed the iterative random function approach to establishing the existence of a stationary measure. Now we will look at this from a more general Markov chain point of view. Roughly speaking there are two approaches; a topological Markov chain approach and a measure theoretic Markov chain approach. The latter generally gives stronger results, but it is based on an irreducibility assumption which is not fulfilled in general for the models (4.1), (4.2), but which is made to be fulfilled in the perturbation approach. Markov theory, as applied to time series, has largely been dominated by the irreducibility approach, but Tweedie (1988) showed that irreducibility can be avoided at a cost. The integer time series theory has given a new impetus to proving existence of a stationary measure using the first approach, and we will look at this in Sections 4.2.2 through 4.2.5, and then use of irreducibility and perturbation in Section 4.2.6 and 4.2.7.

A common condition for both approaches is a form of stability condition. It can be phrased in many ways. We take it from Meyn and Tweedie (2009). In fact a substantial part of our material is based on their classical book. For a Markov process $\{X_t\}$ with transition probability $P(x, dy)$, the drift operator Δ is defined for any nonnegative measurable V by

$$\Delta V(x) = \int_S P(x, dy)V(y) - V(x). \tag{4.12}$$

Stability conditions in terms of the drift operator can be formulated in many ways, for example

$$\Delta V(x) \leq -1 + b1_C(x), \tag{4.13}$$

where $b > 0$ is a constant and $1_C(\cdot)$ is the indicator function of the measurable set C. This entails a strict drift towards the set C.

Geometric drift towards C is stronger. Then there exists a function $V : S \to [1, \infty]$ and a measurable set C such that

$$\Delta V(x) \leq -\beta V(x) + b1_C(x), \quad x \in S \tag{4.14}$$

where $\beta > 0$, $b < \infty$.

The weak Feller property is an important Markov chain property. As can be seen from Theorem 4.1 mentioned later, it can be used as an instrument in establishing the existence of a stationary measure, but not uniqueness. It states that if g is a continuous bounded function, then $E(g(X_{t+1})|X_t = x)$ is a continous bounded function of x. This can be formulated for a locally compact state space S. Then we have the following result.

Theorem 4.1 *(Meyn and Tweedie (2009), Theorem 12.3.4): If $\{X_t\}$ is weakly Feller, the drift condition (4.13) holds with C a compact set, and if there exists an x_0 such that $V(x_0) < \infty$, then there exists an invariant measure for $\{X_t\}$.*

All of the conditions of Theorem 4.1 are weak in the context we are considering. For the linear Poisson system (4.4), (4.5) it is easily seen that if $0 < a + b < 1$, then the drift condition (4.13) and the condition of existence of an x_0 such that $V(x_0) < \infty$ are satisfied by choosing $V(\lambda) = |\lambda| + 1 = \lambda + 1$, and by choosing the support of the compact set C large enough. Indeed, for verifying the weak Feller property, consider a continuous bounded function g and let $h(x) = E(g(\lambda_t)|\lambda_{t-1} = x) = E(g(d + ax + bN_{t-1}(x))$. The boundedness of h follows trivially. Consider $h(x) - h(y)$ and let A be the event that N_t has no jumps in $(y - \eta, y + \eta)$, where η is to be chosen later. Then $P(A) = e^{-2\eta}$, and

$$h(x) - h(y) = E[(g(d + ax + bN_{t-1}(x)) - g(d + ay + bN_{t-1}(y))]1_A$$
$$+ E[g(d + ax + bN_{t-1}(x)) - g(d + ay + bN_{t-1}(y))]1_{A^c} \doteq I + II$$

On A^c we have

$$|II| \leq 2||g||_\infty P(A^c) = 2||g||_\infty(1 - e^{-2\eta})$$

where $||g||_\infty = \sup_x |g(x)|$. This expression can be made arbitrarily small by choosing η small enough. On A the mapping $x \to g(d + ax + bN_{t-1}(x))$ is continuous if x is close to y, and by Lebesgue dominated convergence $|I| \to 0$ as $y \to x$, the proof can be completed by standard arguments. In the nonlinear situation where

$$\lambda_t = d + f_1(\lambda_{t-1}) + f_2(N_{t-1}(\lambda_{t-1})) \tag{4.15}$$

and with f_1 and f_2 positive, monotone and continuous such that for $y < x$, $f_1(x) - f_1(y) \leq a(x - y)$, $f_2(x) - f_2(y) \leq b(x - y)$, then if $a + b < 1$, the stability condition (4.13) and the weak Feller property are both satisfied, so that an invariant measure exists.

The disadvantage of Theorem 4.1 is that uniqueness is not obtained in general. The strong Feller property, where a bounded measurable function g is mapped into a continuous bounded function $h(x) = E(g(X_{t+1})|X_t = x)$ is sufficient to obtain uniqueness, but unfortunately, the Markov chains generated by the models we look at are not strong Feller, so an alternative way has to be found.

There are in fact several ways of doing this. One route goes via the so-called *e*-chains. The "*e*" stands for equicontinuity, and an *e*-chain is a strengthening of the weak Feller property from continuity to equicontinuity. On the other hand, it is sufficient to look at functions g having a compact support, and such that for a given $\varepsilon > 0$, there exists a $\delta > 0$ implying

$$|E(g(X_{t+1})|X_t = x) - E(g(X_{t+1})|X_t = y)| < \varepsilon$$

for all $t \geq 1$ whenever $|x - y| < \delta$. We also need the concept of boundedness in probability on average: First, a sequence of probability measures $\{P_k, k \in \mathcal{N}\}$ is tight if for each $\varepsilon > 0$, there is a compact set $C \subset S$ such that $\liminf_{k\to\infty} P_k(C) \geq 1 - \varepsilon$. The chain $\{X_t\}$ will be said to be bounded in probability on average if for each initial condition $x \in S$, the sequence $\{\bar{P}_k(x; \cdot), k \in \mathcal{N}\}$ is tight, where

$$\bar{P}_k(x; \cdot) = \frac{1}{k}\sum_{t=1}^{k} P^t(x; \cdot).$$

In other words

$$\liminf_{k\to\infty} \frac{1}{k}\sum_{t=1}^{k} P(X_t \in C|X_0 = x) \geq 1 - \varepsilon.$$

Finally, a point $x_0 \in S$ is called reachable if for every open set $O \in \mathcal{B}(S)$, the collection of Borel sets in S, containing x_0 (i.e. for every neighborhood of x_0)

$$\sum_i P^i(O|y) > 0; \quad y \in S.$$

The chain $\{X_t\}$ is said to be open set irreducible if every point is reachable, this being the topological analogy of measure theoretic irreducibility. In the statement of the following theorem we do not need open set irreducibility. It is sufficient that there is one reachable state.

Theorem 4.2 *(Meyn and Tweedie (2009) Theorem 18.0.2 (ii)): Assume that $\{X_t\}$ is a chain on a topological space such that the assumptions of Theorem 4.1 holds and such that in addition*

(i) There exists a reachable state
(ii) $\{X_t\}$ is an e-chain.
(iii) $\{X_t\}$ is bounded in probability on average

Then the invariant probability measure of Theorem 4.1 is unique.

It is possible to state this theorem with weaker conditions; see Meyn and Tweedie (2009) Theorem 18.0.2.

Continuing with the example used to illustrate Theorem 4.1, consider again the linear model (4.4), (4.5). That there is a reachable state for this model is trivial to prove. In fact every state in $\mathcal{R}_1^+$ can be proved to be reachable; see Fokianos et al. (2009). Boundedness in probability on average follows from boundedness in probability, which in turn follows from Chebyshev's inequality (cf. Neumann 2011, eq. (2.5) and the proof of his Theorem 2.1), where

$$E(\lambda_t|x) \leq \sum_{i=1}^{t-1}(a+b)^i d + (a+b)^t x.$$

Proving the *e*-chain property requires a little more work. The proof is by recursion and the reader is referred to Wang et al. (2014). Their proof is for a more general threshold-type model.

For the nonlinear system (4.15) and with the conditions stated after the proof of the weak Feller property, it can be proved using essentially the same method that it is an *e*-chain bounded in probability on average, and that there exists a reachable state. Hence, there exists a unique stationary measure for such a system.

Woodard et al. (2011) also have as their point of departure a general Markov chain. Instead of using the *e*-chain, they use a condition termed "asymptotically strong Feller." This condition is fairly technical, and since we are not using it henceforth, we refer to their paper for a precise formulation. If in addition a reachable point exists, then uniqueness of the stationary distribution is obtained. Douc et al. (2013) use the asymptotic strong Feller property.

4.2.3 Coupling Arguments

An alternative but overlapping approach to proving existence of a stationary measure proceeds via coupling arguments. This is the approach of Neumann (2011). See also Franke (2010). Neumann works with the quite general nonlinear recursive system,

$$Y_t|\mathcal{F}_{t-1}^{Y,\lambda} \equiv \text{Poisson}(\lambda_t), \quad \lambda_t = f(\lambda_{t-1}, Y_{t-1}), \tag{4.16}$$

which clearly is also related to the iterative system (4.10). The function f is supposed to satisfy the contractive condition,

$$|f(\lambda, y) - f(\lambda', y')| \le \alpha_1|\lambda - \lambda'| + \alpha_2|y - y'|, \tag{4.17}$$

so that $\alpha = \alpha_1 + \alpha_2 < 1$. This is essentially the sort of condition mentioned for the nonlinear additive system (4.15), and the same method as was used to prove the weak Feller property for the linear system can be used here, whereas Chebyshev's inequality can be used to prove boundedness in probability (tightness). Theorem 12.1.2 (ii) of Meyn and Tweedie (2009) then implies the existence of a stationary measure. (Note that this is different from Theorem 4.1.)

If we let Q_x^t be the conditional distribution of (Y_t, λ_t) given $\lambda_0 = x$, then $\{Q_x^t, t \in \mathcal{N}\}$ is tight. Hence, there is a subsequence $\{t_k, k \in \mathcal{N}\}$ such that $Q_x^{t_k}$ converges weakly to some probability measure π_x as $k \to \infty$. Coupling arguments are now used to show that this limit does not depend on the starting value x, and that the full sequence $\{Q_x^t, t \in \mathcal{N}\}$ converges. This implies the existence of a unique stationary distribution.

A brief indication of the coupling argument is as follows: In addition to Y_t, we construct another process Y_t' with another starting value y, but other than that its probability generating mechanism is identical to that of Y_t. Using general properties of the Poisson process and the contractivity condition, one obtains

$$E(|\lambda_t - \lambda_t'| \; |x, y) \le \alpha^t |x - y|$$

and

$$E(|Y_t - Y_t'| \; |x, y) \le \alpha^t |x - y|$$

and tightness arguments again can be used to complete the proof. Douc et al. (2013) use coupling techniques in addition to the asymptotic strong Feller criterion of Woodard et al. (2011).

4.2.4 Ergodicity

Existence of a stationary measure is not enough to develop an appropriate law of large numbers and asymptotic distribution theory of parameter estimates. Ergodicity is needed for consistency, and often ergodicity with an additional rate of convergence to the stationary distribution, such as geometric ergodicity, is required. This in turn can be linked to mixing and mixing rates and ultimately to central limit results. Since it turns out that strong mixing cannot in general be established for the models we look at, other mixing concepts such as weak dependence, to be defined in Section 4.2.5, may be useful. Alternatively, perturbation arguments can be used to obtain strong mixing.

In the following, we will just state a few facts about ergodicity and mixing in our context. Ergodicity is of course treated at length in the book by Meyn and Tweedie for many types of Markov chains, but most of this material refers to the measure theoretic framework where irreducibility is available, and not so much to our more general situation, where the discreteness of the observations $\{Y_t\}$ used as innovations creates difficulties. Here, we will look at the models given by (4.1), (4.2), and (4.4), (4.5) and then in Section 4.2.7 the perturbed class of models, where ergodicity is much easier to handle.

It should be noticed that "ergodicity" is used with somewhat different meanings in dynamic system theory and in Markov chain theory. In dynamic system theory a strictly stationary process is ergodic if an invariant set for the shift operator either has probability 0 or 1. This is the sense in which ergodicity is used in Douc et al. (2013), Theorem 3.2. Also, this is how it is mostly used in time series, possibly non-Markovian, and where it is employed to prove almost sure convergence of parameter estimates. The basis is of course Birkhoff's famous ergodic theorem, Birkhoff (1931), providing a law of large numbers for ergodic strictly stationary processes.

In Markov chain theory, however, ergodicity is used to characterize the convergence of the t-step probability transition measure $P^t(x, dy)$ to $\pi(dy)$, assuming that there exists an invariant measure π. Ergodicity for a Markov chain can then be defined as

$$\lim_{t \to \infty} |P^t(x, y) - \pi(y)| = 0$$

pointwise in the countable state space case, and

$$\lim_{t \to \infty} ||P^t(x, \cdot) - \pi(\cdot)|| = 2 \lim_{t \to \infty} \sup_{A \in \mathcal{B}(S)} |P^t(x, A) - \pi(A)| = 0,$$

where $||\cdot||$ is the total variation norm in the general state space case. One can prove ergodicity for (Y_t, λ_t) for the quite general system (4.16). The line of argument is as follows, Neumann (2011): It is assumed that (4.16), (4.17) hold and such that a stationary invariant measure exists and such that $\{Y_t, \lambda_t\}$ is in its stationary regime. The count process $\{Y_t\}$ is absolutely regular. It is well known that this implies strong mixing of $\{Y_t\}$ with a geometric rate. On the other hand from Neumann (2011) or Bradley (2007) Proposition 2.8, absolute regularity of $\{Y_t\}$ implies ergodicity of $\{Y_t\}$. Since we can express $\lambda_t = f_\infty(Y_{t-1}, Y_{t-2}, \cdots)$ almost surely, Proposition 2.10 (ii) in Bradley (2007) implies that the bivariate process $\{Y_t, \lambda_t\}$ is also ergodic. But unfortunately the bivariate process $\{Y_t, \lambda_t\}$ is not strongly mixing in general. See Neumann (2011) Remark 3 for a counter example. This is again caused by the combination of a discrete $\{Y_t\}$ and a continuous-valued $\{\lambda_t\}$. Davis and Liu (2014) have extended this argument and the ergodicity results to their more general setting.

4.2.5 Weak Dependence

We have repeatedly pointed out the difficulties in proving strong mixing for the $\{\lambda_t\}$ process due to the discreteness of the $\{Y_t\}$ process. It is an obstacle to the derivation of an asymptotic theory of parameter estimates. Similar problems occur in other fields of statistics such as bootstrapping where one samples from a discrete distribution. Doukhan and Louhichi (1999) invented the concept of weak dependence to get around these difficulties. The concept of weak dependence was especially designed to handle situations where there is no strong mixing.

The concept of weak dependence is perhaps best approached by consulting Doukhan and Neumann (2008) or the monograph Dedecker et al. (2007), which is partly a survey. In it are also mentioned alternative concepts (Bickel and Bühlmann 1999; Dedecker and Prieur 2005) that have some analogous properties to the weak dependence concept. In Doukhan and Neumann (2008), weak dependence is defined as follows ($\mathcal{Z}$ is the set of all integers, $\mathcal{N}$ is the set of positive integers):

A process $\{Y_t\}$, $t \in \mathcal{Z}$, is called ψ-weakly dependent if there exists a universal null sequence $\{\epsilon_r\}, r \in \mathcal{N}$ such that for any k-tuple $(s_1, \ldots, s_k)$ and any m-tuple $(t_1, \ldots, t_m)$ with $s_1 \leq \cdots \leq s_k < s_{k+r} = t_1 \leq \cdots \leq t_m$ and arbitrary measurable functions $g : \mathcal{R}^k \to \mathcal{R}$, $h : \mathcal{R}^m \to \mathcal{R}$ with $||g||_\infty \leq 1$ and $\|h\|_\infty \leq 1$, the following inequality is fulfilled:

$$|\mathrm{cov}(g(X_{s_1}, \ldots, X_{s_k}), h(X_{t_1}, \ldots, X_{t_m}))| \leq \psi(k, m, \mathrm{Lip}\, g, \mathrm{Lip}\, h)\epsilon_r.$$

Here Lip h denotes the Lipschitz modulus of continuity of h, that is,

$$\mathrm{Lip}\, h = \sup_{x \neq y} \frac{|h(x) - h(y)|}{||x - y||_{l_1}},$$

where $||(z_1, \ldots, z_m)||_{l_1} = \sum_i |z_i|$, and $\psi : \mathcal{N}^2 \times \mathcal{R}_+^2 \to [0, \infty)$ is a function to be specified.

This general definition has subsequently been refined by appropriate choices of ψ yielding variants of weak ψ-dependence as explained in Doukhan and Neumann (2008). It has been shown that weak dependence generalizes classic mixing criteria in a number of situations, and that central-limit-type results can be derived in several cases where there is a discrete innovation process such as for Markov processes driven by discrete innovations and bootstrap versions of linear AR processes.

To show that a certain process exhibits weak dependence, a contraction argument seems generally to be most efficient. In turn the contractivity can often be proved by a coupling device. There are some applications of weak dependence to AR count processes (Franke 2010 and Doukhan et al. 2012, 2013; see also Neumann 2011), and they show weak dependence by such arguments.

The applications to count time series so far have mainly been limited to demonstrating that certain models, linear and some nonlinear, are weakly dependent (Doukhan et al. 2012, 2013, Franke 2010, and the related paper by Neumann 2011). Once weak dependence is proved, it is still not trivial to establish an asymptotic theory that can be used to study asymptotics of parameter estimates for the models we are considering. The paper by Bardet and Wintenberger (2009) could be a useful starting point. This paper uses weak dependence to obtain asymptotic normality of the quasi-maximum likelihood estimator for a multidimensional causal process. See also Doukhan et al. (2012, 2013), and Christou and Fokianos (2014).

Note that unlike much of the theory presented in this paper the concept of weak dependence does not require a Markov framework. This is of course also the case with the mixing concepts. Franke (2010) uses contraction to prove weak dependence for non-Markovian, but Markov type, models where

$$\lambda_t = d + \sum a_i \lambda_{t-i} + \sum b_i Y_{t-i}$$

In Doukhan et al. (2012) it was conjectured that weak dependence also holds for nonlinear ARMA(∞, ∞) models. Unfortunately, as stated in the correction Doukhan et al. (2013) of that paper the attempted proof of this fact in Doukhan et al. (2012) is not correct. Note that ARMA models are also treated by Woodard et al. (2011), who transform them to a Markov framework.

4.2.6 Markov Theory with ϕ-Irreduciblity

Much of the theory of Meyn and Tweedie (2009) is concerned with the measure theoretic aspects of Markov chains, and there is a huge literature on continuous-valued Markov chains where this concept is exploited. Here we just comment very briefly on these developments, partly to put the preceding developments into perspective and partly as a prelude to the theory of the perturbed models presented next. Let us start by formally defining ϕ-irreducibility:

A Markov chain is ϕ-irreducible if there is a nontrivial measure ϕ on $\{S, \mathcal{B}(S)\}$ such that whenever $\phi(A) > 0$, then $P^t(x, A) > 0$ for some $t = t(x, A) \geq 1$ for all $x \in S$. The measure ϕ is usually assumed to be a maximal irreducibility measure, see Meyn and Tweedie (2009).

A few comments about the relationship between ϕ-irreducibility and the other concepts we have discussed are in order. First, from Meyn and Tweedie (2009) Proposition 6.1.5, if $\{X_t\}$ is strong Feller, and S contains one reachable point x_0, then $\{X_t\}$ is ϕ-irreducible with $\phi = P(x_0, \cdot)$. So-called T-chains are somewhat in between weak and strong Feller chains. A T-chain is defined by Meyn and Tweedie (2009), p. 124, as a chain containing a continuous component (made precise by Meyn and Tweedie; see also example in Nummelin 1984, p. 12). The continuous component defines a transition probability denoted by T, and if $\{X_t\}$ is a T-chain and $\{X_t\}$ contains a reachable point x_0, then $\{X_t\}$ is ϕ-irreducible with $\phi = T(x_0, \cdot)$. By perturbing the chains (4.1), (4.2) and (4.4), (4.5) with a random sequence having a density absolutely continuous with respect to Lebesgue measure on some set $A \in \mathcal{B}(S)$ of positive Lebesgue measure, then we essentially obtain a T-chain and hence ϕ-irreducibility, because proving the existence of a reachable point is in general not difficult.

When ϕ-irreducibility holds, one can define recurrence, positive recurrence, Harris recurrence, and geometric ergodicity. Note that geometric ergodicity, in contradistinction to ergodicity discussed earlier, requires ϕ-irreducibility. Appendix A in Meyn and Tweedie (2009) is a useful compressed source for most of these concepts. The difference between just having the existence of a stationary measure and having positive recurrence is highlighted in Theorems 17.1.2 and 17.1.7 in Meyn and Tweedie (2009).

4.2.7 Perturbation Method

The ϕ-irreducible Markov chain theory can now be illustrated on the models (4.1), (4.2), (4.4), (4.5) when they are perturbed as in (4.9), so that the innovations have a continuous component, effectively making this into a T-chain. We will stick to the linear model (4.4), (4.5) in this subsection and consider more general models in the next section on inference. As indicated in the preceding subsection, ϕ-irreducibility cannot be used in (4.1), (4.2) and (4.4), (4.5). For example, if d, a and b in (4.5) are rational numbers, then $\{\lambda_t\}$ will stay on the rational numbers if λ_0 is rational, and if ϕ is taken to be Lebesgue measure, a natural choice, $\{\lambda_t\}$ is not ϕ-irreducible. However, by perturbing it, it may be made into a ϕ-irreducible process. In the linear model case this was done in Fokianos et al. (2009) by adding a continuous perturbation obtaining a new process $\{Y_t^m\}$,

$$Y_t^m = N_t(\lambda_t^m), \quad \lambda_t^m = d + a\lambda_{t-1}^m + bY_{t-1}^m + \varepsilon_{t,m}, \tag{4.18}$$

$$\varepsilon_{t,m} = c_m 1(Y_{t-1}^m = 1)U_t, \quad c_m > 0, \ c_m \to 0, \text{ as } \quad m \to \infty, \tag{4.19}$$

where $1(\cdot)$ is the indicator function, and where $\{U_t\}$ is a sequence of iid uniform variables on [0,1], and such that U_t is independent of $N_t(\cdot)$. One can then prove ϕ-irreducibility and geometric ergodicity as in the proof of Lemma A1 and Proposition 2.1 in Fokianos et al. (2009). The perturbation in (4.18) is a purely auxiliary device to obtain ϕ-irreducibility. The U_ts could be thought of as pseudo-observations generated by the uniform law. The perturbation can be introduced in many other ways. It is enough to let $\{U_t\}$ be an iid sequence of positive random variables possessing a density on the positive real axis and having a bounded support starting at zero. In addition, as will be seen in the next section, the likelihood functions for $\{Y_t\}$ and $\{Y_t^m\}$, as far as dependence on $\{\lambda_t\}$ and $\{\lambda_t^m\}$ is concerned, will be the same for models (4.4), (4.5) and (4.18), (4.19). It will be noted that both $\{Y_t\}$ and $\{Y_t^m\}$ can be identified with the observations in the expression for the likelihood, but they cannot be identified as stochastic variables since they are generated by different models.

The simple condition $0 < a+b < 1$ implies geometric ergodicity of the perturbed process, and hence the existence of a stationary measure. For a nonlinear perturbed process

$$\lambda_t^m = d + f_1(\lambda_{t-1}^m) + f_2(Y_{t-1}^m) + \varepsilon_{t,m}$$

the corresponding simple condition $0 < f_1(\lambda) + f_2(\lambda) < 1$ outside a small (often taken to be compact) set C and bounded inside; that is a nonglobal contraction suffices. On the other hand, it has been seen earlier that typically a global contraction is used to obtain the existence of a stationary measure. Based on this one might think that using the perturbation approach it is possible to get away with weaker conditions when it comes to deriving an asymptotic theory for parameter estimates. Unfortunately, this is not the case, because in the next step we must let $m \to \infty$ and $c_m \to 0$, when we want to approximate (4.4), (4.5) by (4.18), (4.19). Then we must have $\lambda_t^m \to \lambda_t$ and $Y_t^m \to Y_t$ in some sense. The proof of this, see proof of Proposition 3 in Fokianos and Tjøstheim (2012), requires stricter conditions, essentially identical to those required in the nonperturbed approach. In the linear case the global and local conditions are the same.

Almost all of the theory developed so far, may it be for the perturbed or nonperturbed case, has been stated for a first-order model. In principle it can be extended to a higher-order model, but it is not trivial. Ferland et al. (2006) discuss wide sense stationarity for an ARMA-type model

$$\lambda_t = d + \sum_{i=1}^{p} a_i \lambda_{t-i} + \sum_{i=1}^{q} b_i Y_{t-i}. \tag{4.20}$$

Davis and Liu (2014) in their Proposition 5 prove the geometric moment contracting property and the existence of a unique stationary solution under mild regularity conditions. Franke (2010) proves so-called θ-weak dependence and strict stationarity of $\{Y_t\}$ in the nonlinear system

$$Y_t = N_t(\lambda_t), \quad \lambda_t = g(\lambda_{t-1}, \ldots, \lambda_{t-p}, Y_{t-1}, \ldots, Y_{t-q}).$$

When it comes to the perturbed process of (4.20), it can be converted to a Markov vector process. Again, geometric ergodicity can be proved with a one-component perturbation, the appropriate condition being $0 < \sum_i a_i + \sum_i b_i < 1$.

4.3 Statistical Inference

In Section 4.2, I have outlined two main approaches to the theory of INGAR models. In one approach one studies the properties of the model itself, and a number of results on the probabilistic properties, in particular on the existence of a stationary measure, are now available. The lack of ϕ-irreducibility (and hence geometric ergodicity) do create some problems. The following quote by Woodard et al. (2011) may illustrate the problem of bridging the probabilistic results on the existence of a stationary measure with the convergence results needed in asymptotic theory. They state that the existence of a stationary measure "lay the foundation for showing convergence of time averages for a broad class of functions, and asymptotic properties of maximum likelihood estimators. However, these results are not immediate. For instance, laws of large numbers do exist for non-ϕ-irreducible stationary processes (cf. again Meyn and Tweedie 2009, Theorem 17.1.2), and show that the averages of bounded functionals converge. However, the value to which they converge may depend on the initialization of the process. (It may be possible to obtain correct limits of time averages by restricting the class of functions under consideration, or by obtaining additional mixing results for the time series under consideration)."

I am going to review the work on statistical inference for the direct approach and the perturbed approach in Sections 4.3.1 and 4.3.2, respectively, but first let me give some general comments on possible advantages and drawbacks of the two methods.

An obvious advantage of the direct method is that it only uses properties of the process defined by the original model itself. A possible drawback is assuming that the process is in its stationary state, granted that conditions for existence of a stationary state is fulfilled. This makes it possible to extend the process to $0, \pm 1, \pm 2, \ldots$. It is not always straightforward to link a likelihood based on a stationary solution to a likelihood depending on a given initial condition. An approximation argument seems to be needed as in Wang et al. (2014). In the perturbation method the process can be started from an arbitrary initial point, because when the process is perturbed, one typically obtains a ϕ-irreducible geometric ergodic process, and the inference theory for such processes does not require the process to be in a stationary state. The geometric ergodicity drives the process towards its stationary state asymptotically at a geometric rate. A disadvantage of this approach is the mere fact that the perturbed process is just an intermediate step and a different, but in some cases similar, approximation argument to that mentioned earlier for the likelihoods is needed. Both methods require a type of contracting condition. When it comes to the problem of estimating parameters, the maximum likelihood estimates for the two methods are usually identical, but in deriving properties of the estimates the methods may differ and require arguments of different complexity.

4.3.1 Asymptotic Estimation Theory without Perturbation

This is a new topic, so the literature is not extensive. The parameter estimation problem has been treated by Davis and Liu (2014), Wang et al. (2014), Douc et al. (2013), Woodard et al. (2011), and Christou and Fokianos (2014). These authors have somewhat different choices of methods, but the main methodological difference seems to consist in the way the contraction condition is established. Once this is established, fairly standard consistency and likelihood arguments are used. Other aspects of statistical inference are treated by Wu and Shao (2004), Fokianos and Neumann (2013), Fokianos and Fried (2010), Christou and

Fokianos (2013, 2015). I will concentrate on the parameter estimation here, and then briefly mention other aspects.

Davis and Liu (2014) treat a general model

$$Y_t|\mathcal{F}_{t-1} \sim p(y|\eta_t), \quad X_t = g_\theta(X_{t-1}, Y_{t-1})$$

where $\mathcal{F}_t = \sigma\{\eta_1, Y_1, \dots . Y_t\}$, $X_t = E(Y_t|\mathcal{F}_{t-1}) \doteq B(\eta_t)$, and $p(\cdot|\eta)$ is a distribution from a one-parameter exponential family

$$p(y|\eta) = \exp\{\eta y - A(\eta)\}h(y),$$

where η is the natural parameter (e.g., $\ln \lambda$ in a Poisson distribution), and $A(\eta)$ and $h(y)$ are known functions. Many familiar distributions belong to this family, such as the Poisson, negative binomial, Bernoulli, exponential. In Davis and Liu (2014) the Poisson case and the negative binomial with parameters r and p_t are treated. For the latter the integer parameter r is fixed, and the probability parameter p_t is allowed to be stochastic and recursively updated. Using a contraction condition on g and the backward process of Diaconis and Freedman (1999), they establish the existence of a unique stationary measure and prove that the process $\{X_t\}$ is geometric moment contracting, as in Wu and Shao (2004), and this in turn is used to establish that $\{Y_t\}$ is absolutely regular and $\{X_t, Y_t\}$ is ergodic, as did Neumann (2011) using other means. The likelihood function is given by

$$L(\theta|Y_1, \dots, Y_n; \eta_1) = \prod_{t=1}^{n} \exp\{Y_t - A(\eta_t(\theta))\}h(Y_t).$$

Taking logs and differentiating the score function is obtained

$$S_n(\theta) = \frac{\partial l(\theta)}{\partial \theta} = \sum_{t=1}^{n} \{Y_t - B(\eta_t(\theta))\} \frac{\partial \eta_t}{\partial \theta},$$

where B is the derivative of A, that is $B(\eta) = \dot{A}(\eta)$. Note that $\dot{B}(\eta) = \text{var}(Y_t) > 0$ so that $B(\eta)$ is strictly increasing. Moreover, $X_t = B(\eta_t) = E(Y_t|\mathcal{F}_{t-1})$, $\eta_t = B^{-1}(X_t)$, and $X_t = g_\infty^\theta(Y_{t-1}, \dots)$ by expanding $\{Y_t\}$ back to $-\infty$ as explained in Section 4.2.4. If we let $\widehat{\theta}_n$ be a solution of $S_n(\theta) = 0$, then under a string of regularity conditions given in Davis and Liu (2014), strong consistency is obtained and asymptotic normality of $\widehat{\theta}_n$ is demonstrated, that is

$$\sqrt{n}(\widehat{\theta}_n - \theta) \xrightarrow{\mathcal{L}} \mathcal{N}(0, \Omega^{-1}),$$

where $\Omega = E\{\dot{B}(\eta_t)\frac{\partial \eta_t}{\partial \theta}(\frac{\partial \eta_t}{\partial \theta})'\}$ and where $'$ is used to denote the transposed. In the Poisson case $\eta_t = \ln \lambda_t$ and

$$L(\theta) = \prod_t \frac{\exp(-\lambda_t)\lambda_t^{Y_t}(\theta)}{Y_t!}, \quad A = e^{\eta_t(\theta)}, \quad \text{and} \quad \Omega = E\left\{e^{\eta_t(\theta)} \frac{\partial \eta_t(\theta)}{\partial \theta} \left(\frac{\partial \eta_t(\theta)}{\partial \theta}\right)'\right\}$$

which corresponds to the asymptotic distribution in the log linear case in Fokianos and Tjøstheim (2011).

Consistency is proved using standard arguments, whereas asymptotic normality is proved by a linearization argument and the fact that $n^{-1/2}\{Y_t - B(\eta_t(\theta))\}\partial\eta_t/\partial\theta$ is a martingale difference sequence.

Christou and Fokianos (2014) look at the negative binomial case too, and they approach the problem somewhat differently. They use the fact that an appropriate mixture of Poisson distributions yields a negative binomial distribution, and obtain a one-parameter negative binomial distribution in that way. Subsequently, this is used in treating overdispersion in a natural manner.

Somewhat different arguments to those employed by Davis and Liu (2014) were used in Wang et al. (2014) in a threshold-like model. The existence of a stationary measure is established using the weak Feller and *e*-chain properties.

Douc et al. (2013) restrict themselves to proving consistency in a model

$$Y_t|\mathcal{F}_{t-1} \sim H(X_{t-1}, \cdot), \quad X_t = F_{Y_t}(X_{t-1}),$$

where H is a Markov kernel and $\mathcal{F}_t = \sigma(X_s, Y_s, s \leq t, s \in \mathcal{N})$. The authors have a series of regularity conditions, among them the asymptotic strong Feller property (Hairer and Mattingly 2006), existence of a reachable point and a contraction condition set up in a coupling context. They use these conditions to establish an invariant unique measure for the Markov kernel under consideration.

4.3.2 The Perturbation Approach

This approach is carried out in Fokianos et al. (2009), Fokianos and Tjøstheim (2011, 2012) and referred to in Woodard et al. (2011). In those papers it is restricted to the Poisson case for the distribution of the innovations, but I believe that it has wider potential. In the linear case, the essence of the method is to set up a perturbed model (4.18), (4.19) in addition to the original model (4.4), (4.5) to obtain a model that is geometrically ergodic and analogously for the log linear and nonlinear models. Next, this is used to construct two likelihoods, and then to show that their difference tends to zero as the size of the perturbation decreases to zero. This is finally utilized to show that the asymptotics of the parameter estimates of the original model are obtained as limits of the asymptotics of the perturbed model.

We will illustrate this for the nonlinear model of Fokianos and Tjøstheim (2012):

$$Y_t = N_t(\lambda_t), \quad \lambda_t = f_1(\lambda_{t-1}) + f_2(Y_{t-1})$$

with assumptions on f_1 and f_2 as already stated in Section 4.2.7. The corresponding perturbed model is given by

$$Y_t^m = N_t(\lambda_t^m), \quad \lambda_t^m = f_1(\lambda_{t-1}^m) + f_2(Y_{t-1}^m) + \varepsilon_{t,m},$$

$$\varepsilon_{t,m} = c_m 1(Y_t^m = 1)U_t, \quad c_m > 0, \quad U_t \sim \text{iid } U[0,1],$$

where $c_m \to 0$ as $m \to \infty$, and where other possibilities of perturbation exist.

The likelihoods are given by

$$L(\theta) = \prod_{t=1}^{n} \frac{\exp(-\lambda_t(\theta))\lambda_t^{Y_t}(\theta)}{Y_t!}$$

and

$$L^m(\theta) = \prod_{t=1}^{n} \frac{\exp(-\lambda_t^m(\theta))(\lambda_t^m(\theta))^{Y_t^m}}{Y_t^m!} \prod_{t=1}^{n} fu(U_t)$$

by the Poisson assumption and the assumed independence of U_t from $(Y_{t-1}^m, \lambda_{t-1}^m)$ with $fu(u)$ denoting the uniform density. Moreover, the log likelihood and the scores are given by

$$l^m(\theta) = \sum_{t=1}^{n} (Y_t^m \ln \lambda_t^m(\theta) - \lambda_t^m(\theta)) + \sum_{t=1}^{n} \ln fu(U_t) - \sum_{t=1}^{m} \ln Y_t^m$$

and

$$S_n^m(\theta) = \frac{\partial l^m(\theta)}{\partial \theta} = \sum_{t=1}^{n} \left(\frac{Y_t^m}{\lambda_t^m(\theta)} - 1 \right) \frac{\partial \lambda_t^m(\theta)}{\partial \theta}.$$

The Hessian is given by

$$\begin{aligned} H_n^m(\theta) &= -\sum_{t=1}^{n} \frac{\partial^2 l^m(\theta)}{\partial \theta \partial \theta'} \\ &= \sum_{t=1}^{n} \frac{Y_t^m}{(\lambda_t^m(\theta))^2} \frac{\partial \lambda_t^m(\theta)}{\partial \theta} \left(\frac{\partial \lambda_t^m(\theta)}{\partial \theta} \right)' - \sum_{t=1}^{n} \left(\frac{Y_t^m}{\lambda_t^m(\theta)} - 1 \right) \frac{\partial^2 \lambda_t^m(\theta)}{\partial \theta \partial \theta'}. \end{aligned}$$

Using $E(Y_t^m|\mathcal{F}_{t-1}) = \lambda_t^m$, the expectation of $\frac{1}{n}H_n^m$ is

$$G^m(\theta) = E\left(\frac{1}{\lambda_t^m} \left(\frac{\partial \lambda_t^m}{\partial \theta} \right) \left(\frac{\lambda_t^m}{\partial \theta} \right)' \right).$$

Exactly the same calculation can be carried through for the unperturbed system, resulting in

$$G(\theta) = E\left(\frac{1}{\lambda_t} \left(\frac{\partial \lambda_t}{\partial \theta} \right) \left(\frac{\partial \lambda_t}{\partial \theta} \right)' \right).$$

Asymptotic normality for the parameter estimates of the perturbed system follows from standard asymptotic theory. Asymptotic normality with covariance matrix G^{-1} for the unperturbed system will follow from an approximation theorem of Brockwell and Davis (1987), Proposition 6.3.9, if it can be shown that $G^m(\theta)$ converges to $G(\theta)$ when $m \to \infty$ and the perturbation tends to zero as $c_m \to 0$. It is in this argument that the contraction assumption is needed. One needs to evaluate differences

$$E\left\| \frac{\partial \lambda_t^m}{\partial \theta} - \frac{\partial \lambda_t}{\partial \theta} \right\| \quad \text{and} \quad E\left\| \frac{1}{\lambda_t^m} \frac{\partial \lambda_t^m}{\partial \theta} \left(\frac{\partial \lambda_t^m}{\partial \theta} \right)' - \frac{1}{\lambda_t} \frac{\partial \lambda_t}{\partial \theta} \left(\frac{\partial \lambda_t}{\partial \theta} \right)' \right\|$$

and using the regularity conditions of Fokianos and Tjøstheim (2012) these are transformed to showing

$$E|\lambda_t^m - \lambda_t| \to 0, \quad E(\lambda_t^m - \lambda_t)^2 \to 0,$$

$$E|Y_t^m - Y_t| \to 0, \quad E(Y_t^m - Y_t)^2 \to 0,$$

and almost surely

$$|\lambda_t^m - \lambda_t| \to 0, \quad |Y_t^m - Y_t| \to 0.$$

For a proof of these we refer to Fokianos and Tjøstheim (2012).

Similar techniques can be used in the log linear case, where

$$Y_t = N_t(\lambda_t), \;\; \nu_t = d + a\nu_{t-1} + b\ln(Y_{t-1} + 1), \;\; \lambda_t = \exp(\nu_t).$$

In this case there is a large discrepancy between the conditions required for approximating the perturbed system to the unperturbed system and the conditions required for geometric ergodicity of

$$Y_t^m = N_t(\lambda_t), \quad \nu_t^m = d + a\nu_{t-1}^m + b\ln(Y_{t-1}^m + 1) + \varepsilon_{t,m},$$

where ν_0^m, Y_0^m are fixed and $\{\varepsilon_{t,m}\}$ is as in (4.19). For geometric ergodicity it is sufficient that $|a| < 1$ and in addition if $b > 0$, then $|a+b| < 1$, and when $b < 0$, $|a||a+b| < 1$. These conditions are much milder than the corresponding condition $|a + b| < 1$ for the linear system (4.4), (4.5). This is because for the log linear system the process can flip back and forth between the negative and positive domain. To obtain global contraction, where the perturbed likelihood can be compared to the unperturbed one, much stricter conditions are required. In Fokianos and Tjøstheim (2012), a set of sufficient conditions were obtained by requiring $|a + b| < 1$ if a and b have the same sign and $(a^2 + b^2) < 1$ if a and b have different signs. In Douc et al. (2013) who did not consider the asymptotic distribution but did consider contraction, this is improved, and they obtained the sufficient condition $|a + b| \vee |a| \vee |b| < 1$.

The perturbation approach is also possible for non-Poisson processes. For the model considered by Davis and Liu (2014), for example, one must show that the geometric moment contractivity is enough to prove that the unperturbed likelihood can be approximated by a perturbed likelihood. I conjecture that this is the case under weak regularity conditions.

Finally, summing up all this, the perturbation approach and the direct unperturbed approach are not mutually exclusive. The direct approach gives the fundamental probability properties of the process. Geometric ergodicity is not present, and, at least in a time series sense, requires a nonstandard Markov chain theory. The perturbation approach is only used in the inference stage. Asymptotic results are first found for the perturbed system. The geometric ergodicity means that the process is not required to be in its stationary state, and in a time series sense more standard Markov theory can be used.

4.4 Extensions

The area of research reviewed in this paper is quite new, which means that there are many problems that have not been fully explored yet. Here I mention a few possibilities.

4.4.1 Higher-Order Models

Most of the papers written so far have been for the first-order case. This is often stated to be for reasons of convenience. Partly, at least, this seems to be true, but I think it has to be realized that the extension is not as immediate as for the standard autoregressive model. In the linear case there are indications that the theory goes through quite easily, but not all of the details have been written up. In order to state higher-order ARMA models as first-order Markov models, one can introduce a state vector containing present and past observations and innovations, but unlike the ordinary AR case there is very little literature on the theory of multivariate integer time series and virtually nothing for processes with autoregressive dynamics. So far estimation theory for univariate models has been limited to the first-order case. In principle this should be relatively easy to extend to the higher-order case. Proceeding via weak dependence might be a viable alternative. The theory of the log-linear model has only been worked on by Fokianos and Tjøstheim (2011) using perturbations and by Douc et al. (2013) not using weak dependence arguments. It does not seem to be trivial to extend the theory in Fokianos and Tjøstheim (2011) and Douc et al. (2013) to higher-order models because of the much more complex stability theory and the more difficult problem of proving asymptotic proximity for the perturbed model for this kind of processes. A partial analog is the threshold AR model for ordinary time series, where there is a flip-over mechanism like that identified in Section 4.3.2. The problem of finding conditions for geometric ergodicity has been completely solved in the ordinary threshold AR(1) case, but to my knowledge there is not a very satisfying solution even in the second-order threshold case, not to speak of higher-order models. The difficulties involved are outlined already in Tjøstheim (1990) and resemble those being encountered for higher-order log-linear integer AR models.

4.4.2 Vector Models

Such models, not surprisingly, have found useful applications in general integer-valued settings. Various aspects of multivariate modeling is given in Section V of the present volume. Jung et al. (2011) use a factor model on stock market trading activity. Estimates are computed using maximum likelihood, but asymptotic properties are not considered. For ordinary AR processes there is a straightforward extension to vector models. Moreover, the multivariate normal distribution inherits many properties of the univariate case and is the simplest multivariate distribution to work with. All of this ceases to be true for multivariate count processes. First, there is not a simple generalization of the Poisson distribution from the univariate to the multivariate case (Johnson et al. 1997). Second, Heinen and Rengifo (2007) have tried to model dependence between marginal Poisson processes by using a copula directly on the Poisson processes, but it is known that the copula for discrete-valued variables may be difficult to handle because of its ambiguity, although it can be uniquely defined. Fokianos et al. (2014) are trying a more indirect approach. But these models are challenging to handle both theoretically and computationally. As far as I know nobody has tried to implement the log-linear model in this context. Multivariate models are clearly of interest in, for example, finance by studying say the micromarket transaction structure and relationship to volatility. Nonlinear vector models are another step up in difficulty, and so far no work appears to have been done.

4.4.3 Generalizing the Poisson Assumption

The Poisson assumption has an immediate advantage since (in the scalar case) it just depends on one parameter. This is also the case for geometric, binomial, and binary distribution models, and one should think that a meaningful theory can be derived as has been indicated in this paper when formulating the general models (4.1), (4.2). Moysiadis and Fokianos (2014) have recently looked at categorical time series. Alternatives to the Poisson assumption have been examined, such as use of negative binomial or generalized Poisson innovations to allow for over-dispersion in thinning models. Again we refer to the papers by Davis and Liu (2014), Christou and Fokianos (2014), and to Weiß (2008). Another class of distributions is mentioned in Franke (2010) in an integer AR context. Ultimately one would like to establish a theory for a fairly general counting process where the innovation distribution is not explicitly defined, but where one only assumes the existence of a discrete distribution depending on one or several unknown parameters.

4.4.4 Specification and Goodness-of-Fit Testing

It is desirable to derive a suitable model via a specification procedure. Once a model is obtained, one would like to test it via a goodness-of-fit test. Such a test should not only be able to decide between a linear and nonlinear model, but there is also the question of which distribution should be used for the innovations. Not much work has been done, but an important application to predictive model assessment is contained in Czado et al. (2009) pointing out differences in the integer and the continuous-valued cases. Neumann (2011) mentions an application to a goodness-of-fit situation with a much more general treatment given in Fokianos and Neumann (2013). Note also that there are papers by Fokianos and Fried (2010, 2012) on interventions.

Acknowledgement

I am indebted to Richard Davis for a number of useful suggestions on an earlier version of the paper.

References

Bardet, J.-M. and Wintenberger, O. (2009), Asymptotic normality of the quasi-maximum likelihood estimator for multidimensional causal processes, *The Annals of Statistics*, 37, 2730–2759.

Bickel, P. and Bühlmann, P. (1999), A new mixing notion and functional central limit theorems for a sieve bootstrap in time series, *Bernoulli*, 5, 413–446.

Birkhoff, D. (1931), Proof of the ergodic theorem, *Proceedings National Academy of Science*, 17, 656–660.

Bougerol, P. and Picard, N. (1992a), Stationarity of GARCH processes and of some nonnegative time series, *Journal of Econometrics*, 52, 115–127.

Bougerol, P. and Picard, N. (1992b), Strict stationarity of generalized autoregressive processes, *The Annals of Probability*, 4, 1714–1730.

Bradley, R. (2007), *Introduction to Strong Mixing Conditions*, Vol. 1, Kendrick Press, Heber City, Utah.

Brockwell, P. and Davis, R. (1987), *Time Series: Theory and Methods*, Springer Verlag, New York.

Christou, V. and Fokianos, K. (2014), Quasi-likelihood inference for negative binomial time series models, *Journal of Time Series Analysis*, 35, 55–78.

Christou, V. and Fokianos, K. (2013), Estimation and testing linearity for mixed Poisson autoregressions, unpublished manuscript.

Christou, V. and Fokianos, K. (2015), On count time series prediction, *Journal of Statistical Computation and Simulation*, 85, 357–373.

Cox, D. (1981), Statistical analysis of time series: some recent developments [with discussion and reply], *Scandinavian Journal of Statistics*, 8, 93–115.

Czado, C., Gneiting, T., and Held, L. (2009), Predictive model assessment for count data, *Biometrics*, 65, 1254–1261.

Davis, R. and Dunsmuir, W.T.M. (2015). State space models for count time series. In R. A. Davis, S. H. Holan, R. Lund and N. Ravishanker, eds., *Handbook of Discrete-Valued Time Series*, pp. 121–144. Chapman & Hall, Boca Raton, FL.

Davis, R. and Wu, R. (2009), A negative binomial model for time series of counts, *Biometrika*, 96, 735.

Davis, R. A. and Liu, H. (2014), Theory and inference for a class of observation-driven models with application to time series of counts, *arXiv preprint arXiv:1204.3915, Statistica Sinica*, doi:10.5705/ss.2014.145t (to appear).

Dedecker, J., Doukhan, P., Lang, G.R., Leon, J.R., Louhichi, S., and Prieur, C. (2007), *Weak Dependence: With Examples and Applications*, Springer.

Dedecker, J. and Prieur, C. (2005), New dependence coefficients. Examples and applications to statistics, *Probability Theory and Related Fields*, 132, 203–236.

Diaconis, P. and Freedman, D. (1999), Iterated random functions, *SIAM Review*, 41, 45–76.

Douc, R., Doukhan, P., and Moulines, E. (2013), Ergodicity of observation-driven time series models and consistency of the maximum likelihood estimator, *Stochastic Processes and Their Applications*, 123, 2620–2647.

Doukhan, P. (2012), Discussion of "Some recent theory for autoregressive count time series", *TEST*, 21, 447–450.

Doukhan, P., Fokianos, K., and Tjøstheim, D. (2012), On weak dependence conditions for Poisson autoregression, *Statistics and Probability Letters*, 82, 942–948.

Doukhan, P., Fokianos, K., and Tjøstheim, D. (2013), Correction to "On weak dependence conditions for Poisson autoregression", *Statistics and Probability Letters*, 83, 1926–1927.

Doukhan, P. and Louhichi, S. (1999), A new weak dependence condition and applications to moment inequalities, *Stochastic Processes and Their Applications*, 84, 313–342.

Doukhan, P. and Neumann, M. (2008), The notion of ψ-weak dependence and its applications to bootstrapping time series, *Probability Surveys*, 5, 146–168.

Ferland, R., Latour, A., and Oraichi, D. (2006), Integer-valued GARCH process, *Journal of Time Series Analysis*, 27, 923–942.

Fokianos, K. (2012), Count time series models, in *Handbook in Statistics: Time Series-Methods and Aplications*, eds. Rao, C. R. and Subba Rao, T., Elsevier, Vol. 30, Amsterdam.

Fokianos, K. (2015). Statistical analysis of count time series models: A GLM perspective. In R. A. Davis, S. H. Holan, R. Lund and N. Ravishanker, eds., *Handbook of Discrete-Valued Time Series*, pp. 3–28. Chapman & Hall, Boca Raton, FL.

Fokianos, K., Doukhan, P., Støve, B., and Tjøstheim, D. (2014), Approximate multivariate Poisson autoregression, Work in progress.

Fokianos, K. and Fried, R. (2010), Interventions in INGARCH processes, *Journal of Time Series Analysis*, 31, 210–225.

Fokianos, K. and Fried, R. (2012), Interventions in log-linear Poisson autoregression, *Statistical Modelling*, 12, 299–322.

Fokianos, K. and Neumann, M. H. (2013), A goodness-of-fit test for Poisson count processes, *Electronic Journal of Statistics*, 7, 793–819.

Fokianos, K., Rahbek, A., and Tjøstheim, D. (2009), Poisson autoregression, *Journal of the American Statistical Association*, 104, 1430–1439.

Fokianos, K. and Tjøstheim, D. (2011), Log-linear Poisson autoregression, *Journal of Multivariate Analysis*, 102, 563–578.

Fokianos, K. and Tjøstheim, D. (2012), Nonlinear Poisson autoregression, *Annals of the Institute of Statistical Mathematics*, 64, 1205–1225.

Francq, C. and Zakoïan, J. (2004), Maximum likelihood estimation of pure GARCH and ARMA-GARCH processes, *Bernoulli*, 10, 605–637.

Francq, C. and Zakoïan, J.-M. (2011), *GARCH Models: Structure, Statistical Inference and Financial Applications*, Wiley, New York.

Franke, J. (2010), Weak dependence of functional INGARCH processes, Technical report, Working paper, University of Kaiserslautern, Germany.

Hairer, M. and Mattingly, J. (2006), Ergodicity of the 2D Navier-Stokes equations with degenerate stochastic forcing, *Annals of Mathematics*, 164, 993–1032.

Heinen, A. and Rengifo, E. (2007), Multivariate autoregressive modeling of time series count data using copulas, *Journal of Empirical Finance*, 14, 564–583.

Johnson, N., Kotz, S., and Balakrishnan, N. (1997), *Discrete Multivariate Distributions*, Wiley, New York.

Jung, R., Liesenfeld, R., and Richard, J. (2011), Dynamic factor models for multivariate count data: An application to stock-market trading activity, *Journal of Business and Economic Statistics*, 29, 73–85.

Meyn, S. P. and Tweedie, R. L. (2009), *Markov Chains and Stochastic Stability*, Cambridge University Press, Cambridge, U.K.

Moysiadis, T. and Fokianos, K. (2014), On binary and categorical time series models with feedback, *Journal of Multivariate Analysis*, 131, 209–228.

Neumann, M. (2011), Absolute regularity and ergodicity of Poisson count processes, *Bernoulli*, 17, 1268–1284.

Nummelin, E. (1984), *General Irreducible Markov Chains and Non-Negative Operators*, Cambridge University Press, Cambridge, U.K.

Rydberg, T. H. and Shephard, N. (2001), A modelling framework for the prices and times of trades made on the New York stock exchange, *Nonlinear and Nonstationary Signal Processing*, eds. Fitzgerald, W. J., Smith, R. L., Walden, A. T., and Young, P. C., pp. 217–246, Cambridge University Press, Cambridge, U.K.

Tjøstheim, D. (1990), Non-linear time series and Markov chains, *Advances in Applied Probability*, 22, 587–611.

Tjøstheim, D. (2012), Some recent theory for autoregressive count time series (with discussion), *TEST*, 21, 413–476.

Tweedie, R. (1988), Invariant measures for Markov chains with no irreducibility assumptions, *Journal of Applied Probability*, 25, 275–285.

Wang, C. and Li, W. (2011), On the autopersistence functions and the autopersistence graphs of binary autoregressive time series, *Journal of Time Series Analysis*, 32, 639–646.

Wang, C., Liu, H., Yao, J.-F., Davis, R. A., and Li, W. K. (2014), Self-excited Threshold Poisson Autoregression, *Journal of the American Statistical Association*, 109, 777–787.

Weiß, C. (2008), Thinning operations for modeling time series of counts—A survey, *AStA Advances in Statistical Analysis*, 92, 319–341.

Woodard, D., Matteson, D., and Henderson, S. (2011), Stationarity of count-valued and nonlinear time series models, *Electronic Journal of Statistics*, 5, 800–828.

Wu, W. B. and Shao, X. (2004), Limit theorems for iterated random functions, *Journal of Applied Probability*, 41, 425–436.

5

Renewal-Based Count Time Series

Robert Lund and James Livsey

CONTENTS

5.1 Introduction

Some classical methods of generating stationary count time series do not yield a particularly flexible suite of autocovariance structures. For example, integer autoregressive moving-average (INARMA) (Steutel and Van Harn 1979; McKenzie, 1985, 1986, 1988; Al-Osh and Alzaid 1988) and discrete autoregressive moving-average (DARMA) methods (Jacobs and Lewis, 1978a,b) cannot produce negatively correlated series at any lags. This is because mixing ratios or thinning probabilities must always lie in (0,1). Also, INARMA and DARMA models cannot generate series with long memory autocovariances.

This chapter takes a different approach to modeling stationary integer count time series: the renewal paradigm. A renewal sequence is used as a natural model for a stationary autocorrelated binary sequence. Independent copies of the renewal sequence can then be superimposed in various ways to build the desired marginal distribution. Blight (1989) and Cui and Lund (2009) are references for the general idea.

A renewal process is built from a nonnegative integer-valued lifetime L_0 and a sequence of independent and identically distributed (IID) lifetimes $\{L_i\}_{i=1}^{\infty}$. Here, $L_i \in \{1, 2, \ldots\}$ has an aperiodic support (the support set of L is not a multiple of some integer larger than unity) and a finite mean $\mu = E[L_1]$ (notice that the distribution of L_0 may be different than those of the other L_is). Below, L will denote a random draw whose distribution is equivalent to that of any L_i for $i \geq 1$. Define the random walk

$$S_n = L_0 + L_1 + \cdots + L_n, \quad n = 0, 1, \ldots.$$

A renewal is said to occur at time t if $S_n = t$ for some $n \geq 0$. Let X_t be unity if a renewal occurs at time t and zero otherwise. This is the classical discrete-time renewal sequence popularized in Smith (1958), Feller (1968), and Ross (1996) (among others). In fact, $\{X_t\}$ is a correlated binary sequence. Copies of $\{X_t\}$ will be used to build our count series shortly. For notation, let u_t be the probability of a renewal at time t in a nondelayed renewal process (a nondelayed process has $L_0 = 0$). The probabilities of renewal can be recursively calculated via $u_0 = 1$ and

$$u_t = P[L = t] + \sum_{\ell=1}^{t-1} P[L = \ell]u_{t-\ell}, \quad t = 1, 2, \ldots, \tag{5.1}$$

and $E[X_t] = u_t$. The elementary renewal theorem (Smith, 1958) states that $\lim_{t\to\infty} u_t = \mu^{-1}$ when L has a finite mean $E[L] = \mu$ and is aperiodic, both of which we henceforth assume.

While $u_t \to E[L]^{-1}$ as $t \to \infty$, $\{X_t\}$ is not weakly stationary unless the distribution of L_0 is strategically selected. Specifically, if L_0 has the first tail distribution derived from L, viz.,

$$P(L_0 = k) = \frac{P(L > k)}{\mu}, \quad k = 0, 1, \ldots \tag{5.2}$$

then $\{X_t\}$ is covariance stationary. Under this initial distribution, $E[X_t] \equiv \mu^{-1}$ (Ross, 1996) and the autocovariance function of $\{X_t\}$, denoted by $\gamma_X(h) = \mathrm{Cov}(X_t, X_{t+h})$, is

$$\gamma_X(h) = P[X_t = 1 \cap X_{t+h} = 1] - P[X_t = 1]P[X_{t+h} = 1] = \mu^{-1}\left(u_h - \frac{1}{\mu}\right) \tag{5.3}$$

for $h = 0, 1, \ldots$. This calculation uses $P[X_{t+h} = 1 \cap X_t = 1] = P[X_{t+h} = 1 | X_t = 1]$ $P[X_t = 1] = u_h \mu^{-1}$.

As an example, suppose that L is geometric with $P(L = k) = p(1-p)^{k-1}$ for $k = 1, 2, \ldots$ and some $p \in (0, 1)$. Then L is aperiodic and $E[L] = 1/p$. The initial lifetime L_0 in (5.2) that makes this binary sequence stationary has distribution $P(L_0 = k) = p(1-p)^k$ for $k = 0, 1, \ldots$ (this is not geometric as the support set of L_0 contains zero). In the nondelayed case, (5.1) can be solved to get $u_h = p$ for every $h \geq 1$. When L_0 has distribution as in (5.2), $u_h = p$ for every $h \geq 0$. Obviously, $u_h \to 1/E[L]$ as $h \to \infty$. While explicit expressions for u_h are seldom available, this case provides an example of a lifetime where u_h quickly (and exactly) achieves its limit.

The construction described above connects count time series with renewal processes. First, stationary time series knowledge can be applied to discrete renewal theory. As an example, stationary series have a well-developed spectral theory. Specifically, any stationary autocovariance $\gamma(\cdot)$ admits the Fourier representation $\gamma(h) = \int_{(-\pi,\pi]} e^{ih\lambda} dF(\lambda)$ for some nondecreasing right continuous function F satisfying $F(-\pi) = 0$ and $F(\pi) = \gamma(0)$ (Brockwell and Davis, 1991). A spectral representation for the renewal probabilities $\{u_t\}_{t=0}^{\infty}$ immediately follows; specifically,

$$\gamma_X(h) = \frac{1}{\mu}\left(u_h - \frac{1}{\mu}\right) = \int_{(-\pi,\pi]} e^{ih\lambda} dF(\lambda) \tag{5.4}$$

holds for some "CDF like" $F(\cdot)$ supported over $(-\pi, \pi]$. Solving this yields a spectral representation for the renewal probabilities: $u_h = \mu^{-1} + \mu \int_{(-\pi,\pi]} e^{ih\lambda} dF(\lambda)$. Here, F may not be a proper cumulative distribution function (CDF) ($F(\pi) \neq 1$), but rather has total mass $\mu^{-1}(1-\mu^{-1})$. While a different spectral representation for u_h is known from Daley and Vere-Jones (2007) via other methods, the representation here follows from the time series result.

Second, and perhaps more importantly, discrete-time renewal knowledge (which is vast) can now be used in count time series problems. For one example of what can be inferred, it is known (Kendall, 1959; Lund and Tweedie, 1996) that if L has a finite geometric moment ($E[r^L] < \infty$ for some $r > 1$), then the renewal sequence geometrically decays to its limit: $|u_h - \mu^{-1}| \leq \kappa c^{-h}$ for $h = 0, 1, \ldots$ for some finite κ and $c > 1$ (the values of c and r are not necessarily the same). Hence, for lifetimes L with some finite geometric moment, $\{X_t\}$ must have short memory in that $\sum_{h=0}^{\infty} |\gamma(h)| < \infty$. Long memory count time series are explored further in Lund et al. (2015; Chapter 21 in this volume).

5.2 Models for Classical Count Distributions

To demonstrate the flexibility of renewal-based methods, we now build stationary time series models with classical count marginal distributions, including binomial, Poisson, and geometric. For this, let $\{X_{t,1}\}, \{X_{t,2}\}, \ldots$ denote independent copies of the binary discrete-time renewal process $\{X_t\}$.

5.2.1 Binomial Marginals

Constructing a count time series model with binomial marginal distributions with a fixed number of trials M is easy—just add independent Bernoulli renewal processes:

$$Y_t = \sum_{i=1}^{M} X_{t,i} \tag{5.5}$$

as in Blight (1989). The lag h autocovariance is

$$\gamma_Y(h) = \text{Cov}(Y_t, Y_{t+h}) = \frac{M}{\mu}\left(u_h - \frac{1}{\mu}\right), \quad h = 0, 1, \ldots \tag{5.6}$$

and autocorrelations of $\{Y_t\}$ obey

$$\rho_Y(h) = \text{Corr}(Y_t, Y_{t+h}) = \frac{u_h - \mu^{-1}}{1 - \mu^{-1}}, \quad h = 0, 1, \ldots. \tag{5.7}$$

The autocovariances produced by this scheme can be positive or negative. In fact, if L is such that $u_1 = P(L=1) < \mu^{-1}$, then (5.6) implies that $\rho_Y(1) < 0$. Let $\epsilon > 0$ be a small probability and consider a lifetime L that takes values 1, 2, and 3 with probabilities $P[L = 1] = P[L = 3] = \epsilon$ and $P[L = 2] = 1 - 2\epsilon$. Then $E[L] = 2$, $u_1 = \epsilon$, and $\text{Corr}(Y_t, Y_{t+1}) = 2\epsilon - 1$. Letting $\epsilon \downarrow 0$ shows existence of a joint random pair (Y_t, Y_{t+1}) with binomial marginal distributions and a negative correlation arbitrarily close to -1. This is

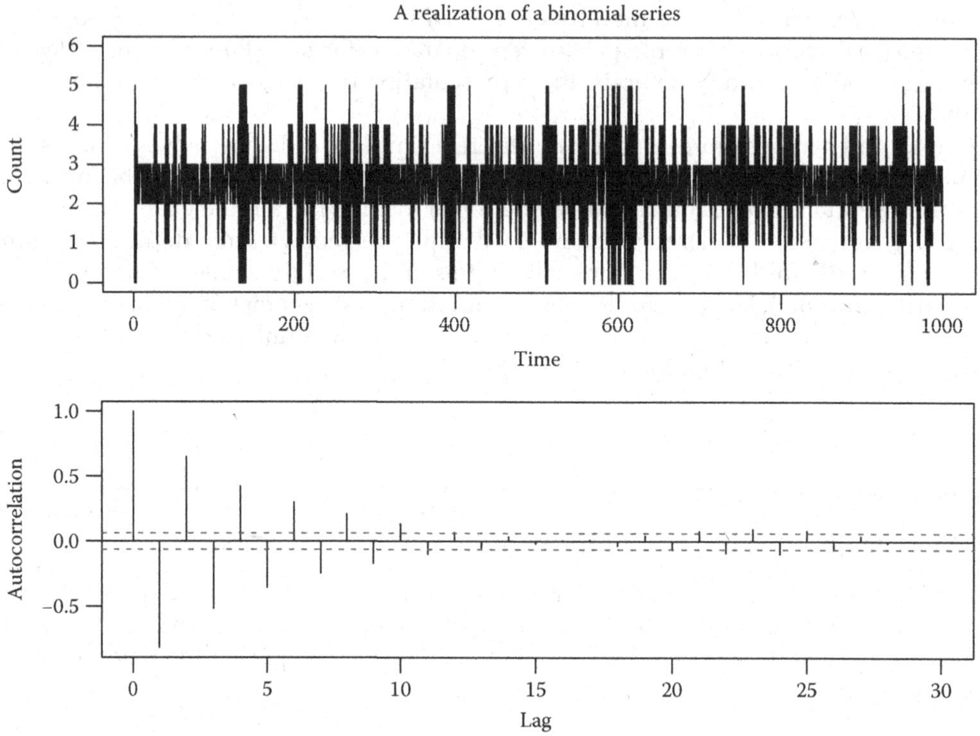

FIGURE 5.1
A realization of length 1000 of a stationary time series with binomial marginal distributions with mean 5 trials and success probability 1/2. Sample autocorrelations are also shown with pointwise 95% critical bounds for white noise. Notice that negative correlations can be achieved.

but one choice for L; other choices yield different autocorrelations. Positively autocorrelated $\{Y_t\}$ are also easily constructed. In fact, any lifetime L generating a renewal sequence with $u_h \geq \mu^{-1}$ for all $h = 0, 1, \ldots$ (these are plentiful) will produce stationary series with nonnegative autocorrelations.

Figure 5.1 shows a realization of length 1000 of a binomial count series with $M = 5$ and L supported on $\{1, 2, 3\}$ with probabilities $P[L=1] = P[L=3] = 1/10$ and $P[L = 2] = 8/10$. This process has a negative lag one correlation of $\rho_Y(1) = -4/5$. The sample autocorrelations of this realization are graphed along with 95% critical bounds under the null hypothesis of IID noise; the plot reveals an oscillating structure, with some negative autocorrelations.

Some properties of this process are worth discussing. First, the distributional relationship

$$Y_{t+h} \stackrel{\mathcal{D}}{=} u_h \circ Y_t + \frac{1 - u_h}{\mu - 1} \circ (M - Y_t), \quad h = 1, 2, \ldots \tag{5.8}$$

holds. Here, $\circ$ denotes binomial thinning, which is defined to operate on a nonnegative integer-valued random variable X via $p \circ X = \sum_{i=1}^{X} B_i$, where $B_1, B_2, \ldots$ are independent Bernoulli trials, independent of X, with the same success probability p. Equation (5.8) is justified as follows. At time t, Y_t of the M Bernoulli processes in the summation in (5.5) are unity and $M - Y_t$ are zero. The distribution of Y_{t+h} is obtained by adding two components: (1) all +1 "particles" at time t that remain +1 at time $t+h$ (this happens with probability $P[X_{t+h} = 1|X_t = 1] = u_h$) and (2) all zero particles that turn to +1 at time $t+h$ (this happens with probability $(1 - u_h)/(\mu - 1)$). This relationship is useful for inference procedures. Cui and Lund (2010) discuss additional properties of this process.

5.2.2 Poisson Marginals

To construct stationary series with Poisson marginal distributions, suppose now that M in (5.5) is random, independent of $\{X_{t,i}\}$ for all $i \geq 1$, and has a Poisson distribution with mean λ. It is easy to check that $\{Y_t\}$ is stationary in time t and has a Poisson marginal distribution with mean $E[Y_t] \equiv \lambda/\mu$. The covariance function of $\{Y_t\}$ is

$$\gamma_Y(h) = \text{Cov}(Y_t, Y_{t+h}) = \frac{\lambda}{\mu} + \frac{\lambda}{\mu}\left(u_h - \frac{1}{\mu}\right), \quad h = 1, 2, \ldots \tag{5.9}$$

with $\gamma_Y(0) = \lambda/\mu$. One may view the covariance in (5.9) as unrealistic since the term λ/μ appears in each and every lag (hence, this series has long memory). A simple modification remedies this—just examine

$$Y_t = \sum_{i=1}^{M_t} X_{t,i}, \tag{5.10}$$

where $\{M_t\}_{t=1}^{\infty}$ is an IID sequence of Poisson random variables each having mean λ. With such $\{M_t\}$, $\{Y_t\}$ is stationary with Poisson marginal distributions and $E[Y_t] \equiv \lambda/\mu$. A calculation shows that $\{Y_t\}$ has $\gamma_Y(0) = \lambda/\mu$ and

$$\gamma_Y(h) = \frac{C(\lambda)}{\mu}\left(u_h - \frac{1}{\mu}\right), \quad h = 1, 2, \ldots, \tag{5.11}$$

where $C(\lambda) = E[\min(M_t, M_{t+h})]$ does not depend on $h \geq 1$. The form of $C(\lambda)$ is identified in the following result.

Lemma 5.1 If M_1 and M_2 are independent and identically distributed Poisson random variables with mean λ, then

$$C(\lambda) = E[\min\{M_1, M_2\}] = \lambda[1 - e^{-2\lambda}\{I_0(2\lambda) + I_1(2\lambda)\}],$$

where $I_j(\cdot)$ is the modified Bessel function

$$I_j(x) = \sum_{n=0}^{\infty} \frac{(x/2)^{2n+j}}{n!(n+j)!}, \quad j = 0, 1.$$

Proof: Observe that

$$E[\min(M_1, M_2)] = \sum_{k=0}^{\infty}\sum_{j=0}^{\infty} \min(k,j)\frac{e^{-\lambda}\lambda^k}{k!}\frac{e^{-\lambda}\lambda^j}{j!}$$

$$= \sum_{k=1}^{\infty}\sum_{j=k}^{\infty} k\frac{e^{-2\lambda}\lambda^{k+j}}{k!j!} + \sum_{k=1}^{\infty}\sum_{j=1}^{k-1} j\frac{e^{-2\lambda}\lambda^{k+j}}{k!j!}.$$

Putting and taking diagonal terms (when $j = k$ in the double summation) gives

$$E[\min(M_1, M_2)] = \sum_{k=1}^{\infty}\sum_{j=k}^{\infty} k\frac{e^{-2\lambda}\lambda^{k+j}}{k!j!} + \sum_{k=1}^{\infty}\sum_{j=1}^{k} j\frac{e^{-2\lambda}\lambda^{k+j}}{k!j!} - \sum_{k=1}^{\infty} k\frac{e^{-2\lambda}\lambda^{2k}}{k!k!}.$$

The two double sums mentioned earlier can be shown to be equal; hence,

$$E[\min(M_1, M_2)] = 2\sum_{k=1}^{\infty}\sum_{j=k}^{\infty} k\frac{e^{-2\lambda}\lambda^{k+j}}{k!j!} - \sum_{k=1}^{\infty} k\frac{e^{-2\lambda}\lambda^{2k}}{k!k!}. \tag{5.12}$$

Dealing with the second summation in (5.12) first, we have

$$\sum_{k=1}^{\infty} \frac{ke^{-2\lambda}\lambda^{2k}}{(k!)^2} = e^{-2\lambda}\sum_{\ell=0}^{\infty} \frac{\lambda^{2\ell+2}}{\ell!(\ell+1)!} = \lambda e^{-2\lambda} I_1(2\lambda).$$

For the first summation in (5.12),

$$2\sum_{k=1}^{\infty}\sum_{j=k}^{\infty} \frac{ke^{-2\lambda}\lambda^{k+j}}{k!j!} = 2\sum_{\ell=0}^{\infty}\sum_{j=\ell+1}^{\infty} \frac{e^{-2\lambda}\lambda^{\ell+j}}{\ell!j!} = 2\lambda\sum_{\ell=0}^{\infty} P(M_1 = \ell \cap M_2 > \ell).$$

Since M_1 and M_2 are independent and identically distributed,

$$2\sum_{k=1}^{\infty}\sum_{j=k}^{\infty} \frac{ke^{-2\lambda}\lambda^{k+j}}{k!j!} = \lambda\sum_{\ell=0}^{\infty} [P(M_1 = \ell \cap M_2 > \ell) + P(M_1 > \ell \cap M_2 = \ell)]$$

$$= \lambda\left(1 - \sum_{\ell=0}^{\infty} P(M_1 = \ell \cap M_2 = \ell)\right).$$

Using $P[M_1 = \ell \cap M_2 = \ell] = P[M_1 = \ell]^2$ and the Poisson probabilities now gives

$$2\sum_{k=1}^{\infty}\sum_{j=k}^{\infty}\frac{ke^{-2\lambda}\lambda^{k+j}}{k!j!} = \lambda[1 - e^{-2\lambda}I_0(2\lambda)]$$

and completes our work.

The autocorrelation function of $\{Y_t\}$ is similar in form to (5.6):

$$\mathrm{Corr}(Y_t, Y_{t+h}) = \frac{C(\lambda)}{\lambda}\left(u_h - \frac{1}{\mu}\right), \quad h = 1, 2, \ldots. \tag{5.13}$$

Series with negative autocorrelations are easily constructed. To see this, again let L be the lifetime supported on $\{1, 2, 3\}$ with probabilities $P[L = 1] = P[L = 3] = \epsilon$ and $P[L = 2] = 1 - 2\epsilon$ for some small ϵ. Then $\mu = 2$, $u_1 = \epsilon$, and the random pair (Y_t, Y_{t+1}) has Poisson marginal distributions with the same mean $\lambda/2$ and

$$\mathrm{Corr}(Y_t, Y_{t+1}) = \frac{C(\lambda)}{\lambda}\left(\epsilon - \frac{1}{2}\right).$$

Letting $\epsilon \downarrow 0$ shows existence of a random pair with Poisson marginals with the same mean $\lambda/2$, whose correlation is arbitrarily close to $-C(\lambda)/(2\lambda)$. This is close to the most negative correlation possible. For recent updates on this problem and bounds, see Shin and Pasupathy (2010) and Yahav and Shmueli (2012).

A classical renewal result (Smith, 1958; Feller, 1968) states that $\sum_{h=0}^{\infty}|u_h - \mu^{-1}| < \infty$ if and only if $E[L^2] < \infty$. Since $\gamma_Y(h)$ is proportional to $u_h - \mu^{-1}$ (see (5.20) below), renewal series will have long memory whenever $E[L^2] = \infty$ (recall that $E[L] < \infty$ is presupposed). Long memory count series are discussed further in Lund et al. (2015; Chapter 21 in this volume). Figure 5.2 displays a sample path of length 500 of a count series with Poisson marginals and its sample autocorrelations. This series was generated by taking $\{M_t\}$ in (5.10) as Poisson with mean $\lambda = 20$ and L as a shifted discrete Pareto variable: $L = 1 + R$, where R is the Pareto random variate satisfying $P(R = k) = ck^{-\alpha}$ for $k \geq 1$ with $\alpha = 2.5$. Here, c is a constant, depending on α, that makes the distribution sum to unity. Since $L \geq 2$, one cannot have a renewal at time 1 in a nondelayed process. Hence, $u_1 = 0$, $u_1 - \mu^{-1}$ is negative, and long memory features have been made in tandem with a negative lag one autocorrelation! Negative autocorrelations at other lags can be devised with non-Pareto choices of L.

The earlier results are useful for multivariate Poisson generation, an active area of research (Karlis and Ntzoufras, 2003). Generating a random pair (X, Y), with X and Y each having a Poisson marginal distribution with the same mean, but with negative correlation, is nontrivial. In fact, suppose that M is a deterministic positive integer and

$$X = \sum_{i=1}^{M} B_i, \quad Y = \sum_{i=1}^{M}(1 - B_i), \tag{5.14}$$

where $\{B_i\}_{i=1}^{M}$ are IID Bernoulli random variables with the same success probability $p = 1/2$. Then X and Y have binomial distributions with M trials and the same mean $M/2$; moreover, $\mathrm{Corr}(X, Y) = -1$. However, should one replace the deterministic M by a Poisson

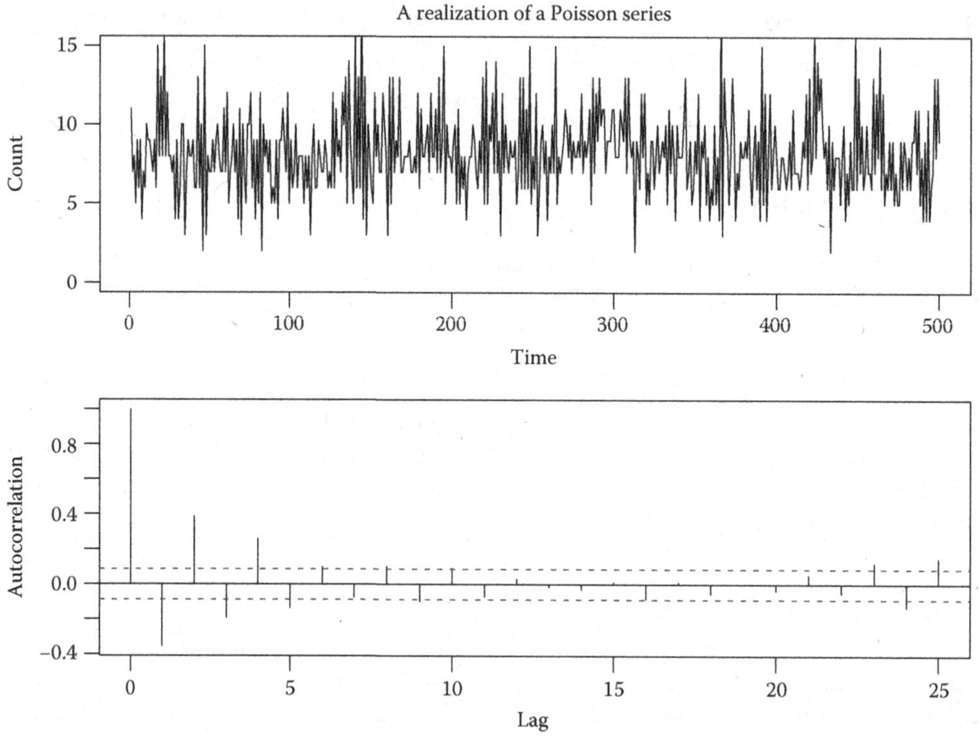

FIGURE 5.2
A realization of length 500 of a long memory stationary time series with Poisson marginal distributions with mean 6.895 (to three decimal places). Sample autocorrelations are also shown with pointwise 95% critical bounds for white noise. Notice that negative correlations can be achieved.

random M with mean λ, then both X and Y have Poisson distributions with mean $\lambda/2$, but now $\text{Corr}(X, Y) = 0$. While there are Poisson partitioning interpretations of the previous (Ross, 1996), the point is that generating negatively correlated Poisson variates can be tricky.

5.2.3 Geometric Marginals

Stationary geometric count series can also be constructed from the renewal class. For concreteness, we clarify geometric and zero-modified geometric distributions. A geometric random variable Z takes values in $\{1, 2, \ldots\}$, with probabilities $P(Z = k) = p(1-p)^{k-1}$ for some $p \in (0, 1)$; the zero-modified geometric distribution has $P(Z = k) = p(1-p)^k$ on the support set $k \in \{0, 1, 2, \ldots\}$, which includes zero. To construct stationary count series $\{Y_t\}$ with geometric marginal distributions with success probability p and generating function

$$E\left[r^{Y_t}\right] = \frac{pr}{1-(1-p)r}, \tag{5.15}$$

let us attempt to write Y_t as the superposition

$$Y_t = 1 + \sum_{i=1}^{M_t} X_{i,t}, \tag{5.16}$$

where $\{M_t\}_{t=1}^{\infty}$ is an IID sequence of random variables to be determined. Computing $E\left[r^{Y_t}\right]$ via (5.16) provides

$$E\left[r^{Y_t}\right] = r\psi_M\left(1 + \frac{r-1}{\mu}\right), \tag{5.17}$$

where $\psi_M(r) = E\left[r^{M_t}\right]$. If such a representation is possible, (5.15) and (5.17) must agree:

$$\frac{p}{1-(1-p)r} = \psi_M\left(1 + \frac{r-1}{\mu}\right).$$

Replacing $1 + (r-1)/\mu$ by z and simplifying gives $\psi_M(z) = p/[p + \mu(1-z)]$. Such a $\psi_M(\cdot)$ is a legitimate generating function. In fact, $\psi_M(\cdot)$ is the probability generating function of a zero-modified geometric random variate with success probability $\alpha = p/[p + (1-p)\mu]$.

The moment structure of $\{Y_t\}$ is $E[Y_t] \equiv 1/p$ and

$$\gamma_Y(h) = \frac{E[\min(M_t, M_{t+h})]}{\mu}\left(u_h - \frac{1}{\mu}\right), \quad h = 1, 2, \ldots.$$

When $h = 0$, $\gamma_Y(0) = (1-p)/p^2$. Evaluating $E[\min(M_t, M_{t+h})]$ is easier than the analogous Poisson computation: when $h > 0$,

$$E[\min(M_t, M_{t+h})] = \sum_{\ell=0}^{\infty} P[\min(M_t, M_{t+h}) > \ell] = \sum_{\ell=0}^{\infty} P[M_t > \ell]^2.$$

Using $P[M_t > k] = (1-\alpha)^{k+1}$ now gives $E[\min(M_t, M_{t+h})] = (1-\alpha)^2/[1-(1-\alpha)^2]$.

Again, negatively correlated series at any lag h can be produced by letting L be such that $u_h < \mu^{-1}$. Let L be the lifetime supported on $\{1, 2, 3\}$ with $P(L = 1) = P(L = 3) = \epsilon$ and $P(L = 2) = 1 - 2\epsilon$. Then $u_1 = \epsilon$ and the correlation at lag 1 can be verified to converge to $-p(1-p)/(4-3p)$ as $\epsilon \downarrow 0$.

Figure 5.3 shows a sample path of a geometric count series of length 1000 along with its sample autocorrelations. Here, we have taken $L = 1 + P$, where P is Poisson with a unit mean ($E[L] = \mu = 2$). The chosen value of α is 1/3, so that $p = 1/2$. Again, note the negative autocorrelations.

Finally, the tactics in (5.14) are not useful in generating negatively correlated geometric variates. Elaborating, set

$$X = 1 + \sum_{i=1}^{M} B_i, \quad Y = 1 + \sum_{i=1}^{M} (1 - B_i),$$

where the individual Bernoulli trials $\{B_i\}_{i=1}^{\infty}$ have mean $E[B_i] = 1/2$ (this corresponds to lifetimes with $\mu = 2$ and implies that X and Y have the same distribution) and M has a zero modified geometric distribution with success probability $\alpha = p/(2-p)$. From the above,

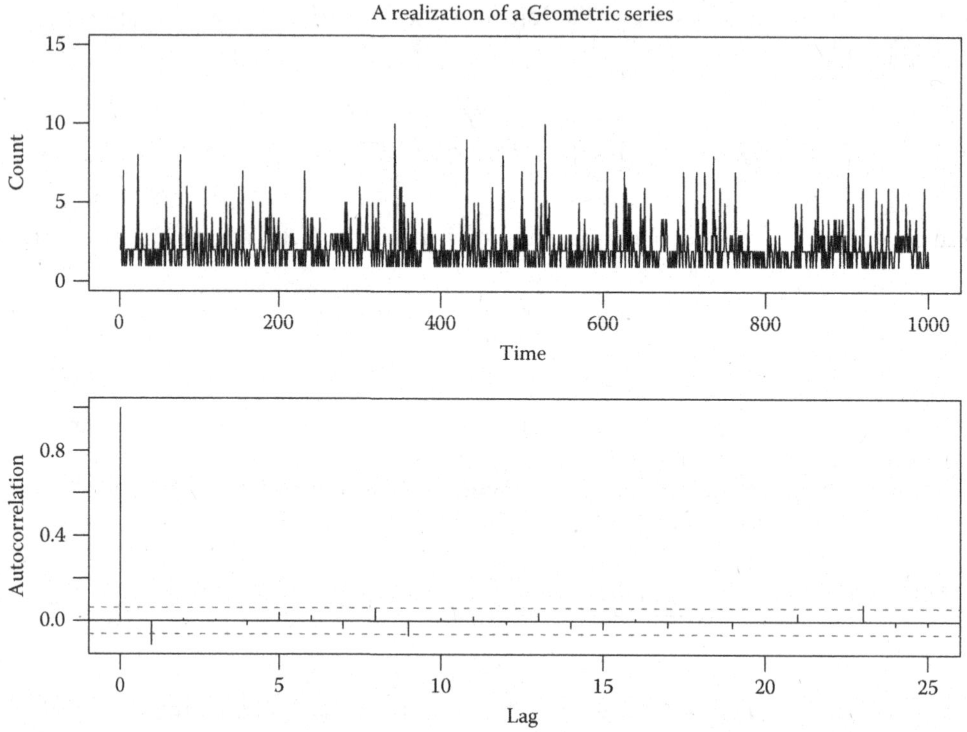

FIGURE 5.3
A realization of length 1000 of a stationary time series with Geometric marginal distributions with success probability 1/2. Sample autocorrelations are also shown along with pointwise 95% critical bounds for white noise. Notice that negative correlations can be achieved.

we know that both X and Y have the same geometric distribution with the success probability p. However, a simple computation gives

$$\text{Corr}(X, Y) = \frac{\sigma_M^2 - \mu_M}{\sigma_M^2 + \mu_M} = \frac{1-p}{1+p},$$

which is positive for any $p \in (0, 1)$.

5.2.4 Generalities

Since any nonnegative integer-valued random variable can be built from Bernoulli trials, any count marginal distribution can be achieved from renewal methods. It may be surprising how easily this is done. Consider trying to build a prespecified marginal distribution for Y_t supported on $\{c, c+1, \ldots\}$ with a renewal superposition of the form

$$Y_t = c + \sum_{i=1}^{M_t} X_{t,i}. \tag{5.18}$$

Here, $\{M_t\}$ is an IID sequence supported on $\{0,1,2,\ldots\}$ and c is some nonnegative integer constant. For simplicity, we assume that $c = 0$. Defined as in (5.18), $\{Y_t\}$ is easily shown to be a strictly stationary sequence.

Specifying the marginal distribution of Y_t is equivalent to specifying the probability generating function $\psi_Y(r) = E\left[r^{Y_t}\right]$. Equation (5.18) gives

$$\psi_Y(r) = \psi_M\left(1 - \frac{1}{\mu} + \frac{r}{\mu}\right), \tag{5.19}$$

where $\psi_M(r) = E\left[r^M\right]$ and M has the same distribution as M_t for any t. Letting $z = r/\mu + (1 - 1/\mu)$ in (5.19) produces

$$\psi_M(z) = \psi_Y\left(1 + \mu(z-1)\right).$$

To verify that $\psi_M(\cdot)$ is a legitimate generating function of a random variable supported on $\{0,1,\ldots\}$, note that the kth derivative of the generating function at zero satisfies $\psi_M^{(k)}(0) = \psi_Y^{(k)}(1-\mu)\mu^k \geq 0$ for $k = 0,1,\ldots$, which is nonnegative if $\psi_Y^{(k)}(1-\mu) \geq 0$ for $k = 0,1,2,\ldots$. Note that $\lim_{z\uparrow 1} \psi_Y(1 + \mu(z-1)) = \psi_Y(1) = 1$ since $\psi_Y(\cdot)$ is a valid probability generating function. Hence, many discrete non-negative integer-valued distributions can be built from the renewal superposition in (5.18).

The autocovariance functions that can be built from (5.18) are less clear. Any solution to (5.18) with $c = 0$ is stationary with mean $E[Y_t] \equiv E[M_1]/\mu$ and autocovariance

$$\gamma_Y(h) = \frac{E[\min(M_1, M_2)]}{\mu}\left(u_h - \frac{1}{\mu}\right), \quad h = 1,2,\ldots. \tag{5.20}$$

The process variance is

$$\gamma_Y(0) = \frac{\mathrm{Var}(M_t)}{\mu^2} + \frac{E[M_t]}{\mu}\left(1 - \frac{1}{\mu}\right).$$

Now suppose that $\{\gamma(h)\}_{h=0}^{\infty}$ is given and is known to be the autocovariance function of some stationary series. We want to construct a renewal count process $\{Y_t\}$ in the form of (5.18) having this autocovariance structure. Taking a power series transform gives

$$\Gamma(z) := \sum_{h=0}^{\infty} \gamma_Y(h) z^h = \frac{\sigma_M^2}{\mu^2} + \frac{\mu_M - E[\min(M_1,M_2)]}{\mu}\left(1 - \frac{1}{\mu}\right)$$
$$+ \frac{E[\min(M_1,M_2)]}{\mu}\sum_{h=0}^{\infty}\left(u_h - \mu^{-1}\right) z^h,$$

where $\mu_M \equiv E[M_t]$ and $\sigma_M^2 \equiv \mathrm{Var}(M_t)$. Let $U(z) = \sum_{h=0}^{\infty} u_h z^h$ and recall the classical relationship $U(z) = 1/[1 - \psi_L(z)]$, where $\psi_L(z) = E[z^L]$ (Smith, 1958). Solving for ψ_L gives

$$\psi_L(z) = 1 - \left[\frac{\mu\Gamma(z)}{E[\min(M_1,M_2)]} - D - \frac{1}{\mu(1-z)}\right]^{-1}, \tag{5.21}$$

where

$$D = \frac{\sigma_M^2 + (\mu - 1)\mu_M}{\mu E[\min(M_1, M_2)]} - \left(1 - \frac{1}{\mu}\right).$$

Unfortunately, without further assumptions, $\psi_L(\cdot)$ in (5.21) may not be a legitimate generating function of a lifetime L supported on $\{1, 2, \ldots\}$. To see an example of such, consider the autocorrelation function

$$\rho_Y(h) = \begin{cases} \phi^{|h/T|}, & h = 0, \pm T, \pm 2T, \ldots \\ 0, & \text{otherwise,} \end{cases}$$

where the parameter $\phi \in (-1, 1)$ and $T > 1$ is some positive integer. This is the autocovariance of a causal Tth order stationary autoregression whose first $T - 1$ autoregressive coefficients are zero.

To see that this autocovariance is not "buildable," the zero autocorrelations at lags $1, 2, \ldots, T-1$ imply that $u_h = \mu^{-1}$ for $h = 1, \ldots, T-1$. Using this in (5.1) with an induction reveals a geometric form over the first $T - 1$ support set values:

$$P(L = k) = \frac{1}{\mu}\left(1 - \frac{1}{\mu}\right)^{k-1}, \quad k = 1, 2, \ldots, T-1.$$

Plugging these into (5.1) gives

$$u_T = P(L = T) + \frac{1}{\mu}\left[1 - \left(1 - \frac{1}{\mu}\right)^{T-1}\right]. \tag{5.22}$$

At lag T, (5.20) requires that

$$\phi = \rho_Y(T) = \frac{E[\min(M_1, M_2)]\left(u_T - \frac{1}{\mu}\right)}{\frac{\sigma_M^2}{\mu} + \mu_M\left(1 - \frac{1}{\mu}\right)}. \tag{5.23}$$

Combining (5.22) and (5.23) now gives

$$P(L = T) = \phi\left[\frac{\frac{\sigma_M^2}{\mu} + \mu_M\left(1 - \frac{1}{\mu}\right)}{E[\min(M_1, M_2)]}\right] + \frac{1}{\mu}\left(1 - \frac{1}{\mu}\right)^{T-1}. \tag{5.24}$$

Using $\mu \geq 1$, a calculus maximization gives the bound $\mu^{-1}(1-\mu^{-1})^{T-1} \leq T^{-1}$. Now define

$$\kappa_M(\mu) = \frac{\frac{\sigma_M^2}{\mu} + \mu_M\left(1 - \frac{1}{\mu}\right)}{E[\min(M_1, M_2)]}$$

and note that $\kappa_M(\mu) > 0$ for every $\mu \geq 1$. Our bounds provide

$$P(L = T) \leq \phi\kappa_M(\mu) + \frac{1}{T},$$

which is negative (a contradiction) when ϕ is chosen close enough to -1 and T large.

To see that $\kappa_M(\mu)$ is bounded away from zero, note that

$$\kappa_M(\mu) > \frac{\min(\sigma_M^2, \mu_M)}{E[\min(M_1, M_2)]},$$

which is strictly positive unless M_t has zero variance. If $\text{Var}(M_t) \equiv 0$, $M_t \equiv M$ where M is a constant. A return to (5.24) now provides

$$P(L = T) = \phi(1-\mu^{-1}) + \mu^{-1}(1-\mu^{-1})^{T-2} \leq \phi(1-\mu^{-1}) + T^{-1},$$

which again can be made negative by choosing T large and ϕ close to -1.

Of course, it is not clear how to build general stationary count series having with a general prespecified autocovariance function and marginal distribution—such processes may not even exist. Notwithstanding, the renewal class seems to be a very flexible class of stationary count time series models.

5.3 Multivariate Series

So far, we have largely concentrated on univariate models. In many applications, multiple count series arise simultaneously and there is a need to develop count process models that allow for correlation between components (cross-correlation). The natural generalization of (5.10) to two dimensions (omitting any additive constants) is

$$\mathbf{Y}_t := \begin{pmatrix} Y_t^{(1)} \\ Y_t^{(2)} \end{pmatrix} = \begin{pmatrix} \sum_{i=1}^{M_t^{(1)}} X_{t,i}^{(1)} \\ \sum_{j=1}^{M_t^{(2)}} X_{t,j}^{(2)} \end{pmatrix}. \tag{5.25}$$

Here, the superscripts (1) and (2) index components, $\mathbf{M}_t = \left(M_t^{(1)}, M_t^{(2)}\right)'$ is a bivariate nonnegative integer-valued pair and $\{\mathbf{X}_{t,i}\} = \left\{\left(X_{t,i}^{(1)}, X_{t,i}^{(2)}\right)'\right\}$ is a two-dimensional binary sequence formed by a bivariate renewal process as described below. We take $\{\mathbf{X}_{t,i}\}$ independent of $\{\mathbf{X}_{t,j}\}$ when $i \neq j$. Here, $X_{t,i}^{(1)}$ is unity if and only if a renewal occurs at time t in the first coordinate of the ith renewal process (similarly for the second coordinate).

Clarifying further, a bivariate renewal process entails a two-dimensional version of the univariate case: bivariate IID lifetimes $\mathbf{L}_i = \left(L_i^{(1)}, L_i^{(2)}\right)_{i=1}^{\infty}$ are added coordinatewise, yielding the random walk

$$\mathbf{S}_n = \mathbf{L}_0 + \mathbf{L}_1 + \mathbf{L}_2 + \cdots + \mathbf{L}_n, \quad n = 0, 1, \ldots.$$

We say that a point occurs in the bivariate random walk at time $\mathbf{t} = (t_1, t_2)'$ when $\mathbf{S}_n = \mathbf{t}$ for some $n \geq 0$. The bivariate walk contains two univariate renewal processes. An important distinction arises from one dimension. Suppose that a point occurs in the bivariate walk at time $(t_1, t_2)'$. Then a renewal occurred at time t_1 in coordinate one and a renewal occurred in coordinate two at time t_2. However, the converse is not true in general. In fact, there are many ways for a renewal to occur in coordinate one at time t_1 and a renewal to occur in coordinate two at time t_2 without having a point in the bivariate walk at $(t_1, t_2)'$. For example, if $\mathbf{L}_0 = (1, 1)'$, $\mathbf{L}_1 \doteq (3, 2)'$, and $\mathbf{L}_2 = (2, 5)'$, then a renewal occurs at time 6 in coordinate one, a renewal occurs in coordinate two at time 3, but there is no point at $(6, 3)'$ in the bivariate walk.

Correlation between $L_j^{(1)}$ and $L_j^{(2)}$ is allowed for a fixed j; however, $L_j^{(1)}$ and $L_k^{(2)}$ are independent when $j \neq k$. The notation $L^{(1)}$ and $L^{(2)}$ will denote generic component lifetimes when the index $i \geq 1$ is of no importance. As in one dimension, an initial lifetime $\mathbf{L}_0$ exists that makes the two-dimensional renewal sequence bivariate stationary. We will not concern ourselves with the form of $\mathbf{L}_0$ here for reasons that follow.

Arguing as in the univariate case, $\{\mathbf{Y}_t\}$ in (5.25) can be shown to be a bivariate stationary (strictly) time series of counts. For autocovariance notation, let $\gamma_Y^{(1,1)}(h) = \text{Cov}\left(Y_t^{(1)}, Y_{t+h}^{(1)}\right)$ and $\gamma_Y^{(2,2)}(h) = \text{Cov}\left(Y_t^{(2)}, Y_{t+h}^{(2)}\right)$ denote coordinatewise lag-h covariances and $\gamma_Y^{(1,2)}(h) = \text{Cov}\left(Y_t^{(1)}, Y_{t+h}^{(2)}\right)$ denote the lag-h cross-covariances. The marginal autocovariances $\gamma_Y^{(1,1)}(\cdot)$ and $\gamma_Y^{(2,2)}(\cdot)$ take the form in (5.20):

$$\gamma_Y^{(1,1)}(h) = \frac{C^{(1)}}{E[L^{(1)}]}\left(u_h^{(1)} - \frac{1}{E[L^{(1)}]}\right), \quad \gamma_Y^{(2,2)}(h) = \frac{C^{(2)}}{E[L^{(2)}]}\left(u_h^{(2)} - \frac{1}{E[L^{(2)}]}\right), \quad h = 1, 2 \ldots$$

Here, $C^{(1)} = E\left[\min\left(M_t^{(1)}, M_{t+h}^{(1)}\right)\right]$ and $C^{(2)} = E\left[\min\left(M_t^{(2)}, M_{t+h}^{(2)}\right)\right]$ for $h > 0$. A detailed calculation shows that, for $h = 0, 1, 2, \ldots$

$$\gamma_Y^{(1,2)}(h) = \frac{\text{Cov}\left(M_t^{(1)}, M_{t+h}^{(2)}\right)}{\mu^{(1)}\mu^{(2)}} + E\left[\min\left\{M_t^{(1)}, M_{t+h}^{(2)}\right\}\right]\left(P\left(X_{t,1}^{(1)} = 1 \cap X_{t+h,1}^{(2)} = 1\right) - \frac{1}{\mu^{(1)}\mu^{(2)}}\right),$$

where $\mu^{(1)} = E\left[L^{(1)}\right]$ and $\mu^{(2)} = E\left[L^{(2)}\right]$. By strict stationarity of the bivariate renewal sequence, $P\left(X_{t,1}^{(1)} = 1 \cap X_{t+h,1}^{(2)} = 1\right) := \Delta_h$ does not depend on t.

An interesting issue now arises. Unless, Corr $\left(L^{(1)}, L^{(2)}\right) = \pm 1$, $\Delta_h = \left(\mu^{(1)}\mu^{(2)}\right)^{-1}$ for all h. While we will not prove this here, the intuition is that at a large time t, the lifetimes in use in coordinates one and two random walks will almost surely be different—and independence is assumed between $\mathbf{L}_i$ and $\mathbf{L}_j$ when $i \neq j$. Elaborating, suppose that at time t in a nondelayed setting $\left(L^{(1)} = L^{(2)} = 0\right)$, the first component is using the ith lifetime $L_i^{(1)}$ and the second component is using the jth lifetime $L_j^{(2)}$. Then when t is large, it is very unlikely that $i = j$; in fact, in the limit as $t \to \infty$, $i \neq j$ with probability one. The implication is that the cross-covariance structure of the process reduces to

$$\gamma_Y^{(1,2)}(h) = \frac{\text{Cov}\left(M_t^{(1)}, M_{t+h}^{(2)}\right)}{\mu^{(1)}\mu^{(2)}}.$$

Hence, if $\{\mathbf{M}_t\}$ is taken as IID as in the univariate case, $\gamma_Y^{(1,2)}(h) = 0$ when $h \geq 1$.

Here are some tactics to induce nonzero cross-correlation between components. The earlier cross-covariance does not assume independence between $M_t^{(1)}$ and $M_{t+h}^{(2)}$. We hence allow them to be dependent in special ways. One way simply links $\{\mathbf{M}_t\}$ to a correlated univariate count series $\{N_t\}$ (say generated by univariate renewal methods) via

$$\mathbf{M}_t = \begin{pmatrix} N_t \\ N_{t-1} \end{pmatrix}.$$

Then $\gamma_Y^{(1,2)}(h) = \gamma_N(h-1)/\left(\mu^{(1)}\mu^{(2)}\right)$ and there can be nonzero correlation between components. A second tactic for inducing cross-correlation is based on copulas. Suppose F_1 and F_2 are CDFs of the desired (prespecified) component marginal distributions. For a Gaussian illustration, suppose that $\mathbf{Z}_t = \left(Z_t^{(1)}, Z_t^{(2)}\right)'$ is multivariate normal with mean $\mathbf{0}$ and covariance matrix $\boldsymbol{\Sigma}$. Now set

$$\mathbf{M}_t = \begin{pmatrix} F_1^{-1}\left(\Phi\left(Z_t^{(1)}\right)\right) \\ F_2^{-1}\left(\Phi\left(Z_t^{(2)}\right)\right) \end{pmatrix}, \tag{5.26}$$

where $\Phi(\cdot)$ is the CDF of the standard normal random variable and $F_i^{-1}(y)$ is the smallest x such that $P(X \leq x) \geq y$ for X distributed as F_i, $i = 1, 2$ (this definition of the inverse CDF has many nice properties—see Theorem 25.6 in Billingsley 1995). The range of possible correlations and characteristics of the transform in (5.26) are discussed in Yahav and Shmueli (2012). If $\{\mathbf{Z}_t\}$ is IID, then $\{\mathbf{M}_t\}$ is also IID and $\gamma_Y^{(1,2)}(h) = 0$ for $h = 1, 2, \ldots$. However, $\gamma_Y^{(1,2)}(0) \neq 0$ and there is nonzero cross-correlation between components of $\mathbf{Y}_t$ (at lag zero) in general. Nonzero cross-covariances can be obtained at lags $h = 1, 2, \ldots$ by allowing $\{\mathbf{Z}_t\}$ to be a stationary bivariate Gaussian process. We will not attempt to derive the autocovariance function of such a series here.

Figure 5.4 shows a realization of (5.25) of length $n = 100$ from Gaussian copula methods that produce Poisson marginal distributions for the two components. Here, $\{\mathbf{M}_t\}$ from (5.26) was generated via a Gaussian copula, with component one of $\{\mathbf{Z}_t\}$ having mean 12 and component two having mean 20. The covariance matrix $\boldsymbol{\Sigma}$ was selected to have ones on

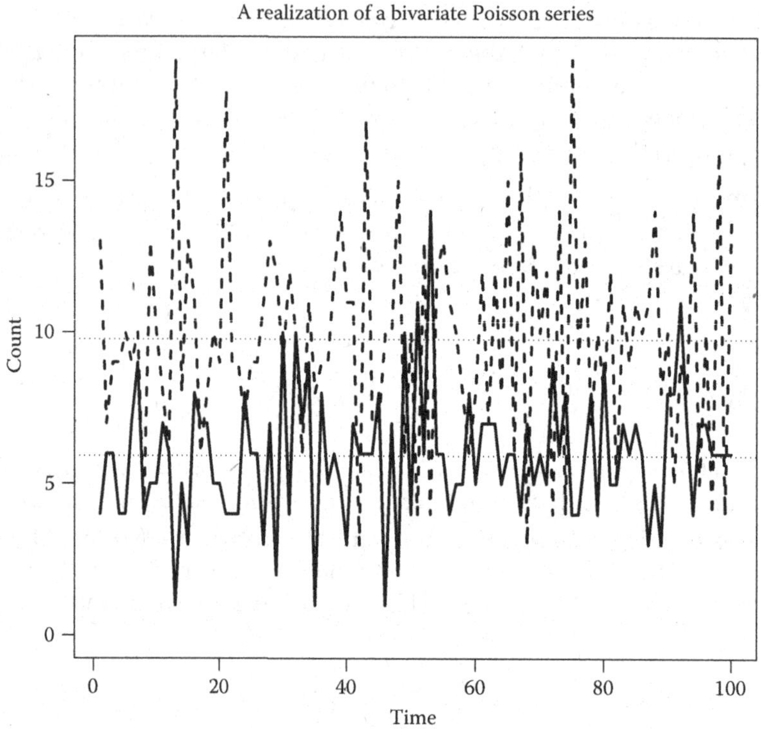

FIGURE 5.4
A realization of length 100 of a bivariate stationary time series, where each coordinate's marginal distribution is Poisson. Negative lag one autocorrelations exist in both coordinates; there is also negative cross-correlation between the components.

its diagonal entries and −0.75 on off-diagonal entries. The renewal lifetimes for the two components were both chosen as the three-point lifetime introduced below (5.13), with $\epsilon = 0.1$. The theoretical means are 6 for component one and 10 for component two. The dotted lines demarcate sample means, which were 6.03 and 9.92 for the first and second components, respectively. This realization has marginal Poisson distributions and negative correlation (at lag zero) between the two components. The negative correlation between components is noticeable in Figure 5.4: large values in coordinate two usually accompany small values of coordinate one, and vice-versa. Because of the lifetimes chosen, there is also negative correlation at some lags in both component series.

5.4 Statistical Inference

Renewal count models are very parsimonious. For a univariate comparison, a Markov chain on S states has $S(S-1)$ free parameters in its one-step-ahead transition matrix. Renewal count models are described entirely through the parameters governing the lifetime L and M_t—this does not change should the state space become infinite. For example, modeling a univariate Poisson series with long memory can be parameterized by a Pareto lifetime L with parameter α and a Poisson $\{M_t\}$, say with marginal mean λ. In this case, only two parameters need to be estimated: α and λ.

Ideally, all parameters would be estimated via maximum likelihood. A true likelihood approach would produce estimators that "feel the joint count distributional structure" (rather than just say process moments). The issue increases in importance with small counts. Unfortunately, likelihood methods have been very difficult to develop for count models as they require calculation of the joint distributional structure. The difficulties encountered can be appreciated in Davis et al. (2003), where likelihood asymptotics are pursued for the Poisson setting.

Kedem and Fokianos (2003) and Thavaneswaran and Ravishanker (2015; Chapter 7 in this volume) have had success in estimating count time series parameters via a quasi-likelihood (Godambe and Heyde, 1987) approach, which is a technique that we use here. Suppose that $Y_1, \ldots, Y_n$ is a sample from a univariate stationary renewal series $\{Y_t\}$. Let θ denote a vector containing all model parameters. One tractable estimation strategy minimizes the sum of squared one-step-ahead prediction errors

$$S(\theta) = \sum_{t=1}^{n} \frac{(Y_t - \hat{Y}_t)^2}{v_t}$$

in θ. Here, $\hat{Y}_t = P(Y_t | 1, Y_1, \ldots, Y_{t-1})$ is the best (minimum mean squared error) linear prediction of Y_t from the process history $Y_1, \ldots, Y_{t-1}$ and a constant (hence the one in the set of predictands mentioned earlier) and $v_t = E[(Y_t - \hat{Y}_t)^2]$ is its unconditional mean squared error. In general, $\hat{Y}_t$ and v_t will depend on θ. Since $\{Y_t\}$ is stationary, v_t converges monotonically downwards to a limit; often this convergence is geometric in t and there is no asymptotic loss of precision in omitting v_t altogether in the sum of squares. While one might try to base inferences on the conditional expectation $\hat{Y}_t = E[Y_t | Y_1, \ldots, Y_{t-1}]$, this quantity also appears intractable in generality.

The stochastic structure of $\{Y_t\}$ in (5.18) is somewhat unwieldy. In general, it is not Markov, conditional Markov, etc. One result that is useful in driving estimating equations is (5.8) in the case where $M_t \equiv M$ is fixed. The scenario gets much more complicated when M_t is allowed to vary; however, the situation can be quantified (we will not do it here) and the results allow us to evaluate $E[Y_t | Y_{t-h}]$ for any $h \geq 1$. This said, pilot computations with Poisson cases indicate that linear predictions from all previous data predict Y_t more accurately than $E[Y_t | Y_{t-1}]$, which is only based on the last observation Y_{t-1}.

As the covariance structure of the process has been identified, best linear one-step-ahead linear predictions are tractable. The one-step-ahead predictions have form

$$P(Y_t | Y_1, \ldots, Y_{t-1}, 1) = \mu_Y + \sum_{t=1}^{n-1} \beta_t (Y_t - \mu_Y),$$

where $\mu_Y \equiv E[Y_t]$ and the coefficients $\beta_1, \ldots, \beta_{t-1}$ are computed from the classical prediction equations. Computation of one-step-ahead linear predictions and their mean squared errors is a well-studied problem (see Chapter 5 of Brockwell and Davis, 1991), which can be done rapidly and recursively in time t.

While no general results have yet to be proven, linear prediction inference methods for count series should yield consistent and asymptotically normal estimators of θ.

Cui and Lund (2010) derive such a theory, replete with an explicit asymptotic information matrix, when $\{Y_t\}$ has binomial marginals. Here, M is constant and known and L is quantified in terms of its hazard rates $h_k = P(L = k|L \geq k), k = 1, 2, \ldots$. Specifically, in the case where h_k is constant for $k \geq 2$, Cui and Lund (2010) establish the joint asymptotic normality of the two estimated hazard rates $\hat{h}_1$ and $\hat{h}_2$ that minimize the prediction sum of squares; viz.,

$$\begin{pmatrix} \hat{h}_1 \\ \hat{h}_2 \end{pmatrix} \sim \text{AN}_2\left(\begin{pmatrix} h_1 \\ h_2 \end{pmatrix}, \frac{\mathbf{R}}{n}\right).$$

The form of $\mathbf{R}$ is explicitly identified in terms of h_1 and h_2. The notation here uses hats to denote both estimators and one-step-ahead predictions (this should not cause confusion). In cases where L has general hazard rates, these methods do not yield an explicit information matrix. Even in simple cases, the computations are intense.

For the more general case, justifying asymptotic normality of the linear prediction estimators may be feasible. If explicit forms for standard errors are not needed, numerical standard error approximations could be obtained by inverting the Hessian matrix associated with the sum of squares at its minimum.

5.5 Covariates and Periodicities

Covariate information often accompanies count data. Frequently, the goal is to explain the counts in terms of the covariates. To modify the renewal paradigm for covariates, one can allow M_t in (5.10) to depend on the covariates. For example, consider the univariate Poisson case and suppose that $C_{1,t}, \ldots, C_{K,t}$ are K covariates at time t. To retain a Poisson marginal distribution, M_t is taken to be Poisson distributed; however, we now allow $E[M_t]$ to vary with the time t covariates via

$$E[M_t] = \exp\left(\beta_0 + \sum_{\ell=1}^{K} \beta_\ell C_{\ell,t}\right),$$

where $\beta_0, \ldots, \beta_K$ are regression coefficients. Of course, such a process has a time-varying mean and is not technically stationary; however, such processes seem useful (Davis et al. 2000). The exponential function is used to keep the Poisson mean positive. While the resulting count series is no longer stationary, it is autocorrelated and can take on a wide and flexible range of covariance structures. It would be desirable to extend the Poisson regression techniques of Davis et al. (2000) to the renewal setting.

Count series with periodicities could be devised in two ways. First, in the univariate case, L could be allowed to depend on the season in which the last renewal occurred. This is done in Fralix et al. (2012), where periodic Markov chains and renewal processes are developed in generality. Second, one could allow M_t to depend on time t in a periodic way. Combinations of both approaches may prove useful.

5.6 Concluding Comments

Renewal superpositioning methods seem to generate more flexible autocovariance structures for count series than traditional ARMA-based approaches. They readily yield stationary series with many marginal integer-valued distribution desired and their autocovariances can be positive and/or negative. Estimation can be conducted by minimizing one-step-ahead prediction errors, which are easily calculated from process autocovariances. Unfortunately, as with other count time series model classes, full likelihood estimation approaches do not appear tractable at this time. Quasilikelihoods, composite likelihood methods, etc. are currently being investigated.

Acknowledgments

Robert Lund's research was partially supported by NSF Award DMS 1407480.

References

Al-Osh, M. and Alzaid, A.A. (1988). Integer-valued moving averages (INMA), *Statistical Papers*, **29**, 281–300.

Blight, P.A. (1989). Time series formed from the superposition of discrete renewal processes, *Journal of Applied Probability*, **26**, 189–195.

Billingsley, P. (1995). *Probability and Measure*, 3rd edn., Springer-Verlag, New York.

Brockwell, P.J. and Davis, R.A. (1991). *Time Series: Theory and Methods*, 2nd edn., Springer-Verlag, New York.

Cui, Y. and Lund, R.B. (2009). A new look at time series of counts, *Biometrika*, **96**, 781–792.

Cui, Y. and Lund, R.B. (2010). Inference for binomial AR(1) models, *Statistics and Probability Letters*, **80**, 1985–1990.

Daley, D.J. and Vere-Jones, D. (2007). *An Introduction to the Theory of Point Processes: Volume II: General Theory and Structure*, Springer, New York.

Davis, R.A., Dunsmuir, W.T.M., and Streett, S.B. (2003). Observation-driven models for Poisson counts, *Biometrika*, **90**, 777–790.

Davis, R.A., Dunsmuir, W.T.M., and Wang, Y. (2000). On autocorrelation in a Poisson regression model, *Biometrika*, **87**, 491–505.

Feller, W. (1968). *An Introduction to Probability Theory and its Applications*, 3rd edn., John Wiley & Sons, New York.

Fralix, B., Livsey, J., and Lund, R.B. (2012). Renewal sequences with periodic dynamics, *Probability in the Engineering and Informational Sciences*, **26**, 1–15.

Godambe, V.P. and Heyde, C.C. (1987). Quasi-likelihood and optimal estimation, *International Statistical Review*, **55**, 231–244.

Jacobs, P.A. and Lewis, P.A.W. (1978a). Discrete time series generated by mixtures I: Correlational and runs properties, *Journal of the Royal Statistical Society, Series B*, **40**, 94–105.

Jacobs, P.A. and Lewis, P.A.W. (1978b). Discrete time series generated by mixtures II: Asymptotic properties, *Journal of the Royal Statistical Society, Series B*, **40**, 222–228.

Karlis, D. and Ntzoufras, I. (2003). Analysis of sports data by using bivariate Poisson models, *Journal of the Royal Statistical Society, Series D*, **52**, 381–393.

Kedem, B. and Fokianos, K. (2003). Regression theory for categorical time series, *Statistical Science*, **18**, 357–376.

Kendall, D.G. (1959). Unitary dilations of Markov transition operators, and the corresponding integral representations for transition-probability matrices. In: *Probability and Statistics*, U. Grenander (ed.), Almqvist & Wiksell, Stockholm, Sweden, 138–161.

Lund, R.B., Holan, S. and Livsey, J. (2015). Long memory discrete-valued time series. In R. A. Davis, S. H. Holan, R. Lund and N. Ravishanker, eds., *Handbook of Discrete-Valued Time Series*, pp. 447–458. Chapman & Hall, Boca Raton, FL.

Lund, R.B. and Tweedie, R. (1996). Geometric convergence rates for stochastically ordered Markov chains, *Mathematics of Operations Research*, **21**, 182–194.

McKenzie, E. (1985). Some simple models for discrete variate time series, *Water Resources Bulletin*, **21**, 645–650.

McKenzie, E. (1986). Autoregressive-moving average processes with negative-binomial and geometric marginal distributions, *Advances in Applied Probability*, **18**, 679–705.

McKenzie, E. (1988). Some ARMA models for dependent sequences of Poisson counts, *Advances in Applied Probability*, **20**, 822–835.

Ross, S.M. (1996). *Stochastic Processes*, 2nd edn., Wiley, New York.

Shin, K. and Pasupathy, R. (2010). An algorithm for fast generation of bivariate Poisson random vectors, *INFORMS Journal on Computing*, **22**, 81–92.

Smith, W.L. (1958). Renewal theory and its ramifications, *Journal of the Royal Statistical Society, Series B*, **20**, 243–302.

Steutel, F.W. and Van Harn, K. (1979). Discrete analogues of self-decomposability and stability, *Annals of Probability*, **7**, 893–899.

Thavaneswaran, A. and Ravishanker, N. (2015). Estimating equation approaches for integer-valued time series models. In R. A. Davis, S. H. Holan, R. Lund and N. Ravishanker, eds., *Handbook of Discrete-Valued Time Series*, pp. 145–164. Chapman & Hall, Boca Raton, FL.

Yahav, I. and Shmueli, G. (2012). On generating multivariate Poisson data in management science applications, *Applied Stochastic Models in Business & Industry*, **28**, 91–102.

6

State Space Models for Count Time Series

Richard A. Davis and William T.M. Dunsmuir

CONTENTS

6.1 Introduction

The family of linear state-space models (SSM), which have been a staple in the time series literature for the past 70 plus years, provides a flexible modeling framework that is applicable to a wide range of time series. The popularity of these models stems in large part from the development of the Kalman recursions, which provides a quick updating scheme for predicting, filtering, and smoothing a time series. In addition, many of the commonly used time series models, such as univariate and multivariate ARMA and ARIMA processes can be embedded in an SSM, and as such can take advantage of fast recursive calculation related to prediction and filtering afforded by the Kalman recursions. Recent accounts of linear state space models can be found in Brockwell and Davis (1991), Brockwell and Davis (2002), Durbin and Koopman (2012), and Shumway and Stoffer (2011).

SSMs can be described via two equations, the *state-equation*, which describes the evolution of the *state*, S_t, of the system, and an *observation equation* that describes the relationship between the observed value Y_t of the time series with the state variable S_t. In this chapter, we shall assume that the state process evolves according to its own probabilistic mechanism; that is, the conditional distribution of S_t given all past states and past observations only depends on the past states. More formally, the conditional distribution of S_t given $S_{t-1}, S_{t-2}, \ldots, Y_{t-1}, Y_{t-2}, \ldots$ is the same as the conditional distribution of S_t given $S_{t-1}, S_{t-2}, \ldots$. The case in which the conditional distribution depends on previous observations is considered in other chapters in this volume (see Tjøstheim [2015; Chapter 4 in this volume]). See also Brockwell and Davis (2002) for a treatment of this more general case.

For the linear SSM specification, the observation equation and its companion state equation that govern the univariate observation process $\{Y_t\}$ and time evolution of the s-dimensional state-variables $\{S_t\}$ are specified via a linear relationship. One of the simplest linear SSM specifications is

$$\begin{array}{ll} \textbf{Observation equation:} & Y_t = GS_t + V_t, \ t = 1, 2, \ldots, \ \{V_t\} \sim \text{IID}(0, \sigma^2), \\ \textbf{State equation:} & S_{t+1} = FS_t + U_t, \ t = 1, 2, \ldots, \ \{U_t\} \sim \text{IID}(0, R), \end{array} \tag{6.1}$$

where

- IID$(0, S)$ denotes an independent and identically distributed sequence of random variables (vectors) with mean 0 and variance (covariance matrix) S,
- G is a $1 \times s$ dimensional matrix,
- F is an $s \times s$ dimensional matrix, and
- The sequence $\{Y_1, (U_t, V_t), t=1, 2, \ldots\}$ is IID where U_t and V_t are allowed to be dependent.

Often, one assumes that the matrix F in the state equation has eigenvalues that are less than one in absolute value, in which case S_t is viewed as a causal vector AR(1) process (see Brockwell and Davis, 1991). In the case when the noise terms $\{U_t\}$ and $\{V_t\}$ are Gaussian, then this state-space system is referred to as a Gaussian SSM.

It is well-known (see, e.g., Proposition 2.1 in Davis and Rosenblatt [1991]) that except in the degenerate case, the marginal distribution of a univariate autoregressive process with iid noise is continuous. So if one considers a stationary solution of (6.1) in the one dimensional case, S_t has a continuous distribution. Consequently, the distribution of the Y_t in (6.1) cannot have a discrete component. As such, this precludes the use of the linear SSM as specified in (6.1) for modeling count time series.

Nonlinear SSMs possess a similar structure as the linear state space model. The linear equations are now replaced by specifying the appropriate conditional distributions. Writing $\mathbf{Y}^{(t)} = (Y_1, \ldots, Y_t)'$ and $\mathbf{S}^{(t)} = (S_1, \ldots, S_t)'$, the observation and state equations become:

Observation equation:

$$p\left(y_t | s_t\right) = p\left(y_t | s_t, \mathbf{s}^{(t-1)}, \mathbf{y}^{(t-1)}\right), \quad t = 1, 2, \ldots \tag{6.2}$$

State equation:

$$p\left(s_{t+1} | s_t\right) = p\left(s_{t+1} | s_t, \mathbf{s}^{(t-1)}, \mathbf{y}^{(t)}\right), \quad t = 1, 2, \ldots, \tag{6.3}$$

where $p(y_t|s_t)$ and $p(s_{t+1}|s_t)$ are prespecified probability density functions*. The joint density for the observations and state components can be rewritten (Brockwell and Davis, 2002, Section 8.8) as

$$p(\mathbf{y}^{(n)}, \mathbf{s}^{(n)}) = \left(\prod_{j=2}^{n} p(s_j|s_{j-1})\right) p_1(s_1),$$

where $p_1(\cdot)$ is a pdf for the initial state vector S_1. Using the fact that $\{S_t\}$ is a Markov process, it follows that

$$p(\mathbf{y}^{(n)}|\mathbf{s}^{(n)}) = \prod_{j=1}^{n} p(y_j|s_j). \tag{6.4}$$

Even though the nonlinear state-space formulation given in (6.2) and (6.3) is quite general, the specification can be limiting in that the state process is required to be Markov. An alternative specification of a general SSM can be fashioned using (6.4) as a starting point. That is, given values for the state process $S_1, \ldots, S_n$, the random variables $Y_1, \ldots, Y_n$ are assumed to be conditionally independent in the sense of (6.4). Using (6.4) and any joint density $p(\mathbf{s}^{(n)})$ for $\mathbf{S}^{(n)}$, the joint density of $\mathbf{Y}^{(n)}$ is then

$$p(\mathbf{y}^{(n)}) = \int_{\mathbb{R}^n} \prod_{j=1}^{n} p(y_j|s_j) p(\mathbf{s}^{(n)}) d\mathbf{s}^{(n)}. \tag{6.5}$$

So in general, the distribution of the state process can be freely specified and, in particular, the state process need not be Markov.

For count time series, the observation probability density function must have support on a subset of $0, 1, 2, \ldots$ and is generally chosen to be one of the common discrete distribution functions (i.e., Poisson, negative-binomial, Bernoulli, geometric, etc.), where the parameter of the discrete distribution is a function of the state S_t. One common choice is to assume that the conditional distribution of Y_t given S_t is a member of a one-dimensional exponential family with canonical parameter s_t, i.e.,

$$p(y_t|s_t) = \exp\{s_t y_t - b(s_t) + c(y_t)\}, \quad y = 0, 1, \ldots, \tag{6.6}$$

where $b(\cdot)$, $c(\cdot)$ are known real functions and s_t is the *canonical parameter* (see McCullagh and Nelder, 1989). Recall that for an exponential family as specified in (6.6), we have the following properties:

- $E(Y_t|S_t = s_t) = b'(s_t)$
- $\text{var}(Y_t|S_t = s_t) = b''(s_t)$,

where b' and b'' refer to the first and second derivatives. The canonical link function g for an exponential family model maps the mean function into the canonical parameter, that is,

* These are probability density functions relative to some dominating measure, which is usually taken to be counting measure on $\{0, 1, \ldots\}$ for $p(y_t|s_t)$ and Lebesgue measure for $p(s_t|s_{t-1})$.

$g(\mu) = (b')^{(-1)}(\mu)$. In other words, the link function is the inverse of the conditional mean, which provides the mapping that expresses the state as a function of the conditional mean.*

Examples:

1. *Poisson:* $p(y_t|s_t) = \exp\{s_t y_t - e^{s_t} - \log(y_t!)\}$, $y_t = 0, 1, \ldots$. In this case $E(Y_t|s_t) = \exp\{s_t\}$ and hence the link function is $g(\mu) = \log \mu$.
2. *Binomial:* Assuming the number of trials in the binomial at time t is m_t, which is known but could vary with t, then $p(y_t|s_t) = \exp\left\{s_t y_t - m_t \log(1 + e^{s_t}) - \log\left(\binom{m_t}{y_t}\right)\right\}$, $y_t = 0, 1, \ldots, m_t$, where the probability of success on a single trial is $p_t = e^{s_t}/(1 + e^{s_t})$. In this case, $b_t(s_t) = m_t \log(1 + e^{s_t})$, $E(Y_t|S_t = s_t) = m_t e^{s_t}/(1 + e^{s_t})$ and the link is the logit function, $g(p_t) = \text{logit}(p_t) = \log(p_t/(1 - p_t))$.
3. *Negative-binomial:* The NegBin(r, p) distribution has density function,

$$p(y; r, p) = \exp\left\{\log(1 - p)y_t + r \log p + \log\binom{y_t + r - 1}{r - 1}\right\}, \quad y_t = 0, 1, 2, \ldots$$

   with mean $r(1 - p)/p$. If we use the natural parameterization where s_t represents the canonical parameter, that is, $s_t = \log(1 - p_t)$, then we see find that $s_t < 0$. From a modeling perspective, this may be too restrictive. Allowing S_t to be a linear time series model that takes both positive and negative values permits greater flexibility in terms of the behavior of S_t. So instead of using the canonical link function $g(\mu) = re^{s_t}/(1 + e^{s_t})$, consider the alternative link function $\tilde{g}(\mu) = -\log(r/\mu) = -\log(p/(1 - p)) = -\text{logit}(p)$. Setting $s_t = \tilde{g}(\mu_t) = -\text{logit}(p_t)$, we obtain $p_t = e^{-s_t}/(1 + e^{-s_t})$ so that the conditional density of Y_t given $S_t = s_t$ becomes

$$p(y_t|s_t) = \exp\left\{-\log(1 + e^{-s_t})y_t + r \log(e^{-s_t}/(1 + e^{-s_t})) + \log\binom{y_t + r - 1}{r - 1}\right\},$$

   $y_t = 0, 1, 2, \ldots$, with conditional mean $E(Y_t|s_t) = re^{-s_t}/(1 + e^{-s_t})$.

To complete the model specification for the count time series after the observation density $p(y_t|s_t)$ has been chosen, it remains to describe the distribution of the state process $\{S_t\}$. Often the primary objective in modeling count time series is to describe associations between observations and a suite of covariates. The covariates may be functions of time (e.g., day of the week effect) or exogenous (e.g., climatic effects). Since the Y_ts are assumed to be conditionally independent given the state process, the mean structure and the dependence in the time series can be modeled entirely through the state process. A natural choice is to use a *regression with time series errors* model for S_t. If $\mathbf{x}_t = (1, x_1 \ldots, x_p)^T$ denotes the vector of covariates associated with the tth observation†, then a regression time series model for S_t is given by

$$S_t = \mathbf{x}_t^T \beta + \alpha_t, \tag{6.7}$$

* One can use other choices for the link function besides the canonical one; as in the negative binomial example. See McCullagh and Nelder (1989) for more details.

† Our covariates always include an intercept term.

where $\boldsymbol{\beta} = (\beta_0, \ldots, \beta_p)^T$ is the vector of regression parameters and $\{\alpha_t\}$ is a strictly stationary time series with zero mean. Sometimes the $\{\alpha_t\}$ process or $\{S_t\}$ itself is referred to as a *latent process* since it is not directly observed. Usually, but not always, one takes $\{\alpha_t\}$ to be a strictly stationary Gaussian time series for which there is an explicit expression for the joint distribution of $\mathbf{S}^{(n)} = (S_1, \ldots, S_n)^T$. In this case, writing $\Gamma_n = \text{cov}(\mathbf{S}^{(n)}, \mathbf{S}^{(n)})$, the joint density of $\mathbf{Y}^{(n)}$ is given by

$$p(\mathbf{y}^{(n)}) = \int_{\mathbb{R}^n} \left(\prod_{t=1}^{n} p(y_t | x_t \boldsymbol{\beta} + \alpha_t) \right) \frac{e^{-\frac{1}{2}(\mathbf{s}^{(n)} - X\boldsymbol{\beta})^T \Gamma_n^{-1} (\mathbf{s}^{(n)} - X\boldsymbol{\beta})} d\mathbf{s}^{(n)}}{(2\pi)^{n/2} |\Gamma_n|^{1/2}}, \tag{6.8}$$

where $X = (\mathbf{x}_1, \ldots, \mathbf{x}_n)^T$ is the design matrix and $\boldsymbol{\alpha}^{(n)} = (\alpha_1, \ldots, \alpha_n)^T$. For estimation purposes, it is convenient to express (6.8) as a likelihood function and writing it in the form

$$L(\boldsymbol{\theta}) = \int_{\mathbb{R}^n} \left(\prod_{t=1}^{n} p(y_t | x_t \boldsymbol{\beta} + \alpha_t) \right) \frac{1}{(2\pi)^{n/2} |\Gamma_n|^{1/2}} e^{-\frac{1}{2}(\boldsymbol{\alpha}^{(n)})^T \Gamma_n^{-1} \boldsymbol{\alpha}^{(n)}} d\boldsymbol{\alpha}^{(n)}, \tag{6.9}$$

where the covariance matrix $\Gamma_n = \Gamma_n(\boldsymbol{\psi})$ depends on the parameter vector $\boldsymbol{\psi}$ and $\boldsymbol{\theta}^T = \left(\boldsymbol{\beta}^T, \boldsymbol{\psi}^T\right)$ denotes the complete parameter vector.

The SSM framework of (6.2) in which the conditional distribution in the observation equation is given by a known family of discrete distributions, such as the Poisson, has a number of desirable features. First, the setup is virtually identical to the starting point of a Bayesian hierarchical model. Conditional on a state-process, which in the Poisson case might be the intensity process, the observations are assumed to be conditionally independent and Poisson distributed. Second, the serial dependence is then modeled entirely through the state-equation. This represents a pleasing physical description for count data, which in the Poisson case, fits under the umbrella of Cox-processes or doubly stochastic Poisson processes. As in most Bayesian modeling settings, the unconditional distribution of the observation, obtained by integrating out the state-variable, rarely has an explicit form. Except in a limited number of cases, it is rare that the unconditional distribution of Y_t is of primary interest. The modeling emphasis in the SSM specification is on choice of conditional distribution in the observation equation and the model for the state process $\{S_t\}$. An overview of tests of the existence of a latent process and estimates of its underlying correlation structure can be found in Davis et al. (1999).

The autocorrelation function (ACF) is the workhorse for describing dependence and model fitting for continuous response data using linear time series models. For nonlinear time series models, including count time series, the ACF plays a more limited role. For example, for financial time series, where now Y_t represents the daily log-returns of an asset or exchange rate on day t, the data are typically uncorrelated and the ACF of the data is not particularly useful. On the other hand, the ACF of the absolute values and squares of the time series $\{Y_t\}$ can be quite useful in describing other types of serial dependence. For time series of counts following the SSM model described above, the ACF can also be used as a measure of dependence but is not always useful. In some cases, the ACF of $\{Y_t\}$ can be expressed explicitly in terms of the ACF of the state-process. For example, if the observation equation is Poisson and $\{S_t\}$ is a stationary Gaussian process with mean β_1 and autocovariance function $\gamma_S(h)$, then (see Davis et al., 2000)

$$\rho_Y(h) = \text{Cor}(Y_t, Y_{t+h}) = \frac{\exp\{\gamma_S(h)\} - 1}{e^{-\beta_1} + (\exp\{\gamma_S(0)\} - 1)}.$$

Moreover, if $\gamma_S(h) > 0$, then

$$0 \le \rho_Y(h) \le \frac{\gamma_S(h)}{\gamma_S(0)} = \rho_S(h).$$

Thus, little or no autocorrelation in the data can mask potential large correlation in the latent process.

In Section 6.2, we consider various methods of estimation for regression parameters and the parameters determining the covariance matrix function for count time series satisfying (6.4) and (6.7). The main issue is that the joint distribution is given by the n-fold integrals in (6.5) or (6.9), which can be difficult to compute numerically for a fixed set of parameter values, let alone maximizing over the parameter space. Strategies for finding maximum likelihood estimators include Laplace-style approximations to the integral in (6.5) and simulation-based procedures using either MCMC or importance sampling. Alternative estimation procedures based on estimating equations and composite likelihood procedures are also discussed in Section 6.2.

Gamerman et al., (2015; Chapter 8 in this volume) consider the formulation and estimation of dynamic Bayesian models. These models combine state-space dynamics with generalized linear models (GLMS). Such models can include covariates where now the coefficients evolve dynamically. Earlier work on dynamic generalized models can be found in Fahrmeir (1992). In the SSM setup of this chapter, the dynamics of the state process $\{S_t\}$ are specified independent of the observed process. The book by Durbin and Koopman (2012) is an excellent reference on SSMs in a general setting.

A different approach, which is discussed by Fokianos (2015; Chapter 1 in this volume) and Tjøstheim (2015; Chapter 4 in this volume), allows the state-process $\{S_t\}$ to be explicit functions of previous observations. Estimation for these models tends to be simpler than the models considered in this chapter. On the other hand, incorporating regressors in a meaningful way in these models as well as underlying theory for these models is far more difficult.

In the remainder of this chapter, estimation procedures are described in Section 6.2 and applied to the Polio data, a classical time series of count data set, in Section 6.3. Forecasting for these models is treated in Section 6.4.

6.2 Approaches to Estimation

Not surprisingly, given the numerical complexity associated with approximating the high dimensional integral required to compute the likelihood (6.9), many different approaches to estimation of nonlinear non-Gaussian models have been used. These include: GLM methods for estimating regression parameters β in which the latent process is ignored, generalized estimating equation (GEE), which uses a working covariance matrix to adjust inference about β for serial dependence effects, composite likelihood methods in which various lower dimensional marginal distributions are combined to define an objective function to be maximized over both β and ψ, approximations to the likelihood such as the Laplace approximation or the penalized quasi likelihood (PQL) method, and use of importance sampling and other Monte Carlo methods. In this section, the main methods are reviewed and compared. Bayesian methods are not reviewed in this chapter but are reviewed elsewhere in this volume (see, for example, Chapters 8 and 11). However, the importance sampling methods to be discussed in this section are also used to implement Bayesian methods.

6.2.1 Estimating Equations

As a precursor to using a fully efficient estimation method for the SSM, inference based on estimating equations can be useful. Zeger (1988) was one of the first to develop GEE methods for Poisson time series (see also Thavaneswaran and Ravishanker [2015; Chapter 7 in this volume]). He proposed finding an estimate of β by solving

$$U_{\text{GEE}}(\beta) = \frac{\partial \mu^{\mathsf{T}}}{\partial \beta} V^{-1}(\beta, \psi)\left(\mathbf{y}^{(n)} - \mu\right) = 0, \tag{6.10}$$

where $\mu^T = \left(\exp(\mathbf{x}_1^T\beta), \ldots, \exp(\mathbf{x}_n^T\beta)\right)$ and $V(\beta, \psi)$ is a working covariance matrix selected to reflect the latent process autocorrelation structure (with parameters ψ) and the variance of the observations.

Zeger (1988) shows that, in the Poisson case, the method is consistent and asymptotically normal with appropriate estimates of the covariance matrix. He applies the method to the Polio data using an AR(1) specification for the lognormal latent process. He also develops a method of moments approach to estimating the variance and autocovariances of the latent process. Davis et al. (1999) develop a bias adjustment for this method based on the GLM estimates of β, that we next discuss.

The most elementary method for estimating β alone uses the GLM estimating equation derived assuming that there is no latent process in (6.7), i.e., it is falsely assumed that $S_t = \mathbf{x}_t^T\beta$ so that the Y_t are independent and not just conditionally so. Therefore, the GLM estimator $\hat{\beta}_n$ of β solves the score equation corresponding to the quasilikelihood under the independence assumption:

$$U_{\text{GLM}}(\beta) = X(\mathbf{y}^{(n)} - \mu) = 0, \tag{6.11}$$

where $\mathbf{x}_t^T$ is the tth row of the design matrix X. See Davis et al. (1999, 2000) for the Poisson case, Davis and Wu (2009) for the negative binomial case in particular and some other members of the exponential family and Wu and Cui (2013) for the binary case. GLM provides preliminary estimates of the regression needed to form assessment of the existence of a latent process and the nature of its autocorrelation structure at least for the Poisson case as discussed in Davis et al. (2000).

The central limit theorem has also been established for GLM estimates for the Poisson case (Davis et al., 2000), the negative binomial case (Davis and Wu, 2009) and the binary case (Wu and Cui, 2013). For the Poisson case, Davis et al. (2000) provide formulae for estimating the asymptotic covariance based on parametric estimators of the latent process model autocovariances. They demonstrate that these are usefully accurate through simulations and applications to data. As pointed out by Wu (2012), use of nonparametric estimates of the autocovariances in a plug-in estimator of the asymptotic covariance will not provide consistent estimators of the asymptotic covariance. Instead, he proposed use of kernel-based estimates and shows they are consistent. It is particularly important that the standard errors produced by GLM (as implemented in standard software packages) not be used because they can be seriously misleading. For example, in the case of the Polio data set to be reviewed in Section 6.3, the standard error reported using GLM for the important linear time trend parameter is 1.40, whereas the correct asymptotic standard error, calculated as just described, results in a standard error of 4.11 (almost 3 times larger).

In the Poisson and negative binomial cases, consistency and asymptotic normality have been established for the GLM estimate $\hat{\beta}_{GLM}$ under suitable conditions on the regressors and log-link function. The regressors x_{nt} are typically defined in terms of a smooth function $f : [0, 1] \to \mathbb{R}^d$ from the unit interval having the form $x_{nt} = f(t/n)$. At least for the Poisson and negative binomial case, this scaling of regression by sample size is required to prevent degeneracy. Regressors can also include harmonic functions (for seasonal components) as well as observed realizations from stationary random processes. In the above treatments, the latent process $\{\alpha_t\}$ is required to be a stationary Gaussian ARMA model or $\epsilon_t = \exp(\alpha_t)$ is required to be a stationary strongly mixing process with finite $(4 + \delta)$th moment. In the Poisson and negative binomial cases with log-link functions, the optimization of the quasi loglikelihood objective function can be performed over all of $\mathbb{R}^d$ since this function is concave.

The same argument can be applied to show that the GLM estimators converge to a limit point and are asymptotically normal when the quasi loglikelihood function is concave. However, in order for the limit point to be the true parameter, the identifiability condition $E\{b'(\mathbf{x}_t^T\beta + \alpha_t)\} = b'(\mathbf{x}_t^T\beta)$ for all t must be met. This is satisfied for some count response distributions including the Poisson and negative binomial distributions with log link functions. However, for the binomial response family, this identity cannot hold for any link function using the normal distribution for the latent process. The same argument can be applied to show that the GEE method proposed by Zeger (1988) will not provide consistent estimates of β in the binomial case.

Correction for the bias in the binary case requires integration over the marginal distribution of α_t, that in turn requires knowledge of its variance, something that is not available from the GLM estimates. Wu and Cui (2013) introduce a modified GLM method that adjusts for the bias in the regression parameter estimates. They introduce a modified estimating equation in which the success probability used in GLM, $\pi_t = 1/(1+\exp(-x_{nt}^T\beta))$, is replaced by the marginal mean $m(x_{nt}^T\beta) = P(Y_t = 1)$. The monotonically increasing function $m(u)$ is estimated nonparametrically using local linear regression. Wu and Cui (2013) prove that this method produces consistent and asymptotically normal estimates of the regression coefficients. They also provide a method for calculating the asymptotic covariance matrix.

6.2.2 Likelihood-based Methods

The main impediments to using maximum likelihood methods for routine analysis are threefold:

Optimizing the likelihood: Calculation of the likelihood in (6.9) cannot be done in closed form for moderate sample sizes let alone large dimensions on the order of magnitude of 1,000 or 10,000. In order to find the maximum likelihood estimator, one typically resorts to numerical optimization of an approximation to $L(\theta)$. These methods often do not rely on derivative information. If a gradient or Hessian is required, then another d or $d(d+1)/2$ integrals need to be approximated, where $d = \dim(\theta)$.

Computing standard errors: Once an approximation $\hat{\theta}$ to the MLE has been produced, standard errors are required for inference purposes. This is especially important in the model fitting stage when perhaps a number of covariates are being considered for inclusion in the model. Because of the large dimensional integrals defining the likelihood, approaches using approximations to the Hessian or scoring algorithms are problematic. Often one resorts to numerical approximations to derivatives of the approximating likelihood evaluated at the estimated value. However, these estimates can be quite variable and numerically sensitive to the choice of tuning parameters in the numerical algorithms. Bootstrap methods, in which each bootstrap replicate would require its own n-dimensional integral to be computed, is one possible workaround for computing more reliable standard errors.

Asymptotic theory: There are currently no proofs that the MLE $\hat{\theta}$ is consistent or asymptotically normal. One would certainly expect these properties to hold, but since the form of the likelihood is rather intractable, the arguments are not standard adaptations of existing proofs. Nonetheless, it is important to have a complete theory worked out for maximum likelihood estimation in these models in order to ensure that inferences about the parameters are justifiable.

As a result of these practical concerns, a large variety of methods have been proposed to approximate the likelihood and, to a lesser extent, derivatives of these approximations. The main approaches in the literature are approximations to the integrand in (6.9), which can be integrated in closed form to get an approximation to the likelihood. Improvements to these approximations are typically based on Monte Carlo methods and importance sampling from approximating importance densities. Quasi Monte Carlo (QMC) methods based on randomly shifted deterministic lattices of points in $\mathbb{R}^n$ have also been applied in recent years. One recent attempt is given in Sinescu et al. (2014) where a Poisson model with a constant mean plus AR(1) state process is considered. While QMC methods hold promise, further development is required before they become competitive with other methods reviewed here.

6.2.3 Earlier Monte Carlo and Approximate Methods

Nelson and Leroux (2006) review various methods in this class for estimating the likelihood function, including the Monte Carlo expectation method first used for the Poisson response AR(1) model in Chan and Ledolter (1995) based on Gibb's sampling for the E-step, a version due to McCulloch based on Metropolis–Hastings sampling, Monte Carlo Newton–Raphson (Kuk and Cheng, 1999) and a modified version of the iterative bias correction method of Kuk (1995). They compare performance of these methods with the original estimating equations approach of Zeger (1988) and the PQL approach (based on a Laplace approximation to the likelihood) of Breslow and Clayton (1993) using simulations and by application to the Polio data in Zeger (1988). Davis and Rodriguez-Yam (2005) also review the Monte Carlo Expectation Maximization and Monte Carlo Newton–Raphson methods.

Apart from the Bayesian method, all the methods reviewed by Nelson and Leroux provide, as a byproduct of the parameter estimation, estimated covariances matrices under the assumption that the estimates satisfy a central limit theorem.

Nelson and Leroux conclude: "The results have clearly shown that the different methods commonly used at present for fitting a log-linear generalized linear mixed model with an autoregressive random effects correlation structure do yield different sets of parameter estimates, in particular, the parameters related to the random effects distribution." We will summarize the main points of differences between the various methods in Section 6.3 when applied to the Polio data set.

6.2.4 Laplace and Gaussian Approximations

Based on (6.5), the likelihood for the unknown parameters is

$$L(\theta) = \int_{\mathbb{R}^n} e^{F(\alpha,\mathbf{y};\theta)}\, d\alpha, \tag{6.12}$$

where

$$F(\alpha,\mathbf{y};\theta) = \sum_{t=1}^{n}\{\log p(y_t|x_{nt}^T\beta + \alpha_t)\} - \frac{1}{2}\alpha^T V\alpha + \frac{1}{2}\log\det(V) - \frac{n}{2}\log 2\pi, \tag{6.13}$$

and $V = \Gamma_n^{-1}$. For many models, the exponent F in (6.12) is unimodal in α. Laplace's approximation replaces the integrand by a normal density that matches that obtained using a second order Taylor series expansion of F around its mode

$$\alpha^* = \arg\max_{\alpha} F(\alpha,\mathbf{y};\theta). \tag{6.14}$$

To find this mode, the Newton–Raphson method has proved effective for the primary model considered here. Let $F'(\alpha,\mathbf{y};\theta)$ denote the first derivative vector and $F''(\alpha,\mathbf{y};\theta)$ the matrix of second derivatives both with respect to α. For the conditionally independent model with a Gaussian latent process of (6.13), it follows that

$$F'(\alpha,\mathbf{y};\theta) = \sum_{t=1}^{n}\frac{\partial}{\partial\alpha}\log p\left(y_t|\mathbf{x}_t^T\beta + \alpha_t\right) - V\alpha \tag{6.15}$$

and

$$F''(\alpha|\mathbf{y};\theta) = -(K+V), \tag{6.16}$$

where, as a result of the conditional independence, K is the diagonal matrix given by

$$K = -\text{diag}\left[\frac{\partial^2}{\partial\alpha_t^2}\log p(y_t|\alpha_t;\theta); t = 1,\ldots,n\right].$$

Let $\alpha^{(k)}$ be the kth iterate (where dependence on $\mathbf{y}$ and θ have been suppressed in the notation). The Newton–Raphson updates are given by

$$\alpha^{(k+1)} = \alpha^{(k)} - F''\left(\alpha^{(k)}\right)^{-1} F'\left(\alpha^{(k)}\right). \tag{6.17}$$

Provided $-F''(\alpha, \mathbf{y}; \theta) = (K + V)$ is positive definite for all α, Newton–Raphson iterates starting at any value $\alpha^{(0)}$ will converge to the unique modal point denoted α^*. A sufficient condition is that $\log p(y|x^T\beta + \alpha)$ be concave in α. This is satisfied for the exponential family with canonical link in which case $K = \text{diag}(b''(X^T\beta + \alpha))$. Note that α^* is a function of θ and $\mathbf{y}$ but we will typically suppress this as given.

Using this expansion, we rewrite F in (6.13) as

$$F(\alpha, \mathbf{y}; \theta) = F_a(\alpha, \mathbf{y}; \theta) + R(\alpha; \alpha^*), \tag{6.18}$$

where

$$F_a(\alpha, \mathbf{y}; \theta) = F(\alpha^*, \mathbf{y}; \theta) - \frac{1}{2}\{(\alpha - \alpha^*)^\top (K^* + V)(\alpha - \alpha^*)\}. \tag{6.19}$$

Ignoring the error term $R(\alpha; \alpha^*)$ provides an approximation to $p(\mathbf{y}, \alpha)$ of the form $p_a(\mathbf{y}, \alpha) = \exp(F_a(\alpha, \mathbf{y}))$, which when integrated over α, gives the Laplace approximation to the marginal distribution of $\mathbf{y}$ and hence to the likelihood (6.12) of the form

$$\begin{aligned} L_a(\theta; \mathbf{y_n}, \alpha^*) &= (2\pi)^{n/2}|K^* + V|^{-1/2}\exp[F(\alpha^*, \mathbf{y}; \theta)] \\ &= \frac{|V|^{1/2}}{|K^* + V|^{1/2}} \exp\left(\log p(\mathbf{y}|X\beta + \alpha^*) - \frac{1}{2}\alpha^{*\top} V \alpha^*\right), \end{aligned} \tag{6.20}$$

and

$$L(\theta) = L_a(\theta)E_a\left[e^{R(\alpha, \alpha^*)}\right], \tag{6.21}$$

where the expectation is with respect to the approximate posterior $p_a(\alpha|\mathbf{y}; \theta) \sim N(\alpha^*, (K^* + V)^{-1})$.

As a result of the very high dimension of the latent process α, Newton–Raphson updates are computationally expensive to implement naively, as each step would cost $O(N^3)$ operations for the calculations involving the inversion of the Hessian matrix. Also, the efficiency of this computation depends very strongly on the form of the autocovariance matrix Γ_n for α. A convenient and flexible choice for the $\{\alpha_t\}$ process is the causal AR(p) models,

$$\alpha_t = \phi_1\alpha_{t-1} + \cdots + \phi_p\alpha_{t-p} + \epsilon_t, \quad \{\epsilon_t\} \sim \text{IID}N(0, \sigma^2), \tag{6.22}$$

in which case V in (6.16) is a banded matrix and is hence sparse. For this class of models, Davis and Rodriguez-Yam (2005) employ the innovations algorithm for computation of the update steps leading to a fast and stable algorithm for obtaining the mode α^* and $|K^* + V|$ required in (6.19)

The above Laplace approximation is equivalent to finding the mode of the smoothing density of the state given the observations. Durbin and Koopman (1997) and Durbin and Koopman (2012) use a modal approximation to obtain the importance density and arrive at an alternative expression to (6.21) shown by Davis and Rodriguez-Yam (2005) to be of the form

$$L(\theta) = L_g(\theta)E_a\left[\frac{p(\mathbf{y}|\alpha)}{g(\mathbf{y}|\alpha)}\right], \tag{6.23}$$

where the expectation is taken with respect to the previously defined Gaussian approximating smoothing density for $\boldsymbol{\alpha}|\mathbf{y} \sim N(\boldsymbol{\alpha}^*, (K^* + V)^{-1})$, $g(\mathbf{y}|\boldsymbol{\alpha})$ is the corresponding Gaussian distribution for $\mathbf{y}|\boldsymbol{\alpha}$, and $L_g(\theta) = A(\boldsymbol{\alpha}^*)L_a(\theta)$, where $A(\boldsymbol{\alpha}^*)$ is defined in Davis and Rodriguez-Yam (2005). As a result of (6.23), the importance sampling method based on the Laplace approximation and that based on the Gaussian approximation of Durbin and Koopmans will give identical results (for the same draws from the approximating conditional of $\boldsymbol{\alpha}|\mathbf{y}$).

Unlike the use of L_a alone to approximate the likelihood, L_g alone cannot be used as an approximation to $L(\theta)$ and simulation is required to adjust it. An alternative approximation to the likelihood, not requiring simulation, is proposed in Durbin and Koopman (1997) mainly for obtaining starting values for $\hat{\theta}$ to optimize the likelihood. This is of the form

$$\log L_{a,DK} = \log L_g + \log \hat{w} + \log\left(1 + \frac{1}{8}\sum_{t=1}^{n} \hat{l}_t^{(4)} v_t^2\right), \tag{6.24}$$

where $\hat{l} = \hat{w} = \log p(\mathbf{y}|\boldsymbol{\alpha}^*) - \log p_a(\mathbf{y}|\boldsymbol{\alpha}^*)$, $v_t = \left[(K^{*-1} + V^{-1})^{-1}\right]_{tt}$, and $\hat{l}_t^{(4)} = b^{(4)}(x_t^{\mathrm{T}}\beta + \alpha_t^*)$ is the fourth derivative of the tth element of $\hat{l}$ with $b^{(4)}$ being the fourth derivative of b. All of the quantities needed for the approximation (6.24) can be calculated readily for the exponential family and require only the quantities needed for the Laplace approximation above.

Shephard and Pitt (1997) and Durbin and Koopman (1997) use Kalman filtering and smoothing to obtain the mode corresponding to an approximating linear, Gaussian state space model. Details are in Davis and Rodriguez-Yam (2005) and Durbin and Koopman (2012). For this procedure to be valid, it is required that $\log p(y|x^T\beta + \alpha)$ be concave in α, which is a stronger condition than is needed for the approach based on the innovations algorithm to obtain the Laplace approximation. In cases where $\log p(y|x^T\beta + \alpha)$ is not log concave, Jungbacker and Koopman (2007) show that the Kalman filter and smoother algorithms of Durbin and Koopman (1997) can still be used in conjunction with a line search version of Newton–Raphson to obtain the required mode for the Laplace approximating density. However, derivation of this result in Jungbacker and Koopman (2007) is based on different arguments than presented in Durbin and Koopman (2012).

We let $\hat{\theta}_a = \arg\max_\theta L_a(\theta)$ and call this the approximate MLE for θ. Use of derivative information can substantially assist numerical optimization algorithms to find the maximum over θ. However, expressions for the first derivative vector involves the third order partial derivatives of $F(\boldsymbol{\alpha}(\theta); \theta)$, which in turn require implicit differentiation to be used. Needless to say the analytic expressions are complex and need to be worked out for each specific model being considered for the latent process autocorrelation structure. Numerical derivatives based on finite differences are often used but these can be slow to execute and inaccurate. An alternative to using numerical derivatives suggested in Skaug (2002) and Skaug and Fournier (2006) is to use automatic differentiation (AD). Sometimes called "algorithmic differentiation", this method takes computer code for calculation of $L_a(\theta; \mathbf{y_n}, \boldsymbol{\alpha}^*)$ in (6.20) and produces new code that evaluates its derivatives. This is not the same as using a symbolic differentiator such as used in Maple or Mathematica.

Davis and Rodriguez-Yam (2005) demonstrate the very good accuracy of Laplace approximation without use of importance sampling for the models considered here. Use of higher order terms (Shun and McCullagh, 1995) in the Taylor series expansion of F in (6.13) would be relatively easy to implement and are likely to improve the performance even further, particular for highly discrete data such as in Bernoulli responses. While the Laplace approximation appears to be very close to the required integral likelihood, convergence of the Laplace approximation for parameter driven processes has not been established, despite many methods that base their analysis on approximating the likelihood with the Laplace approximation and simulating the error such as Durbin and Koopman (2000) and Davis and Rodriguez-Yam (2005).

As yet, there is no useable asymptotic theory for the Laplace approximate MLE. Simulation results in Davis and Rodriguez-Yam (2005) for reasonably simple examples suggest that the bias is small and the distribution is approximately normal.

Despite the drawbacks of asymptotic bias and lack of a CLT for the estimates obtained by maximizing the approximate likelihoods $L_a(\theta)$ in (6.20) or $L_{a,DK}(\theta)$ in (6.24), they could be very useful in the model building phase to decide on appropriate regressors and a latent process model form. Here, the emphasis should be on speed to arrive at a parsimonious model in which relevant regressors are included and serial dependence is adequately taken into account. Approximate standard errors could be obtained using numerical differentiation of the approximate likelihood to obtain an approximate Hessian and hence information matrix. For example, a standard optimizer such as 'optim' in R (R Core Team, 2014), the Hessian at convergence could be requested. Another approach to this is as suggested in Davis and Rodriguez-Yam (2005) and uses bootstrap methods to adjust for bias and to obtain an approximate distribution for the approximate likelihood estimators. Alternatively, information theoretic criteria could be used to compare models of different structures or complexity or informal use of the likelihood ratio test for nested model structures could be employed. Once one or several suitable model structures have been decided upon by this approach, the approximate likelihood methods could be enhanced by importance sampling to improve the accuracy of inference and final model choice.

6.2.5 Importance Sampling

The general use of importance sampling methods to evaluate the likelihood for nonlinear non-Gaussian models is well reviewed in Durbin and Koopman (2012). Importance sampling is used to approximate the expectation of the error term of the Laplace approximation $L_a(\theta)$ to the likelihood $L(\theta)$ in (6.21) and, for the Gaussian approximation, to approximate the expectation term multiplying $L_g(\theta)$ in (6.23). Both of these implementations of importance sampling draw samples from the same modal approximation to the posterior distribution for $\boldsymbol{\alpha}|\mathbf{y}$. An alternative approach to approximating $L(\theta)$ is efficient importance sample (EIS), that uses a global approximation minimizing variance of the log of the importance weights. We discuss these three approaches in more detail here.

6.2.5.1 Importance Sampling based on Laplace Approximation

Based on (6.21), Davis and Rodriguez-Yam (2005) use importance sampling to approximate the expectation as

$$Er_a(\theta) = E_a\left[e^{R(\boldsymbol{\alpha},\boldsymbol{\alpha}^*)}\right] \approx \frac{1}{M}\sum_{i=1}^{M} e^{R(\boldsymbol{\alpha}^{(i)};\boldsymbol{\alpha}^*)} =: e^{\hat{e}(\theta)}, \tag{6.25}$$

where the $\boldsymbol{\alpha}^{(i)}$ are independent draws from $p_a(\boldsymbol{\alpha}|\mathbf{y})$. However, this calculation is slow, needing M Monte Carlo simulations at each step of the optimization of the likelihood.

They also suggest using approximate importance sampling (AIS) in which $e^{\hat{e}(\theta)}$ in (6.25) is approximated using a first order Taylor expansion in θ around $\hat{\theta}_{\mathrm{AL}}$ of the form

$$T_{\hat{e}}(\theta; \hat{\theta}_{\mathrm{AL}}) = \hat{e}(\theta) + q_{\mathrm{AL}}^{\mathrm{T}}(\theta - \hat{\theta}_{\mathrm{AL}}), \tag{6.26}$$

where $q_{\mathrm{AL}} = \frac{\partial}{\partial\theta}\hat{e}(\theta)|_{\hat{\theta}_{\mathrm{AL}}}$ is obtained using numerical differentiation. This method is considerably faster than IS because the approximate AIS approximate likelihood given by

$$L_c(\theta) = L_a(\theta) \exp\left\{T_{\hat{e}}(\theta; \hat{\theta}_{\mathrm{AL}})\right\} \tag{6.27}$$

requires that q_{AL} (which uses importance sampling) only needs to be computed once before $L_c(\theta)$ is optimized.

Numerical experiments in Davis and Rodriguez-Yam (2005) show that this estimator has no significant drop in accuracy compared to traditional importance sampling. However, it is possible to iterate over the AIS approximation of $Er_a(\theta)$ if further accuracy is required, where θ_c is replaced with the updated estimate for θ at each step.

6.2.5.2 Importance Sampling based on Gaussian Approximation

Durbin and Koopman (1997) advocate using the form (6.23) to approximate the likelihood by approximating $E_a[w(\alpha, \mathbf{y})]$, where $w(\alpha, \mathbf{y}) = p(\mathbf{y}, \alpha)/g(\mathbf{y}, \alpha) = p(\mathbf{y}|\alpha)/g(\mathbf{y}|\alpha)$ are the importance weights by

$$E_a[w(\alpha, \mathbf{y})] \approx \frac{1}{M}\sum_{i=1}^{M} w\left(\alpha^{(i)}, \mathbf{y}\right), \tag{6.28}$$

and as before, the $\alpha^{(i)}$ are independent draws from the same $p_a(\alpha|\mathbf{y})$ as used in the Davis and Rodriguez-Yam approach to importance sampling. However, Durbin and Koopman (2012) and Shephard and Pitt (1997) use the simulation smoother of De Jong and Shephard (1995) to draw the samples from this approximating posterior distribution. In order to improve the accuracy and numerical efficiency of the Monte Carlo estimate of the error term, Durbin and Koopman (1997) employ antithetic variables to balance location and scale and a control variable based on a fourth order Taylor expansion of the weight function in the error term. This is not the same as the AIS method of Davis and Rodriguez-Yam (2005).

6.2.5.3 Efficient Importance Sampling

Choice of importance sampling density is critical for efficient approximation of the likelihood by simulation. Modal approximation was used in the methods of the previous two subsections. An alternative choice of importance density leads to the EIS method—see Jung and Liesenfeld (2001), Richard and Zhang (2007), and Jung et al. (2006) for complete details on this approach. For EIS, the importance densities are based on the (approximate) minimization of the variance of the log importance weight $\log w(\alpha, \mathbf{y})$ defined as in the Durbin–Koopman method. Lee and Koopman (2004) compare this with DK procedure and conclude they have similar performance when applied to the stochastic volatility model.

Recently, Koopman et al. (2015) have presented an approach called numerically accelerated importance sampling (NAIS) for the nonlinear non-Gaussian state space models. The NAIS method combines fast numerical integration techniques with the Kalman filter smoothing methods proposed by Shephard and Pitt (1997) and Durbin and Koopman (1997). They demonstrate significant computational speed improvements over standard EIS implementations as well as improved accuracy. Additionally, by using new control variables substantial improvement in the variability of likelihood estimates can be achieved. A key component of the NAIS method is construction of the importance sampling density by numerical integration using Gauss–Hermite quadrature in contrast to using simulated trajectories.

6.2.6 Composite Likelihood

As already noted, the lack of a tractably computable form for the likelihood is one of the drawbacks in using SSMs for count time series. Estimation procedures described in previous sections essentially resort to simulation-based procedures for computing the likelihood and then maximizing it. With any simulation procedure, one cannot be certain that the *simulated likelihood* provides a good facsimile of the actual likelihood or even the likelihood in proximity of the maximum likelihood estimator. An alternative estimation procedure, which has grown recently in popularity, is based on the composite likelihood. The idea is that perhaps one does not need to compute the entire likelihood but only the likelihoods of more manageable subcollections of the data, which are then combined. For example, the likelihood given in (6.5) requires an n-fold integration, which is generally impractical even for moderate sample sizes. However, a similar integral based on only 2 or 3 dimensions can be computed numerically rather accurately. The objective then is to replace a large dimensional integral with many small and manageable integrations.

The special issue of *Statistical Sinica* (Volume 21 (2011)) provides an excellent survey on the use of composite likelihood methods in statistics. In the development below, we will follow the treatment of Davis and Yau (2011) with emphasis on Example 3.4. Ng et al. (2011), also in the special issue, consider composite likelihood for related time series models.

While we will focus this treatment on the *pairwise* likelihood for count time series, extensions to combining weighted likelihoods based on more general subsets of the data are relatively straightforward. The downside in our application is that numerically, it may not be practical to compute density functions of more than just a few observations.

Suppose $Y_1, \ldots, Y_n$ are observations from a time series for which we denote $p(y_{i_1}, y_{i_2}, \ldots, y_{i_k}; \theta)$ as the likelihood for a parameter θ based on the k-tuples of distinct observations $y_{i_1}, y_{i_2}, \ldots, y_{i_k}$. For ease of notation, we have suppressed the dependence of $p(\cdot;\theta)$ on k and the vector $i_1, i_2, \ldots, i_k$. For k fixed, the composite likelihood based on consecutive k-tuples is given by

$$CPL_k(\theta; \mathbf{Y}^{(n)}) = \sum_{t=1}^{n-k} \log p(Y_t, Y_{t+1}, \ldots, Y_{t+k-1}; \theta). \tag{6.29}$$

Normally, one does not take k very large due to the difficulty in computing the k-dimensional joint densities. The composite likelihood estimator $\hat{\theta}$ is found by maximizing CPL_k. Under identifiability, regularity, and suitable mixing conditions, $\hat{\theta}$ is consistent and asymptotically normal. This was established for linear time series models in

Davis and Yau (2011). They also showed in the case $k = 2$ that if one uses all pairs of observations instead of just consecutive pairs of observations, that is, CPL_2 is replaced by the sum of the log likelihood of Y_s, Y_t for all $s < t$, then the composite likelihood estimator need no longer be consistent. Also note that $k = 1$, which corresponds to just marginal distributions, is allowed. While in this case, one might be able to consistently estimate parameters associated with the marginal distribution, there is no hope of estimating dependence parameters since joint distributions are not part of the objective function. In this case, the dependence parameters are not identifiable.

To illustrate the use of the composite likelihood, consider Example 1 from Section 6.1 in which the observational density is Poisson and the state process $\{\alpha_t\}$ follows an AR(1) process. That is, given the state-process $\{\alpha_t\}$, the y_t are independent and Poisson-distributed with mean $\lambda_t = e^{\beta+\alpha_t}$. The SSM is then specified by the equations

$$p(y_t|\alpha_t;\theta) = e^{-e^{\beta+\alpha_t}} e^{(\beta+\alpha_t)y_t}/y_t!,$$

$$\alpha_t = \phi\alpha_{t-1} + \eta_t,$$

where $\eta_t \sim IIDN(0, \sigma^2)$, $|\phi| < 1$, and $\theta = \{\beta, \phi, \sigma^2\}$ is the parameter vector.

Let the observed data be $\mathbf{y}_n = \{y_1, y_2, \ldots, y_n\}$ and set $\boldsymbol{\alpha}_n = \{\alpha_1, \alpha_2, \ldots, \alpha_n\}$. The pairwise log-likelihood (here we are taking $k = 2$), is given by

$$CPL_2(\theta;\mathbf{y}_n) = \log\left[\prod_{t=1}^{n-1}\int\int p(y_t|\alpha_t;\theta)p(y_{t+1}|\alpha_{t+1};\theta)f_\theta(\alpha_t,\alpha_{t+1})d\alpha_t d\alpha_{t+1}\right].$$

So unlike the computation for the full likelihood, that requires the computation of an n-dimensional integral, the pairwise likelihood requires the computation of $(n-1)$ two-dimensional integrals. Each of these integrals can be computed rather quickly using numerical methods such as Gauss–Hermite quadrature.

A comparison of the performance of the composite likelihood relative to the approximate likelihood procedure described in Section 6.2.4 was made via a simulation study in Davis and Yau (2011) (see Table 3 of the paper). The results show that CPL_2 performed comparably to the AIS estimates. It is also worth noting that using higher orders of k, such as $k = 3$ and 4 often gave worse estimates.

Ultimately, the estimation objective is to compute the maximum likelihood estimates and there has been much effort, as described in earlier sections in finding either approximations to the likelihood function of its optimizer. Even if one could compute the MLE directly, the proof of consistency and asymptotic normality has not been fully argued. In contrast, and perhaps one potential advantage in using composite likelihood methods is that one can give a rigorous argument for the consistency and asymptotical normality of such estimates. We give a brief outline of such an argument that follows the lines of the one given in Davis and Yau (2011). For the setup of Example 1, let

$$cpl_t(\theta) = cpl(\theta;y_t,y_{t+1}) = \log\left[\int\int p(y_t|\alpha_t;\theta)p(y_{t+1}|\alpha_{t+1};\theta)f_\theta(\alpha_t,\alpha_{t+1})d\alpha_t d\alpha_{t+1}\right],$$

and note that $CPL_2(\theta; \mathbf{y_n}) = \sum_{t=1}^{n-1} cpl_t(\theta)$. Let θ_0 and $\hat{\theta}$ be the true value and the CPL_2 estimator of the parameter, respectively. Using a Taylor series expansion of $CPL_2'(\hat{\theta}; \mathbf{y_n})$, the derivative of CPL_2, around θ_0 shows that $\sqrt{n}(\hat{\theta} - \theta_0)$ is asymptotically equivalent to

$$-\left(\frac{1}{n}\sum_{t=1}^{n-1} cpl_t''(\theta_0)\right)^{-1} \frac{1}{\sqrt{n}}\sum_{t=1}^{n-1} cpl_t'(\theta). \tag{6.30}$$

Since the process $\{Y_t\}$ is stationary and strongly mixing at a geometric rate, it follows from the ergodic theorem that

$$\frac{1}{n}\sum_{t=1}^{n-1} cpl_t''(\theta_0) \xrightarrow{a.s.} E(cpl_1''(\theta_0)).$$

Moreover, since $\{cpl_t'(\theta_0)\}$ is also a stationary for strongly mixing sequence, a standard central limit theorem for strongly mixing sequences (e.g., Doukhan, 1994), shows the asymptotic normality of $\frac{1}{\sqrt{n}}\sum_{t=1}^{n-1} cpl_t'(\theta)$ with covariance matrix

$$\sum_{h=-\infty}^{\infty} \gamma(h),$$

where $\gamma(h)$ is the autocovariance matrix of $\{cpl_t'(\theta_0)\}$. Hence, $\sqrt{n}(\hat{\theta} - \theta_0)$ is asymptotically normal with mean 0 and covariance matrix given by

$$\Sigma := \left(Ecpl_1''(\theta_0)\right)^{-1} \left(\sum_{h=-\infty}^{\infty} \gamma(h)\right) \left(Ecpl_1''(\theta_0)\right)^{-1}. \tag{6.31}$$

A consistent estimator for Σ is given by

$$\hat{\Sigma}_n = \left(\frac{1}{n}\sum_{t=1}^{n-1} cpl_t''(\hat{\theta})\right)^{-1} \left(\sum_{k=-r_n}^{r_n} \left(1 - \frac{|k|}{n}\right)\hat{\gamma}(k)\right) \left(\frac{1}{n}\sum_{t=1}^{n-1} cpl_t''(\hat{\theta})\right)^{-1}, \tag{6.32}$$

where $r_n \to \infty$, $r_n/n \to 0$, and

$$\hat{\gamma}(k) = \frac{1}{n}\sum_{t=k+1}^{n-1} cpl_t'(\hat{\theta}) cpl_{t-k}'^{T}(\hat{\theta}).$$

The asymptotic variance of a composite likelihood estimator typically has a sandwich-type form as given by (6.31). Such quantities can be difficult to estimate. One approach, in addition to using (6.32), is via the bootstrap for time series. The block bootstrap or stationary bootstrap (see the discussion paper Politis et al. (2003) for a description of these methods) can be used for generating *nonparametric* bootstrap replicates of a stationary time series.

This methodology provides an attractive alternative for computing asymptotic variances of the estimates and for providing approximations to the sampling distribution of $\sqrt{n}(\hat{\theta} - \theta_0)$.

6.3 Applications to Analysis of Polio Data

In this section, we summarize a variety of analyses using the Poisson AR model for the Polio data set consisting of the monthly number of U.S. cases of Poliomyelitis from 1970 to 1983 first analysed by Zeger (1988). We parameterize the model as in Davis and Rodriguez-Yam (2005) for example, in which the distribution of Y_t given the state α_t is Poisson with rate $\lambda_t = e^{\alpha_t + x_t^T \beta}$. Here, $\beta^T := (\beta_1, \ldots, \beta_6)$, $\mathbf{x_t}$ is the vector of covariates given by

$$\mathbf{x_t}^T = (1, t/1000, \cos(2\pi t/12), \sin(2\pi t/12), \cos(2\pi t/6), \sin(2\pi t/6)),$$

and the state process is assumed to follow an AR(1) model. The vector of parameters is $\theta = (\beta_1, \ldots, \beta_6, \phi, \sigma^2)$.

Table 6.1 compiles, from a variety of sources, the estimates and their standard errors for the key parameters in this model, namely the coefficient of the linear time trend, β_2, the serial autocorrelation of the latent process, ϕ, and the innovation variance, σ^2. Note that in some analyses, the parameterization of the latent process variance as $\sigma_\alpha^2 = \sigma^2/(1 - \phi^2)$ is used. The table adjusts these results to the above parameterization. Estimates of this process variance, obtained as $\hat{\sigma}_\alpha^2 = \hat{\sigma}_\epsilon^2/(1 - \hat{\phi}^2)$, are presented as the final column to allow additional comparison between the various model fits.

TABLE 6.1

Estimates and Standard Errors for Key Parameters in Various Methods Applied to the Polio Series

Method (Source)	$\hat{\beta}_2$	se($\hat{\beta}_2$)	$\hat{\phi}$	se($\hat{\phi}$)	$\hat{\sigma}^2$	$\hat{\sigma}_\alpha^2$
MCEM (Chan and Ledolter, 1995)	−4.62	1.38	0.89	0.04	0.09	0.41
MCEM[NL] (McCulloch, 1997)	−4.35	1.96	0.10	0.36	0.50	0.51
Bayes (Oh and Lim, 2001)	−4.24	1.72	0.66	0.16	0.32	0.56
PQL[NL] (Breslow and Clayton, 1993)	−3.46	3.04	0.70	0.13	0.26	0.51
AL (Davis and Rodriguez-Yam, 2005)	−3.81	2.77	0.63	0.23	0.29	0.48
AL-BC (Davis and Rodriguez-Yam, 2005)	−3.96	2.77	0.73	0.23	0.30	0.65
AIS (Davis and Rodriguez-Yam, 2005)	−3.75	2.87	0.66	0.21	0.27	0.48
AIS-BC (Davis and Rodriguez-Yam, 2005)	−3.76	2.87	0.73	0.21	0.30	0.64
MCNR (Kuk and Cheng, 1999)	−3.82	2.77	0.67	0.18	0.27	0.48
EIS (Jung and Liesenfeld, 2001)	−3.61	2.57	0.68	0.15	0.26	0.48
GLM (Davis et al., 2000)	−4.80	4.11	—	—	—	—
GEE (Zeger, 1988)	−4.35	2.68	0.82	—	0.19	0.57
CPL$_2$ (Davis and Yau, 2011)	−4.74	2.54	0.49	0.21	0.37	0.49
IBC[NL] (Kuk, 1995)	−5.01	3.20	0.54	0.28	0.35	0.49

Note: $\hat{\sigma}_\alpha^2 = \hat{\sigma}_\epsilon^2/(1 - \hat{\phi}^2)$.

The origin of the method and results when applied to the Polio data set is listed in parentheses, and additionally, if the results are from application of the method by Nelson and Leroux (2006) these are also indicated by an additional annotation 'NL'. The methods can be roughly partitioned into three groups. Group 1 consists of two implementations of MCEM (Monte Carlo EM) and a Bayes procedure. Group 2, which is essentially approximate likelihood based-methods, consists of PQL (penalized quasilikelihood), AL (approximate likelihood), AL-BC (bias corrected AL), AIS (approximate importance sampling), AIS-BC (bias corrected AIS), MCNR (Monte Carlo Newton–Raphson), and EIS (efficient importance sampling). Note that the first 3 procedures of this group are nonsimulation based, while the last 4 involve some level of simulation. Group 3 consists of nonlikelihood-based procedures: GLM (generalized linear model estimates ignoring the latent process), GEE (generalized estimating equations), CPL_2 (pairwise composite likelihood), and IBC (iterative bias correction using iterative weighted least squares). We exclude from our review the few studies that have used alternative response distributions or latent process distributions for these data so that the methods are compared on the same model.

With the exception of the GLM, GEE, CPL_2, and Bayesian analyses, all other methods aim to obtain approximations to the likelihood estimates and their standard errors. Clearly there are both substantial differences and similarities between the results for various methods, a point also noted in Nelson and Leroux (2006). We now discuss these differences and similarities in more detail in an attempt to draw some general conclusions about which methods may be preferred. Of course, this comparison is only for application to a single data set and much more research is required before general conclusions can be drawn. However, this is the only data set for which all the methods listed have been applied. Unfortunately, simulation evidence comparing the variety of methods is rather limited with the exception of the results in Nelson and Leroux (2006).

6.3.1 Estimate of Trend Coefficient $\hat{\beta}_2$

The GLM, IBC method as implemented by Nelson and Leroux (2006), and CPL_2 give the most negative trend estimates. It would appear as if the IBC method is not adjusting the bias of the GLM estimate sufficiently well and this may be a result of iterative weighted least squares being used as the basis for the bias adjustment simulations. It is likely that these methods are substantially biased. Amongst the remaining methods, there appear to be two groups of values for the trend coefficient estimates: Group 1, the values for both implementations of MCEM and the Bayes fit; and, Group 2 based on approximations to the likelihood with and without importance sampling (PQL, AL, AL-BC, AIS, AIS-BC, MCNR, and EIS). The concordance in Group 2 is perhaps not surprising since they are all aimed at approximating the likelihood. However, it is surprising that the Group 1 do not agree as closely with the Group 2 results. Turning to comparison of the estimated standard errors, those for Group 1 appear to be substantially smaller than those for Group 2, and within this latter group there is considerable agreement. Also note that the MCEM and Bayes methods are biasing the point estimates towards larger negative values and biasing the associated estimated standard errors downwards. The net effect of these two biases would be to increase the ratio of estimate to standard error resulting in a higher chance of concluding that there is a significant downward trend in Polio cases over the time period of observation. On the other hand, for Group 2, these test ratios would all be consistent with a conclusion of no downward trend.

6.3.2 Estimate of Latent Process Parameters

Interestingly, the estimates of overall variance $\hat{\sigma}^2_\alpha$ are remarkably similar for all methods apart from the GEE method, the Bayes method and the bias corrected AL and AIS methods. This suggests that the likelihood-based methods (including the MCEM methods) are all finding the same degree of overall variability in the latent process. However, the two MCEM methods differ substantially in their identification of the source of this latent process variability. The MCEM method of McCulloch (1997) severely underestimates the autocorrelation ϕ, with corresponding larger values for $\hat{\sigma}^2_\epsilon$ when compared with the MCEM method as implemented by Chan and Ledolter (1995) and the other likelihood approximations. The reason for this is not clear. However, since the only difference between the two MCEM methods is that of Gibbs sampler or Metropolis–Hastings, it may be that these are not exploring the sample space sufficiently well when Monte Carlo draws are being generated resulting in what appears to be lack of identifiability between ϕ and σ^2_ϵ. The bias adjusted AL and AIS methods appear to suffer and the CPL method appear to apportion overall variability to autocorrelation and innovation variance differently than the other likelihood approximations (AL, AIS, MCNR, and EIS), which are quite consistent with each other.

Incidentally, the use of AD as in Skaug (2002) gives identical results to the AL method and, because of this, are not recorded in Table 6.1. Further, the use of $M = 100$, $M = 1000$, or $M = 5000$ importance samples as reported independently by Skaug (2002) has very little impact on point estimates.

6.3.3 Comparisons of Computational Speed

Nelson and Leroux (2006) and Skaug (2002) provide some information on comparison of speeds between some of the methods. However, they mix speeds reported by the original authors with those obtained in their applications and simulations acknowledging that different generation computer processors were used. There is no comprehensive comparison of speeds for the models listed in Table 6.1. However, it is clear that the approximate likelihood methods are the fastest overall requiring no simulations or Monte Carlo to obtain estimates.

6.3.4 Some Recommendations

Based on this comparison on a single data set (with all the limitations for generality that implies):

1. Overall, the use of Laplace approximation to the likelihood results in point estimates and standard errors that are sufficiently close to those obtained from the more accurate importance sample augmented approximations. It would appear that, for these data at least, importance sampling is not providing much additional benefit to inference.
2. Use of CPL appears to provide biased results and has no clear computational advantage over the Laplace approximation method.
3. The MCEM method should be avoided until an explanation can be found for the obvious differences between results from two different implementations (Gibbs versus Metropolis–Hastings sampling) and for the clear bias in point estimates and underestimation of standard errors that the method produces.

6.4 Forecasting

For the nonlinear state-space framework of Section 6.1 given in equations (6.2) and (6.3), the forecast density of the next observation Y_{n+1} given the current data $\mathbf{Y}^{(n)}$ can be computed recursively via Bayes' theorem. We follow the development in Section 8.8.1 of Brockwell and Davis (2002). First, using (6.2) and (6.3), the filtering and prediction densities can be recursively obtained via

$$p\left(s_t|\mathbf{y}^{(t)}\right) = \frac{p\left(y_t|s_t\right)p\left(s_t|\mathbf{y}^{(t-1)}\right)}{p\left(y_t|\mathbf{y}^{(t-1)}\right)} \tag{6.33}$$

and

$$p\left(s_{t+1}|\mathbf{y}^{(t)}\right) = \int p\left(s_{t+1}|s_t\right)p\left(s_t|\mathbf{y}^{(t)}\right)d\mu\left(s_t\right), \tag{6.34}$$

where $\mu(\cdot)$ is the dominating measure for the state density function $p\left(s_t|s_{t-1}\right)$. Note that the forecasting density $p\left(y_t|\mathbf{y}^{(t-1)}\right)$ is a normalizing constant that ensures the filtering density integrates to 1, that is, $\int p\left(s_t|\mathbf{y}^{(t)}\right)d\mu\left(s_t\right) = 1$. The updating equation to calculate the forecast density for the observations is then found from

$$p\left(y_{t+1}|\mathbf{y}^{(t)}\right) = \int p\left(y_{t+1}|s_{t+1}\right)p\left(s_{t+1}|\mathbf{y}^{(t)}\right)d\mu\left(s_{t+1}\right). \tag{6.35}$$

In practice, of course, these recursions are not computable in closed form and one needs to resort to Monte Carlo procedures, see Durbin and Koopman (2012). Specifically, if one can generate replicates of S_{n+1} given $\mathbf{Y}^{(n)}$ through MCMC or via importance sampling as described in Section 6.2, then the forecasting density can be computed by averaging $p(y_{n+1}|s_{n+1})$ over those replicates.

In the case that the count time series is modeled under the assumptions as specified in (6.4) with S_t given by (6.7), then one can derive a rather nice expression for $E(Y_{n+1}|\mathbf{Y}^{(n)})$. To see this, condition first on $\mathbf{Y}^{(n)}$ and $\mathbf{S}^{(n+1)}$, which is the same as conditioning on $\boldsymbol{\alpha}^{(n+1)}$, and then using (6.4), we have

$$\begin{aligned} E(Y_{n+1}|\mathbf{Y}^{(n)}) &= E\left(E(Y_{n+1}|S_{n+1}, \mathbf{S}^{(n)}, \mathbf{Y}^{(n)})|\mathbf{Y}^{n)}\right) \\ &= E\left(E(Y_{n+1}|S_{n+1})|\mathbf{Y}^{n)}\right) \\ &= E\left(E(Y_{n+1}|\alpha_{n+1})|\mathbf{Y}^{n)}\right) \\ &= E\left(h(\alpha_{n+1})|\mathbf{Y}^{n)}\right), \end{aligned}$$

where $h\left(\alpha_{n+1}\right) = E(Y_{n+1}|\alpha_{n+1})$. Since $p\left(\mathbf{y}^{(n)}|\alpha_{n+1}, \boldsymbol{\alpha}^{(n)}\right) = p\left(\mathbf{y}^{(n)}|\boldsymbol{\alpha}^{(n)}\right)$, it follows that $p\left(\alpha_{n+1}|\mathbf{y}^{(n)}, \boldsymbol{\alpha}^{(n)}\right) = p\left(\alpha_{n+1}|\boldsymbol{\alpha}^{(n)}\right)$ and hence

$$E\left(Y_{n+1}|\mathbf{Y}^{(n)}\right) = E\left(E(h(\alpha_{n+1})|\boldsymbol{\alpha}^{(n)})|\mathbf{Y}^{n)}\right). \tag{6.36}$$

If the $\{\alpha_t\}$ process is Gaussian, then often $E\left(h\left(\alpha_{n+1}\right)|\boldsymbol{\alpha}^{(n)}\right)$ can be expressed as an explicit function of $\boldsymbol{\alpha}^{(n)}$. Hence, to compute the conditional expectation in (6.36), it is enough to generate a large number of replicates $\tilde{\alpha}_1^{(n)}, \ldots, \tilde{\alpha}_N^{(n)}$ computed from the conditional distribution of $\boldsymbol{\alpha}^{(n)}$ given $\mathbf{Y}^{(n)}$ and then approximate the conditional expectation by

$$E(Y_{n+1}|\mathbf{Y}^{(n)}) \approx \frac{\sum_{i=1}^{N} E\left(h(\alpha_{n+1})|\tilde{\alpha}_i^{(n)}\right)}{N}.$$

The same ideas can be applied for predicting lead times further into the future.

Example: Suppose Y_t given the state S_t is Poisson (e^{S_t}) and that S_t is the linear regression model with Gaussian AR(p) noise as given in (6.7). In this case,

$$h(\alpha_{n+1}) = E(Y_{n+1}|\alpha_{n+1}) = \exp\left\{\mathbf{x}_{n+1}^T \boldsymbol{\beta} + \alpha_{n+1}\right\}.$$

Since $\{\alpha_t\}$ is a stationary Gaussian time series with zero mean,

$$\alpha_{n+1}|\boldsymbol{\alpha}^{(n)} \sim N\left(\gamma_n^T V \boldsymbol{\alpha}^{(n)}, \gamma(0) - \gamma_n^T V \gamma_n\right), \tag{6.37}$$

where $\gamma(h) = \text{cov}(\alpha(0), \alpha(h))$ is the autocovariance function for $\{\alpha_t\}$, Γ_n is the covariance matrix for $\boldsymbol{\alpha}^{(n)}$ and $\gamma = \text{cov}(\alpha(n+1), \boldsymbol{\alpha}^{(n)})$ is a $1 \times n$ covariance vector. Using the $\{\alpha_t\}$ process in (6.37), we have

$$E(e^{\alpha_{n+1}}|\alpha^{(n)}) = \exp\left\{\gamma_n^T V \boldsymbol{\alpha}^{(n)} + \frac{1}{2}\left(\gamma(0) - \gamma_n^T V \gamma_n\right)\right\}$$

and hence

$$E(Y_{n+1}|\mathbf{Y}^{(n)}) = \exp\left\{\mathbf{x}_{n+1}^T \boldsymbol{\beta} + \frac{1}{2}\left(\gamma(0) - \gamma_n^T V \gamma_n\right)\right\} E\left(\exp\left\{\gamma_n^T V \boldsymbol{\alpha}^{(n)}\right\} |\mathbf{Y}^{(n)}\right). \tag{6.38}$$

In order to compute the righthand side of this equation, one needs to integrate $\exp\left\{\gamma_n^T V \boldsymbol{\alpha}^{(n)}\right\}$ relative to the conditional density $\boldsymbol{\alpha}^{(n)}|\mathbf{Y}^{(n)}$, which can be obtained using some of the same methods described in Section 6.2.

The conditional variance $\text{var}(Y_{n+1}|\mathbf{Y}^{(n)})$ can be computed using a similar development. First note that

$$\text{var}(Y_{n+1}|\mathbf{Y}^{(n)}) = E\left(\text{var}(Y_{n+1}|\alpha_{n+1}, \boldsymbol{\alpha}^{(n)}, \mathbf{Y}^{(n)})|\mathbf{Y}^{(n)}\right) + \text{var}\left(E(Y_{n+1}|\alpha_{n+1}, \boldsymbol{\alpha}^{(n)}, \mathbf{Y}^{(n)})|\mathbf{Y}^{(n)}\right). \tag{6.39}$$

Since the conditional mean and variance are the same in this example, the first term in (6.39) coincides with (6.38). As for the second term,

$$\begin{aligned}\text{var}\left(E(Y_{n+1}|\alpha_{n+1}, \boldsymbol{\alpha}^{(n)}, \mathbf{Y}^{(n)})|\mathbf{Y}^{(n)}\right) &= \text{var}\left(\exp\{\mathbf{x}_{n+1}\boldsymbol{\beta} + \alpha_{n+1}\}|\mathbf{Y}^{(n)}\right)\\ &= E\left(\exp\{2\mathbf{x}_{n+1}\boldsymbol{\beta} + 2\alpha_{n+1}\}|\mathbf{Y}^{(n)}\right) - E^2\left(\exp\{\mathbf{x}_{n+1}\boldsymbol{\beta} + \alpha_{n+1}\}|\mathbf{Y}^{(n)}\right)\end{aligned}$$

and hence

$$\begin{aligned}\text{var}(Y_{n+1}|\mathbf{Y}^{(n)}) &= E(Y_{n+1}|\mathbf{Y}^{(n)}) - E^2(Y_{n+1}|\mathbf{Y}^{(n)})\\ &\quad + \exp\left\{2\mathbf{x}_{n+1}^T \boldsymbol{\beta} + (\gamma(0) - \gamma_n^T V \gamma_n)\right\} E\left(\exp\left\{2\gamma_n^T V \boldsymbol{\alpha}^{(n)}\right\} |\mathbf{Y}^{(n)}\right).\end{aligned}$$

Acknowledgments

The work of the first author was supported in part by NSF grant DMS-1107031. Travel funds from a University of New South Wales Faculty of Science Research Fellowship for the first author were used in this collaboration.

References

Breslow, N. E. and Clayton, D. G. (1993). Approximate inference in generalized linear mixed models. *Journal of the American Statistical Association*, 88(421):9–25.

Brockwell, P. J. and Davis, R. A. (1991). *Time Series: Theory and Methods*. Springer, New York.

Brockwell, P. J. and Davis, R. A. (2002). *Introduction to Time Series and Forecasting*. Springer, New York.

Chan, K. and Ledolter, J. (1995). Monte Carlo em estimation for time series models involving counts. *Journal of the American Statistical Association*, 90(429):242–252.

Davis, R. A., Dunsmuir, W., and Wang, Y. (1999). Modeling time series of count data, *Asymptotics, Nonparametrics, and Time Series*, ed. S. Ghosh. Marcel-Dekker, New York.

Davis, R. A., Dunsmuir, W. T., and Wang, Y. (2000). On autocorrelation in a Poisson regression model. *Biometrika*, 87(3):491–505.

Davis, R. A. and Rodriguez-Yam, G. (2005). Estimation for state-space models based on a likelihood approximation. *Statistica Sinica*, 15(2):381–406.

Davis, R. A. and Rosenblatt, M. (1991). Parameter estimation for some time series models without contiguity. *Statistics and Probability Letters*, 11(6):515–521.

Davis, R. A. and Wu, R. (2009). A negative binomial model for time series of counts. *Biometrika*, 96(3):735–749.

Davis, R. A. and Yau, C. Y. (2011). Comments on pairwise likelihood in time series models. *Statistica Sinica*, 21(1):255.

De Jong, P. and Shephard, N. (1995). The simulation smoother for time series models. *Biometrika*, 82(2):339–350.

Doukhan, P. (1994). *Mixing: Properties and Examples*. Springer, New York.

Durbin, J. and Koopman, S. J. (1997). Monte carlo maximum likelihood estimation for non-Gaussian state space models. *Biometrika*, 84(3):669–684.

Durbin, J. and Koopman, S. J. (2000). Time series analysis of non-Gaussian observations based on state space models from both classical and Bayesian perspectives. *Journal of the Royal Statistical Society: Series B (Statistical Methodology)*, 62(1):3–56.

Durbin, J. and Koopman, S. J. (2012). *Time Series Analysis by State Space Methods*. Oxford University Press, Oxford, U.K.

Fahrmeir, L. (1992). Posterior mode estimation by extended Kalman filtering for multivariate dynamic generalized linear models. *Journal of the American Statistical Association*, 87(418): 501–509.

Fokianos, K. (2015). Statistical analysis of count time series models: A GLM perspective. In R. A. Davis, S. H. Holan, R. Lund and N. Ravishanker, eds., *Handbook of Discrete-Valued Time Series*, pp. 3–28. Chapman & Hall, Boca Raton, FL.

Gamerman, D., Abanto-Valle, C. A., Silva, R. S., and Martins, T.G. (2015). Dynamic Bayesian models for discrete-valued time series. In R. A. Davis, S. H. Holan, R. Lund and N. Ravishanker, eds., *Handbook of Discrete-Valued Time Series*, pp. 165–186. Chapman & Hall, Boca Raton, FL.

Jung, R. C., Kukuk, M., and Liesenfeld, R. (2006). Time series of count data: modeling, estimation and diagnostics. *Computational Statistics and Data Analysis*, 51(4):2350–2364.

Jung, R. C. and Liesenfeld, R. (2001). Estimating time series models for count data using efficient importance sampling. *AStA Advances in Statistical Analysis*, 85(4):387–408.

Jungbacker, B. and Koopman, S. J. (2007). Monte Carlo estimation for nonlinear non-Gaussian state space models. *Biometrika*, 94(4):827–839.

Koopman, S. J., Lucas, A., and Scharth, M. (2015). Numerically accelerated importance sampling for nonlinear non-Gaussian state space models. *Journal of Business and Economic Statistics*, 33(1): 114–127.

Kuk, A. Y. (1995). Asymptotically unbiased estimation in generalized linear models with random effects. *Journal of the Royal Statistical Society. Series B (Methodological)*, 57(2):395–407.

Kuk, A. Y. and Cheng, Y. W. (1999). Pointwise and functional approximations in monte carlo maximum likelihood estimation. *Statistics and Computing*, 9(2):91–99.

McCullagh, P. and Nelder, J. A. (1989). *Generalized Linear Models (Monographs on Statistics and Applied Probability 37)*. Chapman & Hall, London, U.K.

McCulloch, C. E. (1997). Maximum likelihood algorithms for generalized linear mixed models. *Journal of the American statistical Association*, 92(437):162–170.

Nelson, K. P. and Leroux, B. G. (2006). Statistical models for autocorrelated count data. *Statistics in Medicine*, 25(8):1413–1430.

Ng, C. T., Joe, H., Karlis, D., and Liu, J. (2011). Composite likelihood for time series models with a latent autoregressive process. *Statistica Sinica*, 21(1):279.

Oh, M.-S. and Lim, Y. B. (2001). Bayesian analysis of time series Poisson data. *Journal of Applied Statistics*, 28(2):259–271.

Politis, D. N. et al. (2003). The impact of bootstrap methods on time series analysis. *Statistical Science*, 18(2):219–230.

R Core Team. (2014). *R: A Language and Environment for Statistical Computing*. R Foundation for Statistical Computing, Vienna, Austria.

Richard, J.-F. and Zhang, W. (2007). Efficient high-dimensional importance sampling. *Journal of Econometrics*, 141(2):1385–1411.

Shephard, N. and Pitt, M. K. (1997). Likelihood analysis of non-Gaussian measurement time series. *Biometrika*, 84(3):653–667.

Shumway, R. H. and Stoffer, D. S. (2011). *Time Series Analysis and Its Applications: With R Examples*. Springer, New York.

Shun, Z. and McCullagh, P. (1995). Laplace approximation of high dimensional integrals. *Journal of the Royal Statistical Society. Series B (Methodological)*, 57(4):749–760.

Sinescu, V., Kuo, F. Y., and Sloan, I. H. (2014). On the choice of weights in a function space for quasi-monte carlo methods for a class of generalised response models in statistics. *Springer Proceedings in Mathematics and Statistics*, nizer) Gerhard Larcher, Johannes Kepler University Linz, Austria Pierre LEcuyer, Université de Montréal, Canada Christiane Lemieux, University of Waterloo, Canada Peter Mathé, Weierstrass Institute, Berlin, Germany, p. 597.

Skaug, H. J. (2002). Automatic differentiation to facilitate maximum likelihood estimation in nonlinear random effects models. *Journal of Computational and Graphical Statistics*, 11(2):458–470.

Skaug, H. J. and Fournier, D. A. (2006). Automatic approximation of the marginal likelihood in non-Gaussian hierarchical models. *Computational Statistics and Data Analysis*, 51(2):699–709.

Thavaneswaran, A. and Ravishanker, N. (2015). Estimating equation approaches for integer-valued time series models. In R. A. Davis, S. H. Holan, R. Lund and N. Ravishanker, eds., *Handbook of Discrete-Valued Time Series*, pp. 145–164. Chapman & Hall, Boca Raton, FL.

Tjøstheim, D. (2015). Count time series with observation-driven autoregressive parameter dynamics. In R. A. Davis, S. H. Holan, R. Lund and N. Ravishanker, eds., *Handbook of Discrete-Valued Time Series*, pp. 77–100. Chapman & Hall, Boca Raton, FL.

Wu, R. (2012). On variance estimation in a negative binomial time series regression model. *Journal of Multivariate Analysis*, 112:145–155.

Wu, R. and Cui, Y. (2013). A parameter-driven logit regression model for binary time series. *Journal of Time Series Analysis*, 35(5):462–477.

Zeger, S. L. (1988). A regression model for time series of counts. *Biometrika*, 75(4):621–629.

7

Estimating Equation Approaches for Integer-Valued Time Series Models

Aerambamoorthy Thavaneswaran and Nalini Ravishanker

CONTENTS

7.1 Introduction

There is considerable current interest in the study of integer-valued time series models, and for time series of counts, in particular. Applications abound in biometrics, ecology, economics, engineering, finance, public health, etc. Given the increase in stochastic complexity and data sizes, there is a need for developing fast and optimal approaches for model inference and prediction. Several observation-driven and parameter-driven (Cox, 1981) modeling frameworks for count time series have been discussed over the past few decades. Further, although there is a large literature for count time series without zero-inflation, including both observation-driven and parameter-driven models, very few papers have been published for modeling time series with excess zeros.

In parameter-driven models, temporal association is modeled indirectly by specifying the parameters in the conditional distribution of the count random variable to be a function of a correlated latent stochastic process (West and Harrison, 1997). In observation-driven models, temporal association is modeled directly via lagged values of the count variable, adopting strategies such as binomial thinning to preserve the integer nature of the data (Al-Osh and Alzaid, 1987; McKenzie, 2003). Davis et al. (2003), Jung and Tremayne (2006), and Neal and Subba Rao (2007), among others, have discussed estimation and inference for these models. Heinen (2003) and Ghahramani and Thavaneswaran (2009b) described autoregressive conditional Poisson (ACP) models. Ferland et al. (2006) and Zhu (2011, 2012a,b) defined classes of integer-valued time series models following different conditional distributions, which they called INGARCH models, and studied the first two process moments. Although these are called INGARCH models, only the conditional mean of the count variable is modeled, and not its conditional variance. In a recent paper, Creal et al. (2013) described generalized autoregressive score (GAS) models to study time-varying parameters in an observation-driven modeling framework, while MacDonald and Zucchini (2015; Chapter 12 in this volume) discussed a hidden Markov modeling framework.

Estimating functions (EFs) have a long history in statistical inference. For instance, Fisher (1924) showed that maximum likelihood and minimum chi-squared methods are asymptotically equivalent by comparing the first order conditions of the two estimation procedures, that is, by analyzing properties of estimators by focusing on the corresponding EFs rather than on the objective functions or estimators themselves. Godambe (1960) and Durbin (1960) gave a fundamental optimality result for EFs for the scalar parameter case. Following Godambe (1985), who first studied inference based on the EF approach for discrete-time stochastic processes, Thavaneswaran and Abraham (1988) described estimation for nonlinear time series models using linear EFs. Naik-Nimbalkar and Rajarshi (1995) and Thavaneswaran and Heyde (1999) studied problems in filtering and prediction using linear EFs in the Bayesian context. Merkouris (2007), Ghahramani and Thavaneswaran (2009a, 2012), and Thavaneswaran et al. (2015), among others, studied estimation for time series via the combined EF approach. Bera et al. (2006) gave an excellent survey on the historical development of this topic.

Except for a few papers, (Dean, 1991), who discussed estimating equations for mixed Poisson models given independent observations, application of the EF approach to count time series is still largely unexplored. In the following sections, we extend this approach for count time series models. For some recently proposed integer-valued time series models (such as the Poisson, generalized Poisson (GP), zero-inflated Poisson, or negative binomial models), the conditional mean and variance are functions of the same parameter. This motivates considering more informative quadratic EFs for joint estimation of the conditional mean and variance parameters, rather than only using linear EFs. It is also possible to derive closed form expressions for the information gain (Thavaneswaran et al., 2015).

In this chapter, we describe a framework for optimal estimation of parameters in integer-valued time series models via martingale EFs and illustrate the approach for some interesting count time series models. The EF approach only relies on a specification of the first few moments of the random variable at each time conditional on its history, and does not require specification of the form of the conditional probability distribution. We start with a brief review of the general theory of EFs in Section 7.2. In Section 7.3, we describe the conditional moment properties for some recently proposed classes of generalized integer-valued models, such as those discussed in Ferland et al. (2006). Specifically, we derive the first four conditional moments, which are typically required for carrying out inference on model parameters using the theory of combined martingale EFs (Liang et al., 2011). Section 7.4 describes the optimal EFs that enable *joint parameter estimation* for such models. We also derive fast, recursive, on-line estimation techniques for parameters of interest and provide examples. In Section 7.5, we describe how hypothesis testing based on optimal estimation facilitates model choice. Section 7.6 concludes with a summary and a brief discussion of parameter-driven doubly stochastic models for count time series.

7.2 A Review of Estimating Functions (EFs)

Godambe (1985) first described an EF approach for stochastic process inference. Suppose that $\{\mathbf{y}_t, t = 1, \ldots, n\}$ is a realization of a discrete time stochastic process, and suppose its conditional distribution depends on a vector parameter $\boldsymbol{\theta}$ belonging to an open subset $\boldsymbol{\Theta}$ of the p-dimensional Euclidean space, with $p \ll n$. Let $(\Omega, \mathcal{F}, P_{\boldsymbol{\theta}})$ denote the underlying probability space, and let $\mathcal{F}_t$ be the σ-field generated by $\{\mathbf{y}_1, \ldots, \mathbf{y}_t, t \geq 1\}$. Let $\mathbf{m}_t = \mathbf{m}_t(\mathbf{y}_1, \ldots, \mathbf{y}_t, \boldsymbol{\theta}), 1 \leq t \leq n$, be specified q-dimensional martingale difference vectors. Consider the class $\mathcal{M}$ of zero-mean, square integrable p-dimensional martingale EFs, viz.,

$$\mathcal{M} = \left\{ \mathbf{g}_n(\boldsymbol{\theta}) : \mathbf{g}_n(\boldsymbol{\theta}) = \sum_{t=1}^{n} \mathbf{a}_{t-1}(\boldsymbol{\theta})\mathbf{m}_t \right\}, \tag{7.1}$$

where $\mathbf{a}_{t-1}(\boldsymbol{\theta})$ are $p \times q$ matrices that are functions of $\boldsymbol{\theta}$ and $\mathbf{y}_1, \ldots, \mathbf{y}_{t-1}$, $1 \leq t \leq n$. It is further assumed that $\mathbf{g}_n(\boldsymbol{\theta})$ are almost surely differentiable with respect to the components of $\boldsymbol{\theta}$, and are such that for each $n \geq 1$, $E\left(\frac{\partial \mathbf{g}_n(\boldsymbol{\theta})}{\partial \boldsymbol{\theta}} \middle| \mathcal{F}_{n-1} \right)$ and $E(\mathbf{g}_n(\boldsymbol{\theta})\mathbf{g}_n(\boldsymbol{\theta})' | \mathcal{F}_{n-1})$ are nonsingular for all $\boldsymbol{\theta} \in \boldsymbol{\Theta}$, where all expectations are taken with respect to $P_{\boldsymbol{\theta}}$. An estimator of $\boldsymbol{\theta}$ is obtained by solving the estimating equation $\mathbf{g}_n(\boldsymbol{\theta}) = \mathbf{0}$. Furthermore, the $p \times p$ matrix $E(\mathbf{g}_n(\boldsymbol{\theta})\mathbf{g}_n(\boldsymbol{\theta})' | \mathcal{F}_{n-1})$ is assumed to be positive definite for all $\boldsymbol{\theta} \in \boldsymbol{\Theta}$. Then, in the class of all zero-mean and square integrable martingale EFs $\mathcal{M}$, the optimal EF $\mathbf{g}_n^*(\boldsymbol{\theta})$ that maximizes, in the partial order of nonnegative definite matrices, the information

$$\mathbf{I}_{\mathbf{g}_n}(\boldsymbol{\theta}) = \left[E\left(\frac{\partial \mathbf{g}_n(\boldsymbol{\theta})}{\partial \boldsymbol{\theta}} \middle| \mathcal{F}_{n-1} \right) \right]' \left[E(\mathbf{g}_n(\boldsymbol{\theta})\mathbf{g}_n(\boldsymbol{\theta})' | \mathcal{F}_{n-1}) \right]^{-1} \left[E\left(\frac{\partial \mathbf{g}_n(\boldsymbol{\theta})}{\partial \boldsymbol{\theta}} \middle| \mathcal{F}_{n-1} \right) \right],$$

is given by

$$\mathbf{g}_n^*(\boldsymbol{\theta}) = \sum_{t=1}^{n} \mathbf{a}_{t-1}^*(\boldsymbol{\theta})\mathbf{m}_t = \sum_{t=1}^{n} \left[E\left(\frac{\partial \mathbf{m}_t}{\partial \boldsymbol{\theta}} \middle| \mathcal{F}_{t-1} \right) \right]' \left[E(\mathbf{m}_t\mathbf{m}_t' | \mathcal{F}_{t-1}) \right]^{-1} \mathbf{m}_t, \tag{7.2}$$

and the corresponding optimal information reduces to

$$\mathbf{I}_{\mathbf{g}_n^*}(\boldsymbol{\theta}) = E(\mathbf{g}_n^*(\boldsymbol{\theta})\mathbf{g}_n^*(\boldsymbol{\theta})' | \mathcal{F}_{n-1}). \tag{7.3}$$

The function $\mathbf{g}_n^*(\boldsymbol{\theta})$ is also called the "quasi-score" and has properties similar to those of a score function: $E(\mathbf{g}_n^*(\boldsymbol{\theta})) = \mathbf{0}$ and $E(\mathbf{g}_n^*(\boldsymbol{\theta})\mathbf{g}_n^*(\boldsymbol{\theta})') = -E(\partial \mathbf{g}_n^*(\boldsymbol{\theta})/\partial \boldsymbol{\theta}')$. This is a general result in that we do not need to assume that the true underlying conditional distribution belongs to an exponential family of distributions. The maximum correlation between the optimal EF and the true unknown score justifies the terminology "quasi-score" for $\mathbf{g}_n^*(\boldsymbol{\theta})$. It is useful to note that the same procedure for derivation of optimal estimating equations may be used when the time series is stationary or nonstationary. Moreover, the finite sample properties of the EFs remain the same, although asymptotic properties will differ. In Chapter 12 of his book, Heyde (1997) discussed general consistency and asymptotic distributional results.

Consider an integer-valued discrete-time scalar stochastic process $\{y_t, t = 1, 2, \ldots\}$ with conditional mean, variance, skewness, and kurtosis given by

$$\mu_t(\theta) = \mathrm{E}\left(y_t|\mathcal{F}_{t-1}\right),$$

$$\sigma_t^2(\theta) = \mathrm{Var}\left(y_t|\mathcal{F}_{t-1}\right),$$

$$\gamma_t(\theta) = \frac{1}{\sigma_t^3(\theta)}\mathrm{E}\left(\left(y_t - \mu_t(\theta)\right)^3|\mathcal{F}_{t-1}\right), \text{ and}$$

$$\kappa_t(\theta) = \frac{1}{\sigma_t^4(\theta)}\mathrm{E}\left(\left(y_t - \mu_t(\theta)\right)^4|\mathcal{F}_{t-1}\right). \tag{7.4}$$

To jointly estimate the conditional mean and variance, which are both functions of θ, Liang et al. (2011) defined optimal combined EFs. We assume that $\mu_t(\theta)$ and $\sigma_t^2(\theta)$ are differentiable with respect to θ, and that the skewness and kurtosis of the standardized y_t do not depend on additional parameters beyond θ. For each data/model combination, our estimation approach for θ requires (1) computation of the first four moments of y_t conditional on the process history, (2) selection of suitable linear and/or quadratic martingale differences, (3) construction of optimal combined EFs, and (4) derivation of recursive estimators of θ when possible. In Section 7.4, we describe optimal estimating equations for θ for some of the integer-valued models discussed in Section 7.3.

7.3 Models and Moment Properties for Count Time Series

Several models have been discussed in the literature for count time series, where parameter estimation using maximum likelihood or Bayesian approaches have been described. For the estimating equations framework described in this chapter, we start from the conditional moments of the process $\{y_t\}$ given the history $\mathcal{F}_{t-1}$. The conditional moments are assumed to be functions of an unknown parameter vector θ and form the basis for constructing the optimal estimating equation. For simplicity, we suppress θ in the notation for the conditional moments and other derived quantities in the following examples. Consider the discrete-time model for μ_t with $P + Q + 1$ parameters defined by

$$\mu_t = \delta + \sum_{i=1}^{P} \alpha_i y_{t-i} + \sum_{j=1}^{Q} \beta_j \mu_{t-j}, \tag{7.5}$$

where $\delta > 0$, $\alpha_i \geq 0$ for $i = 1, \ldots, P$ and $\beta_j \geq 0$ for $j = 1, \ldots, Q$. Let $\theta = (\delta, \boldsymbol{\alpha}', \boldsymbol{\beta}')'$ where $\boldsymbol{\alpha} = (\alpha_1, \ldots, \alpha_P)'$ and $\boldsymbol{\beta} = (\beta_1, \ldots, \beta_Q)'$. We assume that the conditional variance σ_t^2 as well as μ_t depend on θ, and that the conditional skewness γ_t and conditional kurtosis κ_t are available and do not depend on any additional parameters. The higher order conditional moment properties for the models described in Sections 7.3.1 and 7.3.2, especially for the zero-inflated case, are obtained using Mathematica. Section 7.3.3 proposes a model in the framework of the GAS models of Creal et al. (2013).

Equation (7.5) posits an ARMA model for $\{y_t\}$. This ARMA representation is useful for obtaining unconditional moments such as skewness and kurtosis under the stationarity assumption and is often useful in model identification in data analysis. We consider the martingale difference $m_t = y_t - \mu_t$, with conditional mean 0 and conditional variance σ_t^2. Then (7.5) can be written as

$$y_t - m_t = \delta + \sum_{i=1}^{P} \alpha_i y_{t-i} + \sum_{j=1}^{Q} \beta_j (y_{t-j} - m_{t-j}).$$

Rearranging terms and simplifying, we can write

$$\left(1 - \sum_{i=1}^{\max(P,Q)} (\alpha_i + \beta_i) B^i\right) y_t = \delta + \left(1 - \sum_{j=1}^{Q} \beta_j B^j\right) m_t, \text{ or}$$

$$\phi(B) y_t = \delta + \beta(B) m_t,$$

where B denotes the backshift operator. That is, (7.5) can be written as an ARMA model for $\{y_t\}$ with $\phi(B) = 1 - \sum_{i=1}^{\max(P,Q)} \phi_i B^i$, $\phi_i = \alpha_i + \beta_i$, $\beta(B) = 1 - \sum_{i=1}^{Q} \beta_i B^i$, and $\psi(B)\phi(B) = \beta(B)$ with $\psi(B) = 1 + \sum_{i=1}^{\infty} \psi_i B^i$. Similar to the continuous-valued case (Gourieroux, 1997), this model has the same second-order properties as an INARMA($\max(P,Q), Q$) model. When all solutions to $\phi(z) = 0$ lie outside the unit circle, we may write the moving average representation of the causal process as $y_t = \mu + \psi(B) m_t$, where $\psi(B) = \beta(B)/\phi(B)$ and $\mu = \delta/(1 - \phi_1 - \ldots - \phi_{\max(P,Q)})$ is the marginal mean of y_t. The lag k autocovariance and autocorrelation of the process are, respectively, $\gamma_k^{(y)} = E(\sigma_t^2) \sum_{j=0}^{\infty} \psi_j \psi_{j+k}$ and $\rho_k^{(y)} = \gamma_k^{(y)}/\gamma_0^{(y)} = \sum_{j=0}^{\infty} \psi_j \psi_{j+k} / \sum_{j=0}^{\infty} \psi_j^2$, where $E(\sigma_t^2)$ is the unconditional variance of $\{y_t\}$. Note that the temporal correlation $\rho_k^{(y)}$ depends only on the model parameters in (7.5) and not on the conditional distribution of the observed process $\{y_t\}$. Also, the kurtosis of $\{y_t\}$ is given by

$$K^{(y)} = 3 + \frac{(K^{(m)} - 3) \sum_{j=0}^{\infty} \psi_j^4}{\left(\sum_{j=0}^{\infty} \psi_j^2\right)^2}, \tag{7.6}$$

where $K^{(m)} = E(m_t^4)/[E(m_t^2)]^2$. These results follow directly from properties of stationary ARMA processes and often provide guidance in model order choice. By substituting suitable values of ψ_j, we can derive the kurtosis for the integer-valued processes discussed in the following sections.

7.3.1 Models for Nominally Dispersed Counts

Considerable attention has been paid in the literature for modeling count time series via observation-driven models (Zeger and Qaqish, 1988; Davis et al., 2003) and parameter-driven models (Chan and Ledolter, 1995; West and Harrison, 1997). We consider three examples.

Example 7.1

Suppose that the conditional mean, variance, skewness, and kurtosis of y_t are specified as $\mu_t = \lambda_t$, $\sigma_t^2 = \lambda_t$, $\gamma_t = \lambda_t^{-1/2}$, and $\kappa_t = \lambda_t^{-1}$, and suppose that μ_t is modeled by (7.5). These moments match the first four moments of y_t generated by what Ferland et al. (2006) referred to as a Poisson INGARCH process, which assumes that $y_t|\mathcal{F}_{t-1} \sim \text{Poisson}(\lambda_t)$, so that λ_t is the conditional variance as well as the conditional mean. While it seems a misnomer to use the term INGARCH for modeling the conditional mean and not the conditional variance as GARCH models do, the form of (7.5) is similar to the normal-GARCH model (Bollerslev, 1986), where $y_t|\mathcal{F}_{t-1} \sim N(0, \sigma_t^2)$ for all t, and the model for the conditional variance σ_t^2 follows the right side of (7.5), subject to the same conditions on the parameters. For conformity with the literature, we use the term INGARCH in this chapter. The moments of the Poisson INGARCH random variable y_t are easily derived from the probability generating function $G_y(s) = E(\exp(sy_t)|\mathcal{F}_{t-1}) = \exp[\lambda_t(s-1)]$. Implementation of the EF approach does not require that at each time t, y_t has a conditional Poisson distribution, but only requires specification of the conditional moments of $y_t|\mathcal{F}_{t-1}$ for each t. Such moment specifications are also sufficient for the other INGARCH models described in this chapter. □

Example 7.2

Suppose the conditional mean, variance, skewness, and kurtosis of y_t given $\mathcal{F}_{t-1}$ are $\mu_t = \lambda_t^*/(1-\tau) = \lambda_t$, $\sigma_t^2 = \lambda_t^*/(1-\tau)^3 = \tau^{*2}\lambda_t$, $\gamma_t = (1+2\tau)/\sqrt{\lambda_t^*(1-\tau)}$, and $\kappa_t = (1 + 8\tau + 6\tau^2)/[\lambda_t^*(1-\tau)]$, corresponding to moments from the GP INGARCH process (Zhu, 2012a), where $\tau^* = 1/(1-\tau)$. This process is defined as $y_t|\mathcal{F}_{t-1} \sim GP(\lambda_t^*, \tau)$, $\lambda_t^* = (1-\tau)\lambda_t$, $\max(-1, -\lambda_t^*/4) < \tau < 1$, and the conditional mean is again modeled by (7.5). The GP distribution for y_t conditional on $\mathcal{F}_{t-1}$ is

$$P(y_t = k|\mathcal{F}_{t-1}) = \begin{cases} \lambda_t(\lambda_t + \tau k)^{k-1}\exp[-(\lambda_t + \tau k)]/k!, & k = 0,1,2,\ldots \\ 0, & k > m \text{ if } \tau < 0, \end{cases} \tag{7.7}$$

where m is the largest positive integer for which $\lambda_t + \tau m > 0$ when $\tau < 0$. To derive the conditional moments of the GP distribution shown above, we can use the recursive relation for the rth raw moment $\mu(r)$, that is, $(1-\tau)\mu(r) = \lambda_t\mu(r-1) + \lambda_t\frac{\partial\mu(r)}{\partial\lambda_t} + \tau\frac{\partial\mu(r)}{\partial\tau}$, where $\lambda_t > 0$ and $\max(-1, -\lambda_t/m) < \tau < 1$. □

Example 7.3

For $p_t = 1/(1+\lambda_t)$ and $q_t = 1 - p_t$, suppose the conditional mean, variance, skewness, and kurtosis of y_t given $\mathcal{F}_{t-1}$ are $\mu_t = rq_t/p_t = r\lambda_t$, $\sigma_t^2 = rq_t/p_t^2$, $\gamma_t = (2-p_t)/(rq_t)^{1/2}$, and $\kappa_t = (p_t^2 - 6p_t + 6)/rq_t$, which correspond to the moments of a negative binomial INGARCH process, where $y_t|\mathcal{F}_{t-1} \sim NB(r, \lambda_t)$, the conditional mean is modeled by (7.5) as before, and the probability generating function is given by $G_y(s) = p_t^r/(1-q_t s)^r$, and the conditional probability mass function (pmf) of y_t has the form

$$P(y_t = k|\mathcal{F}_{t-1}) = \binom{k+r-1}{r-1} p_t^r q_t^k,\ k = 0,1,2,\ldots. \square \tag{7.8}$$

7.3.2 Models for Counts with Excess Zeros

In several applications, observed counts over time may show an excess of zeros, and the usual Poisson or negative binomial models are inadequate. One example in the area of public health could involve surveillance of a rare disease over time, where the observed counts typically show zero inflation. Yang (2012) studied statistical modeling for time series with excess zeros. We consider two examples.

Example 7.4

When count time series are observed with an excess of zeros in applications, we may assume a specification of the conditional mean, variance, skewness, and kurtosis of y_t given $\mathcal{F}_{t-1}$ as $\mu_t = (1-\omega)\lambda_t$, $\sigma_t^2 = (1-\omega)(1+\omega\lambda_t)\lambda_t$, $\gamma_t = ((1-\omega)\lambda_t)^{-1/2}(1+\omega\lambda_t)^{-3/2}(1+\omega\lambda_t(3+2\omega\lambda_t+\lambda_t))$, and $\kappa_t = [(1-\omega)(1+\omega\lambda_t)^2\lambda_t]^{-1}[1+\omega\lambda_t(7+\lambda_t(-6+12\omega+\lambda_t(1-6(1-\omega)\omega)))]$, where $0 < \omega < 1$, and we assume that the conditional mean is modeled by (7.5). The moments correspond to the first four moments of a zero-inflated Poisson INGARCH process (Zhu, 2012b) given by $y_t|\mathcal{F}_{t-1} \sim ZIP(\lambda_t, \omega)$ defined by

$$P(y_t = k|\mathcal{F}_{t-1}) = \omega\Delta_{k,0} + (1-\omega)\lambda_t^k \exp(-\lambda_t)/k!, \tag{7.9}$$

for $k = 0, 1, 2, \ldots$. When $\omega = 0$, the ZIP INGARCH model reduces to the Poisson INGARCH model. The probability generating function of a ZIP random variable is given by $G_y(s) = \omega + (1-\omega)\exp[\lambda_t(s-1)]$. The conditional variance exceeds the conditional mean, so the ZIP-INGARCH model can handle overdispersion. As mentioned earlier, in this and the following models, we only require specification of the conditional moments of $y_t|\mathcal{F}_{t-1}$ for each t (and no marginal distributional assumptions). □

Example 7.5

An alternate specification for count time series with an excess of zeros corresponds to the conditional moment specifications:

$$\mu_t = \frac{(1-\omega)rq_t}{p_t},$$

$$\sigma_t^2 = \frac{(1-\omega)rq_t(1+r\omega q_t)}{p_t^2},$$

$$\gamma_t = \frac{r\omega q_t(3-rq_t+2\omega rq_t)+2-p_t}{[(1-\omega)rq_t]^{1/2}[1-\omega rq_t]^{3/2}},$$

$$\kappa_t = \frac{1}{(1-\omega)rq_t(1-\omega rq_t)^2}\Big\{\omega rq_t\Big(11-4p_t-6rq_t+r^2q_t^2\Big) + 6\omega^2r^2q_t^2(2-rq_t)+6r^3\omega^3q_t^3+6-6p_t+p_t^2\Big\}.$$

These are the first four moments of a zero-inflated negative binomial INGARCH process (Zhu, 2012b). Here, $y_t|\mathcal{F}_{t-1} \sim ZINB(\lambda_t, \alpha, \omega)$, the conditional mean is again given by (7.5), and $\alpha \geq 0$ is the dispersion parameter. The conditional ZINB distribution is defined by

$$P(y_t = k|\mathcal{F}_{t-1}) = \omega\Delta_{k,0} + (1-\omega)\frac{\Gamma(k+\lambda_t^{1-c}/\alpha)}{k!\Gamma(\lambda_t^{1-c}/\alpha)}\left(\frac{1}{1+\alpha\lambda^c}\right)^{\lambda_t^{1-c}/\alpha}\left(\frac{\alpha\lambda_t^c}{1+\alpha\lambda_t^c}\right)^k,$$

for $k = 0, 1, 2, \ldots$, and where c is an index that assumes the values 0 or 1 and identifies the form of the underlying negative binomial distribution. The probability generating function is $G_y(s) = \omega + (1-\omega)p_t^r/(1-q_ts)^r$. In comparison to the negative binomial distribution shown in (7.8), we have $p_t = 1/\alpha + \lambda_t^c$, $q_t = 1 - p_t$, and $r = \lambda_t/\alpha$. Note that in the limit as $\alpha \to \infty$, the ZINB-INGARCH model reduces to the ZIP-INGARCH model, and when $\omega = 0$, the model reduces to the NB-INGARCH model. □

7.3.3 Models in the GAS Framework

Recently, Creal et al. (2013) proposed a novel observation-driven modeling strategy for time series, that is, the GAS model. Following their approach, we propose an extension of the GAS model for an integer-valued time series $\{y_t\}$ with specified first four conditional moments, and describe the use of estimating equations. Let $\mathbf{f}_t = \mathbf{f}_t(\boldsymbol{\theta})$ denote a vector-valued time-varying function of an unknown vector-valued parameter $\boldsymbol{\theta}$, and suppose that the evolution of $\mathbf{f}_t$ is determined by an autoregressive updating equation with an innovation $\mathbf{s}_t$, which is a suitably chosen martingale difference vector:

$$\mathbf{f}_t = \boldsymbol{\omega} + \sum_{i=1}^{P} \mathbf{A}_i \mathbf{s}_{t-i} + \sum_{j=1}^{Q} \mathbf{B}_j \mathbf{f}_{t-j}. \tag{7.10}$$

Suppose that $\boldsymbol{\omega} = \boldsymbol{\omega}(\boldsymbol{\theta})$, $\mathbf{A}_i = \mathbf{A}_i(\boldsymbol{\theta})$, and $\mathbf{B}_j = \mathbf{B}_j(\boldsymbol{\theta})$. For instance, $\mathbf{f}_t$ could represent the conditional mean μ_t of y_t or its conditional variance σ_t^2.

Suppose a valid conditional probability distribution $p(y_t|\mathcal{F}_{t-1}, \mathbf{f}_t; \boldsymbol{\theta})$ is specified (rather than just assuming the form of the first few moments). Let $\mathbf{s}_t$ correspond to the standardized score function, that is, $\mathbf{s}_t = \mathbf{S}_t \boldsymbol{\nabla}_t$, where

$$\boldsymbol{\nabla}_t = \frac{\partial}{\partial \mathbf{f}_t} \log p(y_t|\mathcal{F}_{t-1}, \mathbf{f}_t; \boldsymbol{\theta}) \text{ and}$$

$$\mathbf{S}_t = -E\left[\frac{\partial}{\partial \mathbf{f}_t \partial \mathbf{f}_t'} \log p(y_t|\mathcal{F}_{t-1}, \mathbf{f}_t; \boldsymbol{\theta})\right]^{-1}.$$

Then (7.10) corresponds to the GAS(P, Q) model discussed by Creal et al. (2013). Alternate specifications have been suggested for $\mathbf{S}_t$ that scale $\boldsymbol{\nabla}_t$ in addition to the inverse information given above. These include the positive square root of the inverse information or the identity matrix.

7.4 Parametric Inference via EFs

In Section 7.4.1, we describe the estimation of the parameter vector $\boldsymbol{\theta}$ in integer-valued time series models via linear estimating equations and give a recursive scheme for fast optimal estimation. In Section 7.4.2, we describe a combined EF approach based on linear and quadratic martingale differences, and show that these combined EFs are more informative when the conditional mean and variance of the observed process depend on the same parameter.

7.4.1 Linear EFs

Consider the class $\mathcal{M}$ of all unbiased EFs $g(m_t(\boldsymbol{\theta}))$ based on the martingale difference $m_t(\boldsymbol{\theta}) = y_t - \mu_t(\boldsymbol{\theta})$. Theorem 7.1 gives the optimal linear estimating equation with corresponding optimal information and the form of an approximate recursive estimator of $\boldsymbol{\theta}$ which is based on a first order Taylor approximation.

Theorem 7.1 *The optimal linear estimating equation and corresponding information are obtained by substituting $m_t(\theta) = y_t - \mu_t(\theta)$ into (7.2) and (7.3). The recursive estimator for θ is given by*

$$\widehat{\theta}_t = \widehat{\theta}_{t-1} + \mathbf{K}_t \mathbf{a}^*_{t-1}(\widehat{\theta}_{t-1}) g(m_t(\widehat{\theta}_{t-1})),$$

$$\mathbf{K}_t = \mathbf{K}_{t-1}\left(\mathbf{I}_p - \left[\mathbf{a}^*_{t-1}(\widehat{\theta}_{t-1})\frac{\partial g(m_t(\widehat{\theta}_{t-1}))}{\partial \theta'} + \frac{\partial \mathbf{a}^*_{t-1}(\widehat{\theta}_{t-1})}{\partial \theta} g(m_t(\widehat{\theta}_{t-1}))\right]\mathbf{K}_{t-1}\right)^{-1}, \quad (7.11)$$

where $\mathbf{I}_p$ is the identity matrix. If $g(x) = x$, then

$$\mathbf{a}^*_{t-1}(\theta) = \frac{\partial \mu_t(\theta)/\partial \theta}{Var(g(m_t(\theta))|\mathcal{F}_{t-1})},$$

while for any other function g (such as the score function),

$$\mathbf{a}^*_{t-1}(\theta) = \frac{[\partial \mu_t(\theta)/\partial \theta][\partial g(m_t(\theta))/\partial m_t(\theta)]}{Var(g(m_t(\theta))|\mathcal{F}_{t-1})}.$$

The proof is similar to that in Thavaneswaran and Heyde (1999) for the scalar parameter case. □

Corollary 7.1 is a special case for the scalar parameter case, where a^*_{t-1} does not depend on θ, while Corollary 7.2 discusses a nonlinear time series model.

Corollary 7.1 (Thavaneswaran and Heyde, 1999). For the class $\mathcal{M}$ of all unbiased EFs $g(m_t(\theta))$ based on the martingale difference $m_t(\theta) = y_t - \mu_t(\theta)$, let $\mu_t(\theta)$ be differentiable with respect to θ. The recursive estimator for θ is given by

$$\widehat{\theta}_t = \widehat{\theta}_{t-1} + \frac{K_{t-1} a^*_{t-1} g(m_t(\theta))}{1 + [\partial \mu_t(\widehat{\theta}_{t-1})/\partial \theta] K_{t-1} a^*_{t-1}}, \text{ where}$$

$$K_t = \frac{K_{t-1}}{1 + [\partial \mu_t(\widehat{\theta}_{t-1})/\partial \theta] K_{t-1} a^*_{t-1}}, \quad (7.12)$$

and a^*_{t-1} is a function of the observations. □

Corollary 7.2 Consider nonlinear time series models of the form

$$y_t = \theta f(\mathcal{F}_{t-1}) + \sigma(\mathcal{F}_{t-1})\varepsilon_t, \quad (7.13)$$

where $\{\varepsilon_t\}$ is an uncorrelated sequence with zero mean and unit variance and $f(\mathcal{F}_{t-1})$ denotes a nonlinear function of $\mathcal{F}_{t-1}$, such as y^2_{t-1}. When $g(x) = x$ in Theorem 7.1, the recursive estimate based on the optimal linear EF $\sum_{t=1}^{n} a^*_{t-1}(y_t - \theta f(\mathcal{F}_{t-1}))$ is given by (Thavaneswaran and Abraham, 1988)

$$\widehat{\theta}_t = \widehat{\theta}_{t-1} + K_t a^*_{t-1}[y_t - \widehat{\theta}_{t-1} f(\mathcal{F}_{t-1})], \text{ where}$$
$$K_t = \frac{K_{t-1}}{1 + f(\mathcal{F}_{t-1}) K_{t-1} a^*_{t-1}}, \tag{7.14}$$

$a^*_{t-1} = -f(\mathcal{F}_{t-1})/\sigma^2(\mathcal{F}_{t-1})$ and $K_t^{-1} = \sum_{t=1}^n a^*_{t-1} f(\mathcal{F}_{t-1})$. After some algebra, (7.14) has the following familiar Kalman filter form:

$$\widehat{\theta}_t = \widehat{\theta}_{t-1} + \frac{K_{t-1} f(\mathcal{F}_{t-1})}{\sigma^2(\mathcal{F}_{t-1}) + f^2(\mathcal{F}_{t-1}) K_{t-1}} [y_t - \hat{\theta}_{t-1} f(\mathcal{F}_{t-1})], \text{ where}$$
$$K_t = K_{t-1} - \frac{(K_{t-1} f(\mathcal{F}_{t-1}))^2}{\sigma^2(\mathcal{F}_{t-1}) + f^2(\mathcal{F}_{t-1}) K_{t-1}}. \quad \square \tag{7.15}$$

The form of the recursive estimate of the fixed parameter θ in (7.14) motivates use of the EF approach for the model in the GAS framework discussed in Section 7.3.3. When the observation at time t comes in, we update the recursive estimate at time t as the sum of the estimate at $t-1$ and the product of the inverse of the information K_t (which is the term S_t in the GAS formulation) and the optimal martingale difference $a^*_{t-1}[y_t - \widehat{\theta}_{t-1} f(\mathcal{F}_{t-1})]$ (which is ∇_t in the GAS formulation).

Estimating equation approaches for recursive estimation of a *time-varying* parameter have not been discussed in the literature. Consider the simple case where $\theta_t = \phi\theta_{t-1}$ for a given ϕ. The following recursions have a Kalman filter form when the state equation has no error:

$$\widehat{\theta}_t = \phi\widehat{\theta}_{t-1} + \frac{\phi K_{t-1} f(\mathcal{F}_{t-1})}{\sigma^2(\mathcal{F}_{t-1}) + f^2(\mathcal{F}_{t-1}) K_{t-1}} \left[y_t - \widehat{\theta}_{t-1} f(\mathcal{F}_{t-1})\right], \text{ where}$$
$$K_t = \phi^2 K_{t-1} - \frac{(\phi K_{t-1} f(\mathcal{F}_{t-1}))^2}{\sigma^2(\mathcal{F}_{t-1}) + f^2(\mathcal{F}_{t-1}) K_{t-1}}. \tag{7.16}$$

7.4.2 Combined EFs

To estimate θ based on the integer-valued data $y_1, \ldots, y_n$, consider two classes of martingale differences for $t = 1, \ldots, n$, viz., $\{m_t(\theta) = y_t - \mu_t(\theta)\}$ and $\{M_t(\theta) = m_t(\theta)^2 - \sigma_t^2(\theta)\}$, see Thavaneswaran et al. (2015). The quadratic variations of $m_t(\theta)$, $M_t(\theta)$, and the quadratic covariation of $m_t(\theta)$ and $M_t(\theta)$ are respectively

$$\langle m\rangle_t = E(m_t^2(\theta)|\mathcal{F}_{t-1}) = \sigma_t^2(\theta),$$
$$\langle M\rangle_t = E(M_t(\theta)^2|\mathcal{F}_{t-1}) = \sigma_t^4(\theta)(\kappa_t(\theta) - 1), \text{ and}$$
$$\langle m, M\rangle_t = E(m_t(\theta) M_t(\theta)|\mathcal{F}_{t-1}) = \sigma_t^3(\theta)\gamma_t(\theta).$$

The optimal EFs based on the martingale differences $m_t(\theta)$ and $M_t(\theta)$ are respectively

$$g_m^*(\theta) = -\sum_{t=1}^{n} \frac{\partial \mu_t(\theta)}{\partial \theta} \frac{m_t}{\langle m \rangle_t} \text{ and}$$

$$g_M^*(\theta) = -\sum_{t=1}^{n} \frac{\partial \sigma_t^2(\theta)}{\partial \theta} \frac{M_t}{\langle M \rangle_t}.$$

The information associated with $g_m^*(\theta)$ and $g_M^*(\theta)$ are respectively

$$\mathbf{I}_{g_m^*}(\theta) = \sum_{t=1}^{n} \frac{\partial \mu_t(\theta)}{\partial \theta} \frac{\partial \mu_t(\theta)}{\partial \theta'} \frac{1}{\langle m \rangle_t} \text{ and}$$

$$\mathbf{I}_{g_M^*}(\theta) = \sum_{t=1}^{n} \frac{\partial \sigma_t^2(\theta)}{\partial \theta} \frac{\partial \sigma_t^2(\theta)}{\partial \theta'} \frac{1}{\langle M \rangle_t}.$$

Theorem 7.2 describes the results for combined EFs based on the martingale differences $m_t(\theta)$ and $M_t(\theta)$ and provides the resulting form of the *recursive* estimator of θ based on a first-order Taylor approximation (Liang et al., 2011; Thavaneswaran et al., 2015.)

Theorem 7.2 *In the class of all combined EFs*

$$\mathcal{G}_C = \left\{ g_C(\theta) : g_C(\theta) = \sum_{t=1}^{n} \left[\mathbf{a}_{t-1}(\theta) m_t(\theta) + \mathbf{b}_{t-1}(\theta) M_t(\theta) \right] \right\},$$

(a) The optimal EF is given by

$$g_C^*(\theta) = \sum_{t=1}^{n} \left[\mathbf{a}_{t-1}^*(\theta) m_t(\theta) + \mathbf{b}_{t-1}^*(\theta) M_t(\theta) \right],$$

where

$$\mathbf{a}_{t-1}^* = \rho_t^2 \left(-\frac{\partial \mu_t(\theta)}{\partial \theta} \frac{1}{\langle m \rangle_t} + \frac{\partial \sigma_t^2(\theta)}{\partial \theta} \eta_t \right) \text{ and} \tag{7.17}$$

$$\mathbf{b}_{t-1}^* = \rho_t^2 \left(\frac{\partial \mu_t}{\partial \theta} \eta_t - \frac{\partial \sigma_t^2}{\partial \theta} \frac{1}{\langle M \rangle_t} \right), \tag{7.18}$$

where $\rho_t^2 = \left(1 - \frac{\langle m, M \rangle_t^2}{\langle m \rangle_t \langle M \rangle_t} \right)^{-1}$ *and* $\eta_t = \frac{\langle m, M \rangle_t}{\langle m \rangle_t \langle M \rangle_t}$;

(b) *The information* $\mathbf{I}_{g^*_C}(\theta)$ *is given by*

$$\mathbf{I}_{g^*_C}(\theta) = \sum_{t=1}^{n} \rho_t^2 \left(\frac{\partial \mu_t(\theta)}{\partial \theta} \frac{\partial \mu_t(\theta)}{\partial \theta'} \frac{1}{\langle m \rangle_t} + \frac{\partial \sigma_t^2(\theta)}{\partial \theta} \frac{\partial \sigma_t^2(\theta)}{\partial \theta'} \frac{1}{\langle M \rangle_t} - \left(\frac{\partial \mu_t(\theta)}{\partial \theta} \frac{\partial \sigma_t^2(\theta)}{\partial \theta'} + \frac{\partial \sigma_t^2(\theta)}{\partial \theta} \frac{\partial \mu_t(\theta)}{\partial \theta'} \right) \eta_t \right);$$

(c) *The gain in information over the linear EF is*

$$\mathbf{I}_{g^*_C}(\theta) - \mathbf{I}_{g^*_m}(\theta) = \sum_{t=1}^{n} \rho_t^2 \left(\frac{\partial \mu_t(\theta)}{\partial \theta} \frac{\partial \mu_t(\theta)}{\partial \theta'} \frac{\langle m, M \rangle_t^2}{\langle m \rangle_t^2 \langle M \rangle_t} + \frac{\partial \sigma_t^2(\theta)}{\partial \theta} \frac{\partial \sigma_t^2(\theta)}{\partial \theta'} \frac{1}{\langle M \rangle_t} - \left(\frac{\partial \mu_t(\theta)}{\partial \theta} \frac{\partial \sigma_t^2(\theta)}{\partial \theta'} + \frac{\partial \sigma_t^2(\theta)}{\partial \theta} \frac{\partial \mu_t(\theta)}{\partial \theta'} \right) \eta_t \right);$$

(d) *The gain in information over the quadratic EF is*

$$\mathbf{I}_{g^*_C}(\theta) - \mathbf{I}_{g^*_M}(\theta) = \sum_{t=1}^{n} \rho_t^2 \left(\frac{\partial \mu_t(\theta)}{\partial \theta} \frac{\partial \mu_t(\theta)}{\partial \theta'} \frac{1}{\langle m \rangle_t} + \frac{\partial \sigma_t^2(\theta)}{\partial \theta} \frac{\partial \sigma_t^2(\theta)}{\partial \theta'} \frac{\langle m, M \rangle_t^2}{\langle m \rangle_t \langle M \rangle_t - \langle m, M \rangle_t^2} - \left(\frac{\partial \mu_t(\theta)}{\partial \theta} \frac{\partial \sigma_t^2(\theta)}{\partial \theta'} + \frac{\partial \sigma_t^2(\theta)}{\partial \theta} \frac{\partial \mu_t(\theta)}{\partial \theta'} \right) \eta_t \right);$$

(e) *The recursive estimate for* θ *is given by*

$$\widehat{\theta}_t = \widehat{\theta}_{t-1} + \mathbf{K}_t \left(\mathbf{a}^*_{t-1}(\widehat{\theta}_{t-1}) m_t(\widehat{\theta}_{t-1}) + \mathbf{b}^*_{t-1}(\widehat{\theta}_{t-1}) M_t(\widehat{\theta}_{t-1}) \right), \tag{7.19}$$

$$\mathbf{K}_t = \mathbf{K}_{t-1} \left(\mathbf{I}_p - \left(\mathbf{a}^*_{t-1}(\widehat{\theta}_{t-1}) \frac{\partial m_t(\widehat{\theta}_{t-1})}{\partial \theta'} + \frac{\partial \mathbf{a}^*_{t-1}(\widehat{\theta}_{t-1})}{\partial \theta} m_t(\widehat{\theta}_{t-1}) + \mathbf{b}^*_{t-1}(\widehat{\theta}_{t-1}) \frac{\partial M_t(\widehat{\theta}_{t-1})}{\partial \theta'} + \frac{\partial \mathbf{b}^*_{t-1}(\widehat{\theta}_{t-1})}{\partial \theta} M_t(\widehat{\theta}_{t-1}) \right) \mathbf{K}_{t-1} \right)^{-1}, \tag{7.20}$$

where $\mathbf{I}_p$ *is the* $p \times p$ *identity matrix and* $\mathbf{a}^*_{t-1}$ *and* $\mathbf{b}^*_{t-1}$ *can be calculated by substituting* $\widehat{\theta}_{t-1}$ *for* θ *in Equations (7.17) and (7.18), respectively;*

(f) *For the scalar parameter case, the recursive estimate of* θ *is given by*

$$\widehat{\theta}_t = \widehat{\theta}_{t-1} + K_t[a^*_{t-1}(\widehat{\theta}_{t-1}) m_t(\widehat{\theta}_{t-1}) + b^*_{t-1}(\widehat{\theta}_{t-1}) M_t(\widehat{\theta}_{t-1})], \text{ where}$$

$$K_t = K_{t-1}\left(1 - \left(a^*_{t-1}(\widehat{\theta}_{t-1})\frac{\partial m_t(\widehat{\theta}_{t-1})}{\partial \theta} + \frac{\partial a^*_{t-1}(\widehat{\theta}_{t-1})}{\partial \theta} m_t(\widehat{\theta}_{t-1})\right.\right.$$
$$\left.\left. + b^*_{t-1}(\widehat{\theta}_{t-1})\frac{\partial M_t(\widehat{\theta}_{t-1})}{\partial \theta} + \frac{\partial b^*_{t-1}(\widehat{\theta}_{t-1})}{\partial \theta} M_t(\widehat{\theta}_{t-1})\right) K_{t-1}\right)^{-1}. \tag{7.21}$$

Since $-E\left(\frac{\partial \mathbf{g}^*_C(\theta)}{\partial \theta}\middle| \mathcal{F}_{t-1}\right)$ *denotes the optimal information matrix based on the first t observations, it follows that* $K_t^{-1} = -\sum_{s=1}^{t} \frac{\partial \mathbf{g}^*_C(\widehat{\theta}_{s-1})}{\partial \theta}$ *can be interpreted as the observed information matrix associated with the optimal combined EF* $\mathbf{g}^*_C(\theta)$. *The proof of this theorem is given in Thavaneswaran et al. (2015).* □

In an interesting recent paper, Fokianos et al. (2009) described estimation for linear and nonlinear autoregression models for Poisson count time series, and used simulation studies to compare conditional least squares estimates with maximum likelihood estimates. Similar to Fisher (1924), we compare the information associated with the corresponding EFs, and show that the optimal EF is more informative than the conditional least squares EF. In the class of estimating functions of the form $\mathcal{G} = \{g_m(\theta) : g_m(\theta) = \sum_{t=1}^{n} \mathbf{a}_{t-1}(\theta) m_t(\theta)\}$, the optimal EF is given by $g^*_m(\theta) = \sum_{t=1}^{n} \mathbf{a}^*_{t-1}(\theta) m_t(\theta)$, where $\mathbf{a}^*_{t-1} = \left(-\frac{\partial \mu_t(\theta)}{\partial \theta}\frac{1}{\langle m \rangle_t}\right)$. The optimal EF and the conditional least squares EF belong to the class $\mathcal{G}$, and the optimal value of $\mathbf{a}_{t-1}$ is chosen to maximize the information. Hence $I^*_{g_m} - I_{g_{CLS}}$ is nonnegative definite. It follows from page 919 of Lindsay (1985) that the optimal estimates are more efficient than the conditional least squares estimates for any class of count time series models.

Note that $\mathbf{g}^*_n = \mathbf{0}$ corresponds in general to a set of nonconvex, nonlinear equations. The formulas for (7.17) through (7.21) may be easily coded as functions in R. For each data/model combination, use of the EF approach in practice requires *soft coding* of the first four conditional moments, derivatives of the first two conditional moments with respect to model parameters, and specification of initial values to start the recursive estimation.

Example 7.6

Consider a zero-inflated regression model. Let $\{y_t\}$ denote a count time series with excess zeros, and assume that the mean, variance, skewness, and kurtosis of y_t conditional on $\mathcal{F}_{t-1}$ are given by

$$\mu_t(\theta) = (1-\omega_t)\lambda_t,$$
$$\sigma_t^2(\theta) = (1-\omega_t)\lambda_t(1+\omega_t\lambda_t),$$
$$\gamma_t(\theta) = \frac{\omega_t(1+2\omega_t)\lambda_t^2 + 3\omega_t\lambda_t + 1}{((1-\omega_t)\lambda_t)^{1/2}(1+\omega_t\lambda_t)^{3/2}}, \text{ and}$$
$$\kappa_t(\theta) = \frac{\omega_t(6\omega_t^2 - 6\omega_t + 1)\lambda_t^3 + 6\omega_t(2\omega_t - 1)\lambda_t^2 + 7\omega_t\lambda_t + 1}{(1-\omega_t)\lambda_t(1+\omega_t\lambda_t)^2},$$

therefore corresponding to the moments of a ZIP(λ_t, ω_t) distribution with pmf

$$p(y_t|\mathcal{F}_{t-1}) = \begin{cases} \omega_t + (1-\omega_t)\exp(-\lambda_t), & \text{if } y_t = 0, \\ (1-\omega_t)\exp(-\lambda_t)\lambda_t^{y_t}/y_t!, & \text{if } y_t > 0. \end{cases}$$

where λ_t is the intensity parameter of the baseline Poisson distribution and ω_t is the zero-inflation parameter. The ZIP model for count time series is an extension of the Poisson autoregression discussed in Chapter 4 of Kedem and Fokianos (2002).

Suppose that λ_t and ω_t are parametrized by $\lambda_t(\boldsymbol{\beta})$ and $\omega_t(\boldsymbol{\delta})$, which are flexible functions of the unknown parameters $\boldsymbol{\beta}$ and $\boldsymbol{\delta}$ and exogenous explanatory variables at time $t-1$, viz., $\mathbf{x}_{t-1}$ and $\mathbf{z}_{t-1}$:

$$\lambda_t(\boldsymbol{\beta}) = \exp(\mathbf{x}'_{t-1}\boldsymbol{\beta}) \text{ and}$$
$$\omega_t(\boldsymbol{\delta}) = \frac{\exp(\mathbf{z}'_{t-1}\boldsymbol{\delta})}{1+\exp(\mathbf{z}'_{t-1}\boldsymbol{\delta})}. \tag{7.22}$$

Let $\boldsymbol{\theta} = (\boldsymbol{\beta}', \boldsymbol{\delta}')'$, which appears in the conditional mean and variance of y_t. Let $m_t = y_t - \mu_t$, $M_t = m_t^2 - \sigma_t^2$, and refer to $\lambda_t(\boldsymbol{\beta})$ and $\omega_t(\boldsymbol{\delta})$ by λ_t and ω_t, respectively. Then

$$\langle m\rangle_t = \lambda_t[1-\omega_t][1+\omega_t\lambda_t],$$
$$\langle M\rangle_t = (1-\omega_t)\lambda_t(\omega_t(4\omega_t^2 - 4\omega_t + 1)\lambda_t^3 + 2\omega_t(4\omega_t - 1)\lambda_t^2 + 5\omega_t\lambda_t + 2),$$
$$\langle m, M\rangle_t = (1-\omega_t)\lambda_t(\omega_t(1+2\omega_t)\lambda_t^2 + 3\omega_t\lambda_t + 1).$$

Also,

$$\frac{\partial\mu_t}{\partial\boldsymbol{\theta}} = \begin{pmatrix} (1-\omega_t)\frac{\partial\lambda_t}{\partial\boldsymbol{\beta}}, \\ -\lambda_t\frac{\partial\omega_t}{\partial\boldsymbol{\delta}} \end{pmatrix} \text{ and}$$
$$\frac{\partial\sigma_t^2}{\partial\boldsymbol{\theta}} = \begin{pmatrix} (1-\omega_t)(1+2\omega_t\lambda_t)\frac{\partial\lambda_t}{\partial\boldsymbol{\beta}} \\ \lambda_t(\lambda_t(1-2\omega_t)-1)\frac{\partial\omega_t}{\partial\boldsymbol{\delta}} \end{pmatrix}.$$

The combined optimal EF based on m_t and M_t is given by

$$\mathbf{g}_C^*(\boldsymbol{\theta}) = \sum_{t=1}^{n}\left(\mathbf{a}_{t-1}^* m_t + \mathbf{b}_{t-1}^* M_t\right),$$

where

$$\mathbf{a}_{t-1}^* = \rho_t^2\left(\begin{pmatrix} -(1-\omega_t)\frac{\partial\lambda_t}{\partial\boldsymbol{\beta}} \\ \lambda_t\frac{\partial\omega_t}{\partial\boldsymbol{\delta}} \end{pmatrix}\frac{1}{\langle m\rangle_t} + \begin{pmatrix} (1-\omega_t)(1+2\omega_t\lambda_t)\frac{\partial\lambda_t}{\partial\boldsymbol{\beta}} \\ \lambda_t(\lambda_t(1-2\omega_t)-1)\frac{\partial\omega_t}{\partial\boldsymbol{\delta}} \end{pmatrix}\eta_t\right),$$
$$\mathbf{b}_{t-1}^* = \rho_t^2\left(\begin{pmatrix} -(1-\omega_t)\frac{\partial\lambda_t}{\partial\boldsymbol{\beta}} \\ \lambda_t\frac{\partial\omega_t}{\partial\boldsymbol{\delta}} \end{pmatrix}\eta_t - \begin{pmatrix} (1-\omega_t)(1+2\omega_t\lambda_t)\frac{\partial\lambda_t}{\partial\boldsymbol{\beta}} \\ -\lambda_t(\lambda_t(1-2\omega_t)-1)\frac{\partial\omega_t}{\partial\boldsymbol{\delta}} \end{pmatrix}\frac{1}{\langle M\rangle_t}\right).$$

The corresponding information matrix is given by

$$\mathbf{I}_C^*(\boldsymbol{\theta}) = \mathbf{I}_m^*(\boldsymbol{\theta}) + \mathbf{I}_M^*(\boldsymbol{\theta})$$
$$= -\sum_{t=1}^{n}\rho_t^2\begin{pmatrix} 2(1-\omega_t)^2(1+2\omega_t\lambda_t)\left(\frac{\partial\lambda_t}{\partial\boldsymbol{\beta}}\right)^2 & \lambda_t(1-\omega_t)(\lambda_t(1-4\omega_t)-2)\left(\frac{\partial\lambda_t}{\partial\boldsymbol{\beta}}\right)\left(\frac{\partial\omega_t}{\partial\boldsymbol{\delta}}\right) \\ \lambda_t(1-\omega_t)(\lambda_t(1-4\omega_t)-2)\left(\frac{\partial\lambda_t}{\partial\boldsymbol{\beta}}\right)\left(\frac{\partial\omega_t}{\partial\boldsymbol{\delta}}\right) & -2\lambda_t^2(\lambda_t(1-2\omega_t)-1)\left(\frac{\partial\omega_t}{\partial\boldsymbol{\delta}}\right)^2 \end{pmatrix}.$$

The recursive estimator for $\boldsymbol{\theta}$ follows from (7.19) and (7.20) as

$$\widehat{\boldsymbol{\theta}}_t = \widehat{\boldsymbol{\theta}}_{t-1} + \mathbf{K}_t \begin{pmatrix} -\dfrac{\frac{\partial}{\partial \beta}\lambda_t(\widehat{\beta}_{t-1})[y_t-(1-\omega_t(\widehat{\delta}_{t-1}))\lambda_t(\widehat{\beta}_{t-1})]}{\lambda_t(\widehat{\beta}_{t-1})(1+\omega_t(\widehat{\delta}_{t-1})\lambda_t(\widehat{\beta}_{t-1}))} \\ \dfrac{\frac{\partial}{\partial \delta}\omega_t(\widehat{\delta}_{t-1})[y_t-(1-\omega_t(\widehat{\delta}_{t-1}))\lambda_t(\widehat{\beta}_{t-1})]}{(1-\omega_t(\widehat{\delta}_{t-1}))(1+\omega_t(\widehat{\delta}_{t-1})\lambda_t(\widehat{\beta}_{t-1}))} \end{pmatrix},$$

and

$$\mathbf{K}_t = \mathbf{K}_{t-1}\left(\mathbf{I}_2 - \left(\frac{1}{1+\omega_t\lambda_t}\begin{pmatrix} \dfrac{\left(\frac{\partial \lambda_t}{\partial \beta}\right)^2(1-\omega_t)}{\lambda_t} & -\dfrac{\partial \lambda_t}{\partial \beta}\dfrac{\partial \omega_t}{\partial \delta} \\ -\dfrac{\partial \lambda_t}{\partial \beta}\dfrac{\partial \omega_t}{\partial \delta} & \dfrac{\left(\frac{\partial \omega_t}{\partial \delta}\right)^2\lambda_t}{1-\omega_t} + \dfrac{(y_t-(1-\omega_t)\lambda_t)}{(1+\omega_t\lambda_t)^2} \end{pmatrix}\right.\right.$$
$$\left.\left.\times \begin{pmatrix} \dfrac{-\frac{\partial^2 \lambda_t}{\partial \beta^2}\lambda_t(1+\omega_t\lambda_t)+\left(\frac{\partial \lambda_t}{\partial \beta}\right)^2(1+2\omega_t\lambda_t)}{\lambda_t^2} & \dfrac{\partial \lambda_t}{\partial \beta}\dfrac{\partial \omega_t}{\partial \delta} \\ -\dfrac{\frac{\partial \lambda_t}{\partial \beta}\frac{\partial \omega_t}{\partial \delta}\omega_t}{(1-\omega_t)} & \dfrac{\frac{\partial^2 \omega_t}{\partial \delta^2}(1-\omega_t)(1+\omega_t\lambda_t)-\left(\frac{\partial \omega_t}{\partial \delta}\right)^2(\lambda_t+1-2\omega_t\lambda_t)}{(1-\omega_t)^2} \end{pmatrix}\right)\mathbf{K}_{t-1}\right)^{-1}.$$

For this example, it is also straightforward to show that the linear optimal EF and the corresponding optimal coefficient are given by

$$\mathbf{g}_m^*(\boldsymbol{\theta}) = \sum_{t=1}^{n}\begin{pmatrix} -\dfrac{\frac{\partial \lambda_t}{\partial \beta}}{\lambda_t(1+\omega_t\lambda_t)} \\ \dfrac{\frac{\partial \omega_t}{\partial \delta}}{(1-\omega_t)(1+\omega_t\lambda_t)} \end{pmatrix} m_t, \text{ and}$$

$$\mathbf{a}_{t-1}^*(\boldsymbol{\theta}) = \begin{pmatrix} -\dfrac{\frac{\partial \lambda_t}{\partial \beta}}{\lambda_t(1+\omega_t\lambda_t)} \\ \dfrac{\frac{\partial \omega_t}{\partial \delta}}{(1-\omega_t)(1+\omega_t\lambda_t)} \end{pmatrix}. \square$$

Example 7.7

Consider an extended GAS(P, Q) model. As discussed in Section 7.3.3, suppose $\{y_t\}$ is an integer-valued time series, its first four conditional moments given $\mathcal{F}_{t-1}$ are available, and $\mathbf{f}_t$ is modeled by (7.10), $\mathbf{s}_t$ being a suitably chosen martingale difference. Suppose the time-varying parameter $\mathbf{f}_t$ corresponds to the conditional mean $\mu_t(\boldsymbol{\theta}) = E(y_t|\mathcal{F}_{t-1})$. Following the discussion in Theorem 7.1, it is natural to choose $\mathbf{s}_t$ as $\mathbf{a}_{t-1}^*(\boldsymbol{\theta})m_t(\boldsymbol{\theta})$. Suppose instead that $\mathbf{f}_t$ corresponds to the conditional variance $\sigma_t^2(\boldsymbol{\theta}) = Var(y_t|\mathcal{F}_{t-1})$; a natural choice of $\mathbf{s}_t$ is $\mathbf{b}_{t-1}^*(\boldsymbol{\theta})M_t(\boldsymbol{\theta})$. When $\mathbf{f}_t$ is modeled by (7.10), the most informative innovation is given by

$$\mathbf{s}_t = \mathbf{K}_t\left(\mathbf{a}_{t-1}^*(\widehat{\boldsymbol{\theta}}_{t-1})m_t(\widehat{\boldsymbol{\theta}}_{t-1}) + \mathbf{b}_{t-1}^*(\widehat{\boldsymbol{\theta}}_{t-1})M_t(\widehat{\boldsymbol{\theta}}_{t-1})\right), \tag{7.23}$$

where $\mathbf{K}_t$ is defined in (7.20) and $\mathbf{a}_{t-1}^*$ and $\mathbf{b}_{t-1}^*$ can be calculated by substituting $\widehat{\boldsymbol{\theta}}_{t-1}$ in equations (7.17) and (7.18) respectively for the fixed parameter $\boldsymbol{\theta}$.

When the form of the conditional distribution of y_t given $\mathcal{F}_{t-1}$ is available, and the score function is easy to obtain, then the optimal choice for $\mathbf{V}_t$ is the score function. However, in situations where we do not wish to assume an explicit form for the conditional distribution, the optimal choice for $\mathbf{V}_t$ is given by components of the optimal linear, or quadratic, or combined EFs.

Consider the nonlinear time series model in (7.13). Based on the optimal EF in the class $\mathcal{G}$ of all unbiased EFs $\mathbf{g} = \sum_{t=1}^{n} b_{t-1} g(y_t - \mu(\theta_t, F_{t-1}^y))$, the form of the extended GAS model for the location parameter θ_t is given by

$$\theta_t = \phi\theta_{t-1} + \frac{\phi K_{t-1} b_{t-1}^*}{1 + [\partial\mu(\hat{\theta}_{t-1}, \mathcal{F}_{t-1})/\partial\theta] K_{t-1} b_{t-1}^*} g(y_t - \mu(\theta_{t-1}, \mathcal{F}_{t-1})), \text{ where}$$

$$K_t = \frac{K_{t-1}}{1 + [\partial\mu(\theta_{t-1}, \mathcal{F}_{t-1})/\partial\theta]\phi^2 K_{t-1} b_{t-1}^*},$$

where b_{t-1}^* is a function of g, θ, and the observations:

$$b_{t-1}^* = \frac{[\partial\mu(\theta_{t-1}, \mathcal{F}_{t-1})/\partial\theta][\partial g(\theta_{t-1}, \mathcal{F}_{t-1})/\partial\mu]}{Var(g|\mathcal{F}_{t-1})}.$$

7.5 Hypothesis Testing and Model Choice

Hypothesis testing situations in stochastic modeling are often tests of linear hypotheses about the unknown parameter $\theta \in \mathcal{R}^p$. Suppose $\theta = (\theta_1, \theta_2)$, and suppose that the optimal EF $g_n^*(\theta)$ and the corresponding unconditional information $F_n(\theta) = E[g_n^*(\theta) g_n^*(\theta)']$ have conformable partitions, that is,

$$g_n^*(\theta) = \begin{pmatrix} g_{n1}^*(\theta) \\ g_{n2}^*(\theta) \end{pmatrix}, \; F_n(\theta) = \begin{pmatrix} F_{n11}(\theta) & F_{n12}(\theta) \\ F_{n21}(\theta) & F_{n22}(\theta) \end{pmatrix}.$$

A test of $H_0 : \theta_2 = \theta_{20}$ versus $H_1 : \theta_2 \neq \theta_{20}$ corresponds to a comparison of a full model versus a nested model when we test $\theta_{20} = \mathbf{0}$. For example, in (7.5), testing $\theta_{20} = \mathbf{0}$ could correspond to testing $\beta = \mathbf{0}$, so that H_0 corresponds to a smaller model with only δ and α as parameters.

Let $\theta_n^* = (\theta_{n1}^*, \theta_{n2}^*)$ denote the optimal estimate of θ unrestricted by H_0, and let $\tilde{\theta}_n = (\tilde{\theta}_{n1}, \tilde{\theta}_{n2})$ denote the optimal estimate under the null hypothesis H_0. As in Thavaneswaran (1991), we propose two test statistics, viz., the Wald-type statistic and the score statistic, as

$$W_n = (\theta_{n2}^* - \theta_{20})' A_{n22}(\theta_{n2}')(\theta_{n2}^* - \theta_{20}), \quad (7.24)$$

$$Q_n = (g_2^*(\tilde{\theta}_n))' A_{n22}^{-1}(\tilde{\theta}_n) g_2^*(\tilde{\theta}_n), \quad (7.25)$$

where

$$A_{n22}(\theta) = F_{n22}(\theta) - F_{n21}(\theta) F_{n11}^{-1}(\theta) F_{n12}(\theta)$$

is the inverse of the second diagonal block in the inverse of the partitioned matrix $F_n(\theta)$.

The Wald and score statistics for testing a general linear hypothesis $H_0 : \mathbf{C}\theta = \mathbf{c}_0$ versus $H_1 : \mathbf{C}\theta \neq \mathbf{c}_0$, where the $r \times p$ matrix $\mathbf{C}$ has full row rank, are

$$\widetilde{W}_n = (\mathbf{C}\theta_n^* - \mathbf{c}_0'[CF_n^{-1}(\theta_n^*)C']^{-1}(\mathbf{C}\theta_n^* - \mathbf{c}_0) \text{ and} \quad (7.26)$$

$$\widetilde{Q}_n = (g_n^*(\tilde{\theta}_n))' F_n^{-1}(\tilde{\theta}_n) g_n^*(\tilde{\theta}_n). \quad (7.27)$$

Thavaneswaran (1991) showed that under certain regularity conditions, the test statistics in (7.26) and (7.27) are asymptotically equivalent, that is, the difference between them converges to zero in probability under H_0 and they have the same limiting null distributions (which is a χ^2_r distribution).

7.6 Discussion and Summary

Interest in developing models for integer valued time series, especially for count time series, is growing. Among these are models discussed in Ferland et al. (2006) and Zhu (2011, 2012a,b), who described classes of INGARCH models with different conditional distributional specifications for the process given its history, and primarily described likelihood based approaches for estimating model parameters, under parametric assumptions such as Poisson, negative binomial, or ZIP for the conditional distributions. Although these models are referred to as INGARCH models in the literature, they model the conditional mean of the time series and not its conditional variance. These models are similar to the ACP models discussed in Heinen (2003) and Ghahramani and Thavaneswaran (2009b). Creal et al. (2013) recently discussed GAS models, while Thavaneswaran and Ravishanker (2015) described models for circular time series.

This chapter considers modeling the conditional mean and conditional variance of an integer-valued time series $\{y_t\}$, where conditional moments are functions of θ. We have described a combined EF approach for estimating θ and have also provided forms for joint recursive estimates for fixed parameters using the most informative combined martingale difference and provided its corresponding information. In Section 7.4.1, we have shown how recursive estimation extends to the case with a time-varying parameter. This approach would be valuable in a study of doubly stochastic models for integer-valued time series, which we briefly discuss below.

Similar to the well-known stochastic volatility (SV) or stochastic conditional duration (SCD) models described in the literature, a general integer-valued doubly stochastic model for y_t with conditional mean $E(y_t|\mathcal{F}_{t-1}) = \mu_t = \exp(\lambda_t)$ is defined via

$$\lambda_t = \delta + \sum_{j=0}^{\infty} \psi_j \varepsilon_{t-j}, \tag{7.28}$$

where δ is a real-valued parameter, $\sum_{j=0}^{\infty} \psi_j^2 < \infty$, $\varepsilon_t|\mathcal{F}_{t-1}$ are independent $N(0, \sigma_\varepsilon^2)$ variables, and ε_s is independent of $y_t|\mathcal{F}_{t-1}$ for all s, t. The conditional moments of y_t may match the moments of a known probability distribution for a count random variable, for example, those corresponding to a Poisson-INDS model, a GP-INDS model, a NB-INDS model, or a ZIP-INDS model. In lieu of (7.28), if λ_t is modeled by $\lambda_t - \delta = (1-B)^{-d}\varepsilon_t = \sum_{j=0}^{\infty} \psi_j \varepsilon_{t-j}$ where $d \in (0, 0.5)$ and $\psi_k = \Gamma(k+d)/[\Gamma(d)\Gamma(k+1)]$, where $\Gamma(\cdot)$ is the gamma function, the model can handle long-memory behavior. We may also define an integer-valued quadratic doubly stochastic (INQDS) model by assuming the first four conditional moments of y_t given $\mathcal{F}_{t-1}$ to be $\mu_t = \exp(a\lambda_t + b\lambda_t^2)$, $\sigma_t^2 = \mu_t$, $\gamma_t = \mu_t^{-1/2}$ and $\kappa_t = \mu_t^{-1}$, which match the first four moments of a Poisson$(\exp(a\lambda_t + b\lambda_t^2))$ process. Naik-Nimbalkar and Rajarshi (1995) and Thompson and Thavaneswaran (1999) studied filtering/estimation for state space models and counting processes in the context of EFs. Thavaneswaran and Abraham (1988) and Thavaneswaran et al. (2015) described combining nonorthogonal EFs following prefiltered estimation.

Acknowledgments

The first author acknowledges support from an NSERC grant.

References

Al-Osh, M. and Alzaid, A. (1987). First-order integer-valued autoregressive (INAR(1)) model. *Journal of Time Series Analysis*, 8:261–275.

Bera, A. K., Bilias, Y., and Simlai, P. (2006). Estimating functions and equations: An essay on historical developments with applications to econometrics. In: T. C. Mills and K. Patterson, (eds.), *Palgrave Handbook of Econometrics*, Palgrave MacMillan: New York, Vol. I, pp. 427–476.

Bollerslev, T. (1986). Generalized autoregressive conditional heteroscedasticity. *Journal of Econometrics*, 31:307–327.

Chan, K. S. and Ledolter, J. (1995). Monte Carlo EM estimation for time series involving counts. *Journal of the American Statistical Association*, 90:242–252.

Cox, D. R. (1981). Statistical analysis of time series: Some recent developments. *Scandinavian Journal of Statistics*, 8:93–115.

Creal, D. D., Koopman, S. J., and Lucas, A. (2013). Generalized autoregressive score models with applications. *Journal of Applied Econometrics*, 28:777–795.

Davis, R. A., Dunsmuir, W. T. M., and Streett, S. B. (2003). Observation-driven models for Poisson counts. *Biometrika*, 90:777–790.

Dean, C. B. (1991). Estimating equations for mixed Poisson models. In: V. P. Godambe, (ed.) *Estimating Functions*, Oxford University Press: Oxford, U.K., pp. 35–46.

Durbin, J. (1960). Estimation of parameters in time-series regression models. *Journal of the Royal Statistical Society Series B*, 22:139–153.

Ferland, R., Latour, A., and Oraichi, D. (2006). Integer-valued GARCH model. *Journal of Time Series Analysis*, 27:923–942.

Fisher, R. A. (1924). The conditions under which χ^2 measures the discrepancy between observation and hypothesis. *Journal of the Royal Statistical Society*, 87:442–450.

Fokianos, K., Rahbek, A., and Tjostheim, D. (2009). Poisson autoregression. *Journal of the American Statistical Association*, 104:1430–1439.

Ghahramani, M. and Thavaneswaran, A. (2009a). Combining estimating functions for volatility. *Journal of Statistical Planning and Inference*, 139:1449–1461.

Ghahramani, M. and Thavaneswaran, A. (2009b). On some properties of Autoregressive Conditional Poisson (ACP) models. *Economic Letters*, 105:273–275.

Ghahramani, M. and Thavaneswaran, A. (2012). Nonlinear recursive estimation of the volatility via estimating functions. *Journal of Statistical Planning and Inference*, 142:171–180.

Godambe, V. P. (1960). An optimum property of regular maximum likelihood estimation. *Annals of Mathematical Statistics*, 31:1208–1212.

Godambe, V. P. (1985). The foundations of finite sample estimation in stochastic process. *Biometrika*, 72:319–328.

Gourieroux, C. (1997). *ARCH Models and Financial Applications*. Springer-Verlag: New York.

Heinen, A. (2003). Modelling time series count data: An Autoregressive Conditional Poisson model. Discussion Paper, vol. 2003/62. Center for Operations Research and Econometrics (CORE), Catholic University of Louvain, Louvain, Belgium.

Heyde, C. C. (1997). *Quasi-Likelihood and its Application: A General Approach to Optimal Parameter Estimation*. Springer-Verlag: New York.

Jung, R. C. and Tremayne, A. R. (2006). Binomial thinning models for integer time series. *Statistical Modelling*, 6:21–96.

Kedem, B. and Fokianos, K. (2002). *Regression Models for Time Series Analysis*. Wiley: Hoboken, NJ.

Liang, Y., Thavaneswaran, A., and Abraham, B. (2011). Joint estimation using quadratic estimating functions. *Journal of Probability and Statistics*, article ID 372512:14 pages.

Lindsay, B. C. (1985). Using empirical partially Bayes inference for increased efficiency. *The Annals of Statistics*, 13:914–931.

MacDonald, I. L. and Zucchini, W. (2015). Hidden Markov models for discrete-valued time series. In R. A. Davis, S. H. Holan, R. Lund and N. Ravishanker, eds., *Handbook of Discrete-Valued Time Series*, pp. 267–286. Chapman & Hall, Boca Raton, FL.

McKenzie, E. (2003). Some simple models for discrete variate time series. In: C. R. Rao and D. N. Shanbhag, (eds.), *Handbook of Statistics-Stochastic models: Modeling and Estimation*, Elsevier Science: Amsterdam, the Netherlands, Vol. 21, pp. 573–606.

Merkouris, T. (2007). Transform martingale estimating functions. *The Annals of Statistics*, 35: 1975–2000.

Naik-Nimbalkar, U. V. and Rajarshi, M. B. (1995). Filtering and smoothing via estimating functions. *Journal of the American Statistical Association*, 90:301–306.

Neal, P. and Subba Rao, T. (2007). MCMC for integer-valued ARMA models. *Journal of Time Series Analysis*, 28:92–110.

Thavaneswaran, A. (1991). Tests based on an optimal estimate. In: V. P. Godambe (ed.), *Estimating Functions*, Clarendon Press: Oxford, U.K., pp. 189–198.

Thavaneswaran, A. and Abraham, B. (1988). Estimation of nonlinear time series models using estimating functions. *Journal of Time Series Analysis*, 9:99–108.

Thavaneswaran, A. and Heyde, C. C. (1999). Prediction via estimating functions. *Journal of Statistical Planning and Inference*, 77:89–101.

Thavaneswaran, A., Ravishanker, N, and Liang, Y. (2015). Generalized duration models and inference using estimating functions. *Annals of the Institute of Statistical Mathematics*, 67:129–156.

Thavaneswaran, A. and Ravishanker, N. (2015). Estimating functions for circular models. Technical Report, *Department of Statistics*, University of Connecticut: Storrs, CT.

Thompson, M. E. and Thavaneswaran, A. (1999). Filtering via estimating functions. *Applied Mathematics Letters*, 12:61–67.

West, M. and Harrison, P. J. (1997). *Bayesian Forecasting and Dynamic Models*. Springer-Verlag: New York.

Yang, M. (2012). Statistical models for count time series with excess zeros. PhD thesis, University of Iowa, Iowa City, IA.

Zeger, S. L. and Qaqish, B. (1988). Markov regression models for time series. *Biometrics*, 44:1019–1031.

Zhu, F. (2011). A negative binomial integer-valued GARCH model. *Journal of Time Series Analysis*, 32:54–67.

Zhu, F. (2012a). Modeling overdispersed or underdispersed count data with generalized Poisson integer-valued GARCH models. *Journal of Mathematical Analysis and Applications*, 389:58–71.

Zhu, F. (2012b). Zero-inflated Poisson and negative binomial integer-valued GARCH models. *Journal of Statistical Planning and Inference*, 142:826–839.

8

Dynamic Bayesian Models for Discrete-Valued Time Series

Dani Gamerman, Carlos A. Abanto-Valle, Ralph S. Silva, and Thiago G. Martins

CONTENTS

8.1 Introduction

State-space models (SSMs) have been discussed in the literature for a number of decades. They are models that rely on a decomposition that separates the observational errors from the temporal evolution. The former usually consists of temporally independent specifications that handle the characteristics of the observational process. The latter is devised to describe the temporal dependence at a latent, unobserved level through evolution disturbances. In the most general form, the observational and evolution disturbances may be related, but in a typical set-up they are independent. SSMs were originally introduced for

Gaussian, hence continuous, time series data, but the above decomposition made it easy to extend them to discrete-valued time series. This chapter describes SSMs with a view towards their use for such data.

The use of SSM by the statistical time series community has become widespread since the books of Harvey (1989) and West and Harrison (1997). These books provided an extensive account of the possibilities of SSM from the classical and Bayesian perspectives, respectively. Another surge of interest has occurred more recently with the development of sequential Monte Carlo (SMC) methods, allowing for approximate online inference; see the seminal paper by Gordon et al. (1993).

The basic framework upon which this chapter lies is called the dynamic generalized linear model (DGLM). It is a special case of SSM, and was introduced by West et al. (1985). Consider a discrete-valued-time series $y_1, \ldots, y_T$ and let $EF(\mu, \phi)$ denote a exponential family distribution with mean μ and variance $\phi c(\mu)$, for some mean function c. The SSM decomposition of DGLM is given, for $t = 1, \ldots, T$, by the equations

$$\text{Observation equation: } y_t \mid x_t, \theta \sim EF(\mu_t, \phi), \tag{8.1}$$

$$\text{Link function: } g(\mu_t) = z_t' x_t, \tag{8.2}$$

$$\text{System equation: } x_t = G_t x_{t-1} + w_t, \quad \text{where } w_t \mid \theta \sim N(0, W), \tag{8.3}$$

where z_t is a known vector (possibly including covariates) at time t, x_t is a time-dependent latent state at time t, and θ is a vector of hyperparameters including ϕ and unknown components of G_t and W. The model is completed with a prior specification for the initial latent state x_0. A Bayesian formulation would also require a prior distribution for the hyperparameter θ. The above model formulation considers only linear models both at the link relation and the system evolution levels. A non-Gaussian evolution with nice integration properties was proposed by Gamerman et al. (2013) to replace (8.3). It includes a few discrete observational models but is not as general as the above formulation.

Usual features of time series can be represented in the above formulation. For example, local linear trends are specified with $z_t = (1, 0)'$, $G_t = \begin{pmatrix} 1 & 1 \\ 0 & 1 \end{pmatrix}$ and $x_t = (\alpha_t, \beta_t)$. In this case, α represents the local level of the series and β represents the local growth in the series. Another common feature of time series is seasonality. There are a few related ways to represent seasonal patterns in time series. Perhaps the simplest representation is the structural form of Harvey (1989) where the seasonal effects s_t are stochastically related via

$$s_t = -(s_{t-1} + s_{t-2} + \cdots + s_{t-p+1}) + \eta_t, \quad \forall\, t, \tag{8.4}$$

for seasonal cycles of length p. Deterministic or static seasonal terms are obtained in the limiting case of $\eta_t = 0$, *a.s.*, thus implying that $\sum_{i=1}^{p} s_{t-i} = 0$, for all p. Evolution 8.3 is recovered by forming the latent component $x_t = (s_t, s_{t-1}, \cdots, s_{t-p+1})'$ with $z_t = (1, 0'_{p-1})'$, $G_t = \begin{pmatrix} 1'_{p-1} & 0 \\ I_{p-1} & 0_{p-1} \end{pmatrix}$ and $w_t = (\eta_t, 0_{p-1})$, where I_m, 1_m, and 0_m are the identity matrix, vector of 1s and vector of 0s of order m, respectively.

By far, the most common discrete-valued specifications are the Poisson and binomial distributions. The Poisson distribution is usually assumed in the analysis of time series of counts. The most popular model for time series of counts is the log-linear dynamic model given by

$$\text{Observation equation: } y_t \mid x_t \sim Poisson(\mu_t), \text{ for } t = 1, \ldots, T \tag{8.5}$$

$$\text{Link function: } \log(\mu_t) = z_t' x_t, \text{ for } t = 1, \ldots, T, \tag{8.6}$$

with system equation (8.3). For binomial-type data, the most popular model is the dynamic logistic regression given by

$$\text{Observation equation: } y_t \mid x_t, \theta \sim Bin(n_t, \pi_t), \text{ for } t = 1, \ldots, T \tag{8.7}$$

$$\text{Link function: } logit(\pi_t) = z_t' x_t, \text{ for } t = 1, \ldots, T, \tag{8.8}$$

with system equation (8.3). Similar models are obtained if the *logit* link is replaced by the probit or complementary log–log links.

A number of extensions/variations can be contemplated:

- Nonlinear models can be considered at the link relation (8.2) and/or at the system evolution (8.3);
- Some components of the latent state x_t may be fixed over time. The generalized linear models (GLM) (Nelder and Wedderburn, 1972) are obtained in the static, limiting case that all components of x_t are fixed;
- The observational equation (8.1) may be robustified to account for overdispersion (Gamerman, 1997);
- The link function (8.2) may be generalized to allow for more flexible forms via parametric (Abanto-Valle and Dey, 2014) or nonparametric (Mallick and Gelfand, 1994) mixtures; and
- The system equation disturbances may be generalized by replacement of Gaussianity by robustified forms (Meinhold and Singpurwalla, 1989) or by skew forms (Valdebenito et al., 2015).

Data overdispersion is frequently encountered in discrete-valued time series observed in human-related studies. It can be accommodated in the DGLM formulation (8.1) through (8.3) via additional random components in the link functions (8.6) and (8.8). These additional random terms cause extra variability at the observational level, forcing a data dispersion larger than that prescribed by the canonical model. These terms may be included in conjugate fashion, thus rendering negative binomial and beta-binomial to replace Poisson and binomial distributions, respectively leading to hierarchical GLM (Lee and Nelder, 1999). Alternatively, random terms may be added to the linear predictors $z_t' x_t$ in the link equations (Ferreira and Gamerman, 2000). The resulting distributions are also overdispersed but no longer available analytically in closed forms. Their main features resemble those of the corresponding negative binomial and beta-binomial distributions, for $N(0, \sigma^2)$ random terms.

Inference can be performed in two different ways: sequentially or in a single block. From a Bayesian perspective, these forms translate into obtaining the sequence of distributions of $[(x_t, \theta) \mid y^t]$, for $t = 1, \ldots, T$ or $[(x_1, \ldots, x_T, \theta) \mid y^T]$, respectively, where

$y^t = \{y^0, y_1, \ldots, y_t\}$ and y^0 represents the initial information. The sequential approach is obtained via iterated use of Bayes' theorem

$$p(x_t, \theta \mid y^t) \propto p(y_t \mid x_t, \theta)\, p(x_t \mid y^{t-1}, \theta) p(\theta \mid y^{t-1}), \tag{8.9}$$

where the first term on the right side is given by (8.1). The second term on the right side is obtained iteratively via

$$p\left(x_t \mid y^{t-1}, \theta\right) = \int p\left(x_t \mid x_{t-1}, y^{t-1}, \theta\right) p\left(x_{t-1} \mid y^{t-1}, \theta\right) dx_{t-1}, \tag{8.10}$$

where the first term in the integrand is given by (8.3).

Single block inference is performed by a single pass of Bayes' theorem as

$$p\left(x_0, x_1, \ldots, x_T, \theta \mid y^T\right) \propto \prod_{i=1}^{T} p(y_t \mid x_t, \theta) \times \prod_{i=1}^{T} p(x_t \mid x_{t-1}, \theta) \times p(x_0) p(\theta), \tag{8.11}$$

where the terms in the products above are respectively given by (8.1) and (8.3).

Inference may also be performed from a classical perspective. In this case, a likelihood approach would use the above posterior distribution as a penalized likelihood, probably with removal of the prior $p(\theta)$. This route is pursued in Durbin and Koopman (2001) (see also Davis and Dunsmuir [2015; Chapter 6 in this volume]). This chapter will concentrate on the Bayesian paradigm.

The resulting distributions are untractable analytically for all cases discussed above and approximations must be used. The following sections describe some of the techniques that are currently being used to approximate the required distributions. They are Markov chain Monte Carlo (MCMC), SMC (or particle filters), and integrated nested Laplace approximations (INLA). These techniques do not exhaust the range of possibilities for approximations, but are the most widely used techniques nowadays. Finally, the techniques are applied to real datasets under a variety of model formulations in order to illustrate the usefulness of the SSM formulation.

8.2 MCMC

We describe MCMC methods in the context of the general SSM given in (8.1)–(8.3). A detailed review on this subject can be found in Fearnhead (2011) and Migon et al. (2005). From the Bayesian perspective, inference in a general SSM targets the joint posterior distribution of parameters and hidden states, $p(\theta, x^T \mid y^T)$, which is given by (8.11). A Markov chain whose state is (θ, x^T) and whose stationary distribution is the joint posterior distribution $p(\theta, x^T \mid y^T)$ is subsequent. A realization of this chain is generated until convergence is reached. After convergence, the following iterations of the chain can be used to form a sample from the posterior distribution. The Gibbs sampler, iteratively drawing samples from $p(x^T \mid y^T, \theta)$ and $p(\theta \mid x^T, y^T)$ is the most popular method to sample from such a posterior distribution. In practice, sampling from $p(\theta \mid x^T, y^T)$ is often easy, whereas designing a sampler for $p(x^T \mid y^T, \theta)$ is trickier due to the high posterior correlation that usually occurs between the states. Next, we will describe approaches to sample from states.

8.2.1 Updating the States

The simplest procedure is to update the components of the states x^T one at time in a single-move fashion (Carlin et al., 1992; Geweke and Tanizaki, 2001). However, due to the severe correlation between states, such a sampler may lead to slow mixing. In such cases it is better to update the states in a multimove fashion as blocks of states $x^{r,s} = (x_r, x_{r+1}, \ldots, x_s)'$, or update the whole state process x^T (Shephard and Pitt, 1997; Carter and Kohn, 1994, 1996).

8.2.1.1 Single-Move Update for the States

Carlin et al. (1992) and Geweke and Tanizaki (2001) introduced the Gibbs sampler and the Metropolis–Hastings algorithms to perform inference for nonnormal and nonlinear SSM in a single-move fashion. For sampling states, a sequential sampler that updates each state conditioning on the rest of the states is used. Such an approach is easy to implement and due to the Markovian evolution of the states, the conditional distribution of each state given all the others reduces to conditioning only on its two adjacent states:

$$p(x_t \mid y_t, x_{t-1}, x_{t+1}, \theta) \propto \begin{cases} p(y_t \mid x_t, \theta)p(x_{t+1} \mid x_t, \theta)p(x_t \mid x_{t-1}, \theta) & t < T \\ p(y_t \mid \theta, x_t)p(x_t \mid x_{t-1}, \theta) & t = T \text{ (end point)}. \end{cases} \tag{8.12}$$

In some situations, we can simulate directly from the full conditional distribution, and such moves will always be accepted. Where this is not possible, a Metropolis–Hastings step within the Gibbs sampler can often be implemented. Geweke and Tanizaki (2001) give a detailed discussion of several proposals in this situation.

8.2.1.2 Multimove Update for the States

While single-move samplers are easy to implement, the resulting MCMC algorithms can mix slowly if there is strong dependence in the state process. Updating the states in a multimove fashion could be an alternative approach to overcome this problem. Ideally, we would update the whole state process in one move. The simulation smoother of de Jong and Shephard (1995) can be used to sample the states.

In situations where it is not possible to update the whole state process, Shephard and Pitt (1997) and Watanabe (2004) propose sampling random blocks of the disturbances $w^{r+1,s} = (w_{r+1}, \ldots, w_s)'$ (equivalently $x^{r+1,s}$ using as proposal transition density a second order Taylor expansion of $l_t = \log p(y_t \mid \zeta_t)$ in the full conditional of $w^{t+1,s}$, where $h(\zeta_t) = d_t = z_t' x_t$. The proposal density is then the multivariate normal with pseudo-observations $\hat{y}_t = z_t' x_t + \hat{V}_t l_t'(\hat{\zeta}_t)$ and $V_t = -l''(\hat{\zeta}_t)$, for $t = r+1, \ldots, s-1$ and $t = T$. For $t = s < T$, we have $\hat{y}_t = \hat{V}_t \left[z_t \{ l'(\hat{\zeta}_t - l''(\hat{\zeta}_t) z_t' \hat{x}_t \} + G_{t+1}' W_{t+1}^{-1} x_{t+1} \right]$ and $V_t = \left[G_{t+1}' W_{t+1}^{-1} G_{t+1} - l''(\hat{\zeta}_t) z_t z_t' \right]^{-1}$. Then the linear SSM with pseudo-observations $\hat{y}_t$ is defined as

$$\hat{y}_t = \begin{cases} z_t' x_t + \nu_t & t = r+1, \ldots, s-1, \text{ and } t = T \\ x_t + \nu_t & t = s < T \end{cases} \tag{8.13}$$
$$x_t = G_t x_{t-1} + w_t, \forall\ t,$$

where ν_t and w_t are all independent and $\nu_t \sim \mathcal{N}(0, \hat{V}_t)$, $w_t \sim \mathcal{N}(0, W_t)$. Notice that sampling from this distribution (g) is the same as sampling $w^{r+1,s}$ given x_r, x_{s+1} and $\hat{y}_{r+1}, \ldots, \hat{y}_s$ in the above model, which is possible by using the de Jong and Shephard (1995) simulation smoother. Since the distribution f of the disturbances is not bounded by g, the Metropolis–Hastings acceptance–rejection algorithm samples from f as recommended by Chib and Greenberg (1995). The expansion blocks $\hat{w}^{r+1,s}$ ($x^{r+1,s}$) are selected as follows. Once an initial expansion block is selected, the auxiliary observations $\hat{y}_t$ are calculated. Next, application of the Kalman filter and a disturbance smoother to the linear Gaussian SSM with the artificial $\hat{y}_t$ yields the mean of $x^{r+1,s}$ conditional on $\hat{x}^{r+1,s}$. By repeating the procedure until the smoothed estimates converge, we obtain the posterior mode $x^{r+1,s}$. According to Shephard and Pitt (1997), the blocks are selected randomly.

Gamerman (1998) suggests the use of a proposal transition density very similar to that of Shephard and Pitt (1997) based on a reparametrization of the model in terms of the system disturbances and sampling from these distributions. The proposal density is the full conditional distribution of w_t in the model with the modified observational equation in (8.13). The reparametrization rewrites the link function in term of the system disturbances as $g(\mu_t) = \alpha_t = z_t' \sum_{j=1}^{t} G^{t-j} w_j$ with $w_t \sim \mathcal{N}(0, W_t)$, $t = 2, \ldots, T$ and $w_1 \sim \mathcal{N}(a_1, R_1)$, if $G_t = G$, $\forall t$.

8.3 Sequential Monte Carlo

Since the seminal work by Gordon et al. (1993), the SMC methods—also known as particle filter algorithms—have gained popularity as generalization of the importance sampling algorithm. The auxiliary particle filter (Pitt and Shephard, 1999) is another generalization of SMC. Since SMC methods are employed for filtering and smoothing given the parameters, the parameters should first be estimated before estimating the state vector. Recently, Andrieu et al. (2010) proposed the particle MCMC algorithm, which combines two algorithms: Metropolis–Hastings and SMC methods. They showed that if the likelihood is unbiasedly estimated by SMC and is plugged into a Metropolis–Hastings algorithm, then the parameters and states can be sampled from the correct posterior distribution. Nonetheless, designing a Metropolis–Hastings algorithm and tuning it can be cumbersome, especially for some state space models. It is possible to use adaptive Metropolis–Hastings sampling schemes as in Pitt et al. (2012) to overcome this problem. The first scheme is the adaptive random walk Metropolis sampling proposed by Roberts and Rosenthal (2009) (see references therein), and the second one is the adaptive independent Metropolis–Hastings sampling algorithm proposed by Giordani and Kohn (2010).

We show below the combination of the particle filter proposed by Gordon et al. (1993) and the adaptive random walk Metropolis sampling proposed by Roberts and Rosenthal (2009), which enables us to draw from the exact posterior distribution of the parameter vector, including the states. We should keep in mind that there are other more efficient combinations of particle filtering and adaptive sampling methods.

8.3.1 Particle Filter

We describe the particle filter by Gordon et al. (1993). Suppose that we have samples $x_{t-1}^k \sim p(x_{t-1}|y^{t-1}, \Theta)$ for $k = 1, \ldots, K$. First, we take the sample $\tilde{x}_t^k \sim p(x_t|x_{t-1}^k)$, for

$k = 1, \ldots, K$, which gives an approximation of $p(x_t|y_1, \ldots, y^{t-1}, \Theta)$—the prediction density. We can compute the corresponding weights and probabilities by

$$\delta_t^k = p\left(y_t|\widetilde{x}_t^k, \Theta\right) \quad \text{and} \quad \omega_t^k = \frac{\delta_t^k}{\sum_{j=1}^{K} \delta_t^k}.$$

The filtering density $p(x_t|y_1, \ldots, y^t, \Theta)$ can be approximated by $\left\{\left(\widetilde{x}_t^k, \omega_t^k\right)\right\}_{k=1}^{K}$, that is,

$$\widehat{p}(x_t|y^t, \Theta) = \sum_{k=1}^{K} \omega_t^k \Delta\left(x_t - \widetilde{x}_t^k\right), \tag{8.14}$$

where Δ is the Dirac function.

The next step is to resample from this mass function to obtain an equally weighted sample, which we call x_t^k for $k = 1, \ldots, K$. We can use the multinomial sampling in this resampling step although stratified sampling reduces the variance of the simulated likelihood. The true likelihood is in fact estimated by the simulated likelihood that is given by

$$\hat{p}(y_t|y_1, \ldots, y^{t-1}, \Theta) = \frac{1}{K}\sum_{k=1}^{K} p\left(y_t|\widetilde{x}_t^k\right) = \frac{1}{K}\sum_{k=1}^{K} \delta_t^k. \tag{8.15}$$

We now move to the next time step.

8.3.2 Adaptive Random Walk Metropolis Sampling

The posterior distribution $p(\boldsymbol{\Theta}|\boldsymbol{Y})$ is our target density from which we wish to draw a sample. However, it is computationally difficult to do so directly and we use the Metropolis–Hastings algorithm. Given an initial $\boldsymbol{\Theta}_0$, we then generate $\boldsymbol{\Theta}_j$ for $j \geqslant 1$ from the proposal density $g_j(\boldsymbol{\Theta}, \boldsymbol{\Theta}^\star)$ where $\boldsymbol{\Theta}^\star$ represents previous iteration values of $\boldsymbol{\Theta}$. Define $\boldsymbol{\Theta}_j^p$ as the proposed value of $\boldsymbol{\Theta}_j$ generated from $g_j(\boldsymbol{\Theta}; \boldsymbol{\Theta}_{j-1})$. We then take $\boldsymbol{\Theta}_j = \boldsymbol{\Theta}_j^p$ with probability

$$\alpha\left(\boldsymbol{\Theta}_{j-1}; \boldsymbol{\Theta}_j^p\right) = \min\left\{1, \frac{p\left(\boldsymbol{\Theta}_j^p|\boldsymbol{Y}\right)}{p\left(\boldsymbol{\Theta}_{j-1}|\boldsymbol{Y}\right)} \frac{g_j\left(\boldsymbol{\Theta}_{j-1}; \boldsymbol{\Theta}_j^p\right)}{g_j\left(\boldsymbol{\Theta}_j^p; \boldsymbol{\Theta}_{j-1}\right)}\right\}, \tag{8.16}$$

and take $\boldsymbol{\Theta}_j = \boldsymbol{\Theta}_{j-1}$ otherwise. Under some regularity conditions (Tierney, 1994), the sequence $\{\boldsymbol{\Theta}_j,\ j = 1, \ldots, n\}$ converges as $n \to \infty$ to draws from the target density $p(\boldsymbol{\Theta}|\boldsymbol{Y})$.

The adaptive random walk Metropolis proposal of Roberts and Rosenthal (2009) is

$$g_j(\boldsymbol{\Theta}; \boldsymbol{\Theta}_{j-1}) = \gamma_j \phi_d(\boldsymbol{\Theta}|\boldsymbol{\Theta}_{j-1}, \eta_1 \boldsymbol{\Omega}_1) + (1 - \gamma_j)\phi_d(\boldsymbol{\Theta}|\boldsymbol{\Theta}_{j-1}, \eta_2 \boldsymbol{\Omega}_{2j}), \tag{8.17}$$

where d is the dimension of $\boldsymbol{\Theta}$ and $\phi_d(\boldsymbol{x}|\boldsymbol{\mu}, \boldsymbol{\Omega})$ is a multivariate d dimensional normal density with mean $\boldsymbol{\mu}$ and covariance matrix $\boldsymbol{\Omega}$. In (8.17), $\gamma_j = 1$ for $j \leqslant k$, with $k \geqslant 1$ representing the initial iterations, $\gamma_j = 0.05$ for $j > k$; $\eta_1 = 0.1^2/d$, which makes the sampler to move locally in small steps; $\eta_2 = 2.38^2/d$, which is optimal (Roberts et al., 1997) when the posterior distribution is a multivariate normal; Ω_1 is a constant covariance matrix that may be derived from an estimate of the parameters or may simply be the identity matrix; Ω_{2j} is the sample covariance matrix of the first $j-1$ iterates (the adaptive step).

On one hand the posterior distribution is considered to be approximated by $\hat{p}(\boldsymbol{\Theta}|\boldsymbol{Y}) \propto \hat{L}(\boldsymbol{Y};\boldsymbol{\Theta})p(\boldsymbol{\Theta})$. On the other hand, we can take all the random variables used in the particle filter as a vector $\boldsymbol{u}$ of uniforms and the posterior distribution in an augmented space as $p(\boldsymbol{\Theta},\boldsymbol{u}|\boldsymbol{Y}) \propto L(\boldsymbol{Y};\boldsymbol{\Theta},\boldsymbol{u})p(\boldsymbol{\Theta})p(\boldsymbol{u})$. We can plug this into a Metropolis–Hastings scheme such as the adaptive random walk Metropolis to draw a sample from $p(\boldsymbol{\Theta}|\boldsymbol{Y}) \propto L(\boldsymbol{Y};\boldsymbol{\Theta})p(\boldsymbol{\Theta})$. See Andrieu et al. (2010) and Pitt et al. (2012) for more details. There are efficient ways to learning sequentially about $\boldsymbol{\Theta}$, if conditional sufficient statistics exist (Carvalho et al., 2010).

8.4 INLA

The INLA, hereafter denoted by INLA, is a deterministic approach developed by Rue et al. (2009) to perform approximate Bayesian inference in the wide class of latent Gaussian models (Rue and Held, 2005). The INLA methodology takes advantage of the hierarchical structure of latent Gaussian models and combines a series of deterministic approximations to obtain posterior marginals of the unknown parameters in the model as well as other summary, predictive, and validation measures of interest.

As mentioned in Martins et al. (2013), implementation of the INLA methodology requires some expertise in numerical methods and computer programming, in order to achieve efficient computing times. The R (R Core Team, 2013) package `INLA`, hereafter denoted as `R-INLA`, was developed to overcome this challenge, and provides a user-friendly interface to the INLA methodology, allowing it to be used routinely, even for those who are not interested in the implementation details behind the program.

The models defined by (8.1)–(8.3) belong to the class of latent Gaussian models for which INLA was originally designed. We refer to Rue et al. (2009) for examples, which are outside this scope. Looking at (8.1 and 8.2), we see that $\pi(y_i|\eta_i(x),\theta_1) = EF(g^{-1}(z_i'x_i),\phi)$, $i = 1,\ldots,T$ with $\theta_1 = \phi$. From (8.3) we can write

$$p(x|\theta_2) = p(x_0|\theta_2)\prod_{i=1}^{T} p(x_i|x_{i-1},\theta_2),$$

where $x_0|\theta_2 \sim N(0,W_0)$ and $x_i|x_{i-1},\theta_2 \sim N(G_i x_{i-1},W)$. Since $x_0|\theta_2$ and $x_i|x_{i-1},\theta_2$, $i = 1,\ldots,T$ are Gaussian, it can be shown that $x|\theta_2 \sim N(0,Q^{-1}(\theta_2))$, with precision matrix given by

$$Q(\theta_2) = \begin{bmatrix} W_0^{-1} & G_1^T W^{-1} & 0 & \cdots & 0 & 0 \\ W^{-1}G_1 & W^{-1}+G_2^T W^{-1}G_2 & G_2^T W^{-1} & \cdots & 0 & 0 \\ 0 & W^{-1}G_2 & W^{-1}+G_3^T W^{-1}G_3 & \ddots & 0 & 0 \\ 0 & 0 & \ddots & \ddots & \ddots & 0 \\ 0 & 0 & \cdots & W^{-1}G_{T-1} & W^{-1}+G_T W^{-1}G_T & G_T^T W^{-1} \\ 0 & 0 & 0 & \cdots & W^{-1}G_T & W^{-1} \end{bmatrix}$$

and θ_2 consists of the unknown parameters within the variance–covariance matrix W and the matrices G_i, $i = 1, \ldots, T$. Therefore, $x|\theta_2$ is a latent Gaussian model with a sparse precision matrix $Q(\theta_2)$, also known as a Gaussian Markov Random Field (GMRF) (Rue and Held, 2005).

Since the dynamic models of interest in this chapter can be written as

$$p(y, x, \theta) = p(y|x, \theta)p(x|\theta)p(\theta),$$

we will describe the INLA methodology in Section 8.4.1, and illustrate how to use the `R-INLA` package through examples in Section 8.4.2.

8.4.1 INLA Methodology

For the hierarchical model described earlier, the joint posterior distribution is given by

$$\begin{aligned} p(x, \theta|y) &\propto p(\theta)p(x|\theta)\prod_{i=1}^{n_d} p(y_i|\eta_i(x), \theta) \\ &\propto p(\theta)|Q(\theta)|^{n/2} \exp\left[-\frac{1}{2}x^T Q(\theta)x + \sum_{i=1}^{n_d} \log\{p(y_i|x_i, \theta)\}\right], \end{aligned}$$

and the posterior marginals of interest can be written as

$$p(x_i|y) = \int p(x_i|\theta, y)p(\theta|y)d\theta, \quad i = 1, \ldots, n, \tag{8.18}$$

$$p(\theta_j|y) = \int p(\theta|y)d\theta_{-j}, \quad j = 1, \ldots, m, \tag{8.19}$$

INLA provides approximations $\tilde{p}(\theta|y)$, $\tilde{p}(x_i|\theta, y)$ to $p(\theta|y)$, and $p(x_i|\theta, y)$, plugs them into (8.18) and (8.19), and uses numerical integration to obtain the approximated posterior marginals $\tilde{p}(x_i|y)$, $\tilde{p}(\theta_j|y)$ of interest.

The approximation used for the joint posterior of the hyperparameters $p(\theta|y)$ is

$$\tilde{p}(\theta|y) \propto \left.\frac{p(x, \theta, y)}{p_G(x|\theta, y)}\right|_{x=x^*(\theta)}, \tag{8.20}$$

where $p_G(x|\theta, y)$ is a Gaussian approximation to the full conditional of x, $p(x|\theta, y)$, obtained by matching the modal configuration and the curvature at the mode, and $x^*(\theta)$ is the mode of the full conditional for x, for a given θ. Expression (8.20) is equivalent to the Laplace approximation of a marginal posterior distribution (Tierney and Kadane, 1986), and it is exact if $p(x|y, \theta)$ is Gaussian.

For $p(x_i|\theta, y)$, three options are available, and they vary in terms of speed and accuracy. The fastest option, $p_G(x_i|\theta, y)$, is to use the marginals of the Gaussian approximation $p_G(x|\theta, y)$, which is already computed when evaluating expression (8.20). The only extra cost in obtaining $p_G(x_i|\theta, y)$ is to compute the marginal variances from the sparse precision matrix of $p_G(x|\theta, y)$, see Rue et al. (2009) for details. The Gaussian approximation often gives reasonable results, but it may contain errors in the location and/or errors due to its lack of skewness (Rue and Martino, 2007). The more accurate approach would be to again use a Laplace approximation, denoted by $p_{LA}(x_i|\theta, y)$, with a form similar to (8.20), that is,

$$p_{LA}(x_i|\theta, y) \propto \left. \frac{p(x, \theta, y)}{p_{GG}(x_{-i}|x_i, \theta, y)} \right|_{x_{-i}=x^*_{-i}(x_i,\theta)}, \tag{8.21}$$

where x_{-i} represents the vector x with its ith element excluded and $p_{GG}(x_{-i}|x_i, \theta, y)$ is the Gaussian approximation to $x_{-i}|x_i, \theta, y$ and $x^*_{-i}(x_i, \theta)$ is the modal configuration. A third option $p_{SLA}(x_i|\theta, y)$, called simplified Laplace approximation, is obtained by doing a Taylor expansion on the numerator and denominator of (8.21) up to third order, thus correcting the Gaussian approximation for location and skewness with a much lower cost when compared to $p_{LA}(x_i|\theta, y)$. We refer to Rue et al. (2009) for a detailed description of the Gaussian, Laplace and simplified Laplace, approximations to $p(x_i|\theta, y)$.

Finally, once we have the approximations $\tilde{p}(\theta|y)$, $\tilde{p}(x_i|\theta, y)$ described earlier, the integrals in (8.18) and (8.19) are numerically approximated by discretizing the θ space through a grid exploration of $\tilde{p}(\theta|y)$. Details about this grid exploration can be found in Martins et al. (2013).

8.4.2 R-INLA through Examples

The syntax for the `R-INLA` package is based on the built-in `glm` function in `R`, and a basic call starts with

```
formula = y ~ a + b + a:b + c*d + f(idx1, model1, ...)
          + f(idx2, model2, ...),
```

where `formula` describes the structured additive linear predictor $\eta(x)$. Here, `y` is the response variable, the term `a + b + a:b + c*d` holds similar meaning as in the builtin `glm` function in `R` and is then responsible for the fixed effects specification. The `f()` terms specify the general Gaussian random effects components of the model. In this case we say that both `idx1` and `idx2` are latent building blocks that are combined together to form a joint latent Gaussian model of interest. Once the linear predictor is specified, a basic call to fit the model with `R-INLA` takes the following form:

```
result = inla(formula, data = data.frame(y, a, b, c, d, idx1, idx2),
              family = "gaussian").
```

After the computations, the list variable `result` will hold an S3 object of class `"inla"`, from which summaries, plots, and posterior marginals can be obtained. We refer to the package website http://www.r-inla.org for more information about model components available to use inside the `f()` functions as well as more advanced arguments to be used within the `inla()` function. In the next section, we show how `R-INLA` can be used to fit dynamic models. Ruiz-Cárdenas et al. (2011) provide more detailed examples.

8.5 Applications

8.5.1 Deep Brain Stimulation

A binary time series of infant sleep status was recorded in a $T = 120$ min electroencephalographic (EEG) sleep pattern study (Stoffer et al., 1998). Let y_t be the indicator of REM sleep cycle. Two time-varying covariates are considered: z_{t1}, the number of body movements during minute t; and, z_{t2}, the number of body movements not due to sucking during minute t. As in Czado and Song (2008), our main objective is to investigate whether or not the probability of being in the REM sleep status is significantly related to the two types of body movements, z_{t1} or z_{t2}. Our analysis considers the probit link and the Student-t link (shown here in the equations) and assumes that

$$y_t \sim \mathcal{B}er(\pi_t) \qquad t = 1, \ldots, T, \tag{8.22}$$

$$\pi_t = T_\nu(\beta_0 + \beta_1 z_{t1} + \beta_1 z_{t2} + x_t), \tag{8.23}$$

$$x_t = \delta x_{t-1} + \tau \eta_t, \tag{8.24}$$

where the innovations η_t are assumed to be mutually independent and normally distributed with mean zero and unit variance and T_ν is the d.f. of the $t_\nu(0, 1)$ distribution. Clearly x_t represents a time-specific effect on the observed process. It follows from the one-to-one relationship (8.23) that $\pi_t = P(y_t = 1 \mid \boldsymbol{\beta}, \mathbf{z}_t, x_t) = T_\nu(\beta_0 + \beta_1 z_{t1} + \beta_1 z_{t2} + x_t)$, where $\boldsymbol{\beta} = (\beta_0, \beta_1, \beta_2)'$, $\mathbf{z}_t = (1, z_{t1}, z_{t2})'$. We also assume that $\mid \delta \mid < 1$, that is, the latent state process is stationary and $x_0 \sim \mathcal{N}(0, \kappa)$, where $\kappa = \tau^2/(1 - \delta^2)$. The model is completed with priors $\delta \sim \mathcal{N}_{(-1,1)}(0.95, 100)$, $\kappa \sim \mathcal{IG}(0.01, 0.01)$, and $\boldsymbol{\beta} \sim \mathcal{N}_3(\boldsymbol{\beta}_0, \Sigma_0)$, where $\boldsymbol{\beta}_0 = \mathbf{0}$ and $\Sigma_0 = 1000\mathbf{I}_3$, $\mathbf{0}$ indicates a 3×1 vector of zeros and $\mathbf{I}_3$ the identity matrix of order 3. For ν, we assume a noninformative prior as in Fonseca et al. (2008).

8.5.1.1 *Computation Details*

As in Abanto-Valle and Dey (2014), we adopt the so-called threshold approach (Albert and Chib, 1993), where y_t is created through dichotomization of a latent continuous process Z_t, given by the one-to-one correspondence

$$y_t = 1 \Leftrightarrow u_t > 0, \qquad t = 1, \ldots, T. \tag{8.25}$$

With the unobservable or latent threshold variable vector $u_t = (u_1, \ldots, u_T)$, equations (8.22) and (8.23) can be rewritten as

$$u_t = \mathbf{z}_t'\beta + x_t + \lambda_t^{-1/2}\epsilon_t, \tag{8.26}$$

$$\lambda_t \sim \mathcal{G}\left(\frac{\nu}{2}, \frac{\nu}{2}\right), \tag{8.27}$$

where the innovations ϵ_t are assumed to be mutually independent and normally distributed with mean zero and unit variance. A similar data augmentation scheme can be used if the link function is logistic instead of probit (Polson et al., 2013). Equations (8.24) and (8.26), conditioned on δ, the vector β, and the mixing variable λ_t, jointly represent a linear state space model, so the algorithm of de Jong and Shephard (1995) is used to simulate the states.

We fit an SSM to the binary observations using probit and Student-t links. For each case, we conducted MCMC simulation for 50,000 iterations. For both cases, the first 10,000 draws were discarded as a burn-in period. In order to reduce the autocorrelation between successive values of the simulated chain, every 20th value of the chain was stored. Posterior means and 95% credible intervals were calculated using the resulting 2000 values.

For SMC, we ran the adaptive random walk Metropolis for 300,000 iterations and discarded the first 100,000 while the particle filter was run with 100 particles only. Note that there are some differences in the estimates, but not much on the persistence parameter ϕ.

The R-INLA commands required for the analysis are described in the Appendix.

8.5.1.2 Results

The results obtained with the different approximation schemes are summarized later in this chapter. They show reasonable agreement between the different schemes for any given model. Results for the Student-t link are not presented for the INLA analysis because this option is not yet available in `R-INLA`.

The main results for the static parameters are summarized in Tables 8.1 and 8.2. In both models, the posterior means of δ are around 0.93–0.95, showing higher persistence of the autoregressive parameter for states variables and thus in the binary time series. The heaviness of the tails is measured by the shape parameter ν in the BSSM-T model. In Table 8.2, the posterior mean of ν is around 10–15. This result seems to indicate that the measurement errors of the u_t threshold variables are better explained by heavy-tailed distributions, as a consequence the t-links could be more convenient than the probit link. We found empirically that the influence of the number of body movements (z_1) is marginal, since the corresponding 95% credible intervals for β_1 contain the zero value. On the other hand, the influence of the number of body movements not due to sucking (x_2) is detected to be statistically significant. The negative value of the posterior mean for β_2 shows that a higher number of body movements not due to sucking will reduce the probability of the infant being in REM sleep.

Figures 8.1 and 8.2 show the posterior smoothed means of the probabilities π_t for both links considered. They show substantial agreement between the different schemes for both models. A more thorough comparison is also possible using fit criteria such as BIC or DIC. We found some differences between the fits from the different models, but in general the results are in accordance with Czado and Song (2008). Such difference between the methods

TABLE 8.1
Probit link: Estimation results for the Infant Sleep data set

	Probit Link		
Parameter	**MCMC**	**SMC**	**INLA**
β_0	−0.2248	0.1043	−0.0315
	(−1.4792, 0.8513)	(−3.0585, 3.8011)	(−2.5043, 2.6736)
β_1	0.2457	0.2532	0.2816
	(−0.1607, 0.8063)	(−0.1061, 0.6129)	(−0.1020, 0.6987)
β_2	−0.5183	−0.4421	−0.5057
	(−1.2111, 0.0002)	(−0.9096, 0.0060)	(−1.0523, −0.0236)
δ	0.9321	0.9481	0.9334
	(0.8029, 0.9913)	(0.8609, 0.9966)	(0.8283, 0.9891)
κ	8.0121	6.9412	7.2467
	(0.7321, 38.6524)	(0.6011, 33.6659)	(0.9710, 31.9357)

Notes: First column, MCMC fit; second column, SMC fit; third column, INLA fit. In each case, we report the posterior mean and the 95% CI.

TABLE 8.2
Student-*t* link: Estimation results for the Infant Sleep data set

	Student-*t* Link	
Parameter	**MCMC**	**SMC**
β_0	−0.3309	0.3949
	(−2.5026, 1.0777)	(−4.6229, 7.5911)
β_1	0.4155	0.8406
	(−0.2594, 1.6057)	(−0.0173, 3.0087)
β_2	−0.7311	−1.2148
	(−2.3480, −0.0042)	(−4.0387, −0.1178)
δ	0.9506	0.9564
	(0.8029, 0.9913)	(0.8693, 0.9973)
κ	18.0723	17.1308
	(0.9921, 128.1231)	(0.7771, 110.7981)
ν	10.8673	13.0285
	(2.1299, 27.1029)	(0.5189, 55.1458)

Notes: First column, MCMC fit; second column, SMC fit. In each case, we report the posterior mean and the 95% CI.

are expected and are mostly within the corresponding Monte Carlo errors associated with the sampling procedures. Error evaluation for the INLA is yet to be derived.

8.5.2 Poliomyelitis in the U.S.

In this section, we consider a time series of monthly counts of cases of poliomyelitis in the United States between January 1970 and December 1983 (Zeger, 1988). The observations are displayed in Figure 8.3. This dataset has been frequently analyzed in the literature,

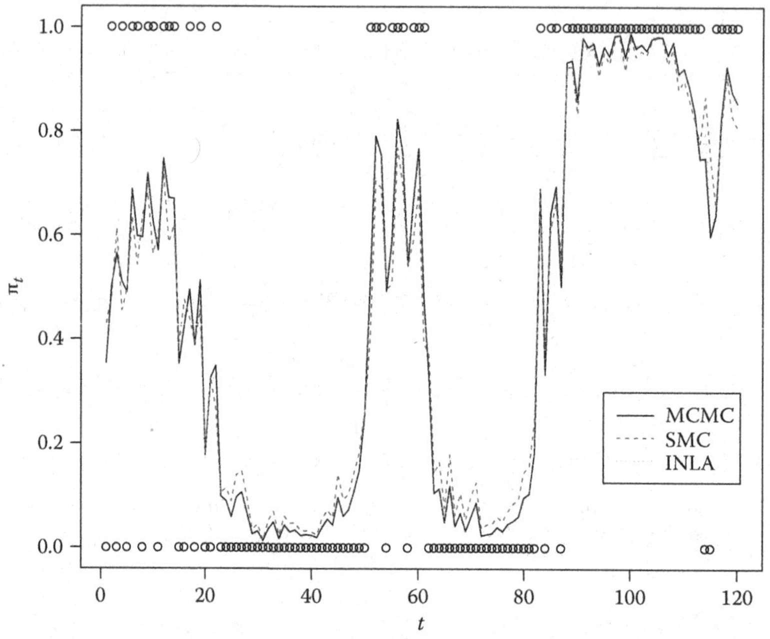

FIGURE 8.1
Probit link: Posterior smoothed mean of π_t applied to the infant sleep status data set.

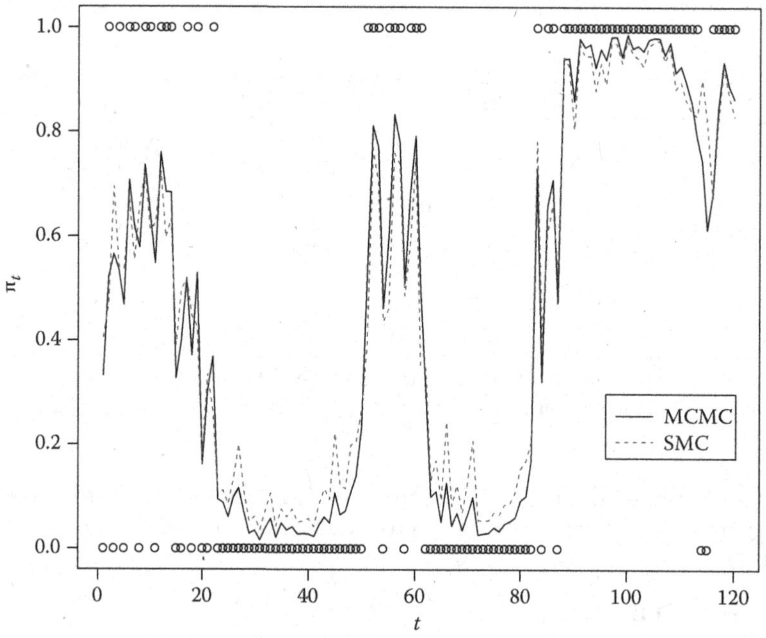

FIGURE 8.2
Student-t link: Posterior smoothed mean of π_t applied to the infant sleep status data set.

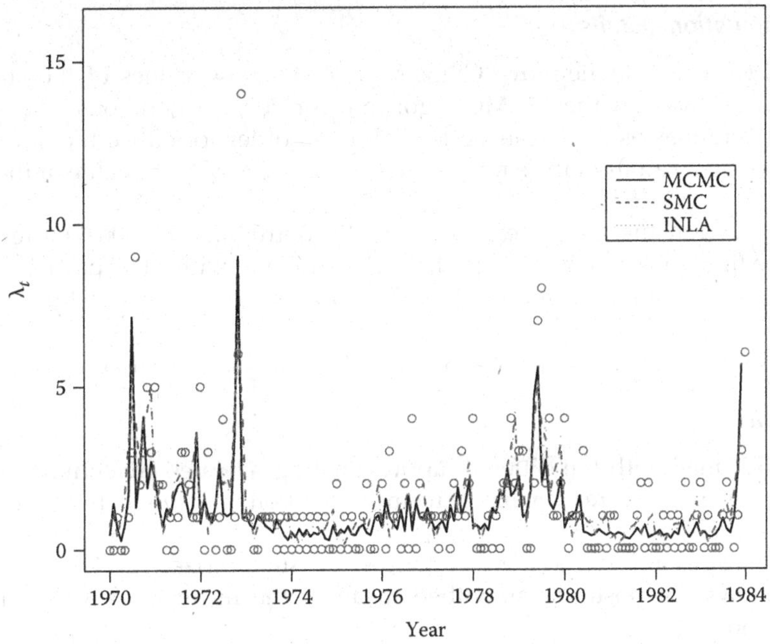

FIGURE 8.3
Polio counts (points) in the United States, January 1970–December 1983, and posterior smoothed mean sequence, λ_t, of the fitted seasonal Poisson SSM without overdispersion.

for example, by Chan and Ledolter (1995), Le Strat and Carrat (1999), and Davis and Rodriguez-Yam (2005). We concentrate below on the estimation of the mean response and the hyperparameters.

We adopt the loglinear seasonal Poisson SSM defined by

$$
\begin{aligned}
y_t &\sim Po(\lambda_t) \\
\ln \lambda_t &= \mu_t + s_t + \gamma_t \\
\mu_t &= \mu_{t-1} + \omega_{1t} \\
s_t &= -(s_{t-1} + \ldots + s_{t-p+1}) + \omega_{2t} \\
\gamma_t &\sim \mathcal{N}(0, W_3),
\end{aligned}
$$

where $\omega_{it} \sim \mathcal{N}(0, W_i)$ independent for $i = 1, 2$ and $p = 12$ is the seasonal period.

We fit the seasonal Poisson SSM without and with overdispersion. First we take $W_3 = 0$ which means $\gamma_t = 0$ for all t. Then, we take γ_t as described above such that the model can handle with overdispersion. We set the prior distributions as $W_i^{-1} \sim$ Gamma $(0.01, 0.01)$ for $i = 1, 2, 3$, $x_0 \sim \mathcal{N}(0, 10{,}000)$ and $s_j \sim \mathcal{N}(0, 10{,}000)$ for $j = -10, \ldots, 0$.

8.5.2.1 Computation Details

For MCMC, we use a single-move Gibbs sampler to draw values of the states λ_t and s_t in both cases. We run the MCMC algorithm for 700,000 iterations. We discard the first 100,000 iterations as a burn-in period. Next, in order to reduce the autocorrelation between successive values of the simulated chain, only every 100th value of the chain was stored.

For SMC, we ran the adaptive random walk Metropolis for 300,000 iterations and discarded the first 100,000, while the particle filter was run with 2,000 particles.

For INLA, the R commands required for the analysis are described in the Appendix.

8.5.2.2 Results

The results obtained with the different approximating schemes are summarized later in this chapter. They show reasonable agreement for both models between the different approximating schemes.

The main results for the system variances are summarized in Tables 8.3 and 8.4. Figure 8.4 shows the posterior smoothed mean of the mean rates λ_t for model with overdisperssion.

TABLE 8.3

Summary of the posterior distribution of the Poisson model without overdispersion applied to the Polio data set

	Poisson without Overdispersion		
Parameter	**MCMC**	**SMC**	**INLA**
W_1	0.0841	0.0889	0.0853
	(0.0201, 0.1953)	(0.0281, 0.2012)	(0.0273, 0.1939)
W_2	0.0209	0.0215	0.0206
	(0.0034, 0.0689)	(0.0039, 0.0679)	(0.0039, 0.0637)

TABLE 8.4

Summary of the posterior distribution of the Poisson model with overdispersion applied to the Polio data set

	Poisson without Overdispersion		
Parameter	**MCMC**	**SMC**	**INLA**
W_1	0.0490	0.0472	0.0460
	(0.0089, 0.1485)	(0.0109, 0.1261)	(0.0107, 0.1187)
W_2	0.0162	0.0157	0.0146
	(0.0028, 0.0601)	(0.0032, 0.0502)	(0.0031, 0.0442)
W_3	0.2526	0.2648	0.2345
	(0.0185, 0.6822)	(0.0256, 0.6022)	(0.0319, 0.5339)

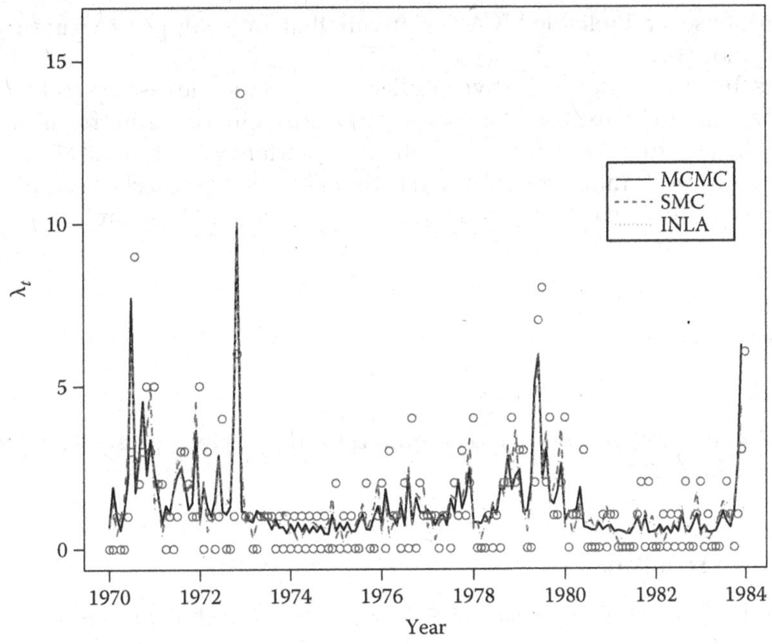

FIGURE 8.4
Polio counts (points) in the United States, January 1970–December 1983, and posterior smoothed mean sequence, λ_t, of the fitted seasonal Poisson SSM with overdispersion.

8.6 Final Remarks

This chapter presents an overview of the possibilities associated with the analysis of discrete time series with SSM under the Bayesian perspective. A number of frequently used model components are described, including autoregressive dependence, seasonality, overdispersion, and transfer functions. The techniques most commonly used nowadays (MCMC, SMC, INLA) to analyze these models are also described and illustrated.

INLA is by far the fastest method for Bayesian data analysis, but suffers from a lack of measures to quantify and assess the errors committed. Also, it is not trivial to set the grid of values for computation of the integrals involved. This shortcoming is largely mitigated by the availability of the software R-INLA, but users become restricted to the options available there and are unable to easily introduce their own building blocks into the software.

SMC enables fairly fast generation of online posterior distributions of the state parameters and the hyperparameters. If one is interested in their smoothed distributions, as we showed in our illustrations, then their processing time becomes nonnegligible. This is still a lively area of research, and we may see in the near future ways to avoid the currently high computing cost of smoothed distributions. Some of the techniques associating SMC with MCMC seem to be promising.

Finally, MCMC is the standard approach for Bayesian data analysis these days. It is time consuming and requires nontrivial tuning when full conditionals are not available, but is well documented and has been used by many users under a large variety of settings.

Further, there are several reliable MCMC software that are equipped to handle a variety of sophisticated models.

The results that we present in the two applications show some discrepancies between the methods. These are to be expected since there are errors involved in the approximations. However, we believe that these differences did not interfere with the overall analysis in any meaningful way. In summary, the message is that all these approaches can be safely used (especially in the linear realm) with the guidance provided in this chapter.

8A Appendix

This Appendix contains the commands required for the analysis of the datasets of Section 5 with the R package `INLA`.

8A.1 Deep Brain Stimulation

The first step is to load the R package `INLA` and set-up a `data.frame` that will be used within the `inla()` function.

```
require(INLA)

# Create data.frame
data_frame = data.frame(y = y, z = z, time = 1:length(y))

head(data_frame, 3)
  y z1 z2 time
1 0  1  1    1
2 1  1  1    2
3 0  3  1    3
```

The response variable `y` takes the value 1 to indicate a REM sleep cycle and zero otherwise. `z1` and `z2` stand for the z covariates described in Section 8.5.1 and `time` assumes the values $t = 1, \ldots, n$, where n is the number of observations, and will be used to model the time-specific effect x_t on the observed process. The model can then be fitted with the command line

```
inla1 <- inla(formula = y ~ 1 + z1 + z2 +
                f(time, model = "ar1", hyper = hyper.ar1),
              family = 'binomial',
              data = data_frame,
              control.fixed = control.fixed(prec.intercept=0.001,
                                            prec = 0.001),
              control.family = list(link = "probit"),
              control.predictor = list(compute = TRUE))
```

The `formula` argument of our model assumes that our linear predictor has an intercept, two covariates `z1` and `z2` and a time-specific effect indexed by `time`. The `control.fixed` argument was used to set the precision of the priors for the fixed-effects equal to 0.001, which leads to a variance equal to 1000. The model for the time-specific effect was set to follow an autoregressive model of order 1 and the priors for its hyperparameters δ and τ were set by `hyper = hyper.ar1` inside the `f()` function, where

```
hyper.ar1 = list(prec = list(prior="loggamma",
                             param = c(0.01, 0.01)),
                 rho = list(prior = table))
```

and `table` was defined in such a way to encode $\delta \sim \mathcal{N}_{(-1,1)}(0.95, 100)$. For detailed information on how to set priors, please visit http://www.r-inla.org/models/priors.

The `family` argument indicates that we are dealing with binomial data. The `control.family` tells INLA that we want a probit link function. Lastly, besides the marginal posterior distribution of the latent field x and the elements of hyperparameters θ, `control.predictor` sets INLA to compute the marginal posterior distribution of the linear predictors $\eta(x)$ as well, which is not done by default.

8A.2 Poliomyelitis in the U.S.

Our `data.frame` contains the following columns: `y`, `time`, `seasonal` and `iid`, which represent the response variable y_t and the indexes for the time-specific effect μ_t, the seasonal effect s_t and the i.i.d. random-effects γ_t, respectively.

```
n = length(y)
data_frame = data.frame(y = y, time = 1:n,
                        seasonal = 1:n, iid = 1:n)
```

Notice that although `time`, `seasonal`, and `iid` assume identical values, the `inla()` function require the definition of a different index for each random-effect defined through the `f()` functions.

We first fit the model without accounting for overdispersion with command line

```
n.seas = 12
formula1 = y ~ -1 +
           f(time, model = "rw1", constr=FALSE, hyper = hyper.rw1) +
           f(seasonal, model = "seasonal", season.length = n.seas,
             hyper = hyper.sea)
inla1 = inla(formula = formula1,
             data = data_frame,
             family="Poisson",
             control.predictor = control.predictor(compute=TRUE))
```

In the model definition, `-1` stands for the absence of an intercept in the model, the time-specific random-effects were chosen to follow a random-walk model of order 1 (rw1) without a sum-to-zero constrain t (`constr=FALSE`) and for the seasonal effect we specified that we were interested in a monthly frequency by setting `season.length = 12`. The priors for the hyperparameter of the rw1 and the seasonal model were defined using a similar syntax as the one displayed in Section 8.4.2. `family="Poisson"` indicates we are dealing with count data and that we will use a Poisson likelihood in our model.

We now proceed to a model that also includes random-effects to account for overdispersion. We have only redefined the arguments that change from one model to the other in order to save space, while the arguments that remain the same as in `inla1` have been represented by `*`. The command line becomes

```
formula2 = y ~ -1 +
               f(time, model = "rw1", constr=FALSE, hyper = hyper.rw1) +
               f(seasonal, model = "seasonal", season.length = n.seas,
                 hyper = hyper.sea) +
               f(iid, model = "iid", hyper = hyper.iid)
inla2 = inla(formula = formula2, *)
```

The only new feature of this model is the addition of the i.i.d. random-effects to account for overdispersion.

Acknowledgments

The authors thank an anonymous reviewer and the editors for their careful reading of this work, and for their comments and suggestions that helped to improve the presentation. Dani Gamerman was supported by a grant from CNPq-Brazil. Carlos A. Abanto-Valle was supported by grants from CNPq-Brazil and FAPERJ. Ralph S. Silva was supported by grants from CNPq-Brazil and FAPERJ.

References

Abanto-Valle, C. A. and Dey, D. K. (2014), State space mixed models for binary responses with scale mixture of normal distributions links, *Computational Statistics & Data Analysis*, 71, 274–287.

Albert, J. and Chib, S. (1993), Bayesian analysis of binary and polychotomous response data, *Journal of the American Statistical Association*, 88, 669–679.

Andrieu, C., Doucet, A., and Holenstein, R. (2010), Particle Markov chain Monte Carlo methods, *Journal of Royal Statistical Society, Series B*, 72, 1–33.

Carlin, B. P., Polson, N. G., and Stoffer, D. S. (1992), A Monte Carlo approach to nonnormal and nonlinear state-space modeling, *Journal of the American Statistical Association*, 87, 493–500.

Carter, C. K. and Kohn, R. (1994), On Gibbs sampler for State space models, *Biometrika*, 81, 541–553.

Carter, C. K. and Kohn, R. (1996), Markov chain Monte Carlo in conditionally Gaussian state space models, *Biometrika*, 81, 589–601.

Carvalho, C., Johannes, M., Lopes, H., and Polson, N. (2010), Particle learning and smoothing, *Statistical Science*, 25, 88–106.

Chan, K. S. and Ledolter, J. (1995), Monte Carlo EM estimation for time series models involving counts, *Journal of the American Statistical Association*, 90, 242–252.

Chib, S. and Greenberg, E. (1995), Understanding the Metropolis-Hastings algorithm, *The American Statistician*, 49, 327–335.

Czado, C. and Song, P. X.-K. (2008), State space mixed models for longitudinal observations with binary and binomial responses, *Statistical Papers*, 49, 691–714.

Davis, R. A. and Dunsmuir, W. T. M. (2015). State space models for count time series. In R. A. Davis, S. H. Holan, R. Lund and N. Ravishanker, eds., *Handbook of Discrete-Valued Time Series*, pp. 121–144. Chapman & Hall, Boca Raton, FL.

Davis, R. and Rodriguez-Yam, G. (2005), Estimation for state-space models based on a likelihood approximation, *Statistica Sinica*, 15, 381–406.

de Jong, P. and Shephard, N. (1995), The simulation smoother for time series models, *Biometrika*, 82, 339–350.

Durbin, J. and Koopman, S. J. (2001), *Time Series Analysis by State Space Methods*, Oxford, U.K.: Oxford University Press.

Fearnhead, P. (2011), MCMC for state space models, in *Handbook of Markov Chain Monte Carlo*, eds. S. Brooks, A. Gelman, G. L. Jones, and X.-L. Meng, New York: CRC, pp. 513–529.

Ferreira, M. A. R. and Gamerman, D. (2000), Dynamic generalized linear models, in *Generalized Linear Models: A Bayesian Perspective*, eds. D. K. Dey, S. K. Ghosh, and B. Mallick, New York: CRC, pp. 57–71.

Fonseca, T. C. O., Ferreira, M. A. R., and Migon, H. S. (2008), Objective Bayesian analysis for the Student-t regression model, *Biometrika*, 95, 325–333.

Gamerman, D. (1997), Efficient sampling from the posterior distribution in generalized linear mixed models, *Statistics and Computing*, 7, 57–68.

Gamerman, D. (1998), Markov chain Monte Carlo for dynamic generalized linear models, *Biometrika*, 85, 215–227.

Gamerman, D., Santos, T. R., and Franco, G. C. (2013), A non-Gaussian family of state-space models with exact marginal likelihood, *Journal of Time Series Analysis*, 35, 625–645.

Geweke, J. and Tanizaki, H. (2001), Bayesian estimation of state-space models using the Metropolis-Hastings algorithm within Gibbs sampling, *Computational Statistics & Data Analysis*, 37, 151–170.

Giordani, P. and Kohn, R. (2010), Adaptive independent Metropolis-Hastings by fast estimation of mixtures of normals, *Journal of Computational and Graphical Statistics*, 19, 243–259.

Gordon, N. J., Salmond, D. J., and Smith, A. F. M. (1993), A novel approach to non-linear and non-Gaussian Bayesian state estimation, *IEE Proceedings of Radar and Signal Processing*, 140, 107–113.

Harvey, A. C. (1989), *Forecasting, Structural Time Series Models and the Kalman Filter*, Cambridge, U.K.: Cambridge University Press.

Lee, Y. and Nelder, J. A. (1999), Hierarchical generalized linear models (with discussion), *Journal of the Royal Statistical Society, Series B*, 58, 619–678.

Le Strat, Y. and Carrat, F. (1999), Monitoring epidemiologic surveillance data using hidden Markov models, *Statistics in Medicine*, 18, 3463–3478.

Mallick, B. and Gelfand, A. (1994), Generalized linear models with unknown link function, *Biometrika*, 81, 237–245.

Martins, T. G., Simpson, D., Lindgren, F., and Rue, H. (2013), Bayesian computing with INLA: New features, *Computational Statistics & Data Analysis*, 67(C), 68–83.

Meinhold, R. J. and Singpurwalla, N. D. (1989), Robustification of Kalman filter, *Journal of the American Statistical Association*, 84, 479–448.

Migon, H. S., Gamerman, D., Lopes, H. F., and Ferreira, M. A. R. (2005), Dynamic models, in *Hand-Book of Statistics - Bayesian Statistics: Modeling and Computation*, eds. C. R. Rao and D. K. Dey, Amsterdam, the Netherlands: Elsevier, pp. 553–588.

Nelder, J. A. and Wedderburn, R. W. M. (1972), Generalized linear models, *Journal of the Royal Statistical Society, Series A*, 135, 370–384.

Pitt, M. K. and Shephard, N. (1999), Filtering via simulation: Auxiliary particle filters, *Journal of the American Statistical Association*, 94, 590–599.

Pitt, M. K., Silva, R. S., Giordani, P., and Kohn, R. (2012), On some properties of Markov chain Monte Carlo simulation methods based on the particle filter, *Journal of Econometrics*, 171(2), 134–151.

Polson, N. G., Scott, J. G., and Windle, J. (2013), "Bayesian inference for logistic models using PólyaGamma latent variables, *Journal of the American Statistical Association*, 108, 1339–1349.

R Core Team (2013), *R: A Language and Environment for Statistical Computing*, R Foundation for Statistical Computing, Vienna, Austria. http://www.R-project.org/.

Roberts, G. O., Gelman, A., and Gilks, W. R. (1997), Weak convergence and optimal scaling of random walk Metropolis algorithms, *Annals of Applied Probability*, 7(1), 110–120.

Roberts, G. O. and Rosenthal, J. S. (2009), Examples of adaptive MCMC, *Journal of Computational and Graphical Statistics*, 18, 349–367.

Rue, H. and Held, L. (2005), *Gaussian Markov Random Fields: Theory and Applications*, Vol. 104 of *Monographs on Statistics and Applied Probability*, London, U.K.: Chapman & Hall.

Rue, H., and Martino, S. (2007), Approximate Bayesian inference for hierarchical Gaussian Markov random field models, *Journal of Statistical Planning and Inference*, 137(10), 3177–3192.

Rue, H., Martino, S., and Chopin, N. (2009), Approximate Bayesian inference for latent Gaussian models by using integrated nested Laplace approximations, *Journal of the Royal Statistical Society: Series B(Statistical Methodology)*, 71(2), 319–392.

Ruiz-Cárdenas, R., Krainski, E., and Rue, H. (2011), Direct fitting of dynamic models using integrated nested Laplace approximations–INLA, *Computational Statistics & Data Analysis*.

Shephard, N. and Pitt, M. K. (1997), Likelihood analysis of non-Gaussian measurement time series, *Biometrika*, 84, 653–667.

Stoffer, D. S., Schert, M. S., Richardson, G. A., Day, N. L., and Coble, P. A. (1998), A Walsh-Fourier analysis of the effects of moderate maternal alcohol consumption on neonatal sleep-state cycling, *Journal of the American Statistical Association*, 83, 954–963.

Tierney, L. (1994), Markov chains for exploring posterior distributions, *Annals of Statistics*, 22(4), 1701–1728.

Tierney, L. and Kadane, J. (1986), Accurate approximations for posterior moments and marginal densities, *Journal of the American Statistical Association*, 81(393), 82–86.

Valdebenito, A., Arellano-Valle, R., Romeo, J.S., Torres-Avilés, F. (2015), A skew-normal dynamic linear model and bayesian forecasting. Universidad de Chile. (Submitted).

Watanabe, T. (2004), A multi-move sampler for estimating non-Gaussian time series models: Comments on Shephard & Pitt (1997), *Biometrika*, 91, 246–248.

West, M. and Harrison, J. (1997), *Bayesian Forecasting and Dynamic Models*, 2nd ed., New York: Springer-Verlag.

West, M., Harrison, P. J., and Migon, H. S. (1985), Dynamic generalized linear models and Bayesian forecasting (with discussion), *Journal of the American Statistical Association*, 80, 73–83.

Zeger, S. (1988), A regression model for time series of counts, *Biometrika*, 75, 621–629.

Section II

Diagnostics and Applications

9

Model Validation and Diagnostics

Robert C. Jung, Brendan P.M. McCabe, and A.R. Tremayne

CONTENTS

9.1 Introduction

Checking the adequacy of a specified model is an important part of any iterative modeling exercise in applied time series analysis. For linear Gaussian time series models, or those based on the framework of generalized linear models, there exist well-developed tools for this purpose that are readily available and routinely employed in applied work. However, for nonlinear time series models for discrete data, this is not the case. Nevertheless, the need to compare two or more competing model specifications, or evaluate the adequacy of fit of a chosen model, is obvious.

To help address this gap, we suggest a range of diagnostic and model validation methods designed to lead to data coherent models that achieve good probabilistic forecasting outcomes. We borrow from the associated literature developed mainly for continuous variables, adapting them where necessary for the discrete context. This leads us to advocate a set of graphical tools and other calibration methods of various kinds.

However, achieving the desired aim may not be straightforward, as the following quote taken from the important paper of Tsay (1992, p. 2) indicates *However, it is well known that the best model with respect to one checking criterion may fare badly with respect to another criterion. ... Consequently, there is a need to specify the objective of data analysis before choosing a checking criterion....Without mentioning objectives, reported model checking statistics are meaningless or could be misleading*. As suggested earlier, our standpoint is that we consider the model class to be introduced next as primarily of use in a probabilistic forecasting sense, including a need to provide not only point forecasts but also good estimates of entire forecast distributions. Hence, we focus our coverage on methods that may help to achieve good outcomes in this respect.

We now briefly introduce the class of integer autoregressive models that will be used as a vehicle to demonstrate the application of the diagnostic tools described in the chapter. When used for other model classes presented in this volume, appropriate adaptations may be necessary. An integer autoregressive process $\{X_t; t = 0, \pm1, \pm2, \ldots\}$ of order p defined on the state space of nonnegative integers is of the form

$$X_t = R_t(\mathcal{F}_{t-1}; \alpha) + \varepsilon_t, \tag{9.1}$$

where $\mathcal{F}_{t-1}$ indicates the relevant past history of X_t to be conditioned on, typically $X_{t-1}, \ldots, X_{t-p}$ in a pth-order model, and ε_t are a sequence of *i.i.d.* discrete random variables. The innovation process ε_t and $\mathcal{F}_{t-1}$ are presumed to be stochastically independent for all points in time. This model specification is inspired by the work of Joe (1996), from which it follows, *inter alia*, that it is often a Markov chain (of some order).

In (9.1), $R_t(\cdot)$ denotes a random operator to be applied at time t (which may differ from specification to specification) that carries the dependence structure and preserves the integer nature of the process. Some practical examples of these random operators are given in the following. Perhaps unsurprisingly, alternative choices of the operator $R_t(\cdot)$ and the innovations ε_t lead to a rich class of models, see, for example, the survey by McKenzie (2003). A variant of (9.1) that is popular in the literature is the following:

$$X_t = \alpha_1 \diamond X_{t-1} + \alpha_2 \diamond X_{t-2} + \cdots + \alpha_p \diamond X_{t-p} + \varepsilon_t \tag{9.2}$$

where, conditional on X_{t-k}, $\alpha_k \diamond X_{t-k}$ is an integer-valued random variable (using operator $\diamond$) with parameter α_k (possibly a vector). The conditional variables $\alpha_k \diamond X_{t-k}$, $k \in \{1, \ldots, p\}$ are mutually independent and independent of the *i.i.d.* innovations sequence ε_t. The operator thus delivers an integer value, and dependence in X_t is induced via the conditioning variables X_{t-k}, $k \in \{1, \ldots, p\}$. The operator used in $\alpha_k \diamond X_{t-k}$ may correspond to binomial thinning and ε_t to a Poisson variable with parameter λ. Then the conditional variables $\alpha_k \diamond X_{t-k}$, $k \in \{1, \ldots, p\}$ have independent binomial distributions with parameters α_k and X_{t-k}. Another possibility is that, conditional on X_{t-k}, $\alpha_k \diamond X_{t-k}$ is beta-binomial, while ε_t is negative binomial. For all these pth order model variants of the form (9.2), the acronym INAR(p) has been introduced.

The special case of (9.1) when $p = 1$ is of importance. Under binomial thinning and Poisson innovations, X_t has a Poisson marginal distribution because closure under convolution applies. This is probably the workhorse model of integer time series modeling. As it can be written in the form of (9.2), it will henceforth be denoted a PINAR(1) model. If, however, ε_t were to be generalized Poisson (GP) random variables, then, to preserve a GP marginal distribution for X_t, the random operator $R_t(\cdot)$, conditional on X_{t-1}, would yield a quasi-binomial distribution; such a model will be denoted GP(1) in the following. Further, if ε_t is negative binomial and, conditional on X_{t-1}, $R_t(\cdot)$ is beta-binomial, then X_t also has a negative binomial marginal distribution (see, e.g., McKenzie, 2003, p. 586).

In what follows we use the PINAR(1) as a first specification to fit to two real life data sets and consider at various subsequent points in the chapter the evidence that this simple specification needs to be elaborated. We seek to reveal data coherent models for each data set in the light of our diagnostic analyses.

Of course, there are a number of avenues that can be used to assess the evidence against the suitability of a specified model. Issues to be considered include (but would not necessarily be limited to) the type of random operator $R_t(\cdot)$ chosen, relevant past history $\mathcal{F}_{t-1}$, the distributional properties of ε_t, and the need to introduce regression effects in some way. Evidently, the third of these might be obviated by using the semiparametric approach of McCabe et al. (2011), but this can introduce added complications when looking at the last and so we do not consider this approach further in our contribution.

The plan of the chapter is as follows. Sections 9.2 through 9.4 provide a description of the diagnostic methods surveyed together with the results of their application to two real data sets. In Section 9.5, we use simulated data to highlight specific properties of the various methods discussed. Finally, Section 9.6 contains concluding remarks.

9.2 Parametric Resampling

A very general informal approach to model diagnostics for time series is proposed by Tsay (1992). He demonstrates the procedure by employing the sample spectral density function of any process as a functional of interest. This is closely related to the (sample) autocorrelation function, (S)ACF, to be used here, since it is a cosine transformation of the spectrum. The flexibility of Tsay's approach stems from the fact that it not only provides an overall evaluation of the fitted model but also can be tailored to meet certain specific needs of the analysis. The procedure is widely applicable and rests on a fairly minimal set of requirements. Although bootstrap methods are ubiquitous, the caveat that they often do depend on asymptotic theory (and sometimes on distributional assumptions) is in order. In our context, the approach emphasizes reproducibility in fitted models and is designed to provide overall evaluation of fit or to check special characteristics of a process. Moreover, the approach can be readily applied to time series models of counts as it is straightforward to implement the data-generating process (DGP) of most of them in standard software packages.

Only the following requirements need to be fulfilled for the implementation of Tsay's proposal: a parametric model of mathematical form with given parameters and a specified distribution for innovations; and one, or more, characteristics or functionals that encapsulate special features of interest. No further restrictions, other than that the model can be used to generate bootstrap samples, apply. Based on artificially generated sample processes, an empirical distribution of the specified functional (in our case, ordinates of sample autocorrelation functions) is obtained. The adequacy of a fitted model is then assessed by comparing this empirical distribution to the corresponding functional quantity of the data itself. A model may be regarded as adequate if it successfully reproduces the observed characteristics of the actual data. Specifically, for each fixed lag of the autocorrelation function, the $100(1-\alpha/2)$ and $100(\alpha/2)$ quantiles (we use $\alpha = 0.05$ for graphical displays in what follows) can be computed to constitute the bounds of an acceptance envelope. If the sample autocorrelations of the data predominantly lie within the acceptance envelopes, the fitted model can be deemed adequate according to the functional chosen. Notice that this is not an interval estimation procedure as such, so one cannot reason that such an envelope will contain the true value of any functional $100(1-\alpha)\%$ of the time in repeated applications; see Tsay (1992, Sec. 2.2) for related discussion.

As we shall use this parametric resampling procedure regularly in this chapter as a tool to assess a fitted model's adequacy, it seems appropriate to first examine how the procedure operates in a stylized setting. We, therefore, conduct pilot Monte Carlo experiments in which the model fitted to artificially generated data is a PINAR(1). The data itself are generated in two ways: first, when the true model is fitted, that is, the data itself follow a PINAR(1) process in truth; and, second, when the true DGP is an INAR(2) of the form (9.2) with Poisson innovations. In the former case, the mean and variance of the marginal distribution of the data are equal and the autocorrelation function is the same as that of the Gaussian AR(1) continuous counterpart. In the second case, the true marginal distribution of the data is not Poisson, there is some overdispersion, and the true autocorrelation function of the process is equivalent to that of a Gaussian AR(2) process. We anticipate that application of the Tsay procedure under the first scenario will indicate no model misspecification and the contrary under the second.

The functionals that we use in this illustrative experiment are as follows: the variance and the first four ordinates of the autocorrelation function. Artificial data are generated from the two specifications and the relevant sample functionals, the sample variance, and the first- through fourth-order sample autocorrelations, denoted SACF(1)–SACF(4) in the following, are calculated for the generated data. A PINAR(1) model is fitted by maximum likelihood (ML) and, using the resultant parameter estimates, B bootstrap samples are generated and the same sample functionals computed. We determine the percentage of times the functionals of the data are covered by a $100(1-\alpha)$ probability interval (for $\alpha = 0.01, 0.05$, and 0.1) constructed from the bootstrap replicates of the resampling procedure. This parallels the procedure described by Tsay (1992, p. 4), and is repeated R times to provide an indication of the performance of the procedure.

In the first experiment, data series of length $T = 500$ are generated from a PINAR(1) model with the following parameter values: $\alpha_1 = 0.8$ and $\lambda = 0.4$. This leads to Poisson distributed count time series with (theoretical) mean and variance of 2 and first four autocorrelations $0.8, 0.64, 0.51$, and $0.8^4 = 0.41$, respectively. The results presented here are based on $R = 1000$ replications, and for each generated series, we perform the parametric resampling procedure as described in the previous paragraph using $B = 5000$ replications.

To provide some information on the sampling variability that can be expected when the true model is fitted to the data, we present the average quantiles for the sample functionals over the 1000 replications for the first experiment in the upper panel of Table 9.1. From this, it is evident that, on average, the sampling distributions of the functionals are centered quite close to the true values used to generate the data. The lower panel of Table 9.1 shows what happens if the sample size is varied from $T = 500$ to $T = 250$. Broadly, increased sampling variability of the anticipated type is seen in the average newly estimated quantiles.

The left panel of Table 9.2 provides the percentages with which the functionals of the data are covered by the three acceptance bounds used in this experiment and a correct model is fitted. It is evident that, in all cases, these percentages show that sample functionals outside the envelopes occur less often than might be expected. We conducted a further experiment to vary the dependence in the generated process (using $\alpha_1 = 0.5$ and $\lambda = 1$) to see if the results were sensitive to this variation, but they were not. The results indicate that the Tsay procedure will generally confirm a correctly specified model.

TABLE 9.1

Average Quantiles from $R = 1000$ Replications for the Monte Carlo Experiment for the Tsay Resampling Procedure When a True PINAR(1) Model is Fitted for $T = 500$ (Upper Panel) and $T = 250$ (Lower Panel)

Quantile (%)	**0.5**	**2.5**	**5**	**50**	**95**	**97.5**	**99.5**
Functional							
				$T = 500$			
Sample variance	1.286	1.417	1.489	1.935	2.537	2.675	2.990
SACF(1)	0.703	0.726	0.737	0.791	0.836	0.844	0.859
SACF(2)	0.482	0.518	0.536	0.624	0.702	0.716	0.742
SACF(3)	0.315	0.359	0.381	0.492	0.549	0.612	0.647
SACF(4)	0.188	0.237	0.261	0.387	0.506	0.527	0.569
				$T = 250$			
Sample variance	1.074	1.228	1.316	1.895	2.767	2.980	3.451
SACF(1)	0.652	0.687	0.704	0.782	0.845	0.856	0.874
SACF(2)	0.403	0.457	0.483	0.610	0.718	0.736	0.770
SACF(3)	0.221	0.285	0.316	0.474	0.614	0.639	0.684
SACF(4)	0.088	0.156	0.190	0.367	0.530	0.559	0.613

TABLE 9.2

Inclusion Rates of Sample Functionals from $R = 1000$ Replications for the Tsay Resampling Procedure When a PINAR(1) Model Is Fitted and the True DGP Includes PINAR(1) (Left Panel) and INAR(2) (Right Panel)

Acceptance Bounds	**90%**	**95%**	**99%**	**90%**	**95%**	**99%**
Functional						
Sample variance	95.20	98.40	99.80	10.30	15.30	30.00
SACF(1)	93.60	98.60	99.90	23.60	36.60	61.00
SACF(2)	93.80	97.60	99.40	0.00	0.00	0.00
SACF(3)	93.20	96.10	99.30	0.00	0.20	0.50
SACF(4)	92.30	96.30	99.10	0.00	0.10	0.80

On the other hand, when an inadequate model is fitted (refer to the right-hand panel Table 9.2), all functionals are able to indicate this on a regular basis. These results are obtained by using an INAR(2) data-generating mechanism with $\alpha_1 = 0.45, \alpha_2 = 0.35$, and $\lambda = 0.4$. This generates data that have a true mean of 2, variance = 2.95, and first four autocorrelation ordinates equal to 0.692, 0.662, 0.540, and 0.475, respectively. Again, $T = 500$, $R = 1000$, and $B = 5000$ are used. Thus, the Tsay procedure does show an ability to detect an incorrectly fitted model, in this pilot experiment at least. In any instance where it indicates a model's inadequacy, a search for a more refined (or different) model specification should be undertaken.

Jung and Tremayne (2011b) applied the method previously with integer time series (though without any examination of its empirical performance, limited evidence on which is provided earlier). Grunwald et al. (1997) report that the procedure is able to discover some surprising results in the context of Bayesian time series models that would have not been detected on the basis of a set of standard residual diagnostics. In addition, Pavlopoulos and Karlis (2008) apply the method by using tail probabilities and log-moments as their functionals of interest. The usefulness of the parametric bootstrap approach is particularly highlighted here, as formulae for both characteristics are generally unavailable for count time series models.

We now introduce two real-life data sets to be employed for illustrative purposes at various points in the chapter. An initial (PINAR(1)) model is fitted to each, and the Tsay procedure of this section is then used to assess a model's suitability.

The first one consists of 120 counts of claimants collecting wage loss benefit for injuries in the workplace at one specific service delivery location of the Workers Compensation Board of British Columbia, Canada. Only injuries due to cuts and lacerations are considered; we refer to these data as the cuts data. The data set is monthly and covers the period from January 1985 to December 1994. It has been analyzed previously by Freeland and McCabe (2004), for example. The sample mean and variance of the data are 6.133 and 11.797, respectively, and Figure 9.1 provides a time series plot together with a summary of sample serial correlation and partial correlation properties. Note that the latter should be thought of as a heuristic for count data, because, to the best of our knowledge, theoretical results for partial autocorrelations with non-Gaussian data have not yet been established.

Our second exemplar data set consists of counts of iceberg orders in the order book on the ask side of Deutsche Telekom stock sampled every 15 min on the XETRA system operated by the Deutsche Börse. Iceberg orders constitute a particular type of order in many limit order markets. Their nomenclature is derived from the fact that only a small part of the order (the tip of the iceberg) is visible in the order book, while the reminder of the order is hidden. Iceberg order data on different stocks from the one considered here have been studied elsewhere, for example, by Jung and Tremayne (2011b) and McCabe et al. (2011). The data set analyzed relates to 32 trading days in the first quarter of 2004. With 8.5 h of trading per day (9 am to 5:30 pm), there are 34 observations per day giving a count time series of 1088 observations. The sample mean and variance are, respectively, 1.39 and 2.08. Figure 9.2 depicts the raw data and its serial correlation structure. These data, which we refer to as the iceberg order data, evidence a dynamic structure that decays more slowly than the cuts data.

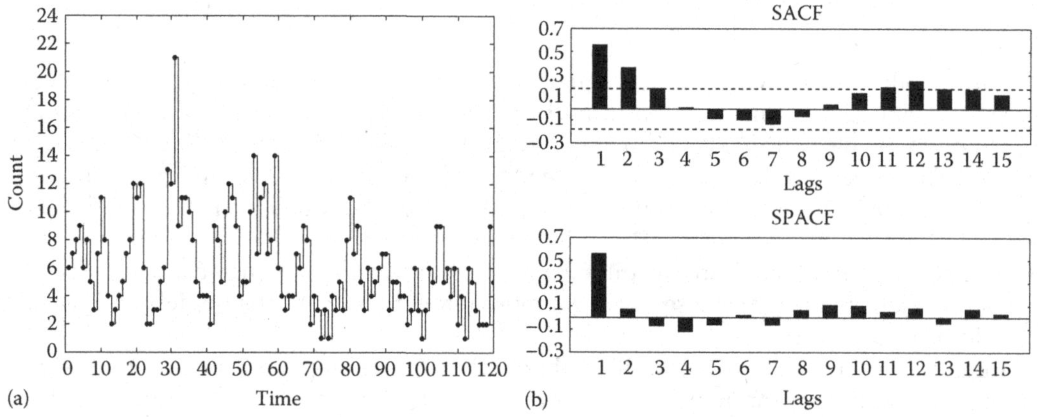

FIGURE 9.1
Time series plot (a) and SACF/SPACF plots (b) of the cuts data.

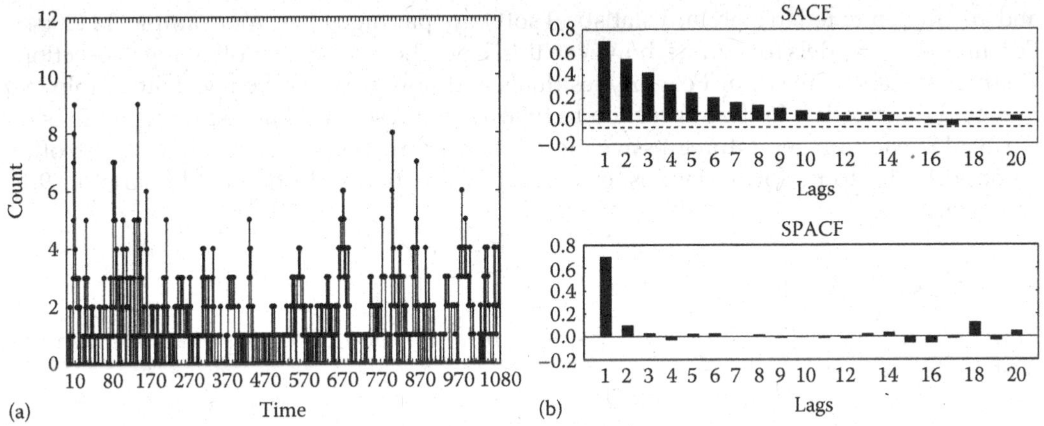

FIGURE 9.2
Time series plot (a) and SACF/SPACF plots (b) of the iceberg order data.

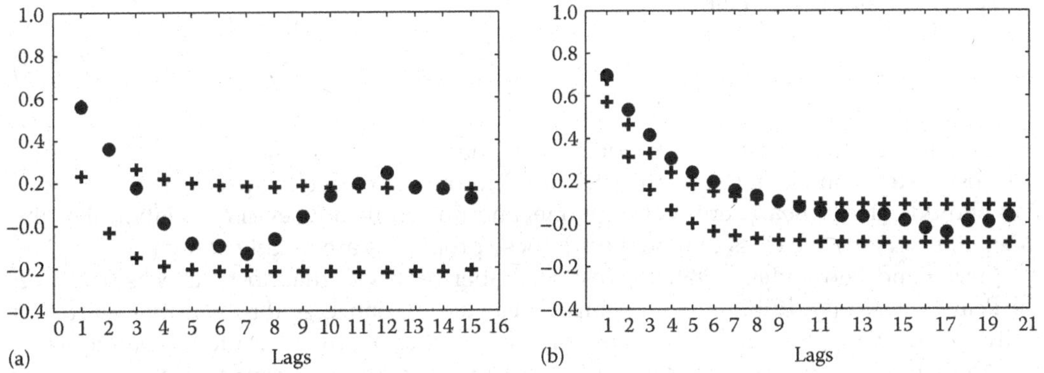

FIGURE 9.3
Parametric resampling diagnostics after fitting a PINAR(1) model to the cuts data (a) and the iceberg order data (b). The 95% acceptance bounds are shown as "+" symbols and the SACF ordinates by "•."

Figure 9.3 provides evidence of the usefulness of the parametric resampling method described earlier and displays relevant graphs after a prototypical PINAR(1) model has been fitted to each data set. The left panel refers to the cuts data and the right panel to the iceberg order data. Both indicate that neither fitted model adequately captures the dynamics in the respective data set, situations that we shall seek to remedy in due course.

9.3 Residual Analysis

Diagnostic checks based on model residuals have a long tradition in time series analysis; see, *inter alia*, discussion in the seminal work of Box and Jenkins (1970). In particular, for linear Gaussian time series models, both formal testing procedures and graphical tools exist and are implemented in standard statistical software packages; see, for example, Li (2004). For time series models for counts, however, this is not the case. In the following subsection, we discuss standardized, or Pearson, residuals and how they can be fruitfully employed in model diagnostics. In the subsection to follow, we present an interesting type of residual available for integer autoregressive models known as component residuals; these offer an opportunity to perform diagnostics on each of the two right-hand side parts of (9.1) separately.

9.3.1 Pearson Residuals

Raw residuals are defined as deviations of X_t from its conditional expectation given the past, that is, for $t = 1, \ldots, T$ as

$$r_t = X_t - E_{t-1}[X_t], \tag{9.3}$$

where E_{t-1} is the expectation taken conditional on $\mathcal{F}_{t-1}$, which contains the relevant past history of the process, including possible covariates. Pearson residuals are defined as the scaled version of the raw residuals

$$e_t = \frac{r_t}{\text{Var}_{t-1}[X_t]^{1/2}}. \tag{9.4}$$

For practical implementation, the population quantities in (9.4) have to be replaced by their estimated counterparts. If a model fitted to data is correctly specified, these residuals should exhibit mean zero, variance one, and no significant serial correlation. For the integer autoregressive class of model (9.1), these properties are readily shown.

Harvey and Fernandes (1989) suggest a number of model diagnostic checks based on the Pearson residuals utilizing their sample mean and variance (and proximity to zero and unity, respectively), together with assessing the presence of any unwanted dynamic structure in them via computing residual autocorrelations at different time lags and depicting them in a residual autocorrelation plot.

Figure 9.4 provides ACF plots of Pearson residuals for the two data sets introduced in Section 9.2 after fitting a PINAR(1) model to each. Note that these plots (as do all other

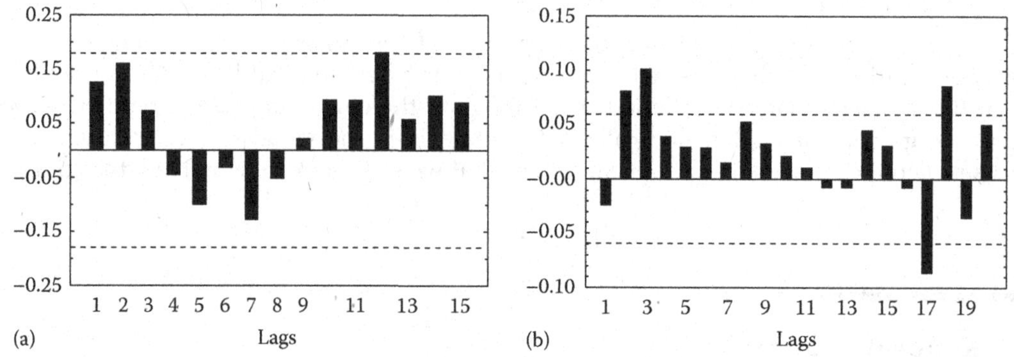

FIGURE 9.4
ACF plots of Pearson residuals after fitting a PINAR(1) model to the cuts data (a) and the iceberg order data (b).

residual ACF plots in the chapter) display dashed lines representing the usual approximate two standard error bounds for departure of the relevant ordinate from zero. The left panel refers to the cuts data set and the right panel to the iceberg order data. Both indicate that this simple model does not adequately capture the dependencies in the data. From the left panel, the dynamic misspecification previously observed in Figure 9.3 is not so evident. However, the residual autocorrelation at lag 12 appears quite large and may suggest a neglected seasonal component. The right panel, associated with the iceberg order data, tells a similar story to that already gleaned from Figure 9.3.

The sample means of the Pearson residuals for both data sets are close to zero. However, their variances are 1.607 and 1.289 for the cuts and iceberg order data, respectively. Both numbers are considerably larger than unity, suggesting potential misspecification of the Poisson innovation distribution specified in the PINAR(1) fitted model.

9.3.2 Component Residuals

At the outset, note that the two parts of the right-hand side of (9.1) and the special case (9.2) are typically unobserved. The first part can be thought of as a specification for the (random) number of survivors from stochastic operations performed at, or prior to, time t, or its complement, the number of departures. The second part reflects the number of new arrivals to the system at time t. It is transparent to derive the concepts in the following for the model variant provided in (9.2). For this case, though each of the $\alpha_k \diamond X_{t-k}$ is unobservable, following Freeland and McCabe (2004), we define a set ($t = 1, \ldots, T$) of departure residuals for each operator in (9.2) ($k \in \{1, \ldots, p\}$) by

$$r_{k,t} = E_t\left[\alpha_k \diamond X_{t-k}\right] - E_{t-1}\left[\alpha_k \diamond X_{t-k}\right], \tag{9.5}$$

where E_t is the expectation conditional on all information up to and including time t. Generally, $E_t\left[\alpha_k \diamond X_{t-k}\right] \neq E_{t-1}\left[\alpha_k \diamond X_{t-k}\right]$ as the conditioning sets are different. Similarly, define the set of arrivals residuals ($t = 1, \ldots, T$) as

$$r_{p+1,t} = E_t\left[\varepsilon_t\right] - E_{t-1}\left[\varepsilon_t\right]. \tag{9.6}$$

By considering the sum of the set of $p+1$ component residuals thereby defined as

$$\begin{aligned}\sum_{k=1}^{p+1} r_{k,t} &= \sum_{k=1}^{p} E_t\left[\alpha_k \diamond X_{t-k}\right] + E_t\left[\varepsilon_t\right] - \left[\sum_{k=1}^{p} E_{t-1}\left[\alpha_k \diamond X_{t-k}\right] + E_{t-1}\left[\varepsilon_t\right]\right] \\ &= E_t\left[X_t\right] - E_{t-1}\left[X_t\right] = X_t - E_{t-1}\left[X_t\right] = r_t,\end{aligned}$$

it is seen that the component residuals add up to the usual raw residuals for model (9.2).

One advantage of sets of residuals being associated with each unobserved part of the model is to offer the potential that they be used to identify the source of problems associated with a fitted model in the following way. Initially, any of this set of $p+1$ residuals may be used to check specification, either informally through the use of time series plots (or other graphical devices) and/or more formally through the construction of statistical specification tests. If some component residual indicates that the corresponding component of the model is not well specified, it may be possible to suggest modifications for improvement. For example, a cyclical pattern in a residual plot may indicate the presence of seasonality and this could formally be tested for using residual autocorrelation methods. Additional lags or covariates could then be added to improve the model. A little care must be taken, however, in examining several sets of residuals as these sets are correlated with one another (as are individual residuals within any set).

Just like the Pearson residuals of the previous subsection, component residuals have to be estimated as their definition requires knowledge of the α_k and whatever parameters are involved in ε_t. But, given estimates, residual sets are typically easy to compute by simply plugging in the estimates for the unknown parameters. As an example, consider the pth order model (9.2) where the thinning operator is the standard binomial thinning one and suppose the arrivals are *i.i.d.* Poisson random variables with mean λ. For this model, the conditional distribution of X_t given the past information $\mathcal{F}_{t-1} = (X_{t-1}, \ldots, X_{t-p})$ is based on the binomial distribution and is given by

$$P(X_t|\mathcal{F}_{t-1}) = \sum_{i_1=0}^{\min[X_{t-1},X_t]} \binom{X_{t-1}}{i_1} \alpha_1^{i_1}(1-\alpha_1)^{X_{t-1}-i_1} \sum_{i_2=0}^{\min[X_{t-2},X_t-i_1]} \binom{X_{t-2}}{i_2} \alpha_2^{i_2}(1-\alpha_2)^{X_{t-2}-i_2}$$

$$\cdots \sum_{i_p=0}^{\min[X_{t-p},X_t-(i_1+\cdots+i_{p-1})]} \binom{X_{t-p}}{i_p} \alpha_p^{i_p}(1-\alpha_p)^{X_{t-p}-i_p} \frac{e^{-\lambda}\lambda^{X_t-(i_1+\cdots+i_p)}}{[X_t-(i_1+\cdots+i_p)]!}. \quad (9.7)$$

The residual sets for $t=1,\ldots,T$ are as introduced in (9.5) and (9.6). A little algebra (see Freeland and McCabe 2004) shows

$$E_t\left[\alpha_k \circ X_{t-k}\right] = \frac{\alpha_k X_{t-k} P(X_t - 1|X_{t-1}, \ldots, X_{t-k}-1, \ldots, X_{t-p})}{P(X_t|\mathcal{F}_{t-1})} \quad (9.8)$$

and

$$E_t[\varepsilon_t] = \frac{\lambda P(X_t - 1|X_{t-1}, \ldots, X_{t-p})}{P(X_t|\mathcal{F}_{t-1})}, \quad (9.9)$$

where we use the convention that $P\left(i|j_1, \ldots, j_p\right) = 0$ if any of i or $j_1, \ldots, j_p$ is less than zero. Thus, given any suitable estimates $\hat{\alpha}_k$ and $\hat{\lambda}$, we can readily compute the estimated residual sets $\left\{\hat{r}_{k,t}; t = 1, \ldots, T\right\}$ using (9.7) in conjunction with (9.8) and (9.9) for $k \in \{1, \ldots, p+1\}$.

To demonstrate the use of component residuals, we compute and analyze them for the cuts data. Figure 9.5 provides ACF plots of the component residuals for the cuts data when the PINAR(1) model has been fitted. It can be compared with the left panel of Figure 9.4. The Pearson residual is decomposed into two components. Both sets of residuals show strong evidence of seasonality based on the graphical and correlation evidence, with much greater variability being seen in the arrivals process. The comparatively large residual ACF ordinates at lags 2 and 12 in Figure 9.4 are seen to be attributable to both departure and arrival residuals with regard to the latter ordinate, but only to the arrival component in the case of the former.

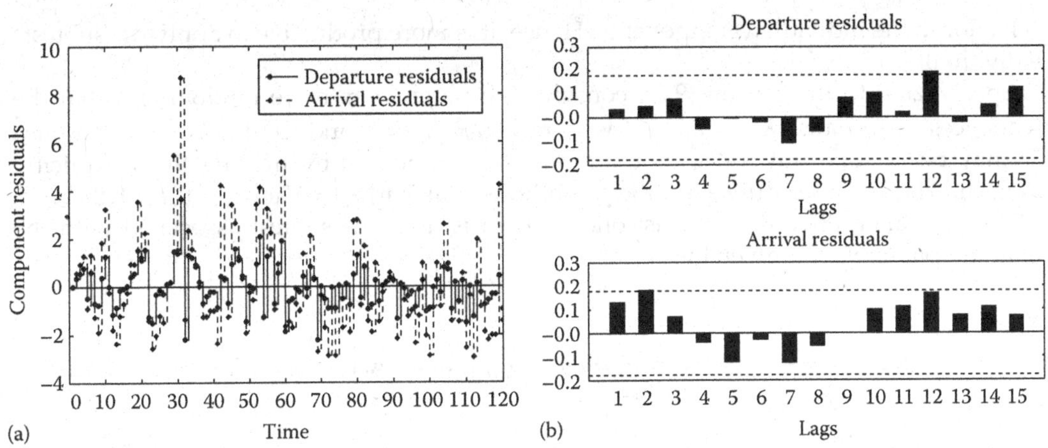

FIGURE 9.5
Time series plot (a) and ACF plots (b) of the component residuals from fitting a PINAR(1) model to the cuts data.

9.3.3 Overdispersion and the Information Matrix Test

A very general procedure to check if a model is correctly specified, the Information Matrix (IM) test, was proposed by White (1982). Basically, the IM test checks to see if the log-likelihood equality

$$-E\left[\frac{\partial^2 \ell}{\partial \theta^2}\right] = \text{Var}\left[\frac{\partial \ell}{\partial \theta}\right]$$

holds, as it should in a well-specified model. Here, ℓ is the log-likelihood of the model and θ is the parameter (which could be a vector). Another motivation for this test was given by Chesher (1983) who considered the parameter θ, under the misspecified alternative, to be random. Hence, the model is well specified when the variance of θ, Var(θ), is zero and checking Var[θ] $= 0$ is essentially the IM test. McCabe and Leybourne (2000) show that such tests, in the multiparameter case, are locally mean most powerful.

Freeland and McCabe (2004) investigated the behavior of the IM test in the context of integer time series with a focus on a random coefficient interpretation. Suppose a PINAR(1) model with fixed coefficients is to be tested against an alternative with random coefficients, where the thinning and arrivals parameters are considered random with beta and gamma distributions, respectively. Then, if the appropriate IM test rejects, a model with greater dispersion would be a natural candidate for consideration as a new specification.

When (as in the example of the previous paragraph) the parameter vector can be partitioned into natural subgroups, it turns out that the IM test can also be decomposed into subtests, each associated with a natural subgrouping of the parameters. Thus, for example, in the case of the INAR(p) model, there is a subtest associated with the parameters of the thinning operators and another associated with the arrivals process. Unfortunately, the combined IM test is often not very useful in practice as it may be difficult to interpret a rejection constructively and, even when it does not reject, it may obscure behavior in the individual components. Hence, it is more productive to apply the subtests individually.

Specializing to the case in (9.2), consider the parameters to be random and that the sequences $\{\alpha_t\}_{t=1}^T$ and $\{\lambda_t\}_{t=1}^T$ are *i.i.d* with means $E[\alpha_t] = \alpha$ and $E[\lambda_t] = \lambda$. Each of α_t and λ_t may be vectors, but we do not complicate the notation by treating them explicitly as such. The IM procedure tests the hypothesis that $\text{Var}[\alpha_t] = \text{Var}[\lambda_t] = 0, t = 1, 2, \ldots, T$, against the alternative that at least one of them is not. The subtest associated with the thinning parameters is given by

$$U_D = \sum_{t=1}^{T} u_{D,t} = \sum_{t=1}^{T} \left\{ \dot{\ell}_{\alpha_t}^2 + \ddot{\ell}_{\alpha_t} \right\}$$

where we denote the score of the model with respect to α_t by $\dot{\ell}_{\alpha_t}$ and the second derivatives of the log-likelihood by $\ddot{\ell}_{\alpha_t}$. We evaluate these derivatives at the mean values, that is, using $\dot{\ell}_{\alpha_t}^2 = \dot{\ell}_{\alpha_t}^2\big|_{\alpha_t=\alpha,\lambda_t=\lambda}$ and $\ddot{\ell}_{\alpha_t}\big|_{\alpha_t=\alpha,\lambda_t=\lambda}$. In the vector parameter case, $\dot{\ell}_{\alpha_t}^2$ and $\ddot{\ell}_{\alpha_t}$ are matrices and their traces should be used in the computations. Since terms like $\dot{\ell}_{\alpha_t}^2$ depend on the parameters, estimates are required to implement these tests. Hence, we construct $\hat{u}_{D,t}$ based on estimated quantities using $\dot{\ell}_{\alpha_t}^2\big|_{\alpha_t=\hat{\alpha},\lambda_t=\hat{\lambda}}$ and $\ddot{\ell}_{\alpha_t}\big|_{\alpha_t=\hat{\alpha},\lambda_t=\hat{\lambda}}$. By means of the martingale properties of the likelihood process, it usually follows, under mild regularity, that $\hat{U}_D = s_D \sum_{t=1}^T \hat{u}_{D,t} \to^d N(0,1)$ where $s_D^2 = \sum_{t=1}^T \hat{u}_{D,t}^2$. Similarly, the subtest associated with the arrivals process is $U_A = \sum_{t=1}^T u_{A,t} = \sum_{t=1}^T \left\{ \dot{\ell}_{\lambda_t}^2 + \ddot{\ell}_{\lambda_t} \right\}$ and this may be implemented with $\hat{u}_{A,t}$ estimated using $\dot{\ell}_{\lambda_t}^2\Big|_{\alpha_t=\hat{\alpha},\lambda_t=\hat{\lambda}}$ and $\ddot{\ell}_{\lambda_t}\big|_{\alpha_t=\hat{\alpha},\lambda_t=\hat{\lambda}}$. It also follows that $\hat{U}_A = s_A \sum_{t=1}^T \hat{u}_{A,t} \to^d N(0,1)$ where $s_A^2 = \sum_{t=1}^T \hat{u}_{A,t}^2$. See Freeland and McCabe (2004), Sec. 5 for further details.

The use of the IM test is again illustrated by reference to a PINAR(1) model fitted to the cuts data. Relevant computations show that the p-values of the IM tests, $\hat{U}_D$ and $\hat{U}_A$, for the departure and arrival processes are 0.0250 and 0.0159, respectively. The low p-values indicate that there is more variation in the data than is described by the model. The relatively larger p-value for the departure process may indicate that a greater problem exists for the arrival process. Note again the pervasive suggestion of seasonality in this monthly data, an issue revisited in Section 9.4.3.

9.4 Analyses Based on the Predictive Distributions

The literature on probabilistic forecasting has developed a number of tools to compare and evaluate predictive distributions, see, for example, Diebold et al. (1998). These tools can also be fruitfully employed in diagnostic checking, as proposed by Gneiting et al. (2007) or Jung and Tremayne (2011a). Following Geweke and Amisano (2010), we first discuss the probability integral transform (PIT) method as a means of assessing the absolute performance of models. Then, in the second subsection, the relative performance of models within a group of competing ones is addressed using scoring rules and information criteria.

9.4.1 PIT Histograms for Discrete Data

The use of the PIT as a method for assessing the adequacy of distributional assumptions for a model dates back at least to the work of Dawid (1984) and exploits Rosenblatt's (1952) transformation of an absolutely continuous (conditional) distribution into a uniform distribution. To be more specific, define the random variable u_t on the basis of the cumulative predictive distribution $F_c(\cdot)$ corresponding to the true DGP $u_t = F_c(X_t|\mathcal{F}_{t-1})$. Under correct specification of the predictive distribution, the series of PIT random variables $\{u_t\}$ are *i.i.d.* standard uniform $(0, 1)$. If a specified model does not correspond to the true DGP and has cumulative predictive distribution $F(X_t|\mathcal{F}_{t-1})$, departures from this behavior can be expected. Diebold et al. (1998) discussed various ways in which the uniformity of $\{u_t\}$ from any model may be assessed. These are divided into two categories: those that are designed to check the unconditional uniformity of the $\{u_t\}$ and those that check whether the PIT series is *i.i.d.* The former is typically checked in an informal way by plotting the empirical cumulative distribution function of $\{u_t\}$ and comparing it to the identity function, or by constructing a histogram of the $\{u_t\}$ and checking for uniformity. Diebold et al. (1998, p. 869) also argued that formal tests of whether the $\{u_t\}$ are *i.i.d.* $U(0, 1)$ using, perhaps, the Kolmogorov–Smirnov or the Cramer–von Mises test are readily available but are nonconstructive and, therefore, of little practical value.

If the $\{u_t\}$ are not uniformly distributed, the nature of the deviation from uniformity can be informative. An obvious tool for analyzing the conformity with the *i.i.d.* assumption for the $\{u_t\}$ is to examine their autocorrelation structure. Incidentally, Berkowitz (2001) suggested transforming the $\{u_t\}$ to standard normal variables by means of an inverse Gaussian distribution. Then a quantile–quantile (QQ) plot against the standard normal distribution assesses the distributional fit of different models, although we do not pursue this proposal here.

In the context of discrete distributions, some modifications to standard methods are required, because predictive cumulative distribution functions are step functions. Two methods have been proposed in the literature. The first of these, suggested by Denuit and Lambert (2005), is the so-called randomized PIT obtained by perturbing the step function nature of the distribution function for discrete random variables, rather in the manner that a hypothesis testing procedure might use a randomization device to yield a test with a prespecified significance level in the context of discrete variables. A random draw, v, from a (standard) uniform distribution is used to construct a randomized PIT based on the distribution function of observed counts x_t from

$$u_t^+ = F(x_t - 1|\mathcal{F}_{t-1}) + v[F(x_t|\mathcal{F}_{t-1}) - F(x_t - 1|\mathcal{F}_{t-1})]. \tag{9.10}$$

Here, $F(\cdot)$ remains the predictive cumulative distribution of the counts from some model and $F(-1) = 0$, compare to Czado et al. (2009). As in the case of a continuous cumulative predictive function, if the model is correctly specified, u_t^+ is a serially independent random variable following a uniform distribution on the interval $[0, 1]$.

The assessment may be carried out by constructing histograms from the $\{u_t^+\}$; see, for example, Jung and Tremayne (2011a) who employ this method in the context of count time series models. They find that, for data sets of small sample size, the shape of the PIT histogram can change appreciably between sets of random draws from v. One approach to overcoming this is to average out this effect by using N independent sets of random uniform numbers to compute N PIT histograms. The average histogram bin heights over these N replications could then be used. But this method is, effectively, indistinguishable from the nonrandomized PIT, to be discussed next, rendering further discussion of the approach redundant.

An alternative method of computing the PIT for a discrete cumulative predictive distribution has been proposed by Czado et al. (2009). It utilizes the distribution function observed at the count x_t via

$$F^{(t)}(u^\star|\mathcal{F}_{t-1}) = \begin{cases} 0, & u^\star \leq F(x_t - 1|\mathcal{F}_{t-1}) \\ \dfrac{u^\star - F(x_t - 1|\mathcal{F}_{t-1})}{F(x_t|\mathcal{F}_{t-1}) - F(x_t - 1|\mathcal{F}_{t-1})}, & F(x_t - 1|\mathcal{F}_{t-1}) \leq u^\star \leq F(x_t|\mathcal{F}_{t-1}) \\ 1, & u^\star \geq F(x_t|\mathcal{F}_{t-1}) \end{cases}. \quad (9.11)$$

The assessment of the distributional assumption can be carried out by aggregating over the set of $T - p$ predictions for all observed counts and comparing the mean PIT

$$F_m(u^\star) = (T-p)^{-1} \sum_{t=p+1}^{T} F^{(t)}(u^\star|\mathcal{F}_{t-1}), \quad 0 \leq u^\star \leq 1,$$

to the cumulative distribution function of a standard uniform random variable. The computation of the relevant predictive cumulative distributions used in (9.10) and (9.11) is straightforward for most time series models for count data.

We operationalize the procedure by plotting a (nonrandomized) PIT histogram obtained by allocating the values to J equally spaced bins $j = 1, \ldots, J$, where the height of the jth bin is computed from $f_j = F_m(j/J) - F_m((j-1)/J)$, and checking for uniformity. We supplement such a graph by incorporating approximate $100(1-\alpha)\%$ confidence intervals (here we use $\alpha = 0.05$) obtained from a standard χ^2 goodness-of-fit test of the null hypothesis that the J bins of the histogram are drawn from a uniform distribution. Under the null hypothesis, the statistic, G, is asymptotically $\chi^2(J-1)$ distributed; we set J to 10. Alternatively, $F_m(u^\star)$ may be plotted against $u^\star$ and checked for deviations from the identity function; this requires no (arbitrary) choice of the number of bins J.

Figure 9.6 provides a summary of the PIT-based diagnostics resulting from fitting a PINAR(1) model to each of the cuts and the iceberg order data sets. Based on the earlier discussion, we present the nonrandomized variant of the PIT histogram in the figure. Both the PIT histogram and the $F_m(u^\star)$ chart for the cuts data suggest misspecification of the distributional assumption. The pronounced U-shaped pattern of the PIT histogram indicates that not enough dispersion has been allowed for in the conditional distribution. This is mildly supported by the results for the uniformity test statistic for the PIT histogram for the cuts data, which is 14.8053 (p-value: 0.0964). In the case of the iceberg order data, the PIT histogram is closer to a uniform distribution, although a slight U-shape pattern is evident. The uniformity test statistic for the iceberg orders is 9.4657 (p-value: 0.3954) and, therefore, does not reject the null hypothesis of a uniform PIT histogram at any conventional significance level. For both data sets, the correlograms of the (centered) $\{u_t^+\}$ provide similar qualitative information to those of the Pearson residuals of Figure 9.4 and are indicative of some misspecification of the dependence structure.

It is, therefore, evident from all the graphs from both residual and predictive distribution analyses for both data sets that the basic PINAR(1) model does not provide a satisfactory fit for either. In a modeling exercise seeking data coherent specifications for these data, some model refinement is clearly called for. This may lead a researcher to need to choose between

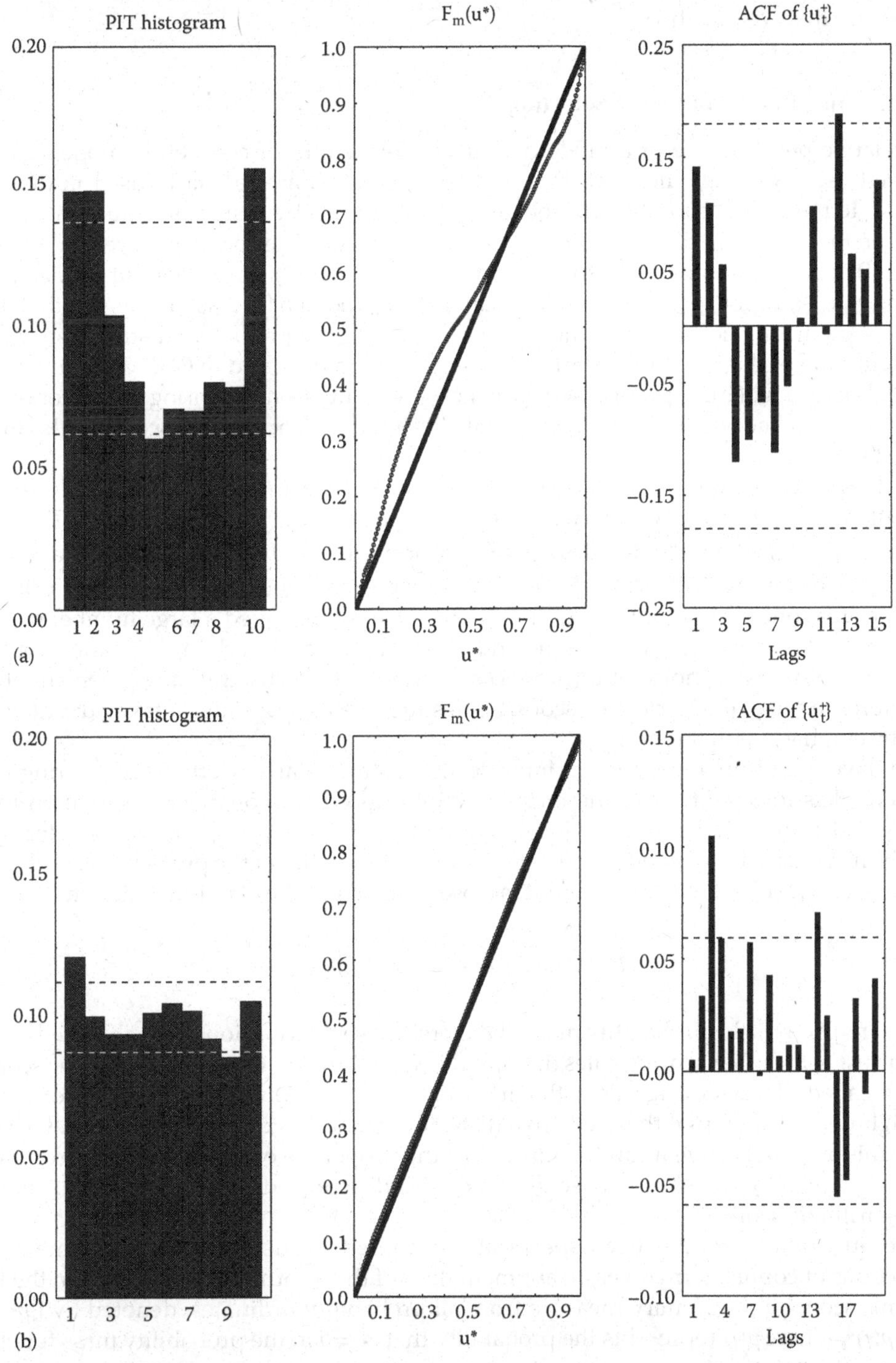

FIGURE 9.6
PIT-based diagnostics after fitting a PINAR(1) model to data: cuts (a) and iceberg order (b).

a multiplicity of different model specifications for the same data set and it is to issues of this nature that we now turn.

9.4.2 Scoring Rules and Model Selection

The relative performance of a model within a given group of competing models can be assessed using scoring rules. Scoring rules, or functions, are regularly used in decision analysis to measure the quality of probabilistic predictions by assigning a numerical score based on the predictive distribution and the observed data. They are closely related to (generalized) entropy measures; see, for example, Jose et al. (2008). An important property of a scoring rule is propriety. A scoring rule is said to be proper if a forecaster achieves their best score by predicting according to their true belief about the predictive distribution. A formal definition of this concept and further discussion can be found in Gneiting and Raftery (2007). Boero et al. (2011) provide a comprehensive evaluation of scoring rules along with some historical background, and Czado et al. (2009) offer an account of scoring rules in the context of count data.

In the present framework, scoring rules are used as a model selection tool and are computed as averages over the relevant set of (in-sample) predictions, say $(T - p)^{-1} \sum_{t=p+1}^{T} s\left[F(x_t)\right]$, where $s\left[\cdot\right]$ denotes a generic scoring rule and observed count x_t and $F(x_t)$ is defined in the text following (9.10). Scoring rules are, generally, negatively oriented penalties that one seeks to minimize. The literature has developed a large number of scoring rules and, unless there is a unique and clearly defined underlying decision problem, there is no automatic choice of a (proper) scoring rule to be used in any given situation. Therefore, the use of a variety of scoring rules may be appropriate to take advantage of specific emphases and strengths.

We have found three proper scoring rules to be particularly useful in comparing time series models for counts, and these are now introduced as scores per observation to be aggregated as indicated in the previous paragraph. The first scoring rule we consider is the logarithmic score. It is defined as the negative of the logarithm of the predictive distribution evaluated at the observed count, and it is closely related to the classical Shannon entropy

$$logs(F(x_t|\mathcal{F}_{t-1})) = -\log p(x_t|\mathcal{F}_{t-1}),$$

where $p(x_t|\mathcal{F}_{t-1})$ is the probability mass of the predictive distribution at the observed count. In contrast to the other scoring rules discussed in the following text, the logarithmic score is what is called a local scoring rule in that it provides a small value if the observed count is in the high-density region of the predictive distribution and large values otherwise. Geweke and Amisano (2011) have a careful analysis of the properties of weighted linear combinations of prediction models and base their model choice procedure on the minimization of the logarithmic score.

The quadratic score has been specifically proposed in the assessment of time series predictions of counts. It involves an augmentation of the information contained in the logarithmic score by a summary measure from all probability ordinates, denoted by $||p||^2 = \sum_{j=0}^{\infty} p(j)^2$, where $p(j)$ represents the probability that $x_t = j$ in the probability mass function of the predictive distribution and is given by

$$qs(F(x_t|\mathcal{F}_{t-1})) = -2p(x_t|\mathcal{F}_{t-1}) + ||p||^2.$$

The quadratic score was proposed by Wecker (1989) in the specific context of predictions for time series of counts.

The final scoring measure we consider is the ranked probability score defined by

$$rps(F(x_t|\mathcal{F}_{t-1})) = \sum_{j=0}^{\infty} [F(j) - \mathbf{1}(x_t \le j)]^2,$$

where $\mathbf{1}(\cdot)$ is an indicator function. This rule assesses the sum of squared differences of the cumulative probabilities using the modeled conditional distribution from the observations. Hence, it penalizes more severely when the predictions are far from the observed outcomes.

Some authors (including Weiß 2009 and Zhu 2011) proposed the use of information criteria, such as the popular Akaike information criterion (AIC), as means of choosing between nonnested time series models for counts, despite the fact that little is known about their ability to do so in this framework. In addition, Psaradakis et al. (2009) examined the ability of some popular information criteria, like AIC, the Bayesian information criterion (BIC) and the Hannan and Quinn criterion (HQ) to distinguish between some nonlinear times series models. They argued that all three criteria have a useful role to play in a time series model selection exercise. Although their study was not based on count time series models directly, it may serve as some justification for using such model selection devices in the present context.

Scoring rules and values for two information criteria for a PINAR(1) model fitted to the cuts data are as follows: logarithmic score 2.4549, quadratic score 0.9001, ranked probability score 1.5932, $AIC = 290.2678$, and $BIC = 294.4490$. For the iceberg order data, the corresponding values are as follows: logarithmic score 1.3014, quadratic score 0.6500, ranked probability score 0.5122, $AIC = 1414.9705$, and $BIC = 1422.4586$. These values are simply reported here, but will be employed in the following section to facilitate comparison of the simple PINAR(1) model to others fitted to each data set.

9.4.3 Cuts and Iceberg Data Revisited

It will now be very clear that the basic PINAR(1) models fitted to the two real life data sets introduced in Section 9.2 have been revealed to be deficient according to a range of criteria. We are now in a position to reconsider specification of appropriate count time series models for these data.

Consider first the cuts data. Diagnostic analyses provided in earlier sections after fitting a simple PINAR(1) model to these data reveal a number of difficulties. First, while the mean of the Pearson residuals is close to zero, their variance is considerably larger than unity (at 1.607). Next, the dependence structure in the data is not well captured by the model. This is evident in the left-hand panels of two figures: from the correlogram of the Pearson residuals in Figure 9.4 and from the parametric resampling exercise depicted in Figure 9.3. Possible evidence of distributional misspecification is available in the PIT in the relevant panel of Figure 9.6. In addition, an analysis of the component residuals of Section 9.3 reveals unaccounted variation in the data and the graphical evidence in Figure 9.5 may be indicative of unmodeled seasonal variation. From a pragmatic point of view, given the lower arrivals p-value for the IM test reported previously and the fact that seasonal arrivals could very well induce seasonal departures, it seems reasonable to account for this variation by modifying the arrival process first.

In seeking to remedy the aforementioned deficiencies in the PINAR(1) model with no covariates, we undertake a limited specification search. This leads us to propose fitting a GP(1) model of the form (9.1) with time-varying innovation rate λ_t to the data. The resultant fitted model is (estimated asymptotic standard errors are given in parentheses below parameter estimates)

$$\hat{X}_t = R_t(X_{t-1}; \underset{(0.072)}{0.478}) + \hat{\varepsilon}, \quad \text{where} \quad \hat{\varepsilon} \sim GP(\hat{\lambda}_t, \underset{(0.066)}{0.165}),$$

$$\text{and} \quad \hat{\lambda}_t = \exp\Big(\underset{(0.190)}{0.942} \; \underset{(0.106)}{-0.216} \sin(2\pi t/12) \; \underset{(0.110)}{-0.333} \cos(2\pi t/12)\Big).$$

It can be seen that estimated coefficients relating to seasonal effects are both statistically different from zero at most conventional significance levels, as is the dispersion parameter of the GP distribution (p-value 0.0125). The values for the various scoring rules and the information criteria are as follows: logarithmic score 2.3252, quadratic score 0.8840, ranked probability score 1.4645, $AIC = 277.0928$, and $BIC = 284.0615$. All of these are lower than their counterparts provided toward the end of Section 9.4.2 for the PINAR(1) model with no covariates. Further summary statistics are as follows: variance of the Pearson residuals: 1.0165 and uniformity test, G, of the PIT histogram: 1.9107 (p-value 0.9928). On the evidence presented, a researcher would clearly prefer the GP(1) model with deterministic seasonality to the original PINAR(1) model.

Some diagnostic plots relating to this new model specification are the subject of Figure 9.7. It is readily seen from all three panels in the figure that there is no evidence of model misspecification. These should be compared and contrasted with comparable panels in Figures 9.3, 9.4, and 9.6. Hence, a simple change in innovation distribution, together with allowing a time-varying innovation mean, leads to a marked improvement in the suitability of the (new) model for which the methods discussed do not reveal any statistical inadequacies.

Turning to the iceberg order data, diagnostic results reported in earlier sections after fitting a PINAR(1) model clearly reject this initial model. However, evidence for distributional misspecification is not as clear-cut as for the cuts data set. The variance of the Pearson residuals is larger than unity at 1.289. But the PIT histogram in the lower left panel of Figure 9.6 displays only limited departure from uniformity and the G-statistic corroborates this by not rejecting the null of a uniform PIT histogram (p-value = 0.395). A misspecification of the dependence structure is evident from the right-hand panels of Figures 9.3, 9.4 and the bottom right panel in Figure 9.6. In contrast to the previous data set, we do not infer that a seasonal pattern is unaccounted for, as some experimentation (not reported here) shows no improvement over the basic PINAR(1) model.

A limited specification search (details of which again go unreported to save space) leads us to propose a model of the form (9.1) with no covariates for the iceberg order data, but with GP innovations and associated random operator of Joe (1996). The proposed DGP sets $p = 2$ and is denoted a GP(2) model. Note that the model cannot be written in the form (9.2), since the random operator $R_t(\mathcal{F}_{t-1}, \boldsymbol{\alpha})$ of (9.1) has two lags in $\mathcal{F}_{t-1}$, but the dependence parameter vector $\boldsymbol{\alpha}$ has three elements. By closure under convolution, this leads to the marginal distribution of the counts being taken to be GP. The resultant fitted model is (estimated asymptotic standard errors again in parentheses)

$$\hat{X}_t = R_t(X_{t-1}; \underset{(0.0129)}{0.1954}, \underset{(0.0139)}{0.046}, \underset{(0.0268)}{0.4671}) + \hat{\varepsilon}, \quad \text{where} \quad \hat{\varepsilon} \sim GP(\underset{(0.0262)}{0.3259}, \underset{(0.0255)}{0.1696}).$$

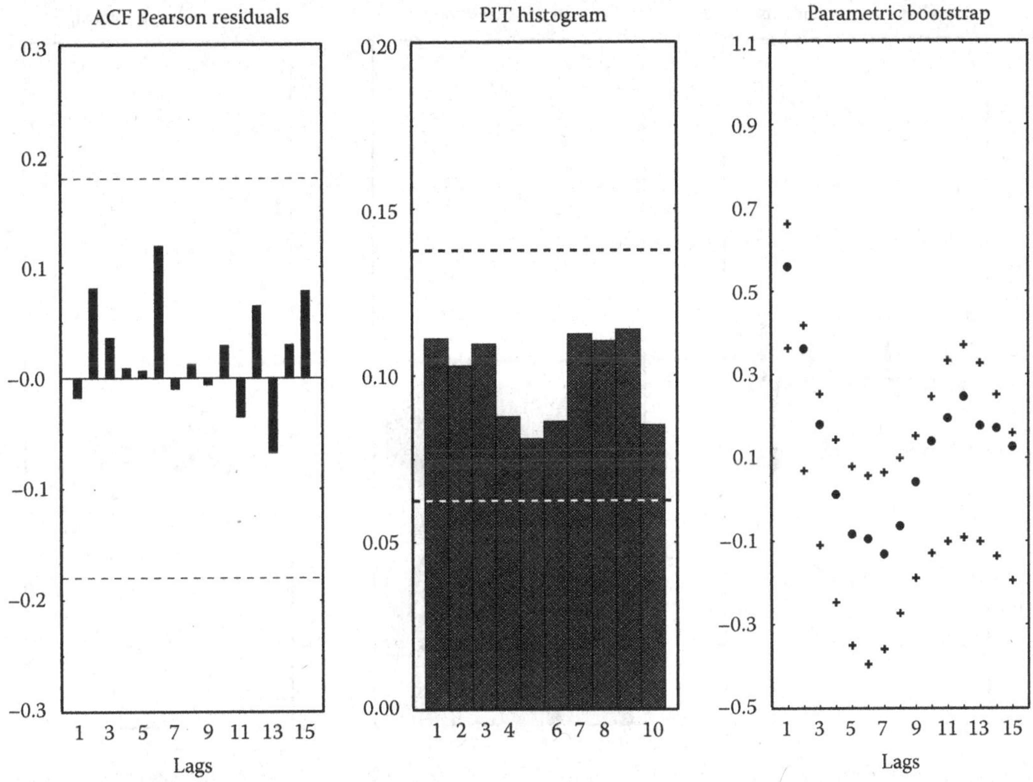

FIGURE 9.7
Diagnostics for a GP(1) model with exogenous variables fitted to the cuts data.

All the parameter estimates in α are positive and significantly different from zero. The overdispersion parameter η is also significant (the p-value is zero to four decimal places). The values for the various scoring rules and information criteria are as follows: logarithmic score 1.2592, quadratic score 0.6376, ranked probability score 0.5024, $AIC = 1370.00$, and $BIC = 1382.48$. The scoring rules and information criteria, as compared to those reported in Section 9.4.2, uniformly favor the GP(2) specification over the PINAR(1) model. Further summary statistics are as follows: variance of the Pearson residuals: 1.0001 and uniformity test of the PIT histogram: 0.9958 (p-value 0.9994). The diagnostic plots are provided in Figure 9.8. None of the three panels indicate model misspecification.

9.5 Evidence with Artifical Data

The evidence on the use of model validation and diagnostic methods provided in Section 9.4.3 relates only to two data sets for which deficiencies in a PINAR(1) model can be highlighted and, perhaps, rectified by simple model respecifications. To further gauge and illustrate the performance of the diagnostic tools presented in previous sections, we report

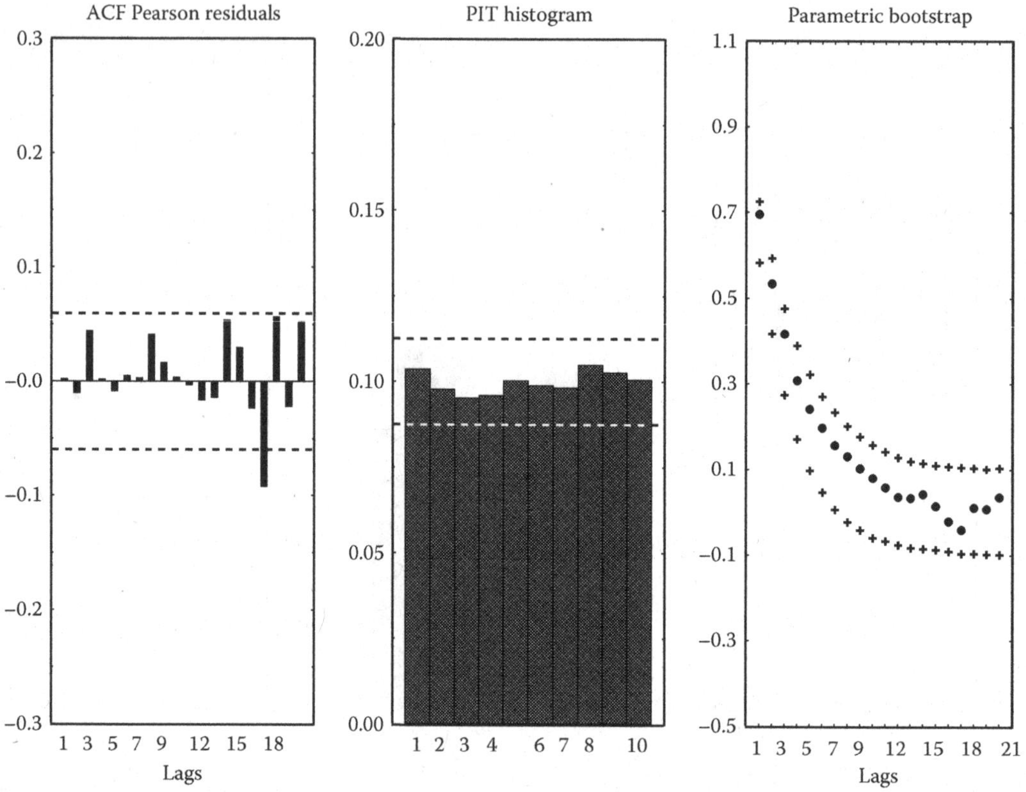

FIGURE 9.8
Diagnostics for a GP(2) model fitted to the iceberg order data.

the results of some simulation experiments seeking to reflect additional common situations that might be faced by applied workers wishing to assess the adequacy of a fitted model for count time series. The purpose of this section is to report the results of four such experiments.

The first experiment aims to analyze the ability of the diagnostic devices, including scoring rules, to detect a misspecification in the distributional assumption of the innovations in a proposed model. In particular, it reflects a common situation in applied work where the count time series exhibits marginal overdispersion reflected by a variance-to-mean ratio greater than one.

Data are generated from a first-order integer autoregressive process (9.1) with innovations $\varepsilon_t \sim GP(\lambda, \eta)$ and the random operator $R_t(\cdot)$ proposed by Joe (1996); this is the setup denoted GP(1) in the Section 9.1. The specific form of the GP distribution we employ is $p(\varepsilon) = \lambda(\lambda + \varepsilon\eta)^{\varepsilon-1}\exp(-\lambda - \varepsilon\eta)/\varepsilon!$, with $\varepsilon = 0, 1, 2, \ldots, \lambda > 0$ and $\eta \in [0, 1)$. Further details can be found, *inter alia*, in Jung and Tremayne (2011b).

For the simulation experiment, we use the following parameter values: $\alpha_1 = 0.6; \lambda = 0.8$, and $\eta = 0.2$. This leads to a GP-distributed count time series with (theoretical) mean and variance of 2.5 and 3.9, respectively, and moderate dependence structure (the τ'th autocorrelation function ordinate $\rho(\tau) = 0.6^{\tau}$, for $\tau = 1, 2, \ldots$). Here, and in conjunction with the next two experiments in this section, we discuss the results from single simulation run where the sample size is set to 50,000 in order to limit the impact of sampling uncertainty on the results. In this case, the generated data are analyzed with two different scenarios: (a) a GP(1) estimated model corresponding to the true DGP and (b) a PINAR(1) estimated model where the innovation distribution is erroneously taken to be Poisson and binomial thinning is assumed.

Figure 9.9 provides three graphs associated with relevant diagnostic tools. Graphs for neither the ACF of the $\{u_t^+\}$ nor the parametric bootstrap are provided, since they are qualitatively similar to the ACF of the Pearson residuals and provide no added insight.

None of the graphical diagnostics depicted in the top panels of Figure 9.9 suggest that the GP(1) model is inadequate for the data. However, this is not the case when attention is focused on the lower row of panels in Figure 9.9. Here, the results of erroneously assuming equidispersed Poisson innovation distribution (and binomial thinning) are clearly evident. The distributional misspecification is seen in the U-shaped PIT histogram and an $F_m(u^\star)$ chart (in the bottom row, third column of the figure), which deviates from the 45° line (compare the corresponding figure in the row above). In addition, the correlogram of the Pearson residuals indicates misspecification with respect to the dynamics of the generated data. This result is explained by the fact that the maximum likelihood estimate for the parameter α_1 in the PINAR(1) model is biased downward. Therefore, the strength of the dependence in the data is underestimated, resulting in residual serial correlation remaining in the Pearson residuals; this is depicted in the left-hand panel in the lower row of Figure 9.9.

A summary of numerical results is given in Table 9.3. It can be seen that, in contrast with that for the correct specification, the sample variance of the Pearson residuals from the PINAR(1) is considerably larger than one (1.3171), indicating that not all the dispersion in the generated data has been accounted for in the fitted specification. Note also that the scoring rules and the information criteria of the two fitted models uniformly indicate a preference for the true GP(1) model for the data over the PINAR(1) one. For some of these statistics, including the variance of the Pearson residuals, both information criteria and the p-values of G, the evidence is emphatic.

The second experiment specifically targets a model misspecification with respect to the predictive distribution. The data are generated using an INAR(1) of the form (9.2) with binomial thinning, but with GP innovations in truth. Such a model is discussed by Jung and Tremayne (2011b), where it is indicated that the resultant counts exhibit marginal overdispersion, but are not GP, because closure under convolution does not apply.

For this simulation experiment, we employ the same set of parameter values for α_1, λ, and η as in the first experiment. Also, we use the same estimated models as mentioned earlier, that is, (1) a GP(1) model based on the random operator $R_t(\cdot)$ of Joe (1996) and (2) a PINAR(1) model based on binomial thinning with the parameters estimated being the dependence parameter and the innovation mean based on a Poisson innovation assumption. Note that neither fitted model is correct. Figure 9.10 provides graphs associated with various diagnostic tools. Summary numerical results are given in Table 9.4.

Note that both fitted models are misspecified, since the first implies a misspecified thinning mechanism (the marginal distribution of the data will be overdispersed but will not be GP) and the second assumes an incorrect innovation, because the likelihood is based on a Poisson assumption for innovations. This is reflected at various junctures in the diagnostic analysis.

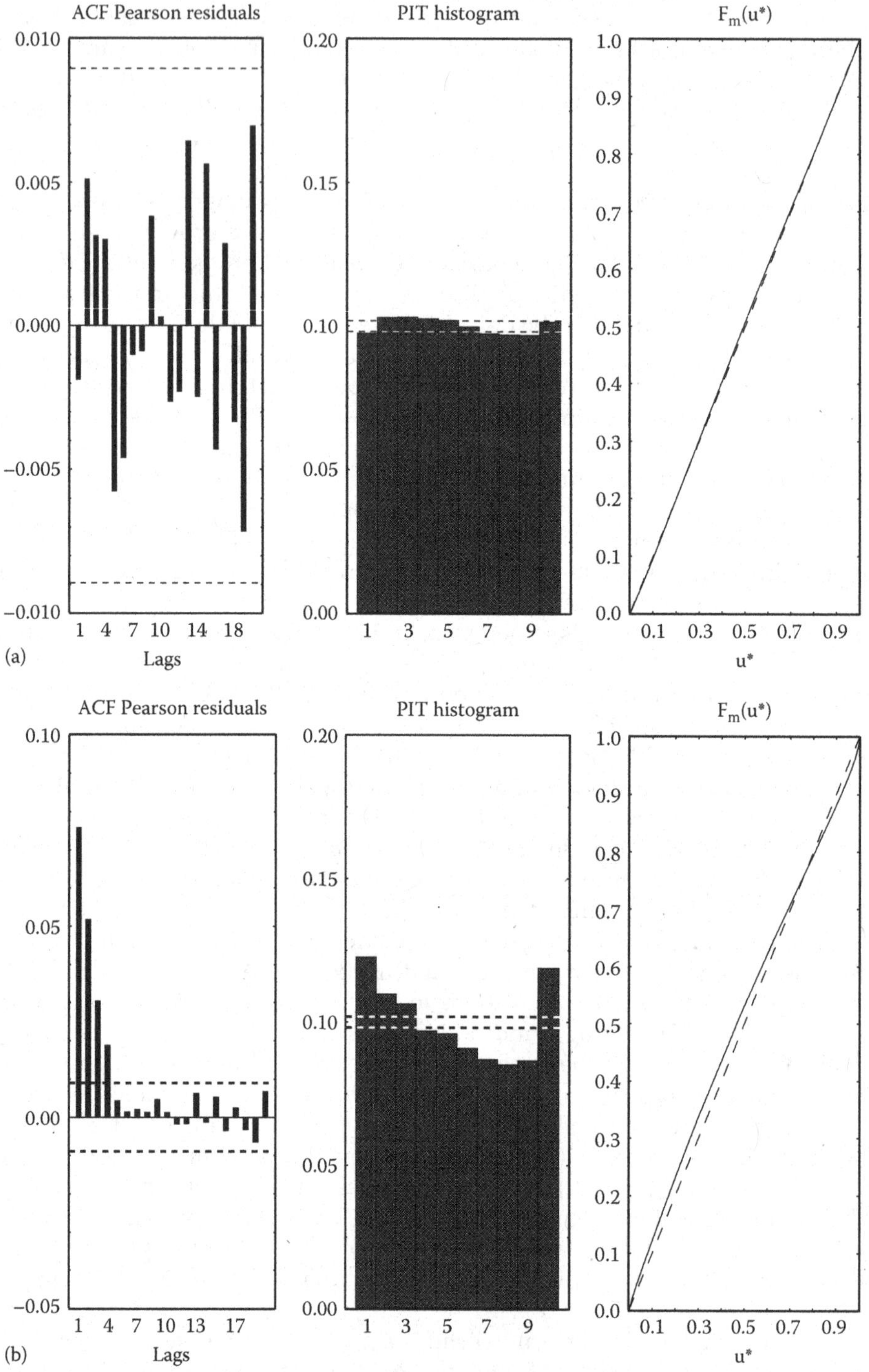

FIGURE 9.9
Graphical results for the first Monte Carlo experiment. (a) GP(1) estimated model and (b) PINAR(1) estimated model.

TABLE 9.3

Summary of Numerical Results for the First Monte Carlo Experiment

	GP(1) Model	PINAR(1) Model
Pearson residual		
Mean	−0.0008	−0.0137
Variance	1.0169	1.3171
Scoring rules		
logs	1.7347	1.7689
qs	0.7816	0.7896
rps	0.8119	0.8227
AIC	86735.49	88444.71
BIC	86748.72	88452.53
G	11.584	777.8477
p-value	(0.2378)	(< 0.000)

Starting with the second set of results related to the PINAR(1) estimated model, we see that, due to a downward bias in the ML estimation of the dependence parameter, there are obvious unwanted spikes in the correlogram of the Pearson residuals. Both the (nonrandomized) PIT histogram and the $F_m(u^{\star})$ chart indicate some misspecification in the distributional assumption. From Table 9.4, it can be seen that the variance of the Pearson residuals (at 1.2547) is considerably larger than unity and the G-statistic decisively rejects uniformity of the PIT histogram.

Interpreting the first set of results related to the GP(1) estimated specification, where the misspecification is essentially due to the thinning operator assumed, is less obvious. We reiterate (Jung and Tremayne 2011b) that the degree of overdispersion in the innovations is attenuated in the true marginal distribution of the observations by the binomial thinning operation used in the data-generating mechanism. The estimated GP(1) model is able to capture the dependence structure in the data, reflected by a Pearson residual correlogram that shows the dependence structure in the data to be adequately modeled (top row, first column of the figure). Also, the variance of the Pearson residuals from the estimated model is larger than one, but only marginally so. Diagnostic results related to other aspects of the specification do tentatively suggest model misspecification in that the (nonrandomized) PIT histogram and the $F_m(u^{\star})$ chart exhibit limited unwanted features. In particular, the former shows some departure from uniformity, a conclusion backed up by the goodness-of-fit statistic G and its associated p-value. Overall, the results displayed in Figure 9.10 and Table 9.4 suggest a preference for the GP(1) specification over the PINAR(1) one, but there may be doubt about whether or not the former is fully data coherent.

The third experiment is designed to reflect underspecification of dynamics in the estimated model. Data is generated using a second-order integer autoregressive model with GP innovations and the operator due to Joe (1996); by closure under convolution, the marginal distribution of the generated counts is GP. See the discussion of the previous subsection relating to the preferred model for the iceberg data set for further information on this specification. The following set of parameters are used in the experiment for the dependence parameter vector $\boldsymbol{\alpha}$, $\alpha_1 = 0.4$; $\alpha_2 = 0.25$; $\alpha_3 = 0.1$; $\lambda = 0.5$; and $\eta = 0.2$ leading to a process mean of 2.5 and first- and second-order autocorrelations of $0.45 + 0.1 = 0.55$ and $0.25 + 0.1 = 0.35$, respectively.

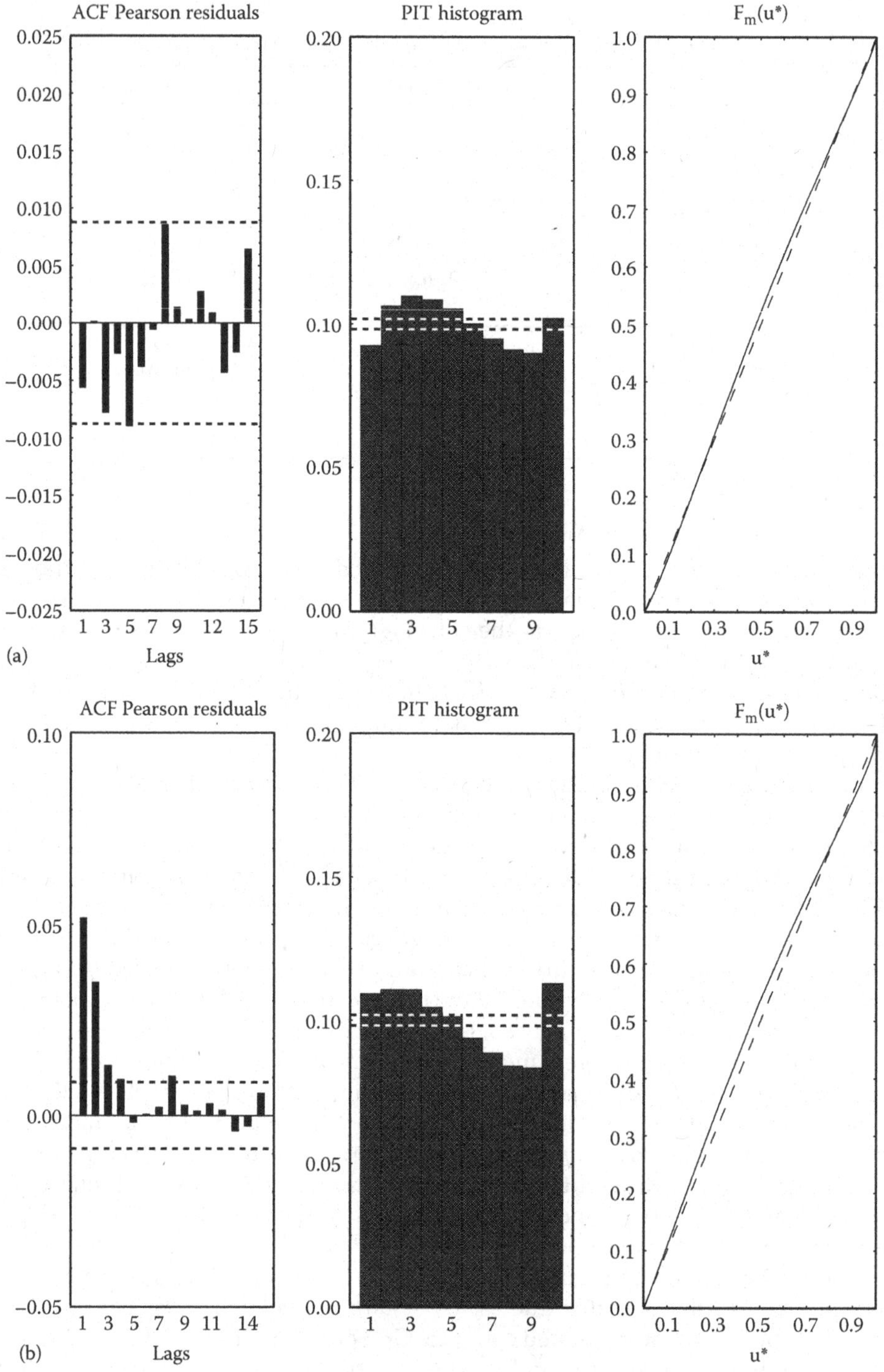

FIGURE 9.10
Graphical results for the second Monte Carlo experiment. (a) GP(1) estimated model and (b) PINAR(1) estimated model.

TABLE 9.4
Summary Numerical Results for the Second Monte Carlo Experiment

	GP(1) Model	PINAR(1) Model
Pearson residual		
Mean	−0.0018	−0.0090
Variance	1.0534	1.2547
Scoring rules		
logs	1.6927	1.7111
qs	0.7724	0.7768
rps	0.7661	0.7704
AIC	84631.63	85550.68
BIC	84644.86	85563.91
G	244.089	611.650
p-value	(< 0.000)	(< 0.000)

Two different models are fitted to the generated data: (1) a GP(2) (correct) estimated model and (2) a misspecified GP(1) model, so both estimated models utilize the thinning operator of Joe (1996). Figure 9.11 displays graphs associated with some of the diagnostic tools discussed for the latter case only, since the fitted GP(2) model evidences no misspecification. Summary numerical results for both estimated models are given in Table 9.5.

It is evident from the (nonrandomized) PIT histogram and the $F_m(u^\star)$ chart that the estimated GP(1) model is able to capture the distributional assumption correctly. However, the correlogram of the Pearson residuals indicates misspecified dynamics in this underspecified fitted model. In particular, it shows Pearson residual autocorrelations of the GP(1) model that decay exponentially (after the third). This arises because the autocorrelations of the data themselves exhibit a more complicated persistence pattern than a first-order model can account for. From the numerical results displayed in Table 9.5, it can be seen, as it is to be expected, that all the scoring rules and the information criteria clearly favor the GP(2) estimated model over the first-order counterpart. Note, however, that the summary statistics relating to the Pearson residuals and the G-statistic do not indicate model misspecification.

Finally, we conduct a fourth experiment by generating data from a second-order integer autoregressive model with Poisson innovations. Data is again generated using the Joe (1996) thinning operator and is of the form (9.1) with $p = 2$. Instead of using a constant innovation rate as in the earlier experiments, we employ a time varying innovation designed to capture (deterministic) seasonality often employed in empirical work. Suppose the innovation rate λ_t to be time varying using harmonics given by

$$\lambda_t = \exp\left(\theta_1 + \theta_2 \sin\left(\frac{2\pi t}{200}\right) + \theta_3 \cos\left(\frac{2\pi t}{200}\right)\right) \tag{9.12}$$

and choose the following set of parameters: $\alpha_1 = 0.4$, $\alpha_2 = 0.25$, $\alpha_3 = 0.1$, $\theta_1 = 0.05$, $\theta_2 = -0.2$, and $\theta_3 = 0.2$. The harmonics introduce additional dynamic effects in the

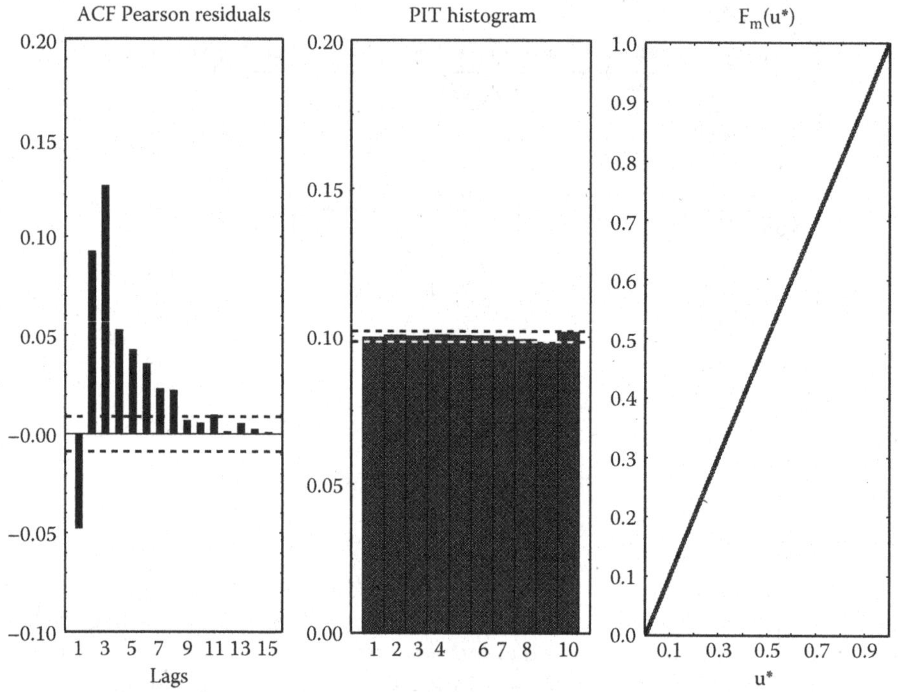

FIGURE 9.11
Graphical results for the third Monte Carlo experiment.

TABLE 9.5
Numerical Results for the Third Monte Carlo Experiment

	GP(2) Model	GP(1) Model
Pearson residual		
Mean	−0.0001	0.0001
Variance	1.0023	0.9999
Scoring rules		
logs	1.7585	1.8363
qs	0.7845	0.8074
rps	0.8515	0.9111
AIC	87924.38	91838.88
BIC	87946.43	91852.11
G	2.551	4.710
p-value	(0.9795)	(0.8588)

generated data and lead to (marginal) overdispersion. The parameters have been chosen such that the data can be fitted to two stationary specifications, namely, the GP(2) and the GP(1) models as used in the previous experiment, as well as to the true specification. From a practical point of view, if the harmonics were "too strong," standard optimization routines may encounter problems fitting the misspecified models to the data and we seek to avoid this unnecessary complication. The idea behind fitting the two incorrect models is, of course, to see which diagnostic tool(s) detect model misspecification when the deterministic seasonality is ignored. In this experiment, we concentrate on the correlogram of the Pearson residuals and the Tsay parametric bootstrap procedure using the sample serial correlations as functional and set the sample size to 10,000.

Figure 9.12 provides a row of panels corresponding to three fitted models: the top row arises from fitting the true model with included regression effects, the second to a misspecified GP(2) model ignoring harmonics, and the final one to a misspecified GP(1) model omitting harmonics. The top row of panels shows that neither the autocorrelation plot of the Pearson residuals nor the parametric bootstrap analysis finds any evidence of misspecification; this is unsurprising, since the true generating process was fitted. The middle panel of the figure based on a GP(2) estimated model shows that both the ACF of the Pearson residuals and the parametric bootstrap diagnostics indicate misspecification of this model. However, the former shows this markedly less than the latter and then only at high lags. As far as the fitted GP(1) specification goes, it is evident that both diagnostic devices indicate indubitably that this model is misspecified.

We conclude from these experiments that the diagnostic methods we present can be fruitfully applied in model validation and diagnostic checking in integer autoregressive models for count time series. Moreover, all four potential problems mentioned in the penultimate paragraph of Section 9.1 may be detectable with one or more of the tools discussed in this chapter.

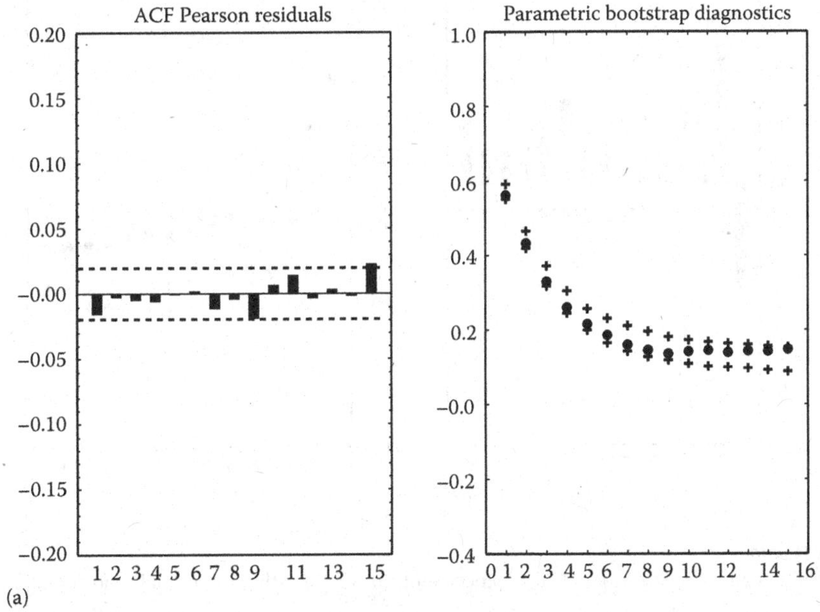

FIGURE 9.12
Graphical results for the fourth Monte Carlo experiment. (a) Estimated model based on the true DGP.

(Continued)

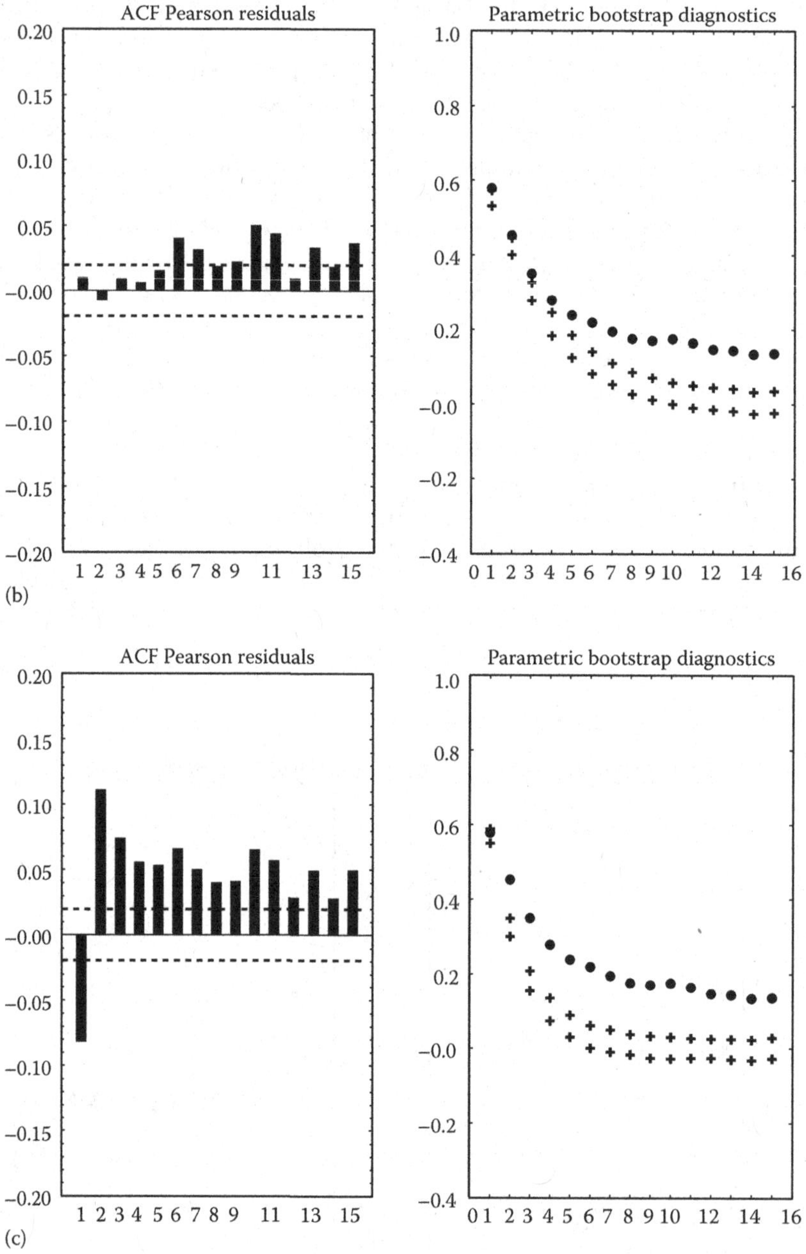

FIGURE 9.12 ***(Continued)***
Graphical results for the fourth Monte Carlo experiment. (b) GP(2) estimated model, and (c) GP(1) estimated model.

9.6 Conclusions

Model validation and the assessment of the adequacy of a fitted specification in discrete-valued time series models has not been the primary concern of many authors in the field so far. However, it should form an integral part of any modern iterative modeling exercise, and so we exposit a number of tools and methods to help fill this gap. Considering these as probabilistic forecasting tools allows us to draw upon a well-developed body of literature and adapt it for our purposes when such adaptation is required.

We focus on a range of diagnostic tools that help to assess the adequacy of fit of a chosen model specification. Many of these facilitate comparison between two or more model specifications. Some methods are graphical in nature, some are scoring rules, and others are based on statistical tests. We demonstrate the applicability of our methods to two example data sets, one each from economics and finance. Further, in a range of carefully designed simulation experiments, we document the ability of these tools and methods to serve our purpose for the popular class of integer autoregressive models.

We try to show how the search for a data coherent model for a count data set can be aided by using some or all of the methods discussed. Despite the caveat quoted from Tsay (1992) in Section 9.1, our experience is that model misspecification in integer time series models will usually be revealed by more than one diagnostic device. Moreover, we would counsel that no model should be used for probabilistic forecasting until it passes satisfactorily all the diagnostic checks advocated, for the subject matter of this chapter remains an area of statistical methodology ripe for further development. From a practical point of view, what we suggest here should be thought of as fairly minimal requirements that ought to be satisfied by a proposed model for data. Further, any chosen model should perform at least comparably with respect to other specifications entertained with regard to summary statistics such as scores. Finally, we hope that this work will achieve two aims: first, to stimulate researchers to improve on what we have put forward and second, to provide applied count time series analysts with model checking techniques that can be routinely applied in their current work.

References

Berkowitz, J. (2001). Testing density forecasts, with applications to risk management. *Journal of Business and Economic Statistics*, 19:465–474.

Boero, G., Smith, J., and Wallis, K. F. (2011). Scoring rules and survey density forecasts. *International Journal of Forecasting*, 27:379–393.

Box, G. and Jenkins, G. (1970). *Time Series Analysis: Forecasting and Control*. Holden-Day, San Francisco, CA.

Chesher, A. (1983). The information matrix test: Simplified calculation via a score test interpretation. *Economics Letters*, 13:45–48.

Czado, C., Gneiting, T., and Held, L. (2009). Predictive model assessment for count data. *Biometrics*, 65:1254–1261.

Dawid, A. P. (1984). Statistical theory: The prequential approach. *Journal of the Royal Statistical Society: Series A*, 147:278–292.

Denuit, M. and Lambert, P. (2005). Constraints on concordance measures in bivariate discrete data. *Journal of Multivariate Analysis*, 93:40–57.

Diebold, F. X., Gunther, T. A., and Tay, A. S. (1998). Evaluating density forecasts with applications to financial risk management. *International Economic Review*, 39:863–883.

Freeland, R. K. and McCabe, B. P. M. (2004). Analysis of count data by means of the Poisson autoregressive model. *Journal of Time Series Analysis*, 25:701–722.

Geweke, J. and Amisano, G. (2010). Comparing and evaluating Bayesian predictive distributions of asset returns. *International Journal of Forecasting*, 26:216–230.

Geweke, J. and Amisano, G. (2011). Optimal prediction pools. *Journal of Econometrics*, 164:130–141.

Gneiting, T., Balabdaoui, F., and Raftery, A. E. (2007). Probabilistic forecasts, calibration and sharpness. *Journal of the Royal Statistical Society: Series B*, 69:243–268.

Gneiting, T. and Raftery, A. E. (2007). Strictly proper scroing rules, prediction, and estimation. *Journal of the American Statistical Association*, 102:359–378.

Grunwald, G., Hamza, K., and Hyndman, R. (1997). Some properties and generalizations of non-negative Bayesian time series models. *Journal of the Royal Statistical Society: Series B*, 59:615–626.

Harvey, A. C. and Fernandes, C. (1989). Time series models for count or qualitative observations. *Journal of Business & Economic Statistics*, 7:407–417.

Joe, H. (1996). Time series models with univariate margins in the convolution-closed infinitely divisible class. *Journal of Applied Probability*, 33:664–677.

Jose, V. R. R., Nau, R. F., and Winkler, R. L. (2008). Scoring rules, generalized entropy, and utility maximization. *Operations Research*, 56:1146–1157.

Jung, R. C. and Tremayne, A. (2011a). Useful models for time series of counts or simply wrong ones? *Advances in Statistical Analysis*, 95:59–91.

Jung, R. C. and Tremayne, A. R. (2011b). Convolution-closed models for count time series with applications. *Journal of Time Series Analysis*, 32:268–280.

Li, W. K. (2004). *Diagnostic Checks in Time Serie*. Chapman & Hall, Boca Raton, FL.

McCabe, B. P. M. and Leybourne, S. J. (2000). A general method of testing for random parameter variation in statistical models. In Heijmans, R., Pollock, D., and Satorra, A., eds., *Innovations in Multivariate Statistical Analysis: A Festschrift for Heinz Neudecker*, pp. 75–85. Kluwer, London, U.K.

McCabe, B. P. M., Martin, G. M., and Harris, D. (2011). Efficient probabilistic forecasts for counts. *Journal of the Royal Statistical Society: Series B*, 73:253–272.

McKenzie, E. (2003). Discrete variate time series. In Shanbhag, D. and Rao, C., eds., *Handbook of Statistics, Vol. 21*, pp. 573–606. Elsevier Science, Amsterdam, the Netherlands.

Pavlopoulos, H. and Karlis, D. (2008). INAR(1) modeling of overdispersed count series with an environmental application. *Environmetrics*, 19:369–393.

Psaradakis, Z., Sola, M., Spagnolo, F., and Spagnolo, N. (2009). Selecting nonlinear time series models using information criteria. *Journal of Time Series Analysis*, 30:369–394.

Rosenblatt, M. (1952). Remarks on a multivariate transformation. *Annals of Mathematical Statistics*, 23:470–472.

Tsay, R. S. (1992). Model checking via parametric bootstraps in time series analysis. *Applied Statistics*, 41:1–15.

Wecker, W. E. (1989). Comment: Assessing the accuracy of time series model forecasts of count observations. *Journal of Business & Economic Statistics*, 7:418–422.

Weiß, C. H. (2009). Modelling time series of counts with overdispersion. *Statistical Methods and Applications*, 18:507–519.

White, H. (1982). Maximum likelihood estimation of misspecified models. *Econometrica*, 50:1–25.

Zhu, F. (2011). A negative binomial integer-valued GARCH model. *Journal of Time Series Analysis*, 32(1):54–67.

10

Detection of Change Points in Discrete-Valued Time Series

Claudia Kirch and Joseph Tadjuidje Kamgaing

CONTENTS

10.1 Introduction

There has recently been a renewed interest in statistical procedures concerned with the detection of structural breaks in time series, for example, the recent review articles by Aue and Horváth [2] and Horváth and Rice [16]. The literature contains statistics to detect simple mean changes, changes in linear regression, changes in generalized autoregressive conditionally heteroscedastic (GARCH) models; from likelihood ratio to robust M methods (see, e.g., Berkes et al. [3], Davis et al. [6], Hušková and Marušiaková [26], and Robbins et al. [31]). While at first sight, the corresponding statistics appear very different, most of them are derived using the same principles. In this chapter, we shed light on those principles, explaining how corresponding statistics and their respective asymptotic behavior under both the null and alternative hypotheses can be derived. This enables us to give a unified presentation of change point procedures for integer-valued time series. Because the methodology considered in this chapter is by no means limited to these situations, it allows for future extensions in a standardized way.

Hudecová [17] and Fokianos et al. [11] propose change point statistics for binary time series models while Franke et al. [13] and Doukhan and Kegne [7] consider changes in Poisson autoregressive models. Related procedures have also been investigated by Fokianos and Fried [9,10] for integer valued GARCH and log-linear Poisson autoregressive time series, respectively, but with a focus on outlier detection and intervention effects rather than change points.

Section 10.2 explains how change point statistics can be constructed and derives asymptotic properties under both the null and alternative hypotheses, based on regularity conditions, which are summarized in Appendix 10.7.1 to lighten the presentation. This methodology is then applied to binary time series in Section 10.3 and to Poisson autoregressive models in Section 10.4, generalizing the statistics already discussed in the literature. In Section 10.5, some simulations as well as applications to real data illustrate the performance of these procedures. A short review of sequential (also called online) procedures for count time series is given in Section 10.6. Finally, the proofs are given in Appendix 10.7.2.

10.2 General Principles of Retrospective Change Point Analysis

Assume that data $Y_1, \ldots, Y_n$ are observed with a possible structural break at the (unknown) change point k_0. We will first look at likelihood ratio tests for structural breaks before explaining how to generalize these ideas. To this end, we assume that the data before and after the change can be parameterized by the same likelihood function L but with different (unknown) parameters $\theta_0,\ \theta_1 \in \Theta \subset \mathbf{R}^d$. A likelihood ratio approach yields the following statistic:

$$\max_{1\leqslant k\leqslant n} \ell(k) := \max_{1\leqslant k\leqslant n} \left(\ell\left((Y_1,\ldots,Y_k),\widehat{\theta}_k\right) + \ell\left((Y_{k+1},\ldots,Y_n),\widehat{\theta}_k^{\circ}\right) - \ell\left((Y_1,\ldots,Y_n),\widehat{\theta}_n\right)\right),$$

where $\ell(\mathbf{Y},\theta)$ is the log-likelihood function and $\widehat{\theta}_k$ and $\widehat{\theta}_k^{\circ}$ are the maximum likelihood estimator based on $Y_1,\ldots,Y_k$ and $Y_{k+1},\ldots,Y_n$, respectively. The maximum over k is due to the fact that the change point is unknown, so the likelihood ratio statistic maximizes over all possible change points. A similar approach based on some asymptotic Bayes statistic leads to a sum-type statistic, where the sum over $\ell(k)$ is considered (see, e.g., Kirch [19]). Davis et al. [6] proposed this statistic for linear autoregressive processes of order p with standard normal errors:

$$Y_t = \beta_0 + \sum_{j=1}^{p} \beta_j Y_{t-j} + \varepsilon_t, \quad \varepsilon_t \overset{i.i.d.}{\sim} N(0,1). \tag{10.1}$$

In this situation (which includes mean changes as a special case ($p=0$)), this maximum likelihood statistic does not converge in distribution to a nondegenerate limit but almost surely to infinity (Davis et al. [6]). Nevertheless, asymptotic level α tests based on this maximum likelihood statistic can be constructed using a Darling–Erdös limit theorem as stated in Theorem 10.1b. In small samples, however, the slow convergence of Darling–Erdös limit theorems often leads to some size distortions.

Similarly, one can construct Wald-type statistics based on maxima or sums of quadratic forms of

$$W(k) := \widehat{\theta}_k - \widehat{\theta}_k^{\circ}, \quad k = 1, \ldots, n.$$

Wald statistics can be generalized to any other estimation procedure for θ and are not restricted to maximum likelihood estimators. However, for both maximum likelihood and Wald statistics, the estimators $\widehat{\theta}_k$ and $\widehat{\theta}_k^{\circ}$ need to be calculated, which can be problematic in nonlinear situations. In such situations, which are typical for integer-valued time series, these estimators are usually not analytically tractable, but need to be calculated using numerical optimization methods. This can lead to additional computational effort to calculate the statistics or large numerical errors. The latter problems can be reduced by using score-type statistics based on maxima or sums of quadratic forms of

$$S(k) := S\left(k, \widehat{\theta}_n\right) = \frac{\partial}{\partial \theta} \ell\left((Y_1, \ldots, Y_k), \theta\right) |_{\theta = \widehat{\theta}_n}, \quad k = 1, \ldots, n.$$

In this case, only the estimator based on the full data set $Y_1, \ldots, Y_n$ needs to be calculated (possibly using numerical methods). The likelihood score statistic for the linear regression model has been investigated in detail by Hušková et al. [27]. Similarly to Wald statistics, score statistics do not need to be likelihood based but can be generalized to different estimators as long as those estimators can be obtained as a solution to

$$S\left(n, \widehat{\theta}_n\right) = \sum_{j=1}^{n} F\left((Y_t, \mathbb{Y}_{t-1}), \widehat{\theta}_n\right) = 0$$

for some estimating function F. Important estimators of this type are M estimators, which have been used in the context of linear regression models to construct score-type change point tests by Antoch and Hušková [1].

In the linear autoregressive situation in (10.1), the likelihood-based statistics of the type mentioned earlier are all equivalent. Specifically, some calculations yield

$$2\ell(k) = W(k)^T C_k C_n^{-1} C_k^{\circ} W(k) = S(k)^T C_k^{-1} C_n \left(C_k^{\circ}\right)^{-1} S(k),$$

$$\text{where} \quad C_k = \sum_{t=1}^{k} \mathbb{Y}_{t-1}\mathbb{Y}_{t-1}^T, \quad C_k^{\circ} = \sum_{t=k+1}^{n} \mathbb{Y}_{t-1}\mathbb{Y}_{t-1}^T, \quad \mathbb{Y}_{t-1} = (1, Y_{t-1}, \ldots, Y_{t-p})^T.$$

As already mentioned, the maximum likelihood statistic (hence the corresponding likelihood Wald and score statistics) does not converge in distribution. Under the null hypothesis, the matrix in the quadratic form of the likelihood score statistic can be approximated asymptotically (as $k \to \infty, n - k \to \infty, n \to \infty$) by

$$C_k^{-1} C_n (C_k^{\circ})^{-1} = \frac{1}{\frac{k}{n}\left(1 - \frac{k}{n}\right)} \frac{1}{n} C^{-1} + o_P(1), \quad C = \mathrm{E}\mathbb{Y}_{t-1}\mathbb{Y}_{t-1}^T. \tag{10.2}$$

Replacing this term by $w(k/n)C_n^{-1}$ for a suitable weight function $w(\cdot)$ leads to a statistic that does converge in distribution. More precisely, we consider

$$\max_{1\leqslant k\leqslant n} w\left(\frac{k}{n}\right) S(k)^T C_n^{-1} S(k),$$

where $w : [0,1] \to \mathbb{R}_+$ is a nonnegative continuous weight function fulfilling

$$\lim_{t\to 0} t^{\alpha} w(t) < \infty, \quad \lim_{t\to 1} (1-t)^{\alpha} w(t) < \infty, \quad \text{for some } 0 \leqslant \alpha < 1$$
$$\sup_{\eta\leqslant t\leqslant 1-\eta} w(t) < \infty \quad \text{for all } 0 < \eta \leqslant \frac{1}{2}. \tag{10.3}$$

Theorem 10.1a shows that this class of statistics converges, under regularity conditions, in distribution to a nondegenerate limit. The following choice of weight function, closely related to the choice of the weights in (10.2), has often been proposed in the literature:

$$w(t) = (t(1-t))^{-\gamma}, \quad 0 \leqslant \gamma < 1,$$

where γ close to 1 detects early or late changes with better power. In the econometrics literature, the following weight functions are often used, which correspond to a truncation of the likelihood ratio statistic and can be viewed as the likelihood ratio statistic under restrictions on the set of admissible change points,

$$w(t) = (t(1-t))^{-1/2} \, 1_{\{\epsilon\leqslant t\leqslant 1-\epsilon\}}$$

for some $\epsilon > 0$. Similarly, if a priori knowledge of the location of the change point is available, one can increase the power of the designed test statistic for such alternatives by choosing a weight function that is larger near these points (Kirch et al. [20]). Nevertheless, these statistics have asymptotic power one for other change locations (See Theorem 10.2).

Additionally, many change point statistics discussed in the literature do not use the full score function but rather a lower-dimensional projection, where C_n is replaced by a lower rank matrix. For linear autoregressive models as in (10.1), for example, Kulperger [24] and Horváth [25] use a partial sum process based on estimated residuals, which corresponds to the first component of the likelihood score vector in this example.

For this reason, in the following, we do not require $S(k,\theta)$ to be the likelihood score (nor even of the same dimension as θ), nor do we assume that $\widehat{\theta}_n$ is the maximum likelihood estimator. In fact, we allow for general score-type statistics that are based on partial sum processes of the type

$$S\left(k,\widehat{\theta}_n\right) = \sum_{j=1}^{k} H\left(\mathbf{X}_j,\widehat{\theta}_n\right), \quad \text{with } S\left(n,\widehat{\theta}_n\right) = 0 \quad \text{and} \quad \widehat{\theta}_n \to \theta_0, \tag{10.4}$$

where θ_0 is typically the correct parameter, $\mathbf{X}_j$ are observations, where, for example, for the autoregressive case of order one, a vector $\mathbf{X}_j = (X_j, X_{j-1})^T$ is used, and H is some function usually of the type AF for some (possibly lower rank) matrix A and an estimating function

F that defines the estimator $\widehat{\theta}_n$ as the unique zero of $\sum_{j=1}^{n} F(\boldsymbol{X}_j, \widehat{\theta}_n) = 0$. Furthermore, it is possible to allow for misspecification, in which case, θ_0 becomes the best approximating parameter in the sense of $EF(\boldsymbol{X}_j, \theta_0) = 0$. More details on this framework in a sequential context can be found in Kirch and Kamgaing [22].

We are now able to derive the limit distribution of the corresponding score-type change point tests under the null hypothesis under the regularity conditions given in Section 10.7.1. These regularity conditions are implicitly shown in the proofs for change point tests of the types mentioned earlier. Examples for integer-valued time series are given in Sections 10.3 and 10.4.

Theorem 10.1 *We obtain the following null asymptotics:*

(a) Let A.1 and A.2 (i) in Section 10.7.1 hold. Assume that the weight function is either a continuous nonnegative and bounded function $w : [0,1] \to \mathbb{R}_+$, or for unbounded functions fulfilling (10.3), let additionally A.2 (ii) in Section 10.7.1 hold. Then:

(i) $\max_{1\leqslant k\leqslant n} \frac{w(k/n)}{n} S\left(k, \widehat{\theta}_n\right)^T \Sigma^{-1} S\left(k, \widehat{\theta}_n\right) \xrightarrow{\mathcal{D}} \sup_{0\leqslant t\leqslant 1} w(t) \sum_{j=1}^{d} B_j^2(t),$

(ii) $\sum_{1\leqslant k\leqslant n} \frac{w(k/n)}{n^2} S\left(k, \widehat{\theta}_n\right)^T \Sigma^{-1} S\left(k, \widehat{\theta}_n\right) \xrightarrow{\mathcal{D}} \int_0^1 w(t) \sum_{j=1}^{d} B_j^2(t)\, dt,$

where $B_j(\cdot), j = 1, \ldots, d$, are independent Brownian bridges and Σ can be replaced by $\widehat{\Sigma}_n$ if $\widehat{\Sigma}_n - \Sigma = o_P(1)$.

(b) Under A.1 and A.3 in Section 10.7.1 it holds

$$P\left(a(\log n) \max_{1\leqslant k\leqslant n} \sqrt{\frac{n}{k(n-k)} S(k, \widehat{\theta}_n)^T \Sigma^{-1} S(k, \widehat{\theta}_n)} - b_d(\log n) \leqslant t\right) \to \exp(-2e^{-t}),$$

where $a(x) = \sqrt{2\log x}$, $b_d(x) = 2\log x + \frac{d}{2}\log\log x - \log\Gamma(d/2)$, $\Gamma(\cdot)$ is the Gamma-function, and d is the dimension of the vector $S(k, \theta)$. Furthermore, Σ can be replaced by an estimator $\widehat{\Sigma}_n$ if $\|\widehat{\Sigma}_n^{-1/2} - \Sigma^{-1/2}\| = o_P((\log\log n)^{-1})$.

The assumption of continuity of the weight function in (b) can be relaxed to allow for a finite number of points of discontinuity, where w is either left or right continuous with existing limits from the other side.

Similarly, under alternatives, we provide some regularity conditions, which ensure that the tests mentioned earlier have asymptotic power one. Additionally, we propose a consistent estimator of the change point in rescaled time.

Theorem 10.2 *Under alternatives with a change point of the form*

$$k_0 = \lfloor \lambda n \rfloor, \quad 0 < \lambda < 1, \tag{10.5}$$

we get the following assertions:

(a) *If Assumptions B.1, B.3, and B.4(i) in Section 10.7.1 hold, then the Darling–Erdős- and max-type statistics for continuous weight functions fulfilling $w(\lambda) > 0$ have asymptotic power one, that is, for all $x \in \mathbb{R}$ it holds that*
(i) $P\left(\max_{1\leqslant k\leqslant n} \frac{w(k/n)}{n} S(k,\widehat{\theta}_n)^T\Sigma^{-1}S(k,\widehat{\theta}_n) \geqslant x\right) \to 1,$

(ii) $P\left(a(\log n) \max_{1\leqslant k\leqslant n} \sqrt{S(k,\widehat{\theta}_n)^T\Sigma^{-1}S(k,\widehat{\theta}_n)} - b_d(\log n) \geqslant x\right) \to 1.$

If B.4(i) is replaced by B.4(ii), then the assertion remains true if Σ is replaced by $\widehat{\Sigma}_n$, a consistent estimator of Σ.

(b) *If additionally B.5 holds, then the sum-type statistics for a continuous weight function $w(\cdot) \neq 0$ fulfilling* (10.3) *has power one, that is, it holds for all $x \in \mathbb{R}$*

$$P\left(\sum_{1\leqslant k\leqslant n} \frac{w(k/n)}{n^2} S(k,\widehat{\theta}_n)^T\Sigma^{-1}S(k,\widehat{\theta}_n) \geqslant x\right) \to 1.$$

If B.4(i) is replaced by B.4(ii), then the assertion remains true if Σ is replaced by $\widehat{\Sigma}_n$, a consistent estimator of Σ.

(c) *Let the change point be of the form $k_0 = \lfloor \lambda n \rfloor$, $0 < \lambda < 1$, and consider*

$$\widehat{\lambda}_n = \frac{\arg\max S(k,\widehat{\theta}_n)^T\Sigma^{-1}S(k,\widehat{\theta}_n)}{n}.$$

Under Assumptions B.1, B.3, and B.4(i), $\widehat{\lambda}_n$ is a consistent estimator for the change point in rescaled time λ, that is,

$$\widehat{\lambda}_n - \lambda = o_P(1).$$

If B.4(i) is replaced by B.4(iii), then the assertion remains true if Σ is replaced by $\widehat{\Sigma}_n$, a consistent estimator of Σ.

Assumption (10.5) is standard in change point analysis but can be weakened for Part (a) of Theorem 10.2. The continuity assumption on the weight function can be relaxed in this situation.

While these test statistics were designed for the situation, where at most one change is expected, they usually also have power against multiple changes. This fact is the underlying principle of binary segmentation procedures (first proposed by Vostrikova [36]), which works as follows: The data set is tested using an at most one change test as given earlier. If that test is significant, the data set is split at the estimated change point and the procedure repeated on both data segments until insignificant. Recently, a randomized version of binary segmentation has been proposed for the simple mean change problem (Fryzlewicz [14]).

The optimal rate of convergence in (b) is usually given by $\widehat{\lambda}_n - \lambda = O_P(1/n)$ but requires a much more involved proof (Csörgő and Horváth [5], Theorem 2.8.1, for a proof in a mean change model).

10.3 Detection of Changes in Binary Models

Binary time series are important in applications, where one is observing whether a certain event has or has not occurred. Wilks and Wilby [40], for example, observe whether it has been raining on a specific day, and Kauppi and Saikkonen [18] and Startz [34] observe whether a recession has occurred or not in a given month. A common binary time series model is given by

$$Y_t \mid Y_{t-1}, Y_{t-2}, \ldots, \mathbb{Z}_{t-1}, \mathbb{Z}_{t-2}, \ldots \sim \text{Bern}(\pi_t(\beta)), \quad \text{with } g(\pi_t(\beta)) = \beta^T \mathbb{Z}_{t-1},$$

where $\mathbb{Z}_{t-1} = (Z_{t-1}(1), \ldots, Z_{t-1}(p))^T$ is a regressor, which can be purely exogenous (i.e., $\{\mathbb{Z}_t\}$ is independent of $\{Y_t\}$), purely autoregressive (i.e., $\mathbb{Z}_{t-1} = (Y_{t-1}, \ldots, Y_{t-p})$) or a mixture of both (in particular, the independence assumption does not need to hold). Similar to generalized linear models, the canonical link function $g(x) = \log(x/(1-x))$ is used and statistical inference is based on the partial likelihood

$$L(\beta) = \prod_{t=1}^{n} \pi_t(\beta)^{y_t} (1 - \pi_t(\beta))^{1-y_t},$$

with corresponding score vector

$$S^{\text{BAR}}(k, \beta) = \sum_{t=1}^{k} \mathbb{Z}_{t-1}(Y_t - \pi_t(\beta)) \tag{10.6}$$

for the canonical link function.

Theorem 10.3 *We get the following assertions under the null hypothesis:*

(a) Let the covariate process $\{\mathbb{Z}_t\}$ be strictly stationary and ergodic with finite fourth moments. Then, under the null hypothesis, A.1 and A.3 (i) in Section 10.7.1 are fulfilled for the partial sum process $S^{BAR}(k, \beta_n)$ and $\widehat{\beta}_n$ defined by

$$S^{BAR}(n, \widehat{\beta}_n) = 0. \tag{10.7}$$

(b) If $(Y_t, Z_{t-1}, \ldots, Z_{t-p})^T$ is also α-mixing with exponential rate, then A.3 (ii) and (iii) in Section 10.7.1 are fulfilled.

In particular, change point statistics based on $S^{BAR}(k, \widehat{\beta}_n)$ *have the null asymptotics as stated in Theorem 10.1 with* $\Sigma = \mathrm{cov}(\mathbb{Z}_{t-1}(Y_t - \pi_t(\beta_0))$*, where* β_0 *is the true parameter under the null hypothesis.*

Remark 10.1 *For* $\mathbb{Z}_{t-1} = (Y_{t-1}, \ldots, Y_{t-p})^T$*,* Y_t *is the standard binary autoregressive model (BAR(p)), for which the assumptions of Theorem 10.3 (b) are fulfilled, see, for example, Wang and Li [37]. However, considering some regularity assumptions on the exogenous process, one can prove that* $(Y_t, \ldots, Y_{t-p+1}, Z_t, \ldots, Z_{t-q})$ *is a Feller chain, for which Theorem 1 of Feigin and Tweedie [8] implies geometric ergodicity (see Kirch and Tadjuidje Kamgaing [21] for details) implying that it is* β*-mixing with exponential rate.*

Remark 10.2 *The mixing concept can be regarded as an asymptotic measure of independence in time between the observations. The reader can refer to Tadjuidje et al. [35], Remark 4, for* α*- and* β*-mixing definitions, as well as a concise summary of their relationship to geometric ergodicity for a Markov chain.*

Instead of using the full vector partial sum process $S^{BAR}(k, \widehat{\beta}_n)$ to construct the test statistics, often lower-dimensional linear combinations are used, such as

$$\tilde{S}^{BAR}(k, \widehat{\beta}_n) = \sum_{t=1}^{k} (Y_t - \pi_t(\widehat{\beta}_n)), \tag{10.8}$$

where $\widehat{\beta}_n$ is defined by (10.7). If the assumptions of Theorem 10.3 are fulfilled, then the null asymptotics as in Theorem 10.1 with $\Sigma = \mathrm{cov}(Y_t - \pi_t(\beta_0))$ hold.

The statistic based on $S^{BAR}(k, \widehat{\beta}_n)$ for $w \equiv 1$ has been proposed by Fokianos et al. [11]; a statistic based on $\tilde{S}^{BAR}(k, \widehat{\beta}_n)$ with a somewhat different standardization in a purely autoregressive setup has been considered by Hudecová [17]. Hudecová's statistic is the score statistic based on the partial likelihood and the restricted alternative of a change only in the intercept.

BAR-Alternative: Let the following assumptions hold:

- H_1(i) The change point is of the form $k_0 = \lfloor \lambda n \rfloor$, $0 < \lambda < 1$.
- H_1(ii) The binary time series $\{Y_t\}$ and the covariate process $\{Z_t\}$ before the change fulfills the assumptions of Theorem 10.3a.
- H_1(iii) The time series after the change as well as the covariate process after the change can be written as $Y_t = \tilde{Y}_t + R_1(t)$ and $\mathbb{Z}_t = \tilde{\mathbb{Z}}_t + \mathbb{R}_2(t)$, respectively, $t > \lfloor n\lambda \rfloor$, where $\{\tilde{Y}_t\}$ is bounded, stationary, and ergodic and $\{\tilde{\mathbb{Z}}_t\}$ is square integrable as well as stationary and ergodic with remainder terms fulfilling

$$\frac{1}{n} \sum_{j=\lfloor \lambda n \rfloor+1}^{n} R_1^2(t) = o_P(1), \quad \frac{1}{n} \sum_{j=\lfloor \lambda n \rfloor+1}^{n} \|\mathbb{R}_2(t)\|^2 = o_P(1).$$

- H_1(iv) $\lambda \mathrm{E}\mathbb{Z}_0(Y_1 - \pi_1(\beta)) + (1-\lambda)\mathrm{E}\tilde{\mathbb{Z}}_{n-1}(\tilde{Y}_n - \tilde{\pi}_1(\beta))$ has a unique zero $\beta_1 \in \Theta$ and Θ is compact and convex with $\widehat{\beta}_n \in \Theta$.

Assumption (i) is standard in change point analysis but could be relaxed, and assumption (ii) states that the time series before the change fulfills the assumption under the null hypothesis. Assumption (iii) allows for rather general alternatives, including situations where starting values from before the change are used resulting in a nonstationary time series after the change. Assumption (iv) guarantees that the estimator $\widehat{\beta}_n$ converges to β_1. Neither Hudecová [17] nor Fokianos et al. [11] have derived the behavior of their statistics under alternatives.

Theorem 10.4 *Let $H_1(i)$–$H_1(iv)$ hold.*

(a) For $S^{BAR}(k, \beta)$ as in (10.6), B.1 and B.2 are fulfilled, which implies B.3. If $k_0 = \lfloor \lambda n \rfloor$, then B.5 is fulfilled with $F_\lambda(\beta) = \lambda \mathbb{E} \mathbb{Z}_0 (Y_1 - \pi_1(\beta))$.

(b) For $\tilde{S}^{BAR}(k, \beta)$ as in (10.8) and if $k_0 = \lfloor \lambda n \rfloor$, then B.5 is fulfilled with $F_\lambda(\beta) = \lambda \mathbb{E} (Y_1 - \pi_1(\beta))$.

B.4 is fulfilled for the full score statistic from Theorem 10.4a if the time series before and after the change are correctly specified binary time series models with different parameters. Otherwise, restrictions apply. Together with Theorem 10.2, this implies that the corresponding change point statistics have power one and the point where the maximum is obtained is a consistent estimator for the change point in rescaled time.

10.4 Detection of Changes in Poisson Autoregressive Models

Another popular model for time series of counts is the Poisson autoregression, where we observe $Y_{1-p}, \ldots, Y_n$ with

$$Y_t \mid Y_{t-1}, Y_{t-2}, \ldots, Y_{t-p} \sim \text{Pois}(\lambda_t), \quad \lambda_t = f_\gamma(Y_{t-1}, \ldots, Y_{t-p}) \tag{10.9}$$

for some d-dimensional parameter vector $\gamma \in \Theta$. If $f_\gamma(\mathbf{x})$ is Lipschitz continuous in $\mathbf{x}$ for all $\gamma \in \Theta$ with Lipschitz constant strictly smaller than 1, then there exists a stationary ergodic solution of (10.9) that is β-mixing with exponential rate (Neumann [28]). For a given parametric model f_θ, this allows us to consider score-type change point statistics based on likelihood equations using the tools of Section 10.2. The mixing condition in connection to some moment conditions typically allows one to derive A.3, while a Taylor expansion in connection with $\sqrt{n}$-consistency of the corresponding maximum likelihood estimator (e.g., derived by Doukhan and Kegne [7], Theorem 3.2) gives A.1 under some additional moment conditions. Related test statistics for independent Poisson data are discussed in Robbins et al. [32]. However, in this chapter, we will concentrate on change point statistics related to those proposed by Franke et al. [13], which are based on least square scores and, as such, do not make use of the Poisson structure of the process. However, the methods described in Section 10.2 can be used to derive change point tests based on the partial likelihood, which can be expected to have higher power if the model is correctly specified.

Consider the least squares estimator $\widehat{\gamma}_n$ defined by

$$S^{\text{PAR}}(n,\widehat{\gamma}_n) = 0, \quad \text{where } S^{\text{PAR}}(k,\gamma_n) = \sum_{t=1}^{k} \nabla f_\gamma((Y_{t-1},\ldots,Y_{t-p}))(Y_t - f_\gamma(Y_{t-1},\ldots,Y_{t-p})), \tag{10.10}$$

where ∇ denotes the gradient with respect to γ. Under the additional assumption $f_\gamma(\mathbf{x}) = \gamma_1 + f_{\gamma_2,\ldots,\gamma_d}(\mathbf{x})$, that is, if γ_1 is an additive constant in the regression function, this implies in particular that

$$\sum_{t=1}^{n} \left(Y_t - f_{\widehat{\gamma}_n}(Y_{t-1},\ldots,Y_{t-p})\right) = 0. \tag{10.11}$$

Assumptions under H_0*:*

H_0(i) $\{Y_t\}$ is stationary and α-mixing with exponential rate such that $\mathbb{E}\sup_{\gamma\in\Theta} f_\gamma^2(Y_t) < \infty$.

H_0(ii) $f_\gamma(\mathbf{x}) = \gamma_1 + f_{\gamma_2,\ldots,\gamma_d}(\mathbf{x})$ is twice continuously differentiable with respect to γ for all $\mathbf{x} \in \mathbb{N}_0^p$ and

$$\mathbb{E}\sup_{\gamma\in\Theta} \|\nabla f_\gamma(\mathbb{Y}_t)\nabla^T f_\gamma(\mathbb{Y}_t)\| < \infty, \quad \mathbb{E}\left(Y_t \sup_{\gamma\in\Theta} \|\nabla^2 f_\gamma(\mathbb{Y}_{t-1})\|\right) < \infty.$$

H_0(iii) $e(\gamma) = \mathbb{E}(Y_t - f_\gamma(\mathbb{Y}_{t-1}))^2$ has a unique minimizer γ_0 in the interior of some compact set Θ such that the Hessian of $e(\gamma_0)$ is positive definite.

As already mentioned, (i) is fulfilled for a large class of Poisson autoregressive processes under mild conditions. Assumption (ii) means that the autoregressive function used to construct the test statistic is linear in the first component guaranteeing (10.11). Note that this assumption does not need to be fulfilled for the true regression function of $\{Y_t\}$ (in fact, Y_t does not even need to be a Poisson autoregressive time series). Assumption (iii) is fulfilled for the true value if $\{Y_t\}$ is a Poisson autoregressive time series with regression function f_γ.

Theorem 10.5 *Let under the null hypothesis* H_0*(i)–(iii) be fulfilled.*

(a) If $\mathbb{E}\|\nabla f_{\gamma_0}(Y_p,\ldots,Y_1)\|^\nu < \infty$ *for some* $\nu > 2$*, then* $\tilde{S}^{PAR}(k,\gamma) = \sum_{j=1}^{k}(Y_t - f_{\widehat{\gamma}_n}(Y_{t-1},\ldots,Y_{t-p}))$ *together with* $\widehat{\gamma}_n$ *as in (10.10), A.1 and A.3 in Section 10.7.1 hold with* $\Sigma = (\sigma^2)$ *the long-run variance of* $Y_t - f_{\gamma_0}((Y_{t-1},\ldots,Y_{t-p}))$.

(b) If $f_\gamma(\mathbf{x}) = \gamma_1 + (\gamma_2,\ldots,\gamma_d)^T\mathbf{x}$ *(*$p = d-1$*) and* $\mathbb{E}|Y_0|^\nu < \infty$ *for some* $\nu > 4$*, then* $S^{PAR}(k,\gamma) = \sum_{j=1}^{k}\mathbb{Y}_{t-1}(Y_t - \widehat{\gamma}_n^T\mathbb{Y}_{t-1})$*, where* $\mathbb{Y}_{t-1} = (1, Y_{t-1},\ldots,Y_{t-d+1})^T$*. Together with* $\widehat{\gamma}_n$ *as in (10.10), A.1 and A.3 in Section 10.7.1 hold, where* Σ *is the long-run covariance matrix of* $\{\mathbb{Y}_{t-1}(Y_t - \gamma_0^T\mathbb{Y}_{t-1})\}$.

In particular, the change point statistics based on $\tilde{S}^{PAR}(k,\widehat{\gamma}_n)$ *in (a) and* $S^{PAR}(k,\widehat{\gamma}_n)$ *in (b) have the null asymptotics as stated in Theorem 10.1.*

Assumptions under H_1:

- H_1(i) The change point is of the form $k_0 = \lfloor \lambda n \rfloor, 0 < \lambda < 1$.
- H_1(ii) For all $\gamma \in \Theta$, Θ is compact and convex, and $f_\gamma(\mathbf{x})$ is uniformly Lipschitz in $\mathbf{x}$ with Lipschitz constant $L_\gamma < 1$.
- H_1(iii) The time series before the change is stationary and ergodic such that $\mathbb{E}\sup_{\gamma\in\Theta} \|Y_{t-j}\nabla f_\gamma(Y_{t-1},\ldots,Y_{t-p})\| < \infty, j = 0,\ldots,p$.
- H_1(iv) The time series after the change fulfills $Y_t = \tilde{Y}_t + R_1(t), t > \lfloor \lambda n \rfloor$, where $\{\tilde{Y}_t\}$ is stationary and ergodic such that $\mathbb{E}\sup_{\gamma\in\Theta} \|\tilde{Y}_{t-j}\nabla f_\gamma(\tilde{Y}_{t-1},\ldots,\tilde{Y}_{t-p})\| < \infty$, $j = 0,\ldots,p$, with the remainder term fulfilling

$$\frac{1}{n}\sum_{j=\lfloor \lambda n \rfloor+1}^{n} R_1^2(t) = o_P(1).$$

- H_1(v) $\lambda\mathbb{E}\nabla f_\gamma((Y_0,\ldots,Y_{1-p}))(Y_1 - f_\gamma((Y_0,\ldots,Y_{1-p}))) + (1-\lambda)\mathbb{E}\nabla f_\gamma(\tilde{Y}_0,\ldots,\tilde{Y}_{1-p})$ $(\tilde{Y}_1 - f_\gamma(\tilde{Y}_0,\ldots,\tilde{Y}_{1-p}))$ has a unique zero $\gamma_1 \in \Theta$ in the strict sense of B.2.

The formulation in H_1(iv) allows for certain deviations from stationarity of the time series after the change which can, for example, be caused by starting values from the stationary distribution before the change, while H_1(v) guarantees that $\widehat{\gamma}_n$ converges to γ_1 under alternatives.

The following theorem extends the results of Franke et al. [13].

Theorem 10.6 *Let assumptions H_1(i)–H_1(iv) be fulfilled.*

(a) For $S^{PAR}(k,\gamma)$ as in (10.10), *B.1 and B.2 are fulfilled.*
(b) For $\tilde{S}^{PAR}(k,\gamma) = \sum_{j=1}^{k}(Y_t - f_\gamma(Y_{t-1},\ldots,Y_{t-p}))$ and if $k_0 = \lfloor \lambda n \rfloor$, B.5 is fulfilled with $F_\lambda(t) = \mathbb{E}(Y_1 - \gamma_1^T \mathbb{Y}_0)$.
(c) For $S^{PAR}(k,\gamma)$ as in (10.10) *with $f_\gamma(\mathbf{x}) = \gamma^T\mathbf{x}$ and if $k_0 = \lfloor \lambda n \rfloor$, then B.5 is fulfilled with $F_\lambda(t) = \mathbb{E}\mathbb{Y}_0(Y_1 - \gamma_1^T\mathbb{Y}_0)$.*

From this, we can give assumptions under which the corresponding tests have asymptotic power one and the point where the maximum is attained is a consistent estimator for the change point in rescaled time by Theorem 10.2.

B.4 is always fulfilled for the full score statistic if the time series before and after the change are correctly specified by the given Poisson autoregressive model. Otherwise, restrictions apply.

Doukhan and Kengne [7] propose to use several Wald-type statistics based on maximum likelihood estimators in Poisson autoregressive models. While their statistics are also designed for the at most one change situation, they explicitly prove consistency under the multiple change point alternative.

10.5 Simulation and Data Analysis

In the previous sections, we have derived the asymptotic limit distribution for various statistics as well as shown that the corresponding tests have asymptotic power one under relatively general conditions. In particular, we have proven the validity of these conditions for two important classes of integer-valued time series: binary autoregressive and Poisson counts. In this section, we give a short simulation study in addition to some data analysis to illustrate the small sample properties of these tests complementing simulations of Hudecová [17] and Fokianos et al. [11]. The critical values are obtained from Monte Carlo experiments of the limit distribution based on 1000 repetitions.

10.5.1 Binary Autoregressive Time Series

In this section, we consider a first-order binary autoregressive time series (BAR(1)) as defined in Section 10.3 with $\mathbb{Z}_{t-1} = (1, Y_{t-1})$. We consider the statistic

$$T_n = \max_{1 \leqslant k \leqslant n} \frac{1}{n} \left(S^{\text{BAR}}\left(k, \widehat{\boldsymbol{\beta}}_n\right)\right)^T \widehat{\boldsymbol{\Sigma}}^{-1} S^{\text{BAR}}\left(k, \widehat{\boldsymbol{\beta}_n}\right),$$

$$\text{where } \widehat{\boldsymbol{\Sigma}} = \frac{1}{n} \sum_{t=1}^{n} \mathbb{Z}_{t-1} \mathbb{Z}_{t-1}^T \pi_t\left(\widehat{\boldsymbol{\beta}}_n\right)\left(1 - \pi_t(\widehat{\boldsymbol{\beta}}_n)\right)$$

and S^{BAR} is as in (10.6) and $\widehat{\boldsymbol{\beta}}_n$ as in (10.7). Since $\widehat{\boldsymbol{\Sigma}}$ consistently estimates $\boldsymbol{\Sigma} = \mathbb{E}(\mathbb{Z}_{t-1}\mathbb{Z}_{t-1}^T \pi_t(\boldsymbol{\beta}_0)(1 - \pi_t(\boldsymbol{\beta}_0))^T$ under the null hypothesis, by Theorem 10.3 and Theorem 10.1, the asymptotic null distribution of this statistic is given by

$$\sup_{0 \leqslant t \leqslant 1} \left(B_1^2(t) + B_2^2(t)\right) \tag{10.12}$$

for two independent Brownian bridges $\{B_1(\cdot)\}$ and $\{B_2(\cdot)\}$ with a simulated 95% quantile of 2.53. Table 10.1 reports the empirical size and power (based on 10,000 repetitions) for various alternatives, where a change always occurred at $n/2$. Figure 10.1 shows one sample path for the null hypothesis and each of the alternatives considered there. The size is always conservative and gets closer to the nominal level with increasing sample size as predicted by the asymptotic results. The power is good and increases with the sample size, where some alternatives have better power than others.

TABLE 10.1

Empirical Size and Power of Binary Autoregressive Model with $\boldsymbol{\beta}_1 = (2, -2)$ (Parameter before the Change)

	H_0			H_1: $\boldsymbol{\beta}_2 = (1, -2)$			H_1: $\boldsymbol{\beta}_2 = (1, -1)$			H_1: $\boldsymbol{\beta}_2 = (2, -1)$		
n	200	500	1000	200	500	1000	200	500	1000	200	500	1000
	0.032	0.040	0.044	0.650	0.985	1.00	0.176	0.520	0.871	0.573	0.961	0.999

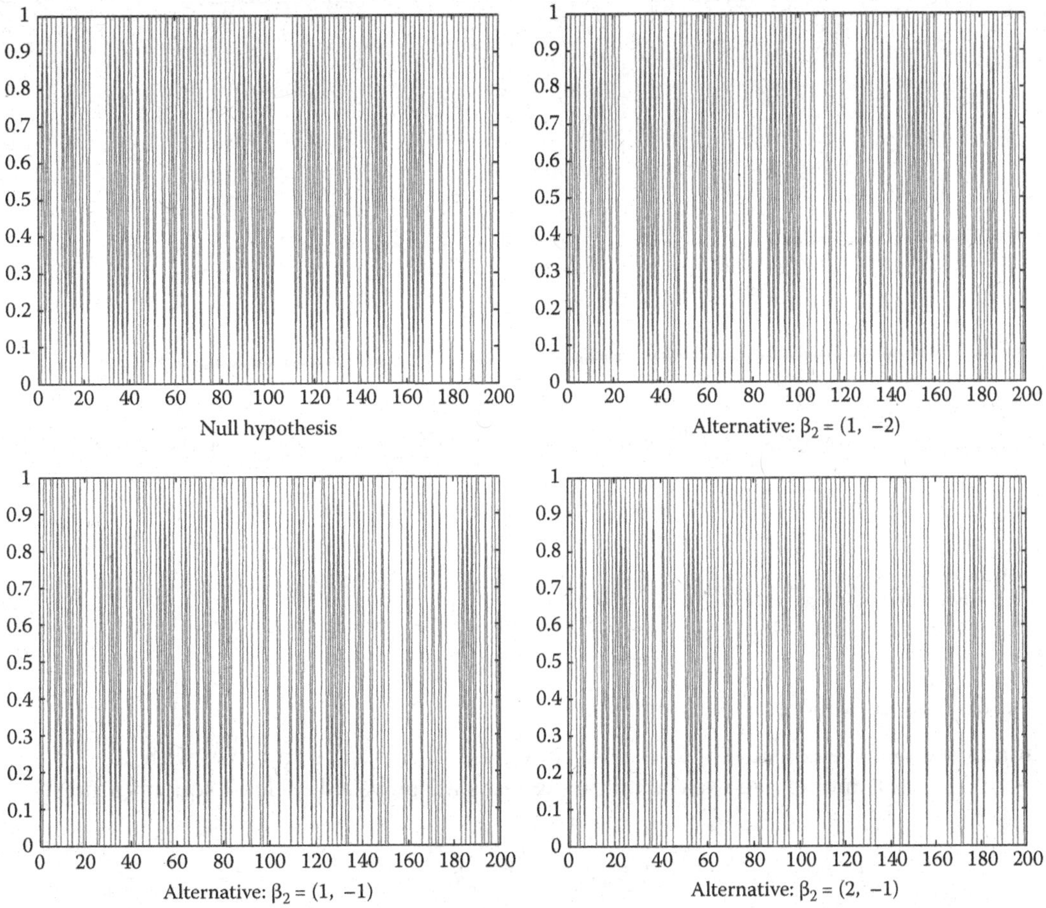

FIGURE 10.1
Sample paths for the BAR(1) model, $\boldsymbol{\beta}_1 = (2, -2)$, $k_0 = 100$, $n = 200$.

10.5.1.1 Data Analysis: U.S. Recession Data

We now apply the test statistic mentioned earlier to the quarterly recession data in Figure 10.2 from the United States for the period 1855–2012*. The datum is 1 if there has been a recession in at least one month in the quarter and 0 otherwise. The data have been previously analyzed by different authors; in particular, they have recently been analyzed in a change point context by Hudecová [17].

We find a change in the first quarter of 1933, which corresponds to the end of the great depression that started in 1929 in the United States and leads to a huge unemployment rate in 1932. If we split the time series at that point and repeat the change point procedure, no further significant change points are found. This is consistent with the findings in Hudecová [17], who applied a different statistic based on a binary autoregressive time series of order 3.

* This data set can be downloaded from the National Bureau of Economic Research at http://research.stlouisfed.org/fred2/series/USREC.

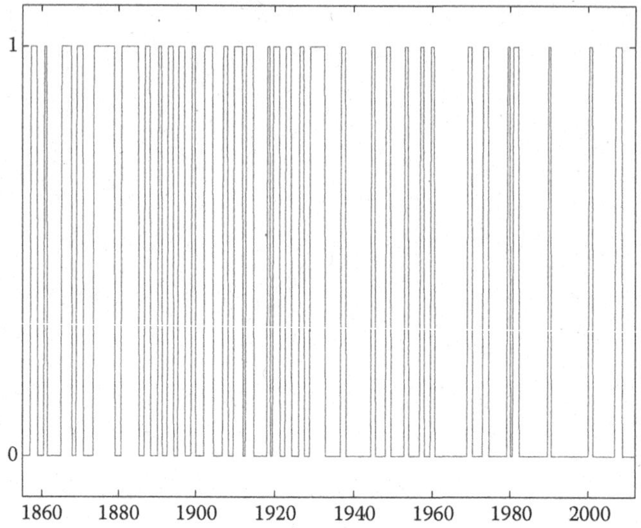

FIGURE 10.2
Quarterly U.S. recession data (1855–2012).

10.5.2 Poisson Autoregressive Models

In this section, we consider a Poisson autoregressive model as in (10.9) with $\lambda_t = \gamma_1 + \gamma_2 Y_{t-1}$. For this model, we use the following test statistic based on least squares scores:

$$T_n = \max_{1\leqslant k\leqslant n} \frac{1}{n} S^{\text{PAR}}(k, \widehat{\gamma}_n)^T \widehat{\boldsymbol{\Sigma}}^{-1} S^{\text{PAR}}(k, \widehat{\gamma}_n),$$

$$\text{where } S^{\text{PAR}}(k, \gamma) = \sum_{t=1}^{k} \mathbb{Y}_{t-1}(Y_t - \lambda_t), \quad \mathbb{Y}_{t-1} = (1, Y_{t-1})^T,$$

and $\widehat{\gamma}_n$ as in (10.10) and $\widehat{\boldsymbol{\Sigma}}^{-1}$ is the empirical covariance matrix of $\{\mathbb{Y}_{t-1}(Y_t - \lambda_t)\}$. By Theorems 10.5b and 10.1, this statistic has the same null asymptotics as in (10.12). Table 10.2 reports the empirical size and power (based on 10,000 repetitions) for various alternatives, where a change always occurred at $n/2$. Figure 10.3 shows one corresponding sample path for each scenario. The test size is always conservative and gets closer to the nominal level with increasing sample size as predicted by the asymptotic results. The power is good

TABLE 10.2
Empirical Size and Power of Poisson Autoregressive Model with $\gamma_1 = (1, 0.75)$ (Parameter before the Change)

	H_0			H_1: $\gamma_2 = (2, 0.75)$			H_1: $\gamma_2 = (2, 0.5)$			H_1: $\gamma_2 = (1, 0.5)$		
n	200	500	1000	200	500	1000	200	500	1000	200	500	1000
	0.028	0.0361	0.036	0.531	0.967	0.999	0.252	0.683	0.968	0.271	0.895	0.999

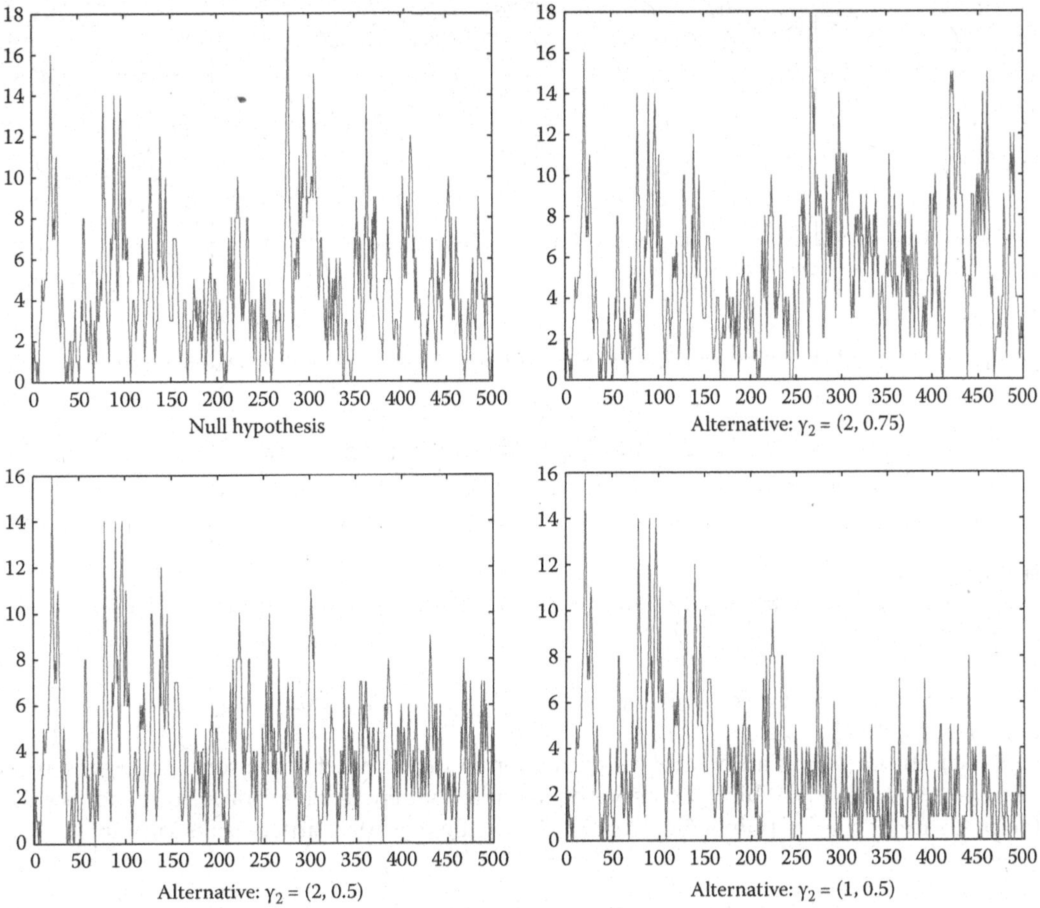

FIGURE 10.3
Sample paths for the Poisson autoregressive model, $\gamma_1 = (1, 0.75)$, $k_0 = 250$, $n = 500$.

and increases with the sample size. Some pilot simulations suggest that using statistics associated with partial likelihood scores can further increase the power. While a detailed theoretic analysis can in principle be done based on the results in Section 10.2, it is beyond the scope of this work.

10.5.2.1 Data Analysis: Number of Transactions per Minute for Ericsson B Stock

In this section, we use the methods given earlier to analyze the data set that consists of the number of transactions per minute for the stock Ericsson B during July 3, 2002. The data set consists of 460 observations instead of 480 for 8 h of transactions because the first 5 min and last 15 min of transactions are ignored. Fokianos et al. [12] have analyzed the transactions count from the same stock on a different day with a focus on forecasting the number of transactions. The data and estimated change points (using a binary segmentation procedure as described in Theorem 10.2) are illustrated in Figure 10.4. The red vertical lines

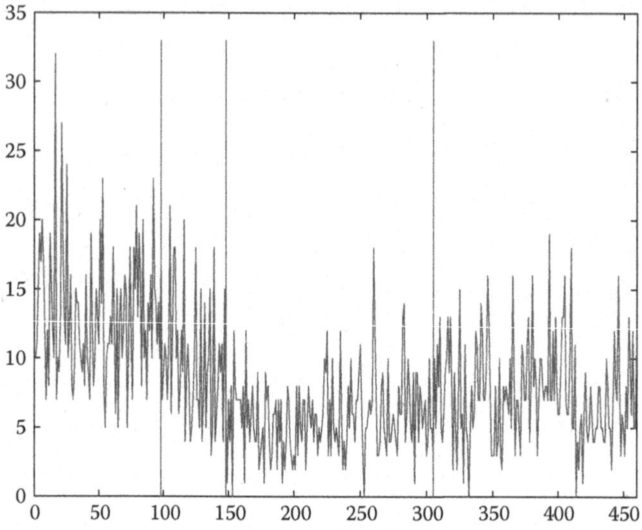

FIGURE 10.4
Transactions per minute for the stock Ericsson B on July 3, 2002, where the red vertical lines are the estimated change points.

indicated estimated change points at the 98th (11:12), 148th (12:02), and 300th (14:39) data points, respectively.

In fact, the empirical autocorrelation function of the complete time series in Figure 10.5 decreases very slowly, which can either be taken as evidence of long-range dependence or the presence of change points. Figure 10.6 shows the empirical autocorrelation function of

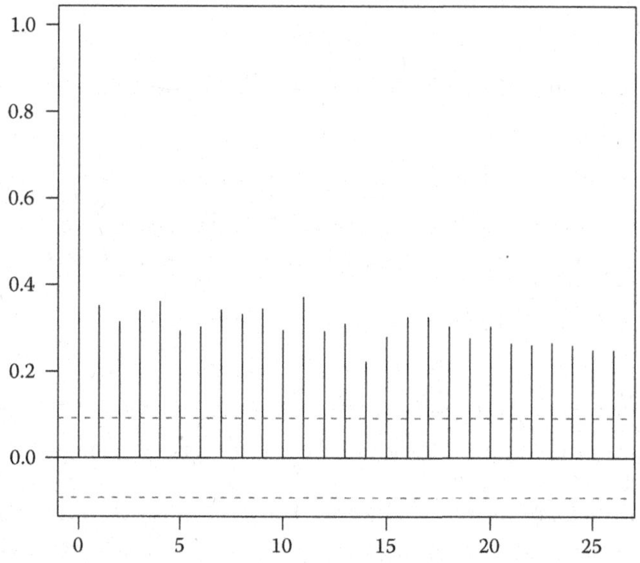

FIGURE 10.5
Sample autocorrelation function.

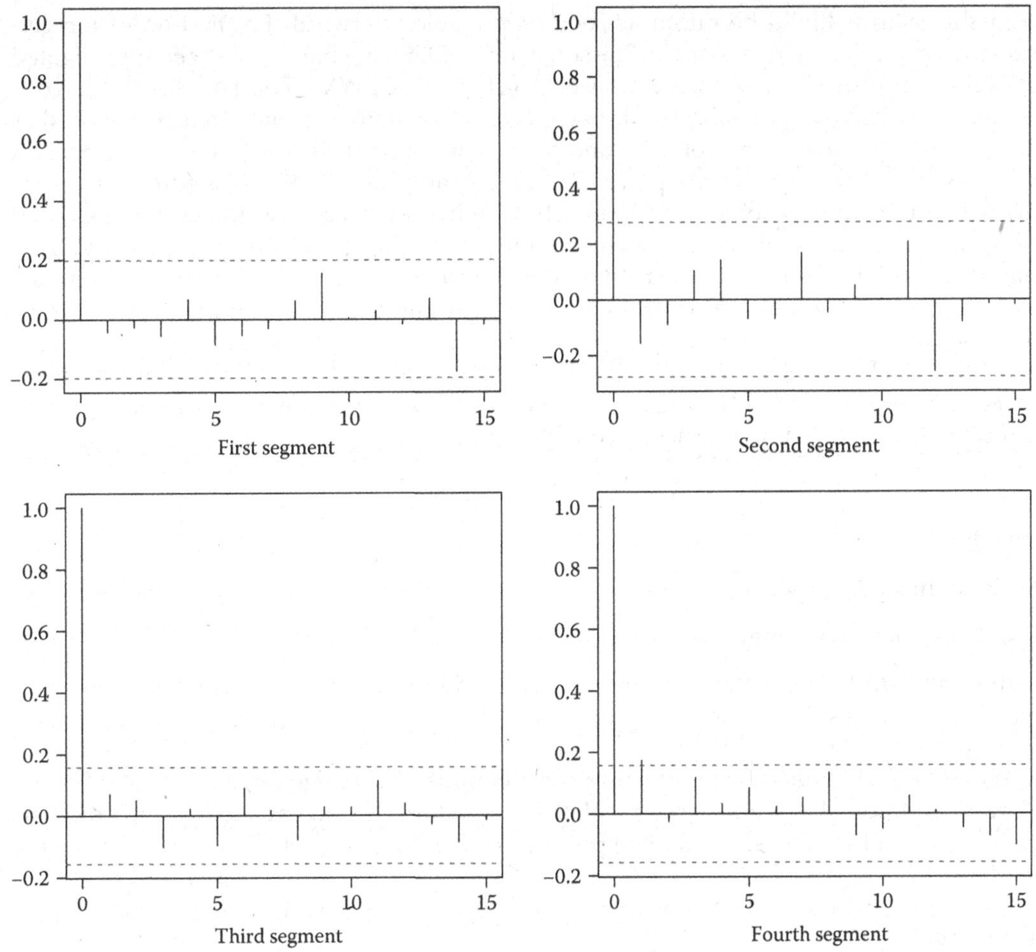

FIGURE 10.6
The empirical autocorrelation of the segmented data, taking into account the estimated change points.

the data in each segment supporting the conjecture that change points, rather than long-range dependence, cause the slow decrease in the empirical autocorrelation function of the full data set.

10.6 Online Procedures

In some situations, statistical decisions on whether a change point has occurred need to be made online as the data arrive point by point. In such situations, classical monitoring charts such as exponentially weighted moving average (EWMA) or cumulative sum (CUSUM) check for changes in a parametric model with known parameters. Unlike in classical statistical testing, it is not the size that is controlled under the null hypothesis but the average run length that should be larger than a prespecified value. On the other hand, if a change occurs an alarm should be raised as soon as possible afterward. For first-order integer-valued autoregressive processes of Poisson counts, CUSUM charts have been investigated by Weiß and Testik [39] as well as Yontay et al. [41] and the EWMA chart by Weiß [38]. A different approach was proposed by Chu et al. [4] in the context of linear regression models that controls the size asymptotically and at the same time tests for changes in a specific model with unknown in-control parameters. If a change occurs, this procedure will eventually reject the null hypothesis. In their setting, a historic data set with no change is used to estimate the in-control parameters before the monitoring of new incoming observations starts. Such a data set usually exists in practice as some data are necessary before any reasonable model building or estimation for statistical inference can take place. Asymptotic considerations based on the length of this data set are used to calibrate the procedure. Kirch and Tadjuidje Kamgaing [22] generalize the approach of Chu et al. [4] to estimating functions in a similar spirit as described in Section 10.2 for the off-line procedure. Examples include integer-valued time series as considered here.

10.7 Technical Appendix

10.7.1 Regularity Assumptions

In this appendix, we summarize the regularity conditions needed to obtain the null asymptotics as well as results under alternatives for change point statistics constructed according to the principles in Section 10.2.

While the techniques used in the proofs are common in change point analysis such regularity conditions have never been isolated to the best of our knowledge and could be quite useful in the analysis of change point tests that have not yet been considered in the literature.

We denote the assumptions under the null hypothesis by A and the assumptions under alternatives by B.

A.1 *Let*

$$\max_{1\leqslant k\leqslant n} \frac{n}{k(n-k)} \left\| S(k,\widehat{\theta}_n) - (S(k,\theta_0) - \frac{k}{n}S(n,\theta_0)) \right\|^2 = o_P(1)$$

for some θ_0.

This assumption allows us to replace the estimator in the statistic by a fixed value θ_0, which is usually given by the true or best approximating parameter of the model. The centering stems from the fact that the estimator $\widehat{\theta}_n$ is the zero of $S(n,\widehat{\theta}_n)$. Typically, this assumption can be derived using a Taylor expansion in addition to the $\sqrt{n}$-consistency of $\widehat{\theta}_n$ for θ_0.

A.2 (*i*) *Let* $\{\frac{1}{\sqrt{n}}S(\lfloor nt\rfloor;\theta_0) : 0 \leqslant t \leqslant 1\}$ *fulfill a functional central limit theorem toward a Wiener process* $\{\mathbf{W}(t) : 0 \leqslant t \leqslant 1\}$ *with regular covariance matrix* Σ *as a limit.*

(ii) *Let both a forward and backward Hájek–Rényi inequality hold true; that is, for all* $0 \leqslant \alpha < 1$, *it holds*

$$\max_{1 \leqslant k \leqslant n} \frac{1}{n^{1-\alpha} k^{\alpha}} \|S(k, \theta_0)\|^2 = O_P(1),$$

$$\max_{1 \leqslant k \leqslant n} \frac{1}{n^{1-\alpha} (n-k)^{\alpha}} \|S(n, \theta_0) - S(k, \theta_0)\|^2 = O_P(1).$$

Both assumptions are relatively weak and are fulfilled by a large class of time series. Hájek–Rényi-type inequalities as given earlier can, for example, be obtained from moment conditions (Appendix B.1 in Kirch [19]).

For the Darling–Erdős-type asymptotics, we need the following stronger assumption.

A.3 (i) *Let* $S(k, \theta_0)$ *fulfill a strong invariance principle, that is, (possibly after changing the probability space) there exists a d-dimensional Wiener process* $\mathbf{W}(\cdot)$ *with regular covariance matrix* Σ *such that*

$$\frac{1}{\sqrt{n}} (S(n, \theta_0) - \mathbf{W}(n)) = o\left((\log \log n)^{-1/2}\right) \quad a.s.$$

(ii) *Let* $\left\{S(n, \theta_0) - S(k, \theta_0) : k = \frac{n}{2}, \ldots, n-1\right\} \stackrel{\mathcal{D}}{=} \left\{\tilde{S}(j, \theta_0) : j = 1, \ldots, n/2\right\}$ *such that*

$$\frac{1}{\sqrt{n}} \left(\tilde{S}(n, \theta_0) - \tilde{\mathbf{W}}(n)\right) = o\left((\log \log n)^{-1/2}\right) \quad a.s.,$$

with $\{\tilde{\mathbf{W}}(\cdot)\} \stackrel{\mathcal{D}}{=} \{\mathbf{W}(\cdot)\}$.

(iii) *Let*

$$\max_{1 \leqslant k \leqslant n/\log n} \frac{1}{k} S(k, \theta_0) \Sigma^{-1} S(k, \theta_0)$$

$$\text{and} \quad \max_{n - n/\log n \leqslant k \leqslant n} \frac{1}{n-k} (S(n, \theta_0) - S(k, \theta_0)) \Sigma^{-1} (S(n, \theta_0) - S(k, \theta_0))$$

be asymptotically independent.

Invariance principles as in (i) have been obtained for different kinds of weak dependence concepts such as mixing to state a classic result (Philipps and Stout [29]), where the rate is typically of polynomial order. Since the definition of mixing is symmetric, the backward invariance principle also follows for such time series. Assumption (iii) is fulfilled by the definition of mixing but can otherwise be difficult to prove. Similarly, Assumption (ii) does not necessarily follow by the same methods as (i) (e.g., the proof of Theorem 10.3, where stronger assumptions are needed to get (ii)). Even with the weaker rate $o(1)$, part (i) implies Assumption A.2 (i) and (ii) for the forward direction (using the Hájek–Rényi inequality for independent random variables) while the backward direction follows from A.3 (ii). Assumption (iii) can be difficult to prove but follows for mixing time series by definition.

We will now state some regularity assumptions under the alternative, for which the statistics given earlier have asymptotic power one.

B.1 *Let* $\sup_{\theta\in\Theta}\left\|\frac{1}{n}S(\lfloor nt\rfloor,\theta)-F_t(\theta)\right\| = o_P(1)$ *uniformly in* $0\leqslant t\leqslant 1$ *for some function* $F_t(\theta)$.

This assumption can be obtained under weak moment assumptions from a strong uniform law of large numbers such as Theorem 6.5 in Ranga Rao [30] if the time series after the change can be approximated by a stationary and ergodic time series and θ comes from a compact parameter set.

B.2 *Let* $S(n,\widehat{\theta}_n) = 0$ *and* θ_1 *the unique zero of* $F_1(\theta)$ *in the strict sense, that is, for every* $\varepsilon > 0$ *there exists a* $\delta > 0$ *such that* $|F_1(\theta)| > \delta$ *whenever* $\|\theta-\theta_1\| > \varepsilon$.

The first assumption is related to the centering in Assumption A.1 and is usually fulfilled by construction. The second assumption guarantees that this estimator converges to θ_1 under alternatives. The strict uniqueness condition given there is automatically fulfilled if F_1 is continuous in θ and Θ is compact.

Proposition 10.1 *Under Assumptions B.1 and B.2, it holds* $\widehat{\theta}_n \xrightarrow{P} \theta_1$.

For the main theorem, we can allow the score function to be different from the one used in the estimation of $\widehat{\theta}_n$. A typical example is a projection into a lower-dimensional space such as only the first component of the vector (see Theorem 10.6b for an example). Assumption B.2 will then typically not be fulfilled as θ_1 can no longer be the unique zero. However, to get the main theorem, it can be replaced by the following:

B.3 $\widehat{\theta}_n \xrightarrow{P} \theta_1$.

B.4 (*i*) $F_\lambda(\theta)\Sigma^{-1}F_\lambda(\theta)\geqslant c>0$ *in an environment of* θ_1.

(*ii*) $\liminf_{n\to\infty} F_\lambda(\theta)\widehat{\Sigma}_n^{-1}F_\lambda(\theta)\geqslant c>0$ *in an environment of* θ_1.

(*iii*) $\widehat{\Sigma}_n\to\Sigma_A$ *with* $F_\lambda(\theta)\Sigma_A^{-1}F_\lambda(\theta)\geqslant c>0$ *in an environment of* θ_1.

This assumption is crucial in understanding which alternatives we can detect. Typically, in the correctly specified model before and after the change, and if the same score vector is used for the change point test as for the parameter estimation, (i) will always be fulfilled. However, if only part of the score vector is used as, for example, in Theorem 10.6b, then this condition describes which alternatives are still detectable. Typically, the power to detect those will be larger than for the full test at the cost of not having power for different alternatives at all. Parts (ii) and (iii) allow one to use estimators of Σ that do not converge or converge to a different limit matrix Σ_A under alternatives than under the null hypothesis.

If this limit matrix Σ_A is positive definite, then this is no additional restriction on which alternatives can be detected. Obviously, (iii) implies (ii). The following additional assumption is also often fulfilled and yields the additional assertions in (b) and (c).

B.5 *$F_t(\cdot)$ is continuous in θ_1 and $F_t(\theta_1) = F_\lambda(\theta_1)g(t)$ with*

$$g(t) = \begin{cases} \frac{1}{\lambda}t, & t \leqslant \lambda, \\ \frac{1}{1-\lambda}(1-t), & t \geqslant \lambda. \end{cases}$$

10.7.2 Proofs

Proof of Theorem 10.1 By Assumption A.1, we can replace $S(k, \widehat{\theta}_n)$ in all three statistics by $S(k, \theta_0)$ without changing the asymptotic distribution, where—for the statistics in (a)—one needs to note that by (10.3)

$$\sup_{1\leqslant k\leqslant n} w\left(\frac{k}{n}\right)\frac{k}{n}\frac{n-k}{n} = O(1).$$

For a bounded and continuous weight function, the assertion then immediately follows by the functional central limit theorem. For an unbounded weight function, note that for any $0 < \tau < 1/2$, it follows from the Hájek–Rényi inequality and (10.3) that

$$\max_{1\leqslant k\leqslant \tau n} \frac{w(k/n)}{n}\left(S(k,\theta_0) - \frac{k}{n}S(n,\theta_0)\right)^T \Sigma^{-1}\left(S(k,\theta_0) - \frac{k}{n}S(n,\theta_0)\right)$$
$$\leqslant \sup_{1-\tau\leqslant t<1} w(t)t^\alpha \|\Sigma^{-1/2}\| \max_{1\leqslant k\leqslant n} \frac{1}{n^{1-\alpha}k^\alpha}\|S(k,\theta_0)\|^2 \xrightarrow{P} 0$$

as $\tau \to 0$ uniformly in n. By the backward inequality and the fact that $S(k,\theta_0) - \frac{k}{n}S(n,\theta_0) = -((S(n,\theta_0) - S(k,\theta_0)) - \frac{n-k}{n}S(n,\theta_0))$, an analogous assertion holds for $\max_{(1-\tau)n\leqslant k\leqslant n}$ as well as for the corresponding maxima over the limit Brownian bridges. Since the functional central limit theorem implies the claimed distributional convergence for $\max_{\tau n\leqslant k\leqslant (1-\tau)n}$, careful arguments yield the assertion. The result for estimated Σ is immediate.

To prove (b), first note that by the invariance principle in A.2 and the law of iterated logarithm, we get

$$\max_{1\leqslant k\leqslant \log n} \frac{n}{k(n-k)}\left(S(k,\theta_0) - \frac{k}{n}S(n,\theta_0)\right)^T \Sigma^{-1}\left(S(k,\theta_0) - \frac{k}{n}S(n,\theta_0)\right) = o_P\left(\left(\frac{b_d(\log n)}{a(\log n)}\right)^2\right).$$

By the invariance principle and Theorem 2.1.4 in Schmitz [33] (the theorem used is for the univariate case but the rates immediately carry over to the multivariate situation here), we also get

$$\max_{n/\log n\leqslant k\leqslant n/2} \frac{n}{k(n-k)}\left(S(k,\theta_0)-\frac{k}{n}S(n,\theta_0)\right)^T \Sigma^{-1}\left(S(k,\theta_0)-\frac{k}{n}S(n,\theta_0)\right)$$
$$= o_P\left(\left(\frac{b_d(\log n)}{a(\log n)}\right)^2\right).$$

The invariance principle in combination with Horvath [15], Lemma 2.2 (in addition to analogous arguments as given earlier), implies that

$$P\left(a(\log n)\max_{\log n\leqslant k\leqslant n/\log n}\sqrt{\frac{n}{k(n-k)}S(k,\widehat{\theta}_n)^T\Sigma^{-1}S(k,\widehat{\theta}_n)}-b_d(\log n)\leqslant t\right)\to \exp(-e^{-t}).$$

By Assumption A.2, the exact same arguments lead to analogous assertions for $k \geqslant n/2$, which imply the assertion by the asymptotic independence guaranteed by Assumption A.3. From this, we also get that $\left\|\sqrt{\frac{n}{k(n-k)}}(S(k,\theta_0)-\frac{k}{n}S(n,\theta_0))\right\| = O_P(\sqrt{\log\log n})$, which implies

$$\sqrt{\log\log n}\sqrt{\frac{n}{k(n-k)}\left(S(k,\theta_0)-\frac{k}{n}S(n,\theta_0)\right)^T \Sigma^{-1}\left(S(k,\theta_0)-\frac{k}{n}S(n,\theta_0)\right)}$$
$$-\sqrt{\log\log n}\sqrt{\frac{n}{k(n-k)}\left(S(k,\theta_0)-\frac{k}{n}S(n,\theta_0)\right)^T \widehat{\Sigma}^{-1}\left(S(k,\theta_0)-\frac{k}{n}S(n,\theta_0)\right)}$$
$$\leqslant \sqrt{\log\log n}\left\|(\Sigma^{-1/2}-\widehat{\Sigma}^{-1/2})\sqrt{\frac{n}{k(n-k)}}\left(S(k,\theta_0)-\frac{k}{n}S(n,\theta_0)\right)\right\|$$
$$= O_P(\log\log n)\left\|\Sigma^{-1/2}-\widehat{\Sigma}^{-1/2}\right\| = o_P(1),$$

showing that the statistic with estimated covariance matrix has the same asymptotics. □

Proof of Proposition 10.1 Using the subsequence principle, it suffices to prove the following deterministic result: Let $\sup_x \|G_n(x)-G(x)\| \to 0$ (as $n\to\infty$). Then it holds for $G_n(x_n) = 0$ and x_1 is the unique zero of $G(x)$ in the strict sense of B.2 that $x_n \to x_1$. To this end, assume that this is not the case. Then there exists $\varepsilon > 0$ and a subsequence $\alpha(n)$ such that $|x_{\alpha(n)} - x_1| \geqslant \varepsilon$. But then since x_1 is a unique zero in the strict sense, $\|G(x_{\alpha(n)})\| \geqslant \delta$ for some $\delta > 0$. However, this is a contradiction as

$$\|G(x_{\alpha(n)})\| = \|G(x_{\alpha(n)}) - G_{\alpha(n)}(x_{\alpha(n)})\| \leqslant \sup_x \|G_{\alpha(n)}(x)-G(x)\| \to 0.$$

□

Proof of Theorem 10.2 B.1, B.3, (10.5), and B.5 (i) imply

$$\frac{1}{n^2}S(k_0,\widehat{\theta}_n)\Sigma^{-1}S(k_0,\widehat{\theta}_n) \geqslant c + o_P(1);$$

hence,

$$\max_{1\leqslant k\leqslant n} \frac{w(k/n)}{n} S(k,\widehat{\theta}_n)^T \Sigma^{-1} S(k,\widehat{\theta}_n) \geqslant n w(\lambda)(c+o_P(1)) \to \infty,$$

$$\frac{a(\log n)}{b_d(\log n)} \max_{1\leqslant k\leqslant n} \frac{n}{k(n-k)} S(k,\widehat{\theta}_n)^T \Sigma^{-1} S(k,\widehat{\theta}_n) \geqslant \frac{a(\log n)}{b_d(\log n)} n \frac{1}{\lambda(1-\lambda)}(c+o_P(1)) \to \infty,$$

which implies assertion (a). If additionally, B.5 holds, we get the assertion for the maximum-type statistic analogously if we replace k_0 by $\lfloor \vartheta n \rfloor$ with $w(\vartheta) > 0$. For the sum-type statistics, we similarly get

$$\frac{1}{n}\sum_{j=1}^{n} w(k/n)\frac{1}{n} S(k,\widehat{\theta}_n)^T \Sigma^{-1} S(k,\widehat{\theta}_n) = nc\left(\int_0^1 w(t) g^2(t)\, dt + o_P(1)\right) \to \infty,$$

since the assumptions on $w(\cdot)$ guarantee the existence of $\int_0^1 w(t)g^2(t)\,dt$ and $\int_0^1 w(t)g^2(t)\,dt \neq 0$. The second assertion follows by standard arguments since λ is the unique maximizer of the continuous function g and by Assumptions B.1 and B.5, it holds

$$\sup_{0\leqslant t\leqslant 1}\left|\frac{1}{n^2} S(\lfloor nt\rfloor,\widehat{\theta}_n)^T \Sigma^{-1} S(\lfloor nt\rfloor,\widehat{\theta}_n) - F_\lambda(\theta_1)\Sigma^{-1}F_\lambda(\theta_1)\, g^2(t)\right| \to 0,$$

where B.4 guarantees that the limit is not zero. The proofs show that the assertions remain true if Σ is replaced by $\widehat{\Sigma}_n$ under the stated assumptions. □

Proof of Theorem 10.3 Assumption A.1 can be obtained by a Taylor expansion, the ergodic theorem, and the $\sqrt{n}$-consistency of the estimator $\widehat{\beta}_n$. The arguments are given in detail in Fokianos et al. [11] (Proof of Proposition 3), where by the stationarity of $\{Z_t\}$ their arguments go through in our slightly more general situation for $k \leqslant n/2$. For $k > n/2$, analogous arguments give the assertion on noting that (with the notation of Fokianos et al. [11])

$$\sum_{t=1}^{k} Z_{t-i}^{(i)} Z_{t-1}^{(j)} \pi_t(\beta)(1-\pi_t(\beta)) - \frac{k}{n}\sum_{t=1}^{n} Z_{t-i}^{(i)} Z_{t-1}^{(j)} \pi_t(\beta)(1-\pi_t(\beta))$$

$$= -\sum_{t=k+1}^{n} Z_{t-i}^{(i)} Z_{t-1}^{(j)} \pi_t(\beta)(1-\pi_t(\beta)) + \frac{n-k}{n}\sum_{t=1}^{n} Z_{t-i}^{(i)} Z_{t-1}^{(j)} \pi_t(\beta)(1-\pi_t(\beta)).$$

Assumption A.3 (i) follows from the strong invariance principle in Proposition 2 of Fokianos et al. [11]. Assumption A.3 (ii) does not follow by the same proof techniques as an autoregressive process in reverse order has different distributional properties than an autoregressive process. However, if the covariate $(Y_t, Z_{t-1}, \ldots, Z_{t-p})^T$ is α-mixing, the same holds true for the summands of the score process (with the same rate). Since the mixing property also transfers to the time-inverse process, the strong invariance principle follows from the invariance principle for mixing processes given by Kuelbs and Philipp [23], Theorem 4. The mixing assumption then also implies A.3 (iii). □

Proof of Theorem 10.4 First note that

$$\sup_{\theta \in \Theta} \|S^{BAR}(k, \beta) - \mathbb{E}S^{BAR}(k, \beta)\| = o_p(n) \tag{10.13}$$

uniformly in $k \leqslant k_0 = \lfloor n\lambda \rfloor$ by the uniform ergodic theorem of Ranga Rao [30], Theorem 6.5. For $k > k_0$, it holds

$$\mathbb{Z}_k Y_k = \tilde{\mathbb{Z}}_k \tilde{Y}_k + \tilde{\mathbb{Z}}_k R_1(t) + \tilde{Y}_k \mathbb{R}_2(t) + R_1(t)\mathbb{R}_2(t),$$

$$\mathbb{Z}_k \pi_k(\beta) = \tilde{\mathbb{Z}}_k \pi_k(\beta) + \mathbb{R}_1(k)\pi_k(\beta) = \tilde{\mathbb{Z}}_k \tilde{\pi}_k(\beta) + O(\tilde{\mathbb{Z}}_k \beta^T \mathbb{R}_1(k)) + O(\mathbb{R}_1(k)),$$

where $g(\tilde{\pi}_k(\beta)) = \beta^T \tilde{\mathbb{Z}}_{t-1}$ and the last line follows from the mean value theorem. An application of the Cauchy–Schwarz inequality together with (iii) and the compactness of Θ shows that the remainder terms are asymptotically negligible, implying that

$$\sup_{\beta} \|S^{BAR}(k, \beta) - S^{BAR}(k_0 \beta) - \mathbb{E}\check{S}^{BAR}(k, \beta) - \mathbb{E}S^{BAR}(k_0, \beta)\| = o_p(n)$$

uniformly in $k > k_0$, where

$$\check{S}^{BAR}(k, \beta) = \sum_{t=1}^{\min(k_0, k)} \mathbb{Z}_{t-1}(Y_t - \pi_t(\beta)) + \sum_{t=k_0+1}^{k} \tilde{\mathbb{Z}}_{t-1}(\tilde{Y}_t - \tilde{\pi}_t(\beta))$$

Together with (10.13), this implies B.1 with

$$F_t(\lambda) = \min(t, \lambda)\, \mathbb{E}\mathbb{Z}_0(Y_1 - \pi_1(\beta)) + (t - \lambda)_+\, \mathbb{E}\tilde{\mathbb{Z}}_{n-1}(\tilde{Y}_n - \tilde{\pi}_n(\beta)).$$

Since $F_t(\beta)$ is continuous in β, B.2 follows from (iv). B.5 follows since by definition of β_1 it holds $\mathbb{E}\tilde{\mathbb{Z}}_{n-1}(\tilde{Y}_n - \tilde{\pi}_n(\beta_1)) = -\lambda/(1-\lambda)\mathbb{E}\mathbb{Z}_0(Y_1 - \pi_1(\beta_1))$. □

Proof of Theorem 10.5 It suffices to show that the assumptions of Theorem 10.1 are fulfilled. Assumption A.1 follows for (a) and (b) analogously to the proof of Lemma 1 in Franke et al. [13]. The invariance principles in A.2 and A.3 then follow from the strong mixing assumption and the invariance principle of Kuelbs and Philipp [23]. The asymptotic independence of A.3 also follows from the mixing condition. □

Proof of Theorem 10.6 This is analogous to the proof of Theorem 10.4, where the mean value theorem is replaced by the Lipschitz assumption, which also implies that $|f_Y(\mathbf{x})| \leqslant |f_Y(0) + \mathbf{x}|$.

□

Acknowledgments

Most of this work was done while the first author was at Karlsruhe Institute of Technology (KIT), where her position was financed by the Stifterverband für die Deutsche Wissenschaft by funds of the Claussen-Simon-trust. The second author was partly supported by the DFG grant SA 1883/4-1. This work was partly supported by DFG grant KI 1443/2-2.

References

1. Antoch, J. and Hušková, M. M-estimators of structural changes in regression models. *Tatra Mt. Math.*, 22:197–208, 2001.
2. Aue, A. and Horváth, L. Structural breaks in time series. *J. Time Ser. Anal.*, 34(1):1–16, 2013.
3. Berkes, I., Horváth, L., and Kokoszka, P. Testing for parameter constancy in GARCH(p,q) models. *Statist. Probab. Lett.*, 70(4):263–273, 2004.
4. Chu, C.-S.J., Stinchcombe, M., and White, H. Monitoring structural change. *Econometrica*, 64:1045–1065, 1996.
5. Csörgő, M. and Horváth, L. *Limit Theorems in Change-Point Analysis*. Wiley, Chichester, U.K., 1997.
6. Davis, R.A., Huang, D., and Yao, Y.-C. Testing for a change in the parameters values and order of an autoregression model. *Ann. Statist.*, 23:282–304, 1995.
7. Doukhan, P. and Kegne, W. Inference and testing for structural change in time series of counts model. *arXiv:1305.1751*, 2013.
8. Feigin, P.D. and Tweedie, R.L. Random coefficient autoregressive processes: A Markov chain analysis of stationarity and finiteness of moments. *J. Time Ser. Anal.*, 6:1–14, 1985.
9. Fokianos, K. and Fried, R. Interventions in INGARCH processes. *J. Time Ser. Anal.*, 31:210–225, 2010.
10. Fokianos, K. and Fried, R. Interventions in log-linear Poisson autoregression. *Statist. Model.*, 12:299–322, 2012.
11. Fokianos, K., Gombay, E., and Hussein, A. Retrospective change detection for binary time series models. *J. Statist. Plann. Inf.*, 145:102–112, 2014.
12. Fokianos, K., Rahbek, A., and Tjostheim, D. Poisson autoregression. *J. Am. Statist. Assoc.*, 104:1430–1439, 2009.
13. Franke, J., Kirch, C., and Tadjuidje Kamgaing, J. Changepoints in times series of counts. *J. Time Ser. Anal.*, 33:757–770, 2012.
14. Fryzlewicz, P. Wild binary segmentation for multiple change-point detection. *Ann. Statist.*, 42(6):2243–2281, 2014.
15. Horváth, L. Change in autoregressive processes *Stoch. Process Appl.*, 44:221–242, 1993.
16. Horváth, L. and Rice, G. Extensions of some classical methods in change point analysis. *TEST*, 23(2):219–255, 2014.
17. Hudeková, S. Structural changes in autoregressive models for binary time series. *J. Statist. Plann. Inf.*, 143(10), 2013.
18. Kauppi, H. and Saikkonen, P. Predicting US recessions with dynamic binary response models. *Rev. Econ. Statist.*, 90:777–791, 2008.
19. Kirch, C. Resampling methods for the change analysis of dependent data. PhD thesis, University of Cologne, Cologne, Germany, 2006. http://kups.ub.uni-koeln.de/volltexte/2006/1795/.
20. Kirch, C., Muhsal, B., and Ombao, H. Detection of changes in multivariate time series with application to EEG data. *J. Am. Statist. Assoc., to appear*, 2014. DOI:10.1080/01621459.2014.957545.
21. Kirch, C. and Tadjuidje Kamgaing, J. Geometric ergodicity of binary autoregressive models with exogenous variables. 2013. http://nbn-resolving.de/urn/resolver.pl?urn:nbn:de:hbz:386-kluedo-36475. (Accessed on 13 November 2013.) Preprint.
22. Kirch, C. and Tadjuidje Kamgaing, J. On the use of estimating functions in monitoring time series for change points. *J. Statist. Plann. Inf.*, 161:25–49, 2015.
23. Kuelbs, J. and Philipp, W. Almost sure invariance principles for partial sums of mixing *B*-valued random variables. *Ann. Probab.*, 8:1003–1036, 1980.

24. Kulperger, R.J. On the residuals of autoregressive processes and polynomial regression. *Stoch. Process Appl.*, 21:107–118, 1985.
25. Horváth, M. Change in autoregressive processes. *Stoch. Process Appl.*, 44:221–242, 1993.
26. Hušková, M. and Marušiaková, M. M-procedures for detection of changes for dependent observations. *Commun. Statist. Simulat. Comput.*, 41:1032–1050, 2012.
27. Hušková, M., Prašková, Z., and Steinebach, J. On the detection of changes in autoregressive time series I. asymptotics. *J. Statist. Plann. Inf.*, 137:1243–1259, 2007.
28. Neumann, M.H. Absolute regularity and ergodicity of Poisson count processes. *Bernoulli*, 17(4):1268–1284, 2011.
29. Philipp, W. and Stout, W. *Almost Sure Invariance Principles for Partial Sums of Weakly Dependent Random Variables*. Memoirs of the American Mathematical Society 161, American Mathematical Society, Providence, RI, 1975.
30. Ranga Rao, R. Relation between weak and uniform convergence of measures with applications. *Ann. Math. Statist.*, 33:659–680, 1962.
31. Robbins, M.W., Gallagher, C.M., Lund, R.B., and Aue, A. Mean shift testing in correlated data. *J. Time Ser. Anal.*, 32:498–511, 2011.
32. Robbins, M.W., Gallagher, C.M., Lund, R.B., and Lu, Q. Changepoints in the North Atlantic tropical cyclone record. *J. Am. Statist. Assoc.*, 106:89–99, 2011.
33. Schmitz, A. Limit theorems in change-point analysis for dependent data. PhD thesis, University of Cologne, Cologne, Germany, 2011. http://kups.ub.uni-koeln.de/4224/.
34. Startz, R. Binomial autoregressive moving average models with an application to US recession. *J. Bus. Econom. Statist.*, 26:1–8, 2008.
35. Tadjuidje Kamgaing, J., Ombao, H., and Davis, R.A. Autoregressive processes with data-driven regime switching. *J. Time Ser. Anal.*, 30(5):505–533, 2009.
36. Vostrikova, L.Y. Detection of 'disorder' in multidimensional random processes. *Sov. Math. Dokl.*, 24:55–59, 1981.
37. Wang, C. and Li, W.K. On the autopersistence functions and the autopersistence graphs of binary autoregressive time series. *J. Time Ser. Anal.*, 32(6):639–646, 2011.
38. Weiss, C.H. Detecting mean increases in Poisson INAR(1) processes with EWMA control charts. *J. Appl. Statist.*, 38:383–398, 2011.
39. Weiß, C.H. and Testik, M.C. The Poisson INAR(1) CUSUM chart under overdispersion and estimation error. *IIE Trans.*, 43(11):805–818, 2011.
40. Wilks, D. and Wilby, R. The weather generation game: A review of stochastic weather models. *Prog. Phys. Geography*, 23:329–357, 1999.
41. Yontay, P., Weiß, C.H., Testik, M.C., and Bayindir, Z.P. A two-sided cumulative sum chart for first-order integer-valued autoregressive processes of Poisson counts. *Qual. Reliab. Eng. Int.*, 29:33–42, 2012.

11

Bayesian Modeling of Time Series of Counts with Business Applications

Refik Soyer, Tevfik Aktekin, and Bumsoo Kim

CONTENTS

11.1 Introduction

Discrete-valued temporal data, and specifically count time series, often arise in numerous application areas including business, economics, engineering, and epidemiology, among others. For instance, observations under study can be the number of arrivals to a bank in a given hour, number of shopping trips of households in a week, number of mortgages defaulted from a particular pool in a given month, number of accidents in a given time interval, or the number of deaths from a specific disease in a given year.

Studies in time series with focus on count data are scarce in comparison to those with continuous data, and of these, many studies consider a Poisson model. A regression model of time series of counts has been introduced by Zeger (1988) where a quasi-likelihood method is used in order to estimate model parameters. Harvey and Fernandes (1989) assume a gamma process on the stochastic evolution of the latent Poisson mean and propose extensions to other count data models such as the binomial, the multinomial, and negative binomial distributions. Davis et al. (2000) develop a method to diagnose the latent factor embedded in the mean of a Poisson regression model and present asymptotic properties of model estimators. Davis et al. (2003) discuss maximum likelihood estimation for a general class of observation-driven models for count data, also referred to as generalized autoregressive moving average models, and develop relevant theoretical properties. Freeland and McCabe (2004) introduce new methods for assessing the fit of the

Poisson autoregressive model via the information matrix equality, provide properties of the maximum likelihood estimators, and discuss further implications in residual analysis.

The Bayesian point of view has also been considered in the time series analysis of count data. Chib et al. (1998) introduce Markov Chain Monte Carlo (MCMC) methods to estimate Poisson panel data models with multiple random effects and use various Bayes factor estimators to assess model fit. Chib and Winkelmann (2001) propose a model that can take into account correlated count data via latent effects and show how the estimation method is practical even with high-dimensional count data. Bayesian state-space modeling of Poisson count data is considered by Frühwirth-Schnatter and Wagner (2006) where MCMC via data augmentation techniques to estimate model parameters is used. Furthermore, Durbin and Koopman (2000) discuss the state-space analysis of non-Gaussian time series models from both the classical and Bayesian perspectives, apply an importance sampling technique for estimation, and illustrate an example with count data. More recently, Santos et al. (2013) consider a non-Gaussian family of state-space models, obtain exact marginal likelihoods, and consider the Poisson model as a special case. Gamerman et al. (2015; Chapter 8 in this volume), have provided an excellent overview of state-space modeling for count time series.

In this chapter, we describe a general class of Poisson time series models from a Bayesian state-space modeling perspective and propose different strategies for modeling the stochastic Poisson rate which evolves over time. Such models are also referred to as parameter-driven time series models as described by Cox (1981) and Davis et al. (2003). The state-space approach allows the modeling of various subcomponents to form an overall time series model as pointed out by Durbin and Koopman (2000). We also present an extension of our framework to multivariate counts where dependence between the individual counts is motivated by a common environment. Specifically, we obtain multivariate negative binomial models and develop Bayesian inference using appropriate MCMC estimation techniques such as the Gibbs sampler, the Metropolis–Hastings algorithm, and the forward filtering backward sampling (FFBS) algorithm. For a good introduction to MCMC, see Gamerman and Lopes (2006), Gilks et al. (1996), Smith and Gelman (1992), and Chib and Greenberg (1995), and for FFBS, see Fruhwirth-Schnatter (1994).

An outline of this chapter is as follows. In Section 11.2, we introduce a Bayesian state-space model for count time series data and show how the parameters can be updated sequentially and forecasting can be performed. We consider a special case excluding covariates where a fully analytically tractable model is presented. In Section 11.3, we introduce MCMC methods to estimate the model parameters. This is followed by Section 11.4 that is dedicated to multivariate extensions of the model. The proposed approach and its extensions are illustrated by three examples from finance, operations, and marketing in Section 11.5. Section 11.6 concludes with some remarks.

11.2 A Discrete-Time Poisson Model

We introduce a Poisson time series model whose stochastic rate evolves over time according to a discrete-time Markov process and covariates. We refer to this model as the basic model. Our model is based on the framework proposed by Smith and Miller (1986) where the likelihood was exponential. Let Y_t be the number of occurrences of an event in a given time interval of length t and let λ_t be the corresponding rate during the same time period.

Given the rate λ_t, we assume that the number of occurrences of an event during period $(0, t)$ (referred to as t for the rest of the chapter) is described by a discrete-time nonhomogeneous Poisson process with probability distribution

$$p(Y_t|\lambda_t) = \frac{\lambda_t^{Y_t} e^{-\lambda_t}}{Y_t!}. \tag{11.1}$$

It is assumed that the Y_ts are conditionally independent given the λ_ts so that the independent increments property holds only conditional on λ_t, whereas unconditionally, the Y_ts are correlated. Equation (11.1) is the measurement equation of a discrete-time Poisson state-space model. It is possible to model the effect of time and of covariates using a multiplicative form for λ_t as

$$\lambda_t = \theta_t e^{\psi z_t}, \tag{11.2}$$

where $\mathbf{z}_t$ is a vector of covariates and ψ is the corresponding parameter vector. Here, θ_t acts like a latent baseline rate that evolves over time but is free of covariate effects.

For the time evolution of the latent rate process, $\{\theta_t\}$, we assume the following Markovian structure:

$$\theta_t = \frac{\theta_{t-1}}{\gamma} \epsilon_t, \tag{11.3}$$

where $(\epsilon_t|D^{t-1}, \psi, \gamma) \sim Beta[\gamma\alpha_{t-1}, (1-\gamma)\alpha_{t-1}]$ with $\alpha_{t-1} > 0$, $0 < \gamma < 1$, and $D^{t-1} = \{Y_1, \mathbf{z}_1, \ldots, Y_{t-1}, \mathbf{z}_{t-1}\}$. In (11.3), γ acts like a discounting factor and its logarithm can be considered to be the first-order autoregressive coefficient for the latent rates, θ_t. It follows from (11.3) that there is an implied stochastic ordering between two consecutive rates, i.e., $\theta_t < \frac{\theta_{t-1}}{\gamma}$. It is also straightforward to show that the conditional distributions of consecutive rates are all scaled Beta densities, $(\theta_t|\theta_{t-1}, D^{t-1}, \psi, \gamma) \sim \text{Beta}\left[\gamma\alpha_{t-1}, (1-\gamma)\alpha_{t-1}; (0, \theta_{t-1}/\gamma)\right]$, and are given by

$$p(\theta_t|\theta_{t-1}, D^{t-1}, \psi, \gamma) = \frac{\Gamma(\alpha_{t-1})}{\Gamma(\gamma\alpha_{t-1})\,\Gamma(\{1-\gamma\}\alpha_{t-1})} \left(\frac{\gamma}{\theta_{t-1}}\right)^{\alpha_{t-1}-1} \theta_t^{\gamma\alpha_{t-1}-1} \times \left(\frac{\theta_{t-1}}{\gamma} - \theta_t\right)^{(1-\gamma)\alpha_{t-1}-1}. \tag{11.4}$$

The state equation (11.3) also implies that $E(\theta_t|\theta_{t-1}, D^{t-1}, \psi, \gamma) = \theta_{t-1}$, that is, a random walk type of evolution in the expected Poisson rates.

Given these measurement and state equations, it is possible to develop a conditionally analytically tractable sequential updating of the model if we assume that at time 0, $(\theta_0|D^0)$ is a gamma distribution, that is,

$$(\theta_0|D^0) \sim \text{Gamma}(\alpha_0, \beta_0), \tag{11.5}$$

where θ_0 is assumed to be independent of the covariate parameter vector ψ at time 0. Given the inductive hypothesis

$$(\theta_{t-1}|D^{t-1},\psi,\gamma) \sim \text{Gamma}(\alpha_{t-1},\beta_{t-1}), \tag{11.6}$$

a recursive updating scheme can be developed as follows. Using (11.4) and (11.6), we can obtain the distribution of θ_t given D^{t-1} as

$$(\theta_t|D^{t-1},\psi,\gamma) \sim \text{Gamma}(\gamma\alpha_{t-1},\gamma\beta_{t-1}). \tag{11.7}$$

It follows that $E(\theta_t|D^{t-1},\psi,\gamma) = E(\theta_{t-1}|D^{t-1},\psi,\gamma)$ and $Var(\theta_t|D^{t-1},\psi,\gamma) = Var(\theta_{t-1}|D^{t-1},\psi,\gamma)/\gamma$. In other words, as we move forward in time, our uncertainty about the rate increases as a function of γ. Given the prior (11.6) and the Poisson observation model (11.1), we obtain the filtering distribution of $(\theta_t|D^t,\psi,\gamma)$ using Bayes' rule as

$$p(\theta_t|D^t,\psi,\gamma) \propto p(Y_t|\lambda_t)p(\theta_t|D^{t-1},\psi,\gamma), \tag{11.8}$$

which implies that

$$p(\theta_t|D^t,\psi,\gamma) \propto \left(\theta_t e^{\psi' z_t}\right)^{\gamma\alpha_{t-1}+Y_t-1} e^{-(\gamma\beta_{t-1}+1)\theta_t e^{\psi' z_t}}.$$

The filtering distribution of the latent rate at time t is a gamma density

$$(\theta_t|D^t,\psi,\gamma) \sim \text{Gamma}(\alpha_t,\beta_t), \tag{11.9}$$

where the model parameters are recursively updated by $\alpha_t = \gamma\alpha_{t-1} + Y_t$ and $\beta_t = \gamma\beta_{t-1} + e^{\psi' z_t}$. This updating scheme implies that as we learn more about the count process over time, we update our uncertainty about the Poisson rate as a function of both counts over time and of covariate effects via β_t. In addition, the conditional one-step-ahead forecast or predictive distribution of counts at time t given D^{t-1} can be obtained via

$$p(Y_t|D^{t-1},\psi,\gamma) = \int_0^\infty p(Y_t|\lambda_t)p(\theta_t|D^{t-1},\psi,\gamma)d\theta_t, \tag{11.10}$$

where $(Y_t|\lambda_t) \sim \text{Poisson}(\lambda_t)$ and $(\theta_t|D^{t-1},\psi,\gamma) \sim \text{Gamma}(\gamma\alpha_{t-1},\gamma\beta_{t-1})$. Therefore,

$$p(Y_t|D^{t-1},\psi,\gamma) = \binom{\gamma\alpha_{t-1}+Y_t-1}{Y_t}\left(1-\frac{1}{\gamma\beta_{t-1}+e^{\psi' z_t}}\right)^{\gamma\alpha_{t-1}}\left(\frac{1}{\gamma\beta_{t-1}+e^{\psi' z_t}}\right)^{Y_t}, \tag{11.11}$$

which is a negative binomial model denoted as

$$(Y_t|D^{t-1},\psi,\gamma) \sim \text{Negbin}(r_t,p_t), \tag{11.12}$$

where $r_t = \gamma\alpha_{t-1}$ and $p_t = \gamma\beta_{t-1}/\gamma\beta_{t-1} + e^{\psi' z_t}$. Given (11.12), one can carry out one-step-ahead predictions and compute prediction intervals in a straightforward manner. The conditional mean of $(Y_t|D^{t-1}, \psi, \gamma)$ can be computed via

$$E(Y_t|D^{t-1}, \psi, \gamma) = \frac{\alpha_{t-1}}{\beta_{t-1}} e^{\psi' z_t}, \tag{11.13}$$

which implies that given the covariates and the counts up to time $t-1$, the forecast for time t is a function of the observed default counts up to time $t-1$ adjusted by the corresponding covariates.

11.2.1 Special Case: Model with No Covariates and Its Properties

Here, we consider a simple but analytically fully tractable version of the basic model presented earlier. We assume that the arrival rate in the Poisson model is not influenced by covariates. In other words, we define $\lambda_t = \theta_t$ in (11.2). In doing so, we can provide analytical results for updating and forecasting conditional on γ by simply replacing $e^{\psi' z_t}$ with 1 and θ_t by λ_t in the previous section.

As earlier, let Y_t as the number of occurrences of an event in a given time interval t; we have the same time evolution of the Poisson rate, λ_t, as earlier given by (11.3):

$$\lambda_t = \frac{\lambda_{t-1}}{\gamma} \epsilon_t, \tag{11.14}$$

where $(\epsilon_t|D^{t-1}, \gamma) \sim Beta[\gamma\alpha_{t-1}, (1-\gamma)\alpha_{t-1}]$ with $\alpha_{t-1} > 0$, $0 < \gamma < 1$, and $D^{t-1} = \{Y_1, \ldots, Y_{t-1}\}$.

Using the same prior at time 0, for $(\lambda_0|D^0)$ as

$$(\lambda_0|D^0) \sim \text{Gamma}(\alpha_0, \beta_0), \tag{11.15}$$

we can develop an analytically tractable sequential updating of Poisson rates over time. It can be shown that

$$(\lambda_t|D^{t-1}, \gamma) \sim \text{Gamma}(\gamma\alpha_{t-1}, \gamma\beta_{t-1}), \tag{11.16}$$

and the filtering distribution of the Poisson rate at time t is a gamma density

$$(\lambda_t|D^t, \gamma) \sim \text{Gamma}(\alpha_t, \beta_t), \tag{11.17}$$

where the model parameters are recursively updated by $\alpha_t = \gamma\alpha_{t-1} + Y_t$ and $\beta_t = \gamma\beta_{t-1} + 1$.

The one-step-ahead forecast distribution of counts at time t given D^{t-1}, γ is again given by a negative binomial model as

$$(Y_t|D^{t-1}, \gamma) \sim \text{Negbin}(r_t, p_t), \tag{11.18}$$

where $r_t = \gamma\alpha_{t-1}$ and $p_t = \gamma\beta_{t-1}/\gamma\beta_{t-1} + 1$. The one-step-ahead forecast is obtained as

$$E(Y_t|D_{t-1},\gamma) = \frac{\alpha_{t-1}}{\beta_{t-1}}.$$

An interesting property of the model is the long-run behavior of its one-step-ahead forecasts. As t gets large, using $\beta_t = \gamma\beta_{t-1} + 1$, we can show that β_t approaches $1/(1-\gamma)$ and we obtain

$$E(Y_t|D_{t-1},\gamma) = (1-\gamma)Y_{t-1} + (1-\gamma)\gamma Y_{t-2} + \ldots + (1-\gamma)\gamma^t\alpha_0,$$

which is an exponentially weighted average of the observed counts.

Although an analytic expression is not available for the k-step-ahead predictive density, the k-step-ahead predictive means can be easily obtained. Using a standard conditional expectation argument one can obtain $E(Y_{t+k}|D^t,\gamma)$ as

$$E(Y_{t+k}|D^t,\gamma) = E_{\lambda_{t+k},\gamma}\{E(Y_{t+k}|\lambda_{t+k},D^t)\} = E(\lambda_{t+k}|D^t,\gamma). \tag{11.19}$$

Furthermore, using the state equation (11.4), we have

$$E(\lambda_{t+k}|D^t,\gamma) = E(\lambda_t|D^t,\gamma)\prod_{n=t+1}^{t+k}\frac{E(\epsilon_n|D^t)}{\gamma} = E(\lambda_t|D^t,\gamma) = \frac{\alpha_t}{\beta_t}, \tag{11.20}$$

where $E(\epsilon_n|D^t) = \gamma$ for any n. Therefore, combining (11.19) and (11.20), we obtain the k-step-ahead forecasts given data up to time t as

$$E(Y_{t+k}|D^t,\gamma) = E(\lambda_{t+k}|D^t,\gamma) = \frac{\alpha_t}{\beta_t}. \tag{11.21}$$

Note that in the case of the model with covariates, it can be shown that

$$E(Y_{t+k}|D^t,z_{t+k},\psi,\gamma) = E(\lambda_{t+k}|D^t,z_{t+k},\psi,\gamma) = \frac{\alpha_t}{\beta_t}e^{\psi' z_{t+k}}. \tag{11.22}$$

If we treat the discount factor γ as a random variable, we lose the analytical tractability of the model described earlier. However selecting a prior distribution for γ can be handled fairly easily. Given D^t, the likelihood function of γ is given by

$$L(\gamma;D^t) = \prod_{i=1}^{t} p(Y_i|D^{i-1},\gamma), \tag{11.23}$$

where $p(Y_i|D^{i-1},\gamma)$ is negative binomial as in (11.18). The posterior distribution of γ can then be obtained as

$$p(\gamma|D^t) \propto \prod_{i=1}^{t} p(Y_i|D^{i-1},\gamma)p(\gamma). \tag{11.24}$$

For some priors for $p(\gamma)$ in (11.24), the posterior distribution will not be available in closed form. However, we can always sample from the posterior using an MCMC method such as the Metropolis–Hastings algorithm. Alternatively, a discrete uniform prior can be a reasonable choice.

11.3 Markov Chain Monte Carlo (MCMC) Estimation of the Model

Since all conditional distributions previously introduced are all dependent on the parameter vectors ψ and γ, we need to discuss how to obtain the joint posterior densities of ψ and γ that cannot be obtained in closed form; therefore, we can use MCMC methods to generate the required samples. Our objective in this section is to obtain the joint posterior distribution of the model parameters given observed counts up to time t, that is, $p(\theta_1, \ldots, \theta_t, \psi, \gamma | D^t)$. We use a Gibbs sampler to generate samples from the full conditionals of $p(\theta_1, \ldots, \theta_t | \psi, \gamma, D^t)$ and $p(\psi, \gamma | \theta_1, \ldots, \theta_t, D^t)$, none of which are available as standard densities.

For notational convenience, we define $\omega = \{\psi, \gamma\}$. The conditional posterior distribution of ω given the latent rates, $(\theta_1, \ldots, \theta_t)$ is

$$p(\omega | \theta_1, \ldots, \theta_t, D^t) \propto \prod_{i=1}^{t} \frac{exp\{\theta_i e^{\psi' z_i}\} (\theta_i e^{\psi' z_i})^{Y_i}}{Y_i!} p(\omega), \tag{11.25}$$

where $p(\omega)$ is the joint prior for ψ and γ. Regardless of the prior selection for ω, (11.25) will not be a standard density. We use an MCMC algorithm such as the Metropolis–Hastings to generate samples from $p(\omega | \theta_1, \ldots, \theta_t, D^t)$. In our numerical examples, we assume flat but proper priors for ψ and γ with $\psi_i \sim Normal(0, 1000)$ for all i and $\gamma \sim Uniform(0, 1)$. Following Chib and Greenberg (1995), the steps in the Metropolis–Hastings algorithm can be summarized as follows:

1. Assume the starting points $\omega^{(0)}$ at $j = 0$.
   Repeat for $j > 0$,
2. Generate ω^* from $q(\omega^* | \omega^{(j)})$ and u from $U(0, 1)$.
3. If $u \leq f(\omega^{(j)}, \omega^*)$ then set $\omega^{(j)} = \omega^*$; else set $\omega^{(j)} = \omega^{(j)}$ and $j = j + 1$,

where

$$f(\omega^{(j)}, \omega^*) = \min \left\{ 1, \frac{\pi(\omega^*) q(\omega^{(j)} | \omega^*)}{\pi(\omega^{(j)}) q(\omega^* | \omega^{(j)})} \right\}. \tag{11.26}$$

In (11.26), $q(.|.)$ is the multivariate normal proposal density and $\pi(.)$ is given by (11.25) which is the density we need to generate samples from. If we repeat this a large number of times, we can obtain samples from $p(\omega | \theta_1, \ldots, \theta_t, D^t)$.

Generation of samples from the full conditional distribution, $p(\theta_1, \ldots, \theta_t|D^t, \omega)$, using the FFBS algorithm as described in Fruhwirth-Schnatter (1994) requires the smoothing distribution of θ_ts, which enable retrospective analysis. In other words, given that we have observed the count data, D^t at time t, we will be interested in the distribution of $(\theta_{t-k}|D^t, \omega)$ for all $k \geq 1$.

We can write

$$p(\theta_{t-k}|D^t, \omega) = \int p(\theta_{t-k}|\theta_{t-k+1}, D^t, \omega)p(\theta_{t-k+1}|D^t, \omega)d\theta_{t-k+1}, \tag{11.27}$$

where $p(\theta_{t-k}|\theta_{t-k+1}, D^t, \omega)$ is obtained via Bayes' rule as

$$\begin{aligned} p(\theta_{t-k}|\theta_{t-k+1}, D^t, \omega) &= \frac{p(\theta_{t-k}|\theta_{t-k+1}, D^{t-k}, \omega)p(Y^*_{(t,k)}|\theta_{t-k}, \theta_{t-k+1}, D^{t-k}, \omega)}{p(Y^*_{(t,k)}|\theta_{t-k+1}, D^{t-k}, \omega)} \\ &= p(\theta_{t-k}|\theta_{t-k+1}, D^{t-k}, \omega), \end{aligned}$$

where $Y^*_{(t,k)} = \{Y_{t-k+1}, \ldots, Y_t\}$. Given θ_{t-k+1}, $Y^*_{(t,k)}$ is independent of θ_{t-k}. In other words, $p\left(Y^*_{(t,k)}|\theta_{t-k}, \theta_{t-k+1}, D^{t-k}, \omega\right) = p\left(Y^*_{(t,k)}|\theta_{t-k+1}, D^{t-k}, \omega\right)$. Thus, (11.27) reduces to

$$p(\theta_{t-k}|D^t, \omega) = \int p(\theta_{t-k}|\theta_{t-k+1}, D^{t-k}, \omega)p(\theta_{t-k+1}|D^t, \omega)d\theta_{t-k+1}. \tag{11.28}$$

Although we cannot obtain (11.28) analytically we can use Monte Carlo methods to draw samples from $p(\theta_{t-k}|D^t, \omega)$. Due to the Markovian nature of the state parameters, we can rewrite $p(\theta_1, \ldots, \theta_t|D^t, \omega)$ as

$$p(\theta_t|D^t, \omega)p(\theta_{t-1}|\theta_t, D^{t-1}, \omega) \ldots p(\theta_1|\theta_2, D^1, \omega). \tag{11.29}$$

We note that $p(\theta_t|D^t, \omega)$ is available from (11.9) and $p(\theta_{t-1}|\theta_t, D^{t-1}, \omega)$ for any t as

$$p(\theta_{t-1}|\theta_t, D^{t-1}, \omega) \propto p(\theta_t|\theta_{t-1}, D^{t-1}, \omega)p(\theta_{t-1}|D^{t-1}, \omega), \tag{11.30}$$

where the first term is available from (11.4) and the second term from (11.6). It is straightforward to show that

$$(\theta_{t-1}|\theta_t, D^{t-1}, \omega) \sim ShGamma[(1-\gamma)\alpha_{t-1}, \beta_{t-1}; (\gamma\theta_t, \infty)],$$

which is a shifted gamma density defined over $\gamma\theta_t < \theta_{t-1} < \infty$.

Therefore, given (11.29) and the posterior samples generated from the full conditional of ω, we can obtain a sample from $p\left(\theta_1, \ldots, \theta_t|\omega, z_t, D^t\right)$ by sequentially simulating the individual latent rates as follows:

1. Assume the starting points $\theta_1^{(0)}, \ldots, \theta_t^{(0)}$ at $j = 0$.
   Repeat for $j > 0$,

2. Using the generated $\omega^{(j)}$, sample $\theta_t^{(j)}$ from $(\theta_t|\omega^{(j)}, D^t)$.
3. Using the generated $\omega^{(j)}$, for each $n=t-1,\ldots,1$ generate $\theta_n^{(j)}$ from $\left(\theta_n|\theta_{n+1}^{(j)}, \omega^{(j)}, D^n\right)$ where $\theta_{n+1}^{(j)}$ is the value generated in the previous step.

If we repeat this a large number of times, we can obtain samples from the full conditional of the latent rates. Consequently, we can obtain samples from the joint density of the model parameters by iteratively sampling from the full conditionals, $p(\omega|\theta_1,\ldots,\theta_t,D^t)$ and $p(\theta_1,\ldots,\theta_t|\omega,D^t)$, via the Gibbs sampler. Once we have the posterior samples from $p(\theta_1,\ldots,\theta_t,\omega|D^t)$ we can also obtain the posterior samples of λ_ts in a straightforward manner using the identity $\lambda_t = \theta_t e^{\psi' z_t}$.

11.4 Multivariate Extension

It is possible to consider several extensions of the basic model to analyze multivariate count time series. For instance, the observations of interest can be the number of occurrences of an event during day t of year j. Another possibility is to consider the analysis of J different Poisson time series. For instance, for a given year, the weekly spending habits of J different households which can exhibit dependence can be modeled using such a structure. Several extensions have been proposed by Aktekin and Soyer (2011), where multiplicative Poisson rates for (11.3) are considered. An alternate approach for modeling multivariate time series of counts is described by Ravishanker, Venkatesan, and Hu (2015; Chapter 20 in this volume).

In what follows, we present a model for J Poisson time series that are assumed to be affected by the same environment. We assume that

$$Y_{jt} \sim Pois(\lambda_{jt}), \text{ for } j = 1,\ldots,J, \tag{11.31}$$

where $\lambda_{jt} = \lambda_j\theta_t, \lambda_j$ is the arrival rate specific to the jth series and θ_t is the common term modulating λ_j. For example, in the case where Y_{jt} is the number of grocery store trips of household j at time t, λ_j is the household-specific rate and we can think of θ_t as the effect of a common economic environment that the households are exposed to at time t. The values of $\theta_t > 1$ represent a more favorable economic environment than usual, implying higher shopping rates.

This is analogous to the concept of an accelerated environment for operating conditions of components used by Lindley and Singpurwalla (1986) in life testing. Our case can be considered as a dynamic version of their setup since we have the Markovian evolution of θ_ts as

$$\theta_t = \frac{\theta_{t-1}}{\gamma}\epsilon_t, \tag{11.32}$$

where, as earlier, $\left(\epsilon_t|D^{t-1},\lambda_1,\ldots,\lambda_J\right) \sim Beta\,[\gamma\alpha_{t-1},(1-\gamma)\alpha_{t-1}]$ with $\alpha_{t-1} > 0, 0 < \gamma < 1$, and $D^{t-1} = \{D^{t-2}, Y_{1(t-1)},\ldots,Y_{J(t-1)}\}$. Furthermore, we assume that

$$\lambda_j \sim \text{Gamma}(a_j, b_j), \text{ for } j = 1,\ldots,J, \tag{11.33}$$

and a priori, λ_js are independent of each other as well as of θ_0. Given θ_ts and λ_js, Y_{jt}s are conditionally independent. In other words, all J series are affected by the same common environment and given that we know the uncertainty about the environment, they will be independent.

At time 0, we assume that $(\theta_0|D^0) \sim \text{Gamma}(\alpha_0, \beta_0)$, and by induction we can show that

$$\left(\theta_{t-1}|D^{t-1}, \lambda_1, \ldots, \lambda_J\right) \sim \text{Gamma}\left(\alpha_{t-1}, \beta_{t-1}\right), \tag{11.34}$$

and

$$\left(\theta_t|D^{t-1}, \lambda_1, \ldots, \lambda_J\right) \sim \text{Gamma}\left(\gamma\alpha_{t-1}, \gamma\beta_{t-1}\right). \tag{11.35}$$

In addition, the filtering density at time t can be obtained as

$$\left(\theta_t|D^t, \lambda_1, \ldots, \lambda_J\right) \sim \text{Gamma}\left(\alpha_t, \beta_t\right), \tag{11.36}$$

where $\alpha_t = \gamma\alpha_{t-1} + Y_{1t} + \ldots + Y_{Jt}$ and $\beta_t = \gamma\beta_{t-1} + \lambda_1 + \ldots + \lambda_J$. Consequently, the marginal distributions of Y_{jt} for any j can be obtained as

$$p\left(Y_{jt}|\lambda_j, D^{t-1}\right) = \binom{\gamma\alpha_{t-1} + Y_{jt} - 1}{Y_{jt}} \left(1 - \frac{\lambda_j}{\gamma\beta_{t-1} + \lambda_j}\right)^{\gamma\alpha_{t-1}} \left(\frac{\lambda_j}{\gamma\beta_{t-1} + \lambda_j}\right)^{Y_{jt}}, \tag{11.37}$$

which is a negative binomial model as earlier. The multivariate distribution of $(Y_{1t}, \cdots, Y_{Jt})$ can be obtained as

$$p\left(Y_{1t}, \ldots, Y_{Jt}|\lambda_1, \ldots, \lambda_J, D^{t-1}\right) = \frac{\Gamma\left(\gamma\alpha_{t-1} + \sum_j Y_{jt}\right)}{\Gamma\left(\gamma\alpha_{t-1}\right)\prod_j \Gamma\left(Y_{jt} + 1\right)} \prod_j \left(\frac{\lambda_j}{\gamma\beta_{t-1} + \sum_j \lambda_j}\right)^{Y_{jt}} \times \left(\frac{\gamma\beta_{t-1}}{\gamma\beta_{t-1} + \sum_j \lambda_j}\right)^{\gamma\alpha_{t-1}}, \tag{11.38}$$

which is a dynamic multivariate distribution of negative binomial type. The bivariate distribution $p(Y_{it}, Y_{jt}|\lambda_i, \lambda_j, D^{t-1})$ can be obtained as

$$\frac{\Gamma\left(\gamma\alpha_{t-1} + Y_{it} + Y_{jt}\right)}{\Gamma\left(\gamma\alpha_{t-1}\right)\Gamma\left(Y_{it} + 1\right)\Gamma\left(Y_{jt} + 1\right)} \left(\frac{\gamma\beta_{t-1}}{\lambda_i + \lambda_j + \gamma\beta_{t-1}}\right)^{\gamma\alpha_{t-1}} \left(\frac{\lambda_i}{\lambda_i + \lambda_j + \gamma\beta_{t-1}}\right)^{Y_{it}} \times \left(\frac{\lambda_j}{\lambda_i + \lambda_j + \gamma\beta_{t-1}}\right)^{Y_{jt}}, \tag{11.39}$$

which is a bivariate negative binomial distribution for integer values of $\gamma\alpha_{t-1}$. This distribution is the dynamic version of the negative binomial distribution proposed by Arbous and Kerrich (1951) for modeling accident numbers.

The conditionals of Y_{jt} will also be negative binomial-type distributions. The dynamic conditional mean (or regression) of Y_{jt} given Y_{jt} can be obtained as

$$E[Y_{jt}|Y_{it}, \lambda_i, \lambda_j, D^{t-1}] = \frac{\lambda_j(\gamma\alpha_{t-1} + Y_{it})}{(\lambda_i + \gamma\beta_{t-1})}, \tag{11.40}$$

which is linear in Y_{it}. It can be easily seen that the bivariate counts are positively correlated and the correlation is given by

$$Cor(Y_{it}, Y_{jt}|\lambda_i, \lambda_j, D^{t-1}) = \sqrt{\frac{\lambda_i\lambda_j}{(\lambda_i + \gamma\beta_{t-1})(\lambda_j + \gamma\beta_{t-1})}}. \tag{11.41}$$

Other properties of the dynamic multivariate distribution are given in Aktekin et al. (2014).

The estimation of this model using MCMC would be straightforward using the FFBS algorithm for θ_ts in conjunction with a Gibbs sampler step for the λ_js whose full conditionals are given by

$$p(\lambda_j|\theta_1, \ldots, \theta_t, D^t) \sim \text{Gamma}(a_{jt}, b_{jt}), \tag{11.42}$$

where $a_{jt} = a_j + Y_{j1} + \cdots + Y_{jt}$ and $b_{jt} = b_j + \theta_1 + \cdots + \theta_t$. By iteratively sampling from the conditional distributions of $(\theta_1, \ldots, \theta_t|\lambda_1, \ldots, \lambda_J, D^t)$ using the FFBS algorithm and $(\lambda_j|\theta_1, \ldots, \theta_t, D^t)$ for all j, one can obtain samples from $(\theta_1, \ldots, \theta_t, \lambda_1, \ldots, \lambda_J|D^t)$.

11.5 Business Applications

In order to show how the models are applied to count time series in business applications, we have used three data sets. Example 11.1 consists of time series counts of the number of calls arriving to a call center in a given time interval. Example 11.2 consists of the number of people who defaulted in a given mortgage pool. Example 11.3 consists of the number of weekly grocery store visits for households. We discuss the implementation and the estimation of the proposed Poisson time series models using these three examples.

11.5.1 Example 11.1: Call Center Arrival Count Time Series Data

To show the use of the basic model without any covariates, we consider the time series of counts of call center arrivals during different intervals of 164 days from an anonymous U.S. commercial bank as discussed in Aktekin and Soyer (2011). Each day consists of 169 time intervals each of which has a duration of 5 min. On a given day, the call center is operational between 7:00 AM and 9:05 PM.

We have only used the first week of the data for illustration purposes and have provided within-day updating and forecasting results separately for Monday–Friday of the week. Such an approach would be of interest to call center managers who would like to be able to determine staff schedules in advance for different time intervals on a given day.

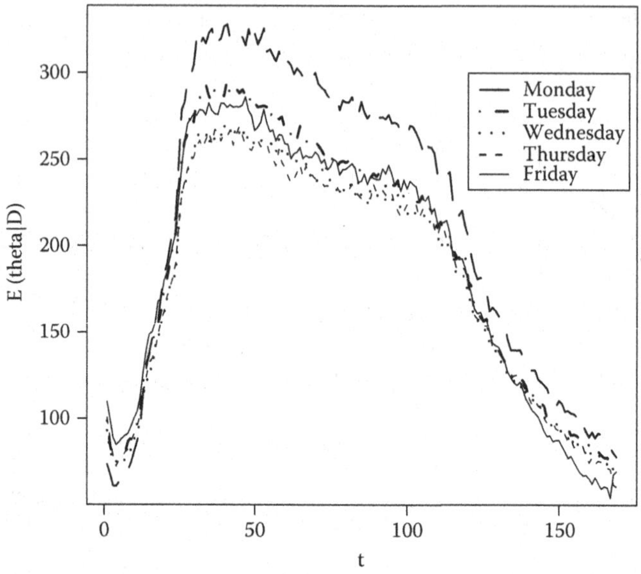

FIGURE 11.1
Posterior arrival rates for different days of the week.

Given the filtering distribution, we obtained the means of the latent arrival rates for a particular time interval for each day given information on the entire data consisting of 164 days each with 169 time intervals. These are shown in Figure 11.1 from which a certain type of ordering between the days of the week can be inferred. As such, we set the initial prior parameters from (11.5) for the arrival rate as $\alpha_0^i = \alpha_0 = 0.001$ and $\beta_0^i = \beta_0 = 0.001$ for all i, with i representing a specific day of the week.

Furthermore, summary statistics for the posterior discounting factor, γ, for each day of the week are shown in Table 11.1. Discounting occurs on the sum of the previously observed values of the call arrivals for a given period and is therefore a function of the data dimension used. Each day seems to exhibit a slightly different discount behavior. The fact that the posterior means of the discounting terms are getting smaller as we observe more data indicates that the model emphasizes arrival counts observed during the within-day interval of interest (say t) more than the previously observed arrival counts (say $t-1, \ldots, 1$).

TABLE 11.1
Posterior means and standard deviations of γ for different days

Day	Mean	St. Dev
Mondays	0.066	0.0025
Tuesdays	0.046	0.0016
Wednesdays	0.092	0.0042
Thursdays	0.084	0.0039
Fridays	0.075	0.0032

11.5.2 Example 11.2: Mortgage Default Count Time Series Data

In illustrating the use of the basic model with covariates, we use data provided by the Federal Housing Administration (FHA) of the U.S. Department of Housing and Urban Development. These data have been analyzed in detail by Aktekin et al. (2013). In our analysis, we use 144 monthly defaulted FHA insured single-family 30-year fixed rate mortgage loans from 1994 in the Atlanta region. In addition, we make use of covariates such as the regional conventional mortgage home price index (CMHPI), federal cost of funds index (COFI), the homeowner mortgage financial obligations ratio (FOR), and regional unemployment rate (Unemp). A time series plot of the monthly mortgage count data is shown in Figure 11.2, where a nonstationary behavior that can be captured by our Poisson state-space models is observed.

In analyzing the default count data, the discounting factor γ introduced in (11.3) is assumed to follow a discrete uniform distribution defined over $(0, 1)$ in order to keep the updating/filtering tractable. The posterior distribution of γ is obtained via (11.24) and is shown in the left panel of Figure 11.3. Thus, given the posterior of γ and the FFBS algorithm, it is possible to obtain the retrospective fit of counts. An overlay plot of the mean default rates and the actual data is shown in the right panel of Figure 11.3. The availability of the joint distribution of the default rate over time, that is, $p(\theta_1, \ldots, \theta_t|D^t)$, would be of interest to institutions that are managing the loans for the purposes of risk management. Furthermore, the Bayesian approach allows direct comparisons of probabilities involving Poisson rates (in this case default rates) during different time periods. For instance, it would be straightforward to compute the probability that the default rate during the second month is greater than that of the first month for a given cohort, that is, $p(\theta_2 \geq \theta_1|D^{144})$.

In order to take into account the effects of covariates (macroeconomic variables in this case) on the default rate, we have used the model with covariates. In doing so, we assume the prior of γ to be continuous uniform over $(0, 1)$ and those of the covariate coefficients, ψ, to be independent normal distributions. The MCMC algorithm was run for 10,000 iterations

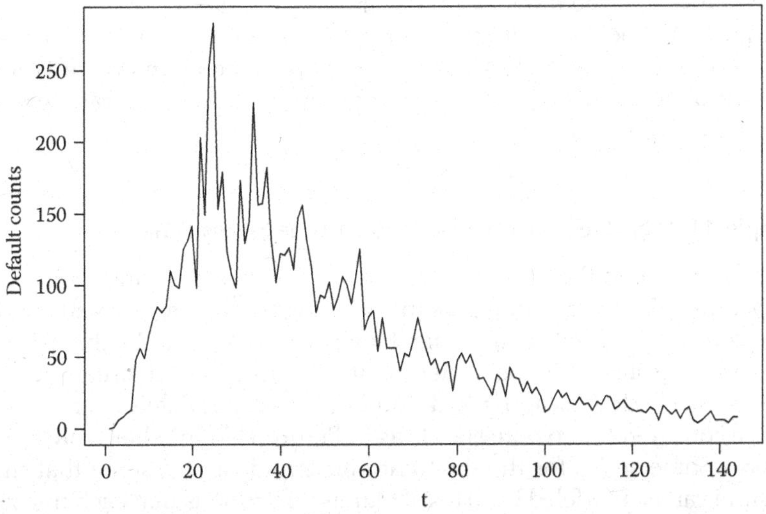

FIGURE 11.2
Time series plot of monthly default counts.

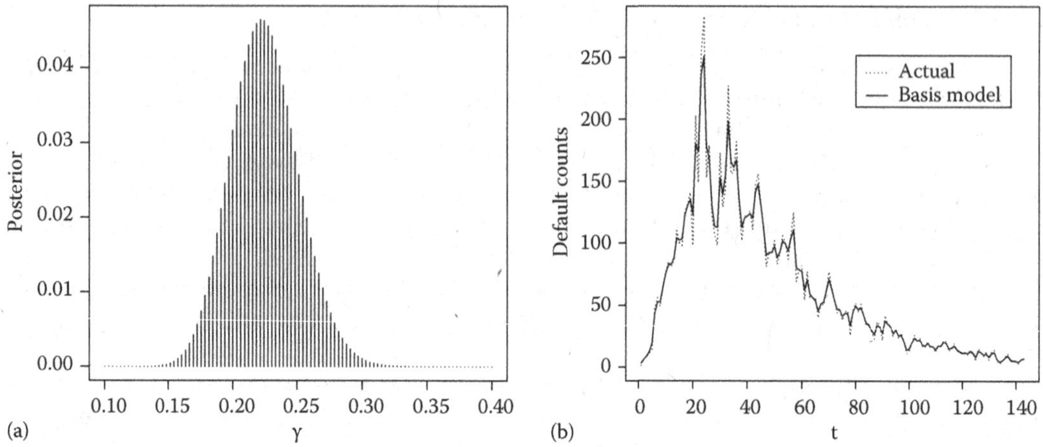

FIGURE 11.3
Posterior γ (a) and the retrospective fit to count data (b).

with a burn-in period of 2,000 iterations with no convergence issues. The posterior density plots of ψ and γ are shown in Figure 11.4 where γ exhibits similar behavior to the posterior discounting term obtained for the basic model as in the left panel of Figure 11.3.

Table 11.2 shows the posterior summary statistics for the covariates. All macroeconomic variables seem to have fairly significant effects on the default rate. CMHPI, COFI, and Unemp have positive effects on default counts. For instance, as unemployment tends to go up, the model suggests that the number of people defaulting tends to increase for the cohort under study. On the other hand, the homeowner FOR seems to decrease the expected number of defaults as it goes up, namely, as the burden of repayment becomes relatively easier, homeowners are less likely to default.

Figure 11.5 shows that the model with covariates provides a reasonably good fit to the data. Furthermore, the behavior of the latent default rates, θ_ts, can be described via their joint distribution $p(\theta_1, \ldots, \theta_{144}|D^{144})$. A boxplot of θ_ts is shown in Figure 11.6, which provides insights into the stochastic and temporal behavior of the latent rates given the count data and the relevant covariates.

11.5.3 Example 11.3: Household Spending Count Time Series Data

Our final example utilizes the multivariate extension of the basic model in the context of household spending. In order to illustrate the workings of the multivariate model in a simple setup, we consider bivariate count data. However, we emphasize that the multivariate count model can be applied to higher orders relatively easily. The data consist of the weekly grocery store visits of 540 Chicago-based households accumulated over 104 weeks, from which we have considered two households as in Figure 11.7. In other words, we have two different Poisson time series for the different households and assume that their visits to the grocery store can be modeled by (11.31). Our assumption is that each household's visit to the grocery store is affected by the same environment, that is, the economic situation, weather, and so on. That is, we assume that the grocery store arrival process of a household in Chicago will exhibit behavior similar to that of any other household.

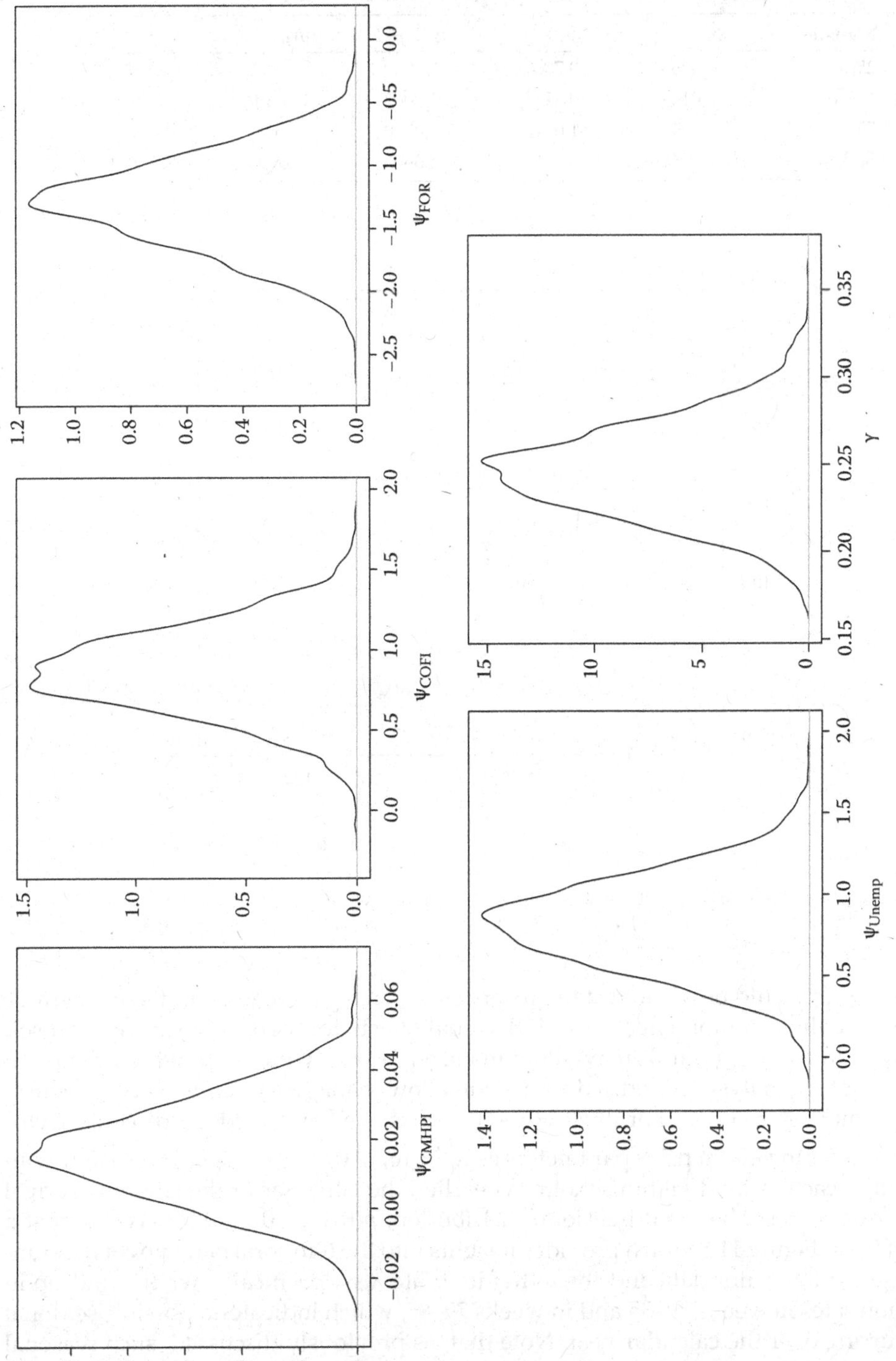

FIGURE 11.4
Posterior density plots of ψ and γ for the model with covariates.

TABLE 11.2

Posterior statistics for ψ and γ for the model with covariates

Statistics	ψ_{CMHPI}	ψ_{COFI}	ψ_{FOR}	ψ_{Unemp}	γ
25th	0.0063	0.7003	−1.5430	0.6252	0.2281
Mean	0.0160	0.8717	−1.3002	0.8191	0.2466
75th	0.0256	1.0510	−1.0550	1.0117	0.2643
St. Dev	0.0141	0.2663	0.3606	0.2826	0.0270

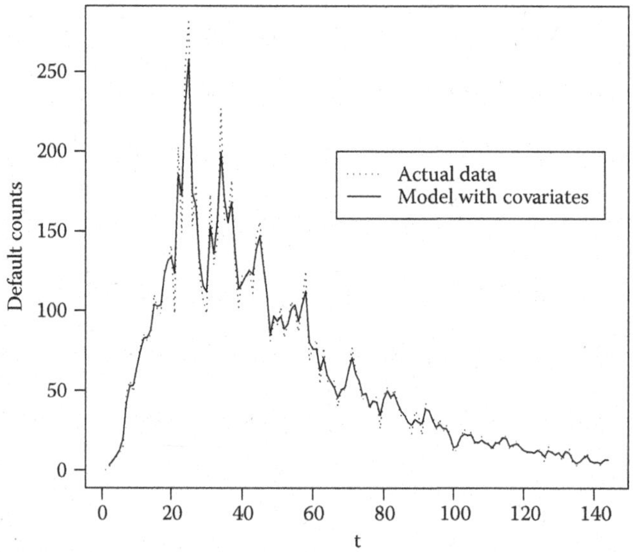

FIGURE 11.5
Retrospective fit of the model with covariates to data.

This approach would be of interest to grocery store managers who would like to be able to differentiate the common effect from individual effects for store promotion purposes. For example, inference about θ_t may allow managers to carry out store-wide promotion activities, whereas analysis on individual λ_js will allow managers to target specific households to promote store visits. For illustrative purposes, we have fixed the discount factor at $\gamma = 0.5$ and set the initial prior parameters as $\alpha_0^j = \alpha_0 = 0.001$ and $\beta_0^j = \beta_0 = 0.001$ with j representing each individual household. As earlier, the behavior of the common arrival rates, θ_ts, can be described via their joint distribution, $p(\theta_1, \ldots, \theta_{104}|D^{104})$. We present a boxplot of θ_ts in Figure 11.8 which provides insights into the temporal behavior of the common rate given the count data and the individual rate λ_js. Specifically, we find a drop in the common rates in weeks 29–35 and in weeks 79–85, which indicates a possible seasonal effect occurring over the calendar year. Note that, as previously discussed, such seasonal effects can be easily incorporated into the model as covariates.

Furthermore, the posterior density plots of λ_js are shown in Figure 11.9. Clearly, household 1 can be characterized by a higher rate compared to household 2. Store managers may

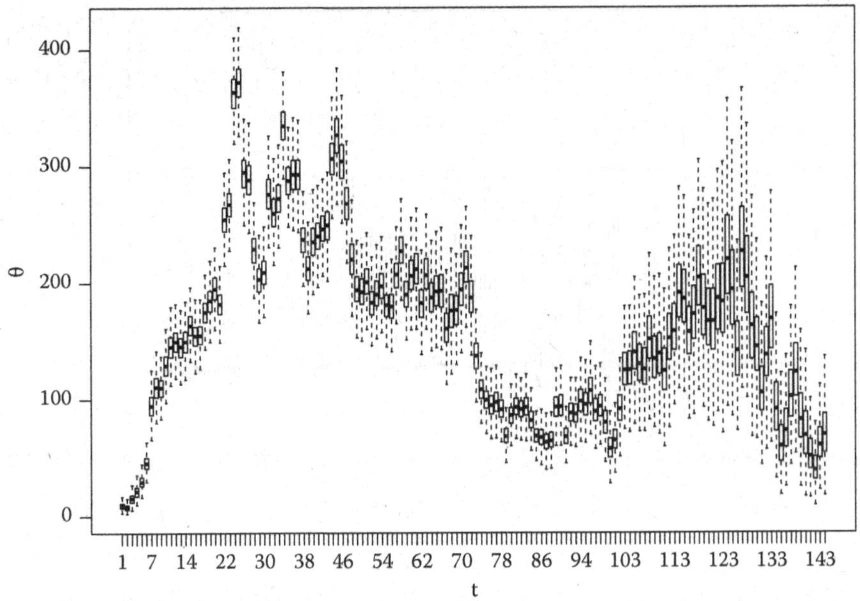

FIGURE 11.6
Boxplots for the latent rates, θ_ts from $p(\theta_1, \ldots, \theta_{144}|D^{144})$.

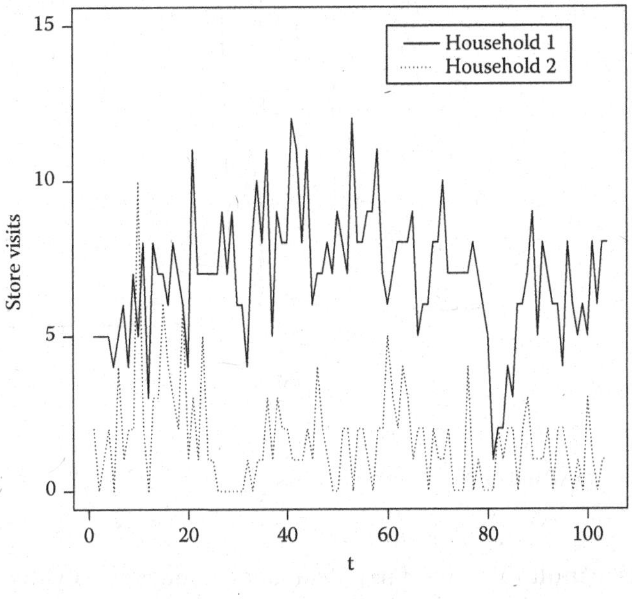

FIGURE 11.7
Time series plot of weekly grocery store visits.

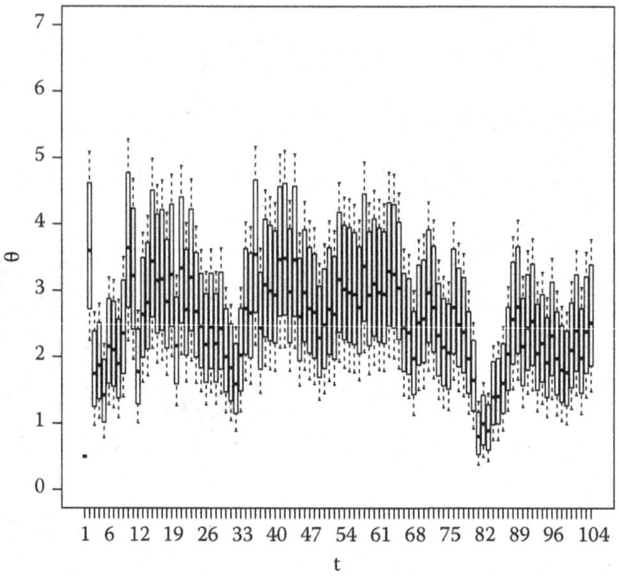

FIGURE 11.8
Boxplots for the common rates, θ_ts from $p(\theta_1, \ldots, \theta_{104}|D^{104})$.

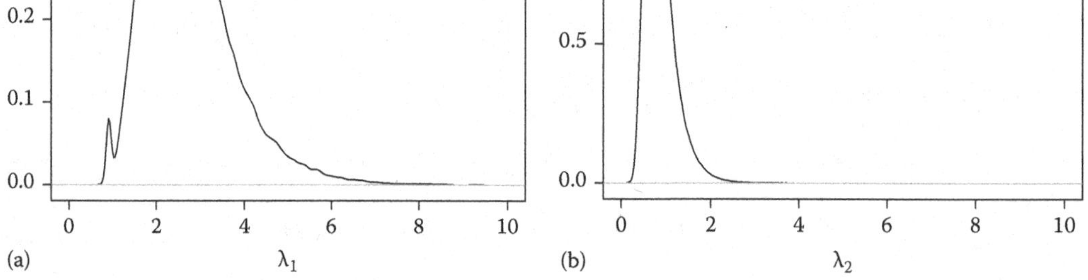

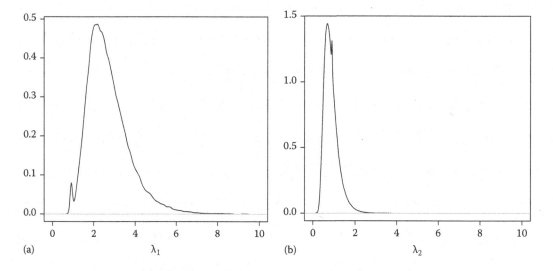

FIGURE 11.9
Posterior density plots of (a) λ_1 and (b) λ_2.

utilize our findings in multiple ways. First, the manager may identify the season associated with low common arrival rates and run a store-wide promotion strategy to lure customers. In addition, store managers faced with a limited budget may also wish to focus their promotion efforts on household 2 with a lower arrival rate rather than on household 1. Finally, one may also extend the model to include covariates to identify the reasons behind differences in individual household λ_js.

11.6 Conclusion

In this chapter, we have introduced a general class of Poisson time series models. We first discussed univariate discrete-time state-space models with Poisson measurements and their Bayesian inference via MCMC methods. Furthermore, we discussed issues of sequential updating, filtering, smoothing, and forecasting. We also introduced modeling strategies for multivariate extensions. In order to show the implementation of the proposed models, we have used real count data from different disciplines such as finance, operations, and marketing.

We believe that several future directions can be pursued as a consequence of this study. One such area is to treat γ as a time-varying or a series-specific discount factor which could potentially create challenges in parameter estimation. Another possibility for estimation purposes is to investigate the implementation of sequential particle filtering methods instead of MCMC that are widely used for state-space applications.

References

Aktekin, T., Kim, B., and Soyer, R. (2014). Dynamic multivariate distributions for count data: A Bayesian approach. Technical report, Institute for Integrating Statistics in Decision Sciences, The George Washington University, Washington, DC.

Aktekin, T. and Soyer, R. (2011). Call center arrival modeling: A Bayesian state-space approach. *Naval Research Logistics*, 58(1):28–42.

Aktekin, T., Soyer, R., and Xu, F. (2013). Assessment of mortgage default risk via Bayesian state space models. *Annals of Applied Statistics*, 7(3):1450–1473.

Arbous, A. G. and Kerrich, J. (1951). Accident statistics and the concept of accident proneness. *Biometrics*, 7:340–432.

Chib, S. and Greenberg, E. (1995). Understanding the Metropolis-Hasting algorithm. *The American Statistician*, 49(4):327–335.

Chib, S., Greenberg, E., and Winkelmann, R. (1998). Posterior simulation and Bayes factors in panel count data models. *Journal of Econometrics*, 86:33–54.

Chib, S. and Winkelmann, R. (2001). Markov chain Monte Carlo analysis of correlated count data. *Journal of Business and Economic Statistics*, 19(4):428–435.

Cox, D. R. (1981). Statistical analysis of time series: Some recent developments. *Scandinavian Journal of Statistics*, 8:93–115.

Davis, R. A., Dunsmuir, W. T. M., and Streett, S. B. (2003). Observation-driven models for Poisson counts. *Biometrika*, 90(4):777–790.

Davis, R. A., Dunsmuir, W. T. M., and Wang, Y. (2000). On autocorrelation in a Poisson regression model. *Biometrika*, 87(3):491–505.

Durbin, J. and Koopman, S. (2000). Time series analysis of non-Gaussian observations based on state space models from both classical and Bayesian perspectives. *Journal of the Royal Statistical Society. Series B*, 62(1):3–56.

Freeland, R. K. and McCabe, B. P. M. (2004). Analysis of low count time series data by Poisson autocorrelation. *Journal of Time Series Analysis*, 25(5):701–722.

Fruhwirth-Schnatter, S. (1994). Data augmentation and dynamic linear models. *Journal of Time Series Analysis*, 15:183–202.

Frühwirth-Schnatter, S. and Wagner, H. (2006). Auxiliary mixture sampling for parameter-driven models of time series of counts with applications to state space modelling. *Biometrika*, 93(4):827–841.

Gamerman, D., Abanto-Valle, C. A., Silva, R. S., and Martins, T. G. (2015). Dynamic Bayesian models for discrete-valued time series. In R. A. Davis, S. H. Holan, R. Lund and N. Ravishanker, eds., *Handbook of Discrete-Valued Time Series*, pp. 165–186. Chapman & Hall, Boca Raton, FL.

Gamerman, D. and Lopes, H. F. (2006). *Markov Chain Monte Carlo Stochastic Simulation for Bayesian Inference*. Chapman & Hall/CRC Press, Boca Raton, FL.

Gilks, W. R., Richardson, S., and Spiegelhalter, D. J. (1996). *Markov Chain Monte Carlo in Practice*. Chapman & Hall/CRC Press, Boca Raton, FL.

Harvey, A. C. and Fernandes, C. (1989). Time series models for count or qualitative observations. *Journal of Business and Economic Statistics*, 7(4):407–417.

Lindley, D. V. and Singpurwalla, N. D. (1986). Multivariate distributions for the life lengths of components of a system sharing a common environment. *Journal of Applied Probability*, 23:418–431.

Ravishanker, N., Venkatesan, R., and Hu, S. (2015). Dynamic models for time series of counts with a marketing application. In R. A. Davis, S. H. Holan, R. Lund and N. Ravishanker, eds., *Handbook of Discrete-Valued Time Series*, pp. 425–446. Chapman & Hall, Boca Raton, FL.

Santos, T. R. D., Gamerman, D., and Franco, G. C. (2013). A non-Gaussian family of state-space models with exact marginal likelihood. *Journal of Time Series Analysis*, 34(6):625–645.

Smith, A. F. M. and Gelman, A. E. (1992). Bayesian statistics without tears: A sampling perspective. *The American Statistician*, 46(2):84–88.

Smith, R. and Miller, J. (1986). A non-Gaussian state space model and application to prediction of records. *Journal of the Royal Statistical Society. Series B*, 48(1):79–88.

Zeger, S. L. (1988). A regression model for time series of counts. *Biometrika*, 75(4):621–629.

Section III

Binary and Categorical-Valued Time Series

12

Hidden Markov Models for Discrete-Valued Time Series

Iain L. MacDonald and Walter Zucchini

CONTENTS

12.1 Structure and Components of a Hidden Markov Model

In the search for useful models for discrete-valued time series, one possible approach is to take a standard model for continuous-valued series, for example, a Gaussian autoregressive moving-average process, and to modify or adapt it in order to allow for the discrete nature of the data. Another approach, the one followed here, is to start from a model for discrete data which assumes independence and then to relax the independence assumption by allowing the distribution of the observations to switch among several possibilities according to a latent Markov chain. To take a simple example, a sequence of independent Poisson random variables with common mean would be inappropriate for a series of unbounded counts displaying significant autocorrelation. But a model that allowed the observations to be Poisson with mean either λ_1 or λ_2, the choice being made by a discrete-time Markov chain, might well be adequate; a model of this kind, it will be seen, allows for both serial dependence and overdispersion relative to the Poisson.

Such a latent Markov chain is a simple device by which any model X_t ($t = 1, 2, \ldots, T$) for a series of independent observations may be generalized to yield a model that allows for serial dependence. A hidden Markov model (HMM) relaxes the assumption of independence; however, we assume that the observations are *conditionally independent*, given the states occupied by an unobserved finite state-space Markov chain. Here, we confine ourselves to discrete data, but the observations could more generally be discrete or continuous, a mixture of those two, univariate, multivariate, circular in nature, spatial images—any kind of observation X_t, in fact, for which it is possible to propose a probabilistic model. The case of discrete data also includes several possibilities: univariate unbounded counts (i.e., Poisson-like observations), univariate bounded counts (i.e., binomial-like observations) including binary observations, observations of categories, and multivariate versions of these.

HMMs consist of two parts: first, an unobserved "parameter process" $\{C_t : t \in \mathcal{N}\}$ that is a Markov chain on $\{1, 2, \ldots, m\}$, and second, an observed process $\{X_t : t \in \mathcal{N}\}$ such that the distribution of X_t is determined only by the current state C_t, irrespective of all previous states and observations. (The symbol $\mathcal{N}$ denotes the natural numbers.) This structure is represented by the directed graph in Figure 12.1 and summarized by the following equations, in which $\mathbf{C}^{(t)}$ and $\mathbf{X}^{(t)}$ denote the "histories" of the processes $\{C_t\}$ and $\{X_t\}$ from time 1 to time t:

$$\Pr(C_t \mid \mathbf{C}^{(t-1)}) = \Pr(C_t \mid C_{t-1}), \; t = 2, 3, \ldots \tag{12.1}$$

$$\Pr(X_t \mid \mathbf{X}^{(t-1)}, \mathbf{C}^{(t)}) = \Pr(X_t \mid C_t), \; t \in \mathcal{N}. \tag{12.2}$$

The Markov chain is assumed here to be irreducible, aperiodic, and (unless it is stated otherwise) homogeneous. The transition probability matrix (t.p.m.) is denoted by $\mathbf{\Gamma} = (\gamma_{ij})$, and the initial distribution, that of C_1, by the row vector $\boldsymbol{\delta}^*$. (All vectors are row vectors.) The stationary distribution of the Markov chain is denoted by $\boldsymbol{\delta}$ and if, as is often assumed, $\boldsymbol{\delta}^* = \boldsymbol{\delta}$, then the Markov chain is stationary. Unless otherwise indicated, the "state-dependent probability" $p_i(x) = \Pr(X_t = x \mid C_t = i)$ is assumed to be independent of t,

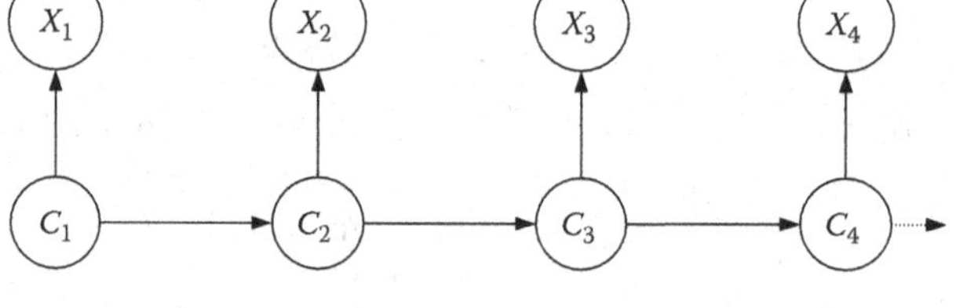

FIGURE 12.1
Directed graph of basic HMM.

but if necessary this assumption can be relaxed. It is notationally convenient to assemble the m state–dependent probabilities $p_i(x)$ into the $m \times m$ diagonal matrix

$$\mathbf{P}(x) = \text{diag}(p_1(x), p_2(x), \ldots, p_m(x)).$$

From this definition, it will be seen that HMMs are "parameter-driven" models, in the sense used by Cox (1981). They are also state-space models whose latent process is discrete-valued.

HMMs have a history going back at least to the work of Leonard Baum and co-authors (see, e.g., Baum et al., 1970; Welch, 2003), although special cases were certainly considered earlier than that. Such models are well known for their applications in speech recognition (Juang and Rabiner, 1991; Rabiner, 1989) and bioinformatics (Durbin et al., 1998, Chapter 3), but we focus here on their use as general-purpose models for discrete-valued time series. HMMs go under a variety of other names or descriptions as well: latent Markov models, hidden Markov processes, Markov-dependent mixtures, models subject to Markov regime, and Markov-switching models, the last being more general in that the conditional independence assumption (12.2) is relaxed.

12.2 Examples of Hidden Markov Time Series Models

We describe here a selection of the ways in which HMMs can be used as models for discrete-valued time series. Some of these models are univariate, some are multivariate, but all have the structure specified by (12.1) and (12.2). Furthermore, the process of estimating parameters by numerical maximization of likelihood will be essentially the same for all; see Section 12.4.1.

12.2.1 Univariate Models for Discrete Observations

The Poisson–HMM is the simple model in which, conditional on a latent Markov chain $\{C_t\}$ on $\{1, 2, \ldots, m\}$, the observation X_t has a Poisson distribution with mean λ_i when $C_t = i$. If the Markov chain is assumed stationary, that is, if $\boldsymbol{\delta}^* = \boldsymbol{\delta}$, the process $\{X_t\}$ is (strictly) stationary and there are m^2 parameters to be estimated: the m state–dependent means λ_i and all but one of the transition probabilities in each row of $\boldsymbol{\Gamma}$, for example, the $m^2 - m$ off-diagonal entries. If the Markov chain is not assumed to be stationary, then $\boldsymbol{\delta}^*$ must also be estimated (see, e.g., Leroux and Puterman, 1992). The number of parameters of an HMM, being of the order m^2, limits the number of states that it is feasible to use in practice. Alternatively, one can reduce the number of parameters by structuring the t.p.m. in some parsimonious way; see Section 12.8 for references.

If we assume instead that the distribution in state i is binomial with parameters n_t (the number of trials at time t) and π_i (the "success probability"), we have a model for a time series of bounded counts. The special case $n_t = 1$ yields a model for binary time series. An alternative to the Poisson in the case of unbounded counts that exhibit overdispersion is the negative binomial. (By "negative binomial" we mean here the version of that distribution that has the nonnegative integers as support.) Indeed, it is not essential to use distributions from the same family in all the states of an HMM. We could use a Poisson distribution in one state of a two-state HMM and a negative binomial in the second state. If instead we use a degenerate distribution at zero in the second state, the resulting marginal distribution of the observations is zero-inflated Poisson.

12.2.2 Multivariate Models: Example of a Bivariate Model for Unbounded Counts

In order to specify an HMM for (say) bivariate observations (X_t, Y_t) of unbounded counts, we need to specify the conditional distribution of (X_t, Y_t) given C_t. One can specify an appropriate bivariate distribution for X_t and Y_t for each of the m states of the Markov chain, but this is not necessarily an easy task, as there is (for instance) no canonical bivariate Poisson distribution. A simple and very convenient alternative that works surprisingly often is to assume that, given C_t, the random variables X_t and Y_t are conditionally independent; that is, they are "contemporaneously conditionally independent." This assumption leads to (unconditional) dependence between X_t and Y_t, very much in the same way as the assumption of conditional independence in a univariate series, given the Markov chain, induces unconditional serial dependence.

As an example, assume that, conditional on the Markov chain $\{C_t\}$, X_t and Y_t have *independent* Poisson distributions and that the pairs (X_t, Y_t) are also conditionally independent, the latter assumption being the usual "longitudinal conditional independence" of an HMM. The dependence structure of such a bivariate model is represented by the directed graph in Figure 12.2. The resulting model consists of serially dependent observations and, in addition, X_t and Y_t are *dependent*. The marginal distributions of X_t and Y_t are mixtures of Poissons.

12.2.3 Multivariate Models: Series of Multinomial-Like Observations

Suppose that at time t there are n_t trials, each of which can result in one of $q \geq 2$ outcomes. To allow for serial dependence, we can simply assume that, conditional on a latent Markov chain, the observations are independent multinomial, with the conditional distribution at time t being multinomial with number of trials n_t and some vector π_i of "success probabilities," i being the current state of the Markov chain. (This vector π_i can be allowed to depend on t, if necessary.) The unconditional distribution of the observations at time t is then a mixture of multinomials.

An important special case of such a model is that in which $n_t = 1$ for all t; this provides a model for categorical series such as DNA base sequences ($q = 4$) or amino acid sequences ($q = 20$). The former example goes back at least to Churchill (1989).

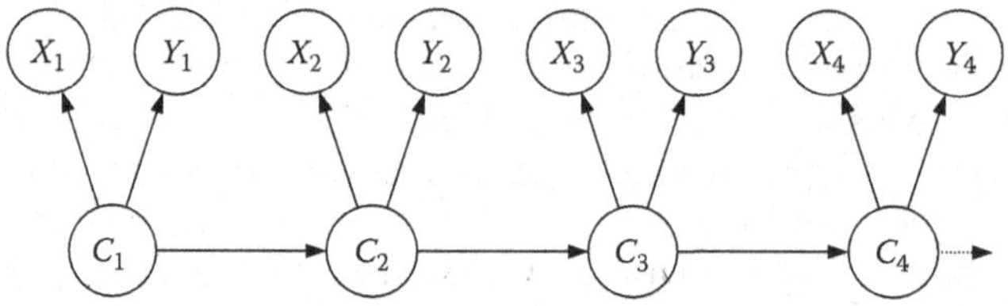

FIGURE 12.2
Directed graph of bivariate HMM assuming contemporaneous conditional independence.

12.2.4 Models with Covariates

A simple way to allow for the effect of covariates, including time-dependent covariates, in (say) a Poisson–HMM, is to leave the Markov chain unchanged but to model the log of ${}_t\lambda_i$ (the mean in state i at time t) as a linear function of a (row) vector of covariates, $\mathbf{y}_t$:

$$\ln {}_t\lambda_i = \boldsymbol{\beta}_i \mathbf{y}_t'.$$

This makes it possible, for example, to build time trend or seasonal components into the state-dependent distributions. A slightly more more difficult, but sometimes more appropriate, way of allowing for covariates is to model the entries of $\boldsymbol{\Gamma}$ as functions of the covariates, in such a way of course as to respect the row-sum and nonnegativity constraints on the transition probabilities.

12.3 Properties of HMMs

12.3.1 Computing the Likelihood

An important property of HMMs is that the likelihood is easily computed. Consider an HMM with initial distribution $\boldsymbol{\delta}^*$, t.p.m. $\boldsymbol{\Gamma}$ and state-dependent distributions $\mathbf{P}(x)$. The likelihood, that is, the joint probability of the observations $x_1, x_2, \ldots, x_T$ is a multiple sum:

$$\begin{aligned} L_T &= \Pr\left(\mathbf{X}^{(T)} = \mathbf{x}^{(T)}\right) \\ &= \sum_{c_1=1}^{m} \sum_{c_2=1}^{m} \cdots \sum_{c_T=1}^{m} \Pr\left(\mathbf{C}^{(T)} = \mathbf{c}^{(T)}\right) \Pr\left(\mathbf{X}^{(T)} = \mathbf{x}^{(T)} \mid \mathbf{C}^{(T)} = \mathbf{c}^{(T)}\right). \end{aligned}$$

That is,

$$L_T = \sum_{c_1, c_2, \ldots, c_T = 1}^{m} \Pr(C_1 = c_1) \prod_{k=2}^{T} \Pr(C_k = c_k \mid C_{k-1} = c_{k-1}) \prod_{k=1}^{T} \Pr(X_k = x_k \mid C_k = c_k). \tag{12.3}$$

It can be shown that this sum can be given as a matrix product:

$$L_T = \boldsymbol{\delta}^* \mathbf{P}(x_1) \boldsymbol{\Gamma} \mathbf{P}(x_2) \boldsymbol{\Gamma} \mathbf{P}(x_3) \cdots \boldsymbol{\Gamma} \mathbf{P}(x_T) \mathbf{1}', \tag{12.4}$$

the vector $\mathbf{1}$ being a row vector of ones. (Justification for this and other unproved assertions can be found in Zucchini and MacDonald (2009).) If the Markov chain is assumed to be stationary, $\boldsymbol{\delta}^* = \boldsymbol{\delta}$ is computed from $\boldsymbol{\Gamma}$; that is, as the solution of the linear system $\boldsymbol{\delta\Gamma} = \boldsymbol{\delta}$ subject to $\boldsymbol{\delta}\mathbf{1}' = 1$.

Thus, L_T can be computed by using the following simple recursion:

$$\alpha_1 = \delta^* \mathbf{P}(x_1);$$
$$\alpha_t = \alpha_{t-1}\mathbf{\Gamma}\mathbf{P}(x_t) \quad \text{for } t = 2, 3, \ldots, T;$$
$$L_T = \alpha_T \mathbf{1}'.$$

The number of operations involved in this computation is of order Tm^2, which is a great improvement on the $O(Tm^T)$ operations required to evaluate L_T using (12.3). There is still the minor complication that the recursion given earlier often results in numerical underflow, but that can be remedied as follows. For each t, scale up the vector α_t so that its m components add to 1, keep track of the sum of the logs of all the scale factors thus applied, and adjust the resulting value of the log-likelihood by this sum.

The recursive algorithm given earlier is called the forward algorithm and the m components of the vector α_t are called the forward probabilities, the jth component being the probability $\Pr(\mathbf{X}^{(t)} = \mathbf{x}^{(t)}, C_t = j)$. The algorithm, which requires a single pass through the data, is fast enough to enable us to apply direct numerical maximization to estimate the parameters of the model.

12.3.2 Marginal and Joint Distributions

The matrix expression for the likelihood provides a simple means of computing various marginal and other distributions of interest. For instance, for a stationary Poisson–HMM (i.e., if $\delta^* = \delta$) it follows that

1. $\Pr(X_t = x) = \delta\mathbf{P}(x)\mathbf{1}'$;
2. $\Pr(X_t = v, X_{t+k} = w) = \delta\mathbf{P}(v)\mathbf{\Gamma}^k\mathbf{P}(w)\mathbf{1}'$;
3. if $\mathbf{X}^{(-t)}$ denotes the observations at all times $\in \{1, 2, \ldots, T\}$ other than t, which lies strictly between 1 and T,

$$\Pr(\mathbf{X}^{(-t)} = \mathbf{x}^{(-t)}) = \delta\mathbf{P}(x_1)\mathbf{\Gamma}\mathbf{P}(x_2)\cdots\mathbf{\Gamma}\mathbf{P}(x_{t-1})\mathbf{\Gamma}^2\mathbf{P}(x_{t+1})\cdots\mathbf{\Gamma}\mathbf{P}(x_T)\mathbf{1}';$$

4. if a and b are nonnegative integers and t lies strictly between 1 and T,

$$\Pr(\mathbf{X}^{(-t)} = \mathbf{x}^{(-t)}, a \le X_t \le b) = \delta\mathbf{P}(x_1)\left(\prod_{k=2}^{t-1}\mathbf{\Gamma}\mathbf{P}(x_k)\right)\mathbf{\Gamma}\left(\sum_{x_t=a}^{b}\mathbf{P}(x_t)\right)\left(\prod_{k=t+1}^{T}\mathbf{\Gamma}\mathbf{P}(x_k)\right)\mathbf{1}'.$$

Statements 3 and 4 generalize in straightforward fashion to cases in which there are several missing observations, or several interval-censored observations, or both.

Statement 4 can be shown by summing the full likelihood

$$L_T = \delta\mathbf{P}(x_1)\left(\prod_{k=2}^{T}\mathbf{\Gamma}\mathbf{P}(x_k)\right)\mathbf{1}'$$

over the appropriate range of x_t-values. Statement 3 then follows from $\sum_{x_t=0}^{\infty} \mathbf{P}(x_t) = \mathbf{I}_m$. The probability $\Pr(\mathbf{X}^{(-t)} = \mathbf{x}^{(-t)})$ is an example of what Little (2009, p. 411) terms an "ignorable likelihood," which can be used to estimate parameters when the missingness mechanism is ignorable. Similarly, likelihoods of the form shown in statement 4 can be used to estimate parameters when there is interval censoring but one can assume that the censoring mechanism is ignorable.

12.3.3 Moments

The moments and autocorrelation function (ACF) of $\{X_t\}$ are also available. Let the distribution p_i have mean μ_i and variance σ_i^2 (both finite) and let $\mathbf{M}$ be the diagonal matrix having $\boldsymbol{\mu} = (\mu_1, \mu_2, \ldots, \mu_m)$ on the diagonal. Then if the Markov chain is stationary, we have the following:

- $\mathrm{E}(X_t) = \boldsymbol{\delta\mu}'$;
- $\mathrm{Var}(X_t) = \sum_{i=1}^{m} \delta_i(\sigma_i^2 + \mu_i^2) - (\boldsymbol{\delta\mu}')^2$;
- for positive integers k, $\mathrm{E}(f(X_t, X_{t+k})) = \sum_{i,j=1}^{m} \delta_i\, (\boldsymbol{\Gamma}^k)_{ij} f(i,j)$;
- for positive integers k, $\mathrm{Cov}(X_t, X_{t+k}) = \boldsymbol{\delta}\mathbf{M}\boldsymbol{\Gamma}^k\boldsymbol{\mu}' - (\boldsymbol{\delta\mu}')^2$;
- for positive integers k, the ACF of $\{X_t\}$ is given by

$$\rho_k = \mathrm{Corr}(X_t, X_{t+k}) = \frac{\boldsymbol{\delta}\mathbf{M}\boldsymbol{\Gamma}^k\boldsymbol{\mu}' - (\boldsymbol{\delta\mu}')^2}{\mathrm{Var}(X_t)}.$$

If the eigenvalues of $\boldsymbol{\Gamma}$ are distinct, or more generally if $\boldsymbol{\Gamma}$ is diagonalizable, then the ACF is a linear combination of the kth powers of the eigenvalues other than 1.

If the state-dependent distributions are all Poisson, it follows that

$$\mathrm{Var}(X_t) = \mathrm{E}(X_t) + \sum_{i<j} \delta_i\delta_j(\mu_i - \mu_j)^2,$$

which displays the overdispersion of the marginal distribution of X_t relative to the Poisson. A Poisson–HMM can, therefore, represent both overdispersion and serial dependence in a series of unbounded counts. The variance and covariance of a two-state Poisson–HMM are

$$\mathrm{Var}(X_t) = \delta_1\mu_1 + \delta_2\mu_2 + \delta_1\delta_2(\mu_1 - \mu_2)^2$$

and, for all positive integers k,

$$\mathrm{Cov}(X_t, X_{t+k}) = \delta_1\delta_2(\mu_1 - \mu_2)^2(1 - \gamma_{12} - \gamma_{21})^k.$$

The ACF is then of the form Aw^k, where (apart from degenerate cases) $A \in (0,1)$ and $w \in (-1,1)$.

12.4 Parameter Estimation by Maximum Likelihood

We describe two methods for computing maximum likelihood estimates of the parameters of an HMM: numerical maximization of likelihood by a general-purpose optimizer and the expectation-maximization (EM) algorithm. For Bayesian estimation, see Scott (2002), Frühwirth-Schnatter (2006), Rydén (2008), or Zucchini and MacDonald (2009, Chapter 7).

12.4.1 Maximum Likelihood via Direct Numerical Maximization

Since likelihood evaluation is straightforward and requires only $O(Tm^2)$ operations, one obvious route to maximum likelihood estimates is numerical maximization by means of a general-purpose numerical optimizer. There are of course constraints on the parameters which the optimization must respect. For an m-state Poisson–HMM there are nonnegativity constraints, both on the m state–dependent means λ_i and on the transition probabilities γ_{ij}, and m row-sum constraints, $\sum_{j=1}^{m} \gamma_{ij} = 1$. Ignoring constraints can cause the optimizer to fail.

One strategy is to use a *constrained* optimizer such as the NAG routine E04UCF or `constrOptim` in R. However, it has been our experience that, in the case of HMMs, convergence is usually faster if one uses an *unconstrained* optimizer. The necessary constraints can easily be imposed indirectly by reparametrizing the model in terms of unconstrained parameters. As an illustration we give a mapping that transforms the constrained "natural parameters" λ_i and γ_{ij} of a (stationary) m-state Poisson–HMM to unconstrained "working parameters" η_i and τ_{ij}. First, for $i = 1, 2, \ldots, m$, set $\eta_i = \ln(\lambda_i)$. Second, for $i \neq j$ set

$$\tau_{ij} = \ln\left(\gamma_{ij}/(1 - \sum_{k\neq i} \gamma_{ik})\right) = \ln\left(\gamma_{ij}/\gamma_{ii}\right).$$

The transformations in the opposite direction are $\lambda_i = \exp(\eta_i)$ and

$$\gamma_{ij} = \exp(\tau_{ij})/\left(1 + \sum_{k\neq i} \exp(\tau_{ik})\right).$$

Given any values of the working parameters, we can find the corresponding natural parameters and evaluate the (log-)likelihood. It is, therefore, possible to maximize the log-likelihood as a function of the working parameters by using an unconstrained optimizer; the maximizing values of the natural parameters are then easily deduced. This process does confine the estimates of the natural parameters to the interior of the parameter space. Put differently, it will not result in transition probabilities or state-dependent means that are *exactly* zero.

An advantage of direct numerical maximization is that the model being fitted can be changed without this causing much change to the estimation procedure. The general matrix expression (12.4) for the likelihood can be used for univariate or multivariate observations, for models with or without covariates, for models in which not all the state-dependent distributions are of the same type, etc.

12.4.2 Maximum Likelihood via the EM Algorithm

There is a strong historical link between HMMs and the EM algorithm, as the Baum–Welch algorithm for finding maximum likelihood estimators in such a model is an important forerunner and special case of EM (Dempster et al., 1977; Welch, 2003). The EM algorithm is therefore regarded by many as the "method of choice" for finding maximum likelihood estimates in HMMs. Indeed, in a very thorough comparative review of the use of EM and Markov chain Monte Carlo in HMMs (Rydén, 2008), there appears to be no mention of direct numerical maximization of likelihood as an alternative to EM in finding MLEs.

We now describe the application of the EM algorithm to an HMM. We do not assume here that the Markov chain is stationary, but we indicate in Section 12.4.3 the modification that will be needed if that assumption is made. In order to apply the EM algorithm to an HMM we need both the forward probabilities (defined in Section 12.3.1) and the "backward probabilities"

$$\beta_t(j) = \Pr(X_{t+1} = x_{t+1}, \ldots, X_T = x_T \mid C_t = j).$$

The latter probabilities are found by the following backward recursion:

$$\boldsymbol{\beta}_T = \mathbf{1},$$
$$\boldsymbol{\beta}'_t = \boldsymbol{\Gamma}\mathbf{P}(x_{t+1})\boldsymbol{\beta}'_{t+1}, \qquad \text{for } t = T-1, T-2, \ldots, 1.$$

For the EM algorithm, we need the complete-data log-likelihood (CDLL), $\Pr(\mathbf{x}^{(T)}, \mathbf{c}^{(T)})$. Defining

$$u_j(t) = 1 \text{ if and only if } c_t = j, \quad (t = 1, 2, \ldots, T)$$

and

$$v_{jk}(t) = 1 \text{ if and only if } c_{t-1} = j \text{ and } c_t = k \quad (t = 2, 3, \ldots, T),$$

we can show that the CDLL of the model is

$$\ln\left(\Pr(\mathbf{x}^{(T)}, \mathbf{c}^{(T)})\right)$$
$$= \sum_{j=1}^{m} u_j(1) \ln \delta_j^* + \sum_{j=1}^{m}\sum_{k=1}^{m}\left(\sum_{t=2}^{T} v_{jk}(t)\right) \ln \gamma_{jk} + \sum_{j=1}^{m}\sum_{t=1}^{T} u_j(t) \ln p_j(x_t) \tag{12.5}$$
$$= \text{term 1} + \text{term 2} + \text{term 3}.$$

The EM algorithm for HMMs is as follows.

- **E step** In the CDLL, replace the quantities $v_{jk}(t)$ and $u_j(t)$ by their conditional expectations given the observations $\mathbf{x}^{(T)}$ and given the current parameter estimates:

$$\widehat{u}_j(t) = \Pr(C_t = j \mid \mathbf{x}^{(T)}) = \alpha_t(j)\beta_t(j)/L_T; \tag{12.6}$$

and

$$\widehat{v}_{jk}(t) = \Pr(C_{t-1} = j, C_t = k \mid \mathbf{x}^{(T)}) = \alpha_{t-1}(j)\, \gamma_{jk}\, p_k(x_t)\, \beta_t(k)/L_T. \tag{12.7}$$

- **M step** Having replaced $v_{jk}(t)$ and $u_j(t)$ by $\widehat{v}_{jk}(t)$ and $\widehat{u}_j(t)$, maximize the CDLL, expression (12.5), with respect to the three sets of parameters: the initial distribution $\boldsymbol{\delta}^*$, the t.p.m. $\boldsymbol{\Gamma}$, and the parameters of the state-dependent distributions (e.g., $\lambda_1, \ldots, \lambda_m$ in the case of a simple Poisson–HMM).

Examination of (12.5) reveals that the M step splits here into three separate maximizations.

1. Set $\delta_j^* = \widehat{u}_j(1)/\sum_{j=1}^{m} \widehat{u}_j(1) = \widehat{u}_j(1)$.
2. Set $\gamma_{jk} = f_{jk}/\sum_{k=1}^{m} f_{jk}$, where $f_{jk} = \sum_{t=2}^{T} \widehat{v}_{jk}(t)$.
3. The maximization of the third term may be easy or difficult, depending on the nature of the state-dependent distributions assumed. It is essentially the standard problem of maximum likelihood estimation for the distributions concerned. In the case of Poisson and normal distributions, closed-form solutions are available. In some other cases, for example, the gamma distributions and the negative binomial, numerical maximization—or some modification of EM—will be necessary to carry out this part of the M step.

One starts by giving initial estimates of the model parameters. The E and M steps are then repeated until convergence is achieved. As in the case of likelihood evaluation via the forward recursion, precautions have to be taken to avoid underflow.

12.4.3 Remarks

It seems clear from Sections 12.4.1 and 12.4.2 that, for such an HMM, direct numerical maximization of the observed data likelihood is at least conceptually simpler than EM. To carry out the former we need only a likelihood evaluator and a general-purpose optimizer such as `nlm` in R. But in order to apply EM, we first compute the forward probabilities (which are all that is needed to evaluate the likelihood) and then do considerably more, including computation of the backward probabilities and the quantities $\widehat{u}_j(t)$, $\widehat{v}_{jk}(t)$, and f_{jk}. In both methods, it is possible to become trapped in a local maximum which is not the global maximum, although EM seems less likely to fail in this way than direct numerical maximization; see Bulla and Berzel (2008, Table 2). However, in HMMs there are in our view several advantages of numerical maximization over EM, as there seem to be in some other contexts as well; see MacDonald (2014).

The assumption of stationarity often seems appropriate in time series applications. However, for HMMs fitted by EM it is almost never assumed that the underlying (homogeneous) Markov chain is stationary, that is, that $\boldsymbol{\delta}^* = \boldsymbol{\delta}$. (The MLE of $\boldsymbol{\delta}^*$ then turns out to be a unit vector: one element is 1 and the others 0.) One obvious reason for not assuming stationarity is convenience. For stationary series, there is no explicit formula for the matrix $\boldsymbol{\Gamma}$ which maximizes term 1 + term 2 of the CDLL in the M step; see, for example, Bulla and Berzel (2008, Section 2.3). It is not difficult to carry out the maximization numerically, but that implies a numerical optimization within each M step, that is, a maximization loop nested within the EM loop. There is no such complication if direct numerical maximization of the observed data likelihood is used; only one numerical optimization is needed, irrespective of whether stationarity is assumed.

Second, in models of relatively complex structure, for example, those of Zucchini et al. (2008), it is a clear advantage of direct numerical maximization that we do not have to derive and code the E and M steps, only a likelihood evaluator. This advantage is particularly marked if details of the model are repeatedly modified in a search for suitable structure.

12.5 Forecasting and Decoding

The matrix product expression (12.4) for the likelihood makes it possible to compute various conditional distributions associated with the observations X_t and latent variables C_t. We list several of these distributions here and discuss their uses in forecasting and "decoding." We do not assume in this section that the Markov chain is stationary.

12.5.1 Forecast Distributions

First note that the forecast distribution (for h periods ahead) is a ratio of likelihoods:

$$\Pr(X_{T+h} = x \mid \mathbf{X}^{(T)} = \mathbf{x}^{(T)}) = \frac{\boldsymbol{\alpha}_T \boldsymbol{\Gamma}^h \mathbf{P}(x)\mathbf{1}'}{L_T}.$$

Then write $\boldsymbol{\phi}_T = \boldsymbol{\alpha}_T / \boldsymbol{\alpha}_T \mathbf{1}' = \boldsymbol{\alpha}_T / L_T$, getting

$$\Pr(X_{T+h} = x \mid \mathbf{X}^{(T)} = \mathbf{x}^{(T)}) = \boldsymbol{\phi}_T \boldsymbol{\Gamma}^h \mathbf{P}(x)\mathbf{1}', \tag{12.8}$$

which is a mixture of the m state–dependent distributions p_i. As h increases, $\boldsymbol{\Gamma}^h$ approaches $\mathbf{1}'\boldsymbol{\delta}$ and the probability (12.8) therefore tends to $\boldsymbol{\delta}\mathbf{P}(x)\mathbf{1}'$. The speed of this convergence is determined by the second largest eigenvalue modulus of $\boldsymbol{\Gamma}$.

12.5.2 State Probabilities, State Prediction, and Decoding

It can be shown that, for $t = 1, 2, \ldots, T$,

$$\alpha_t(i)\beta_t(i) = \Pr(\mathbf{X}^{(T)} = \mathbf{x}^{(T)}, C_t = i),$$

from which it follows that

$$\Pr(C_t = i \mid \mathbf{X}^{(T)} = \mathbf{x}^{(T)}) = \frac{\alpha_t(i)\beta_t(i)}{L_T}. \tag{12.9}$$

This gives the conditional distribution (under the model) of the state at time t, given all the observations.

Similarly, it can be shown that, for $h \in \mathcal{N}$,

$$\Pr(C_{T+h} = i \mid \mathbf{X}^{(T)} = \mathbf{x}^{(T)}) = \boldsymbol{\alpha}_T \boldsymbol{\Gamma}^h \mathbf{e}_i' / L_T = \boldsymbol{\phi}_T \boldsymbol{\Gamma}^h \mathbf{e}_i', \tag{12.10}$$

where $\mathbf{e}_i = (0, \ldots, 0, 1, 0, \ldots, 0)$ has a one in the ith position only. This enables us to forecast the (latent) state h periods ahead. If h tends to infinity the probability (12.10) tends to δ_i:

$$\lim_{h \to \infty} \boldsymbol{\phi}_T \boldsymbol{\Gamma}^h \mathbf{e}_i' = \boldsymbol{\phi}_T \mathbf{1}' \boldsymbol{\delta} \mathbf{e}_i' = 1 \times \boldsymbol{\delta} \mathbf{e}_i' = \delta_i.$$

The speed of convergence is once again determined by the second largest eigenvalue modulus of $\boldsymbol{\Gamma}$.

By using Equation (12.9) we can find that state $i \in \{1, 2, \ldots, m\}$ which is at time t the most likely, given all the observations. This is known as local decoding. If we do that for all times t, the resulting path may in some circumstances be useful, but it is not in general the most likely path followed by the Markov chain, that is, it is not in general that sequence of states $c_1, c_2, \ldots, c_T$ which maximizes the conditional probability

$$\Pr(\mathbf{C}^{(T)} = \mathbf{c}^{(T)} \mid \mathbf{X}^{(T)} = \mathbf{x}^{(T)}).$$

That maximizing path (which need not be unique) can be found by an application of dynamic programming known as the Viterbi algorithm (Viterbi, 1967, 2006), and the process is known as global decoding. The results of local and global decoding are often very similar but not identical.

12.6 Model Selection and Checking

A question that often arises when we seek to use an HMM in practice is: How many states should there be? A common question of another type is: Should we use, for example, Poisson state–dependent distributions or negative binomial? In the absence of useful subject-matter information, model selection questions such as these are not easy to answer completely satisfactorily, but there are some simple tools that can certainly help us to select from a group of competing models. An introduction to model selection may be found in Zucchini (2000). But once we have chosen the "best" model according to some criterion, we still have to consider the question of whether the chosen model can be regarded as adequate. In this section, we describe the model selection criteria: Akaike's information criterion (AIC) and Bayesian information criterion (BIC) and the use of pseudo-residuals (also known as quantile residuals) in order to check the adequacy of the model.

12.6.1 AIC and BIC

Two widely used model selection criteria are the AIC and the BIC. The AIC selects that model which, of those under consideration, has the smallest value of the quantity

$$\text{AIC} = -2 \ln L + 2p.$$

Here, $\ln L$ is the log-likelihood of the fitted model and p denotes the number of parameters of the model. The first term measures the lack of fit, and the second term is a penalty which increases with the number of parameters. The criterion BIC differs from AIC in the penalty term only:

$$\text{BIC} = -2\ln L + p\ln T.$$

The penalty term of BIC is greater than that of AIC if $T > e^2$, that is, for $T \geq 8$. Thus, the BIC generally tends to select models with fewer parameters than does the AIC.

12.6.2 Model Checking by Pseudo-Residuals

There are simple informal checks that can be made on the adequacy of an HMM. One can (assuming stationarity) compare sample and model quantities such as the mean, variance, ACF, and the distribution of the observations. If, for example, the data were to display marked overdispersion but the model did not, we would discard or modify the model.

But there are additional systematic checks that can be performed. We describe here the use of "pseudo-residuals." For $t = 1, 2, \ldots, T$, define

$$u_t^- = \Pr(X_t < x_t \mid \mathbf{X}^{(-t)} = \mathbf{x}^{(-t)}), \quad z_t^- = \Phi^{-1}(u_t^-)$$

and

$$u_t^+ = \Pr(X_t \leq x_t \mid \mathbf{X}^{(-t)} = \mathbf{x}^{(-t)}), \quad z_t^+ = \Phi^{-1}(u_t^+),$$

with Φ denoting the standard normal distribution function. The interval $[u_t^-, u_t^+]$ (on a probability scale) or $[z_t^-, z_t^+]$ (on a "normal" scale) gives an indication of how extreme x_t is relative to its conditional distribution given the other observations. The conditional distribution of one observation given the others is therefore needed, but it is a ratio of likelihoods and can be found in very much the same way as were the conditional distributions in Section 12.5.1.

There are several ways in which such pseudo-residual "segments" $[z_t^-, z_t^+]$ can be used. We give one here. If the observations were continuous there would be a single quantity z_t and not an interval, which would have a standard normal distribution if the model were correct, and so a quantile–quantile (QQ) plot could be used to assess the adequacy of the model. For discrete-valued series, we use the "mid-pseudo-residual" $z_t^m = \Phi^{-1}((u_t^- + u_t^+)/2)$ to sort the pseudo-residuals in order to produce the QQ plot. Although we can claim no more than approximate normality for mid-pseudo-residuals, they are nevertheless useful for identifying poor fits. An example of their application is given in Section 12.7; see Figure 12.5.

12.7 An Example: Weekly Sales of a Soap Product

We describe here the fitting of HMMs to a series of counts and demonstrate briefly the use of the forecasting, decoding, model selection, and checking techniques introduced in Sections 12.5 and 12.6. Consider the series of weekly sales (in integer units, 242 weeks) of a

TABLE 12.1

Weekly Sales of the Soap Product; to Be Read across Rows

1	6	9	18	14	8	8	1	6	7	3	3	1	3	4	12	8	10	8	2
17	15	7	12	22	10	4	7	5	0	2	5	3	4	4	7	5	6	1	3
4	5	3	7	3	0	4	5	3	3	4	4	4	4	4	3	5	5	5	7
4	0	4	3	2	6	3	8	9	6	3	4	3	3	3	3	2	1	4	5
5	2	7	5	2	3	1	3	4	6	8	8	5	7	2	4	2	7	4	15
15	12	21	20	13	9	8	0	13	9	8	0	6	2	0	3	2	4	4	6
3	2	5	5	3	2	1	1	3	1	2	6	2	7	3	2	4	1	5	6
8	14	5	3	6	5	11	4	5	9	9	7	9	8	3	4	8	6	3	5
6	3	1	7	4	9	2	6	6	4	6	6	13	7	4	8	6	4	4	4
9	2	9	2	2	2	13	13	4	5	1	4	6	5	4	2	3	10	6	15
5	9	9	7	4	4	2	4	2	3	8	15	0	0	3	4	3	4	7	5
7	6	0	6	4	14	5	1	6	5	5	4	9	4	14	2	2	1	5	2
6	4																		

particular soap product in a supermarket, as shown in Table 12.1 and Figure 12.3. The data were downloaded in 2007 from http://gsbwww.uchicago.edu/kilts/research/db/dominicks, the Dominick's Finer Food database at the Kilts Center for Marketing, Graduate School of Business of the University of Chicago.* (The product was "Zest White Water 15 oz.," with code 3700031165 and store number 67.)

The data display considerable overdispersion relative to Poisson: the sample mean is 5.44, the sample variance 15.40. In addition, the sample ACF (Figure 12.4) shows strong evidence of serial dependence. Any satisfactory model has to mimic these properties.

Stationary Poisson–HMMs with one to four states were fitted to the soap sales by direct numerical maximization of the log-likelihood; the number of parameters, minus log-likelihood, AIC and BIC values are shown in Table 12.2. Also given are the results for two- to four-state models in which stationarity is not assumed. A one-state model (a sequence of independent Poisson random variables) shows up as completely inadequate,

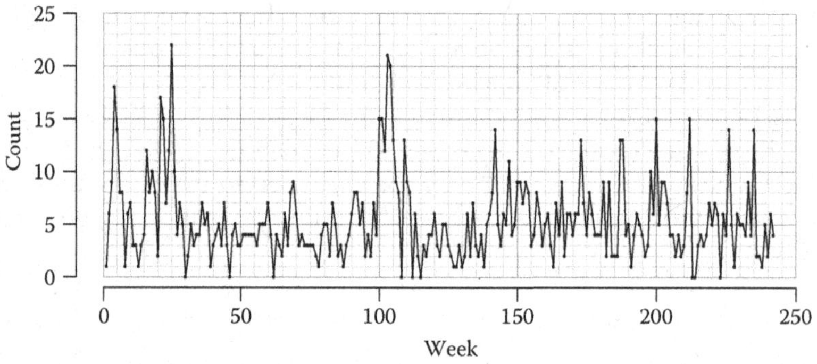

FIGURE 12.3
Series of weekly sales of the soap product.

* That database is now (April 28, 2015) at http://research.chicagobooth.edu/kilts/marketing-databases/dominicks (University of Chicago Booth School of Business, 2015).

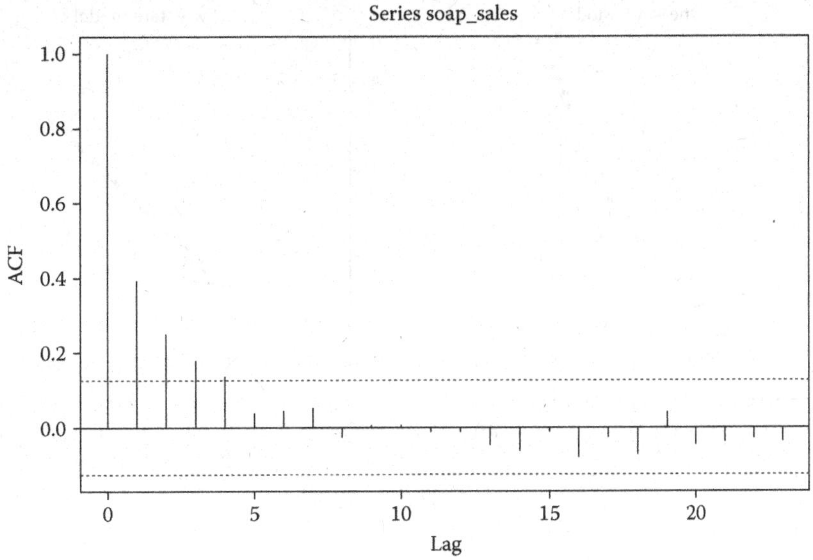

FIGURE 12.4
Autocorrelation function of the weekly soap sales.

TABLE 12.2
Soap Sales Data: Comparison, by AIC and BIC, of Poisson–HMMs with 1–4 States

Type of Model	No. of States	No. of Parameters	$-\ln L$	AIC	BIC
Stationary	1	1	711.8	1425.5	1429.0
	2	4	618.7	1245.3	**1259.3**
	3	9	610.5	**1239.0**	1270.4
	4	16	604.2	1240.4	1296.2
Nonstationary	2	5	618.5	1246.9	**1264.4**
	3	11	610.2	**1242.4**	1280.8
	4	19	602.8	1243.5	1309.8

Note: Bold type indicates the relevant minima.

but this is not surprising in view of the overdispersion and autocorrelation already noted. For both the stationary and the nonstationary case, the AIC is minimized by a three-state model and the BIC by a two-state model.

Figure 12.5 displays QQ plots of (mid-)quantile residuals for stationary models with 1–3 states. The plots indicate that the one-state model is clearly inferior to the two- and three-state models and that there is little to choose between the other two.

In passing, it is interesting to note that, in the three- and four-state cases, it is not difficult to find, both by direct numerical maximization and by EM, a local maximum of the log-likelihood that is not a global maximum. This phenomenon underscores the importance of trying several sets of starting values for the iterations.

The best stationary three-state model found has state-dependent means given by $\widehat{\lambda} = (3.74, 8.44, 14.93)$, t.p.m.

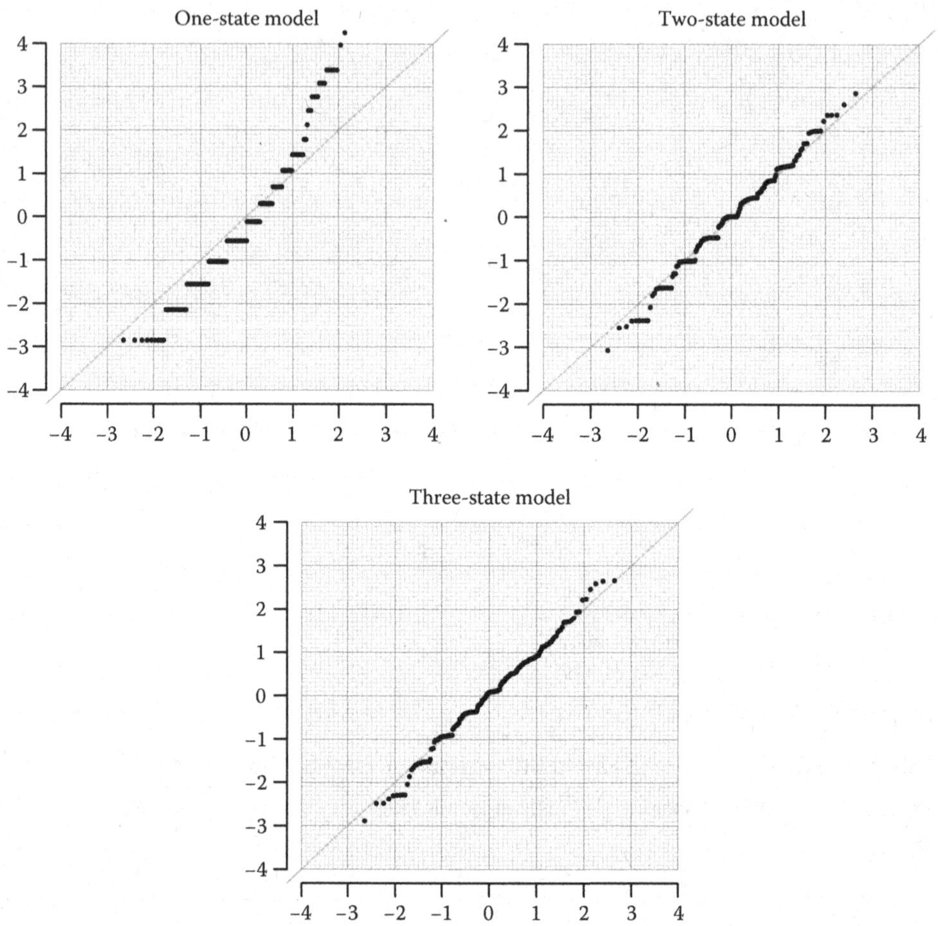

FIGURE 12.5
Soap sales: normal QQ plot of the (mid-)quantile residuals for the one-, two- and three-state stationary models.

$$\widehat{\boldsymbol{\Gamma}} = \begin{pmatrix} 0.864 & 0.117 & 0.019 \\ 0.445 & 0.538 & 0.017 \\ 0.000 & 0.298 & 0.702 \end{pmatrix},$$

and stationary distribution $\widehat{\boldsymbol{\delta}} = (0.722, 0.220, 0.058)$.

The mean (5.42) and variance (14.72) implied by the model certainly reflect the observed overdispersion. The implied ACF (see Section 12.3.3) is given by

$$\widehat{\rho}_k = 0.5392 \times 0.6823^k + 0.0926 \times 0.4220^k.$$

(The non-unit eigenvalues of $\widehat{\boldsymbol{\Gamma}}$ are 0.6823 and 0.4220.) This ACF is close to the sample ACF for the first four lags; see Table 12.3.

Global decoding and local decoding were carried out, and the results are shown in Figure 12.6. The 7 weeks (out of 242) in which global decoding and local decoding differ are indicated there.

TABLE 12.3
Three-State Stationary Poisson–HMM for Soap Sales Data: Comparison of Sample and Model Autocorrelations

Lag	1	2	3	4	5	6
Sample ACF	0.392	0.250	0.178	0.136	0.038	0.044
Model ACF	0.407	0.268	0.178	0.120	0.081	0.055

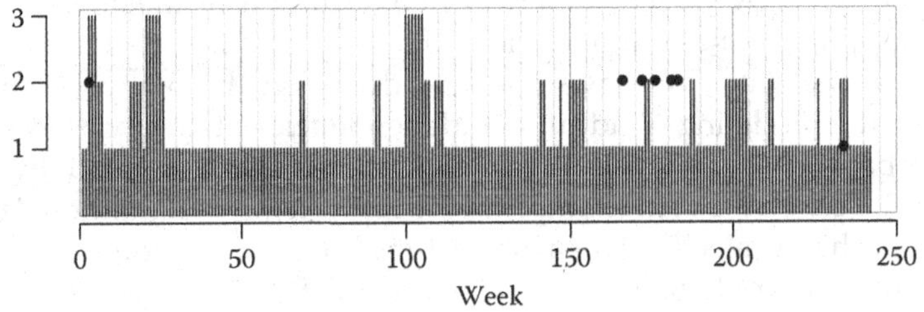

FIGURE 12.6
Global decoding of soap sales: the sequence of states, that is, *a posteriori* the most likely. The black dots indicate the seven occasions on which local decoding led to different states being identified as most likely.

Figure 12.7 displays forecast distributions under this model for weekly sales 1 and 2 weeks into the future, plus the corresponding stationary distribution.

Finally, under this model the state prediction probabilities for the next 3 weeks, compared to the estimated stationary distribution $\widehat{\delta}$, are indicated here.

Week	State 1	2	3
243	0.844	0.138	0.019
244	0.790	0.178	0.031
245	0.763	0.198	0.040
$\widehat{\delta}$	0.722	0.220	0.058

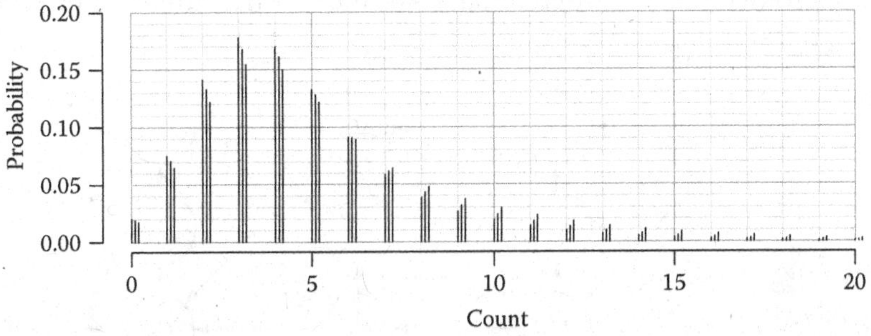

FIGURE 12.7
Forecast distributions for weekly counts of soap sales: one-step-ahead (left vertical lines), two-step-ahead (middle vertical lines), stationary distribution (right vertical lines).

In this case, the convergence to the stationary distribution is relatively fast, which is not surprising as the second largest eigenvalue of $\widehat{\mathbf{\Gamma}}$ (0.6823) is not close to 1.

12.8 Extensions and Concluding Remarks

One of the principal advantages of the use of HMMs as time series models, in particular if they are fitted by direct numerical maximization of likelihood, is the ease of extending or adapting the basic models in order to accommodate known or suspected special features of the data. We have not here dwelt on the many variations that are possible, such as the modeling of additional dependencies at observation level, at latent process level, or between these levels. A selection of possibilities is given by Zucchini and MacDonald (2009, Section 8.6). An example of the last category of additional dependencies is the model of Zucchini et al. (2008) for a binary time series $\{X_t\}$ of animal feeding behavior, which is depicted in Figure 12.8. In that model only the feeding behavior $\{X_t\}$ is observed, and the "nutrient levels" $\{N_t\}$ (an exponentially smoothed version of feeding behavior) are permitted to influence the transition probabilities governing the latent process $\{C_t\}$.

Other important topics not discussed in this chapter, or described only briefly, are the use of HMMs as models for longitudinal data, that is, multiple time series; the incorporation of covariates; the use of Bayesian estimation methods; the structuring of the t.p.m. to reduce the number of parameters required by an HMM; and the construction of HMMs that (accurately) approximate less tractable models having a continuous-valued latent Markov process.

For HMMs as models for longitudinal data, see Altman (2007), Maruotti (2011, 2015), Schliehe-Diecks et al. (2012), and Bartolucci et al. (2013). For examples of models with covariates, see Zucchini and MacDonald (2009, Chapter 14). For Bayesian methods, see the works cited in Section 12.4. For structuring of the transition probability matrix, see, for example, Cooper and Lipsitch (2004) and Langrock (2011). For discretization of continuous-valued latent processes and the resulting application of HMM methods, see Langrock (2011).

To conclude, we suggest that many discrete-valued time series can be usefully modeled by HMMs or variations thereof, and the models relatively easily fitted by direct numerical maximization of likelihood; EM and Bayesian methods are obvious alternatives. The unity—across various types of discrete data—of model structure, and of techniques for

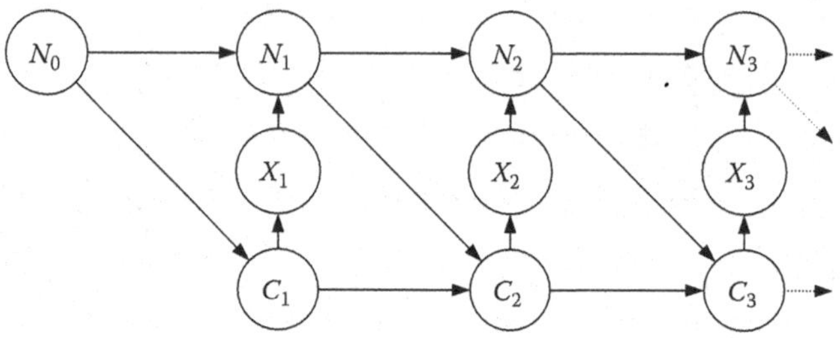

FIGURE 12.8
Directed graph of animal feeding model of Zucchini et al. (2008).

estimation, forecasting, and diagnostic checking, makes HMMs a promising set of models for a wide variety of discrete-valued time series.

Acknowledgments

The James M. Kilts Center, University of Chicago Booth School of Business, is thanked for making available the data analyzed in Section 12.7. The reviewer is thanked for constructive comments and suggestions.

References

Altman, R. M. (2007). Mixed hidden Markov models: An extension of the hidden Markov model to the longitudinal data setting. *Journal of the American Statistical Association*, 102:201–210.

Bartolucci, F., Farcomeni, A., and Pennoni, F. (2013). *Latent Markov Models for Longitudinal Data*. Chapman & Hall/CRC Press, Boca Raton, FL.

Baum, L. E., Petrie, T., Soules, G., and Weiss, N. (1970). A maximization technique occurring in the statistical analysis of probabilistic functions of Markov chains. *Annals of Mathematical Statistics*, 41:164–171.

Bulla, J. and Berzel, A. (2008). Computational issues in parameter estimation for stationary hidden Markov models. *Computational Statistics*, 23:1–18.

Churchill, G. A. (1989). Stochastic models for heterogeneous DNA sequences. *Bulletin of Mathematical Biology*, 51:79–94.

Cooper, B. and Lipsitch, M. (2004). The analysis of hospital infection data using hidden Markov models. *Biostatistics*, 5:223–237.

Cox, D. R. (1981). Statistical analysis of time series: Some recent develpoments. *Scandivanian Journal of Statistics*, 8:93–115.

Dempster, A. P., Laird, N. M., and Rubin, D. B. (1977). Maximum likelihood from incomplete data via the EM algorithm (with discussion). *Journal of the Royal Statistical Society Series B*, 39:1–38.

Durbin, R., Eddy, S. R., Krogh, A., and Mitchison, G. (1998). *Biological Sequence Analysis: Probabilistic Models of Proteins and Nucleic Acids*. Cambridge University Press, Cambridge, U.K.

Frühwirth-Schnatter, S. (2006). *Finite Mixture and Markov Switching Models*. Springer, New York.

Juang, B. H. and Rabiner, L. R. (1991). Hidden Markov models for speech recognition. *Technometrics*, 33:251–272.

Langrock, R. (2011). Some applications of nonlinear and non-Gaussian state-space modelling by means of hidden Markov models. *Journal of Applied Statistics*, 38(12):2955–2970.

Leroux, B. G. and Puterman, M. L. (1992). Maximum-penalized-likelihood estimation for independent and Markov-dependent mixture models. *Biometrics*, 48(2):545–558.

Little, R. J. A. (2009). Selection and pattern-mixture models. In Fitzmaurice, G., Davidian, M., Verbeke, G., and Molenberghs, G., editors, *Longitudinal Data Analysis*, pp. 409–431. Chapman & Hall/CRC, Boca Raton, FL.

MacDonald, I. L. (2014). Numerical maximisation of likelihood: A neglected alternative to EM? *International Statistical Review*, 82(2):296–308.

Maruotti, A. (2011). Mixed hidden Markov models for longitudinal data: An overview. *International Statistical Review*, 79(3):427–454.

Maruotti, A. (2015). Handling non-ignorable dropouts in longitudinal data: A conditional model based on a latent Markov heterogeneity structure. *TEST*, 24:84–109.

Rabiner, L. R. (1989). A tutorial on hidden Markov models and selected applications in speech recognition. *IEEE Transactions on Information Theory*, 77(2):257–284.

Rydén, T. (2008). EM versus Markov chain Monte Carlo for estimation of hidden Markov models: A computational perspective (with discussion). *Bayesian Analysis*, 3(4):659–688.

Schliehe-Diecks, S., Kappeler, P., and Langrock, R. (2012). On the application of mixed hidden Markov models to multiple behavioural time series. *Interface Focus*, 2:180–189.

Scott, S. L. (2002). Bayesian methods for hidden Markov models: Recursive computing in the 21st century. *Journal of the American Statistical Association*, 97:337–351.

University of Chicago Booth School of Business (2015). http://research.chicagobooth.edu/kilts/marketing-databases/dominicks. Accessed date April 28, 2015.

Viterbi, A. J. (1967). Error bounds for convolutional codes and an asymptotically optimal decoding algorithm. *IEEE Transactions on Information Theory*, 13:260–269.

Viterbi, A. J. (2006). A personal history of the Viterbi algorithm. *IEEE Signal Processing Magazine*, 23:120–142.

Welch, L. R. (2003). Hidden Markov models and the Baum–Welch algorithm. *IEEE Information Theory Society Newsletter*, 53(1): 10–13.

Zucchini, W. (2000). An introduction to model selection. *Journal of Mathematical Psychology*, 44(1):41–61.

Zucchini, W. and MacDonald, I. L. (2009). *Hidden Markov Models for Time Series: An Introduction Using R*. Chapman & Hall/CRC, London, U.K./Boca Raton, FL.

Zucchini, W., Raubenheimer, D., and MacDonald, I. L. (2008). Modeling time series of animal behavior by means of a latent-state model with feedback. *Biometrics*, 64:807–815.

13

Spectral Analysis of Qualitative Time Series

David Stoffer

CONTENTS

13.1 Introduction

Qualitative-valued time series are frequently encountered in diverse applications such as economics, medicine, psychology, geophysics, and genomics, to mention a few. The fact that the data are categorical does not preclude the need to extract pertinent information in the same way that is done with quantitative-valued time series. One particular area that was neglected was the frequency domain, or spectral analysis, of categorical time series. In this chapter, we explore an approach based on scaling and the spectral envelope, which was introduced by Stoffer et al. (1993a).

First, we discuss the concept of scaling categorical variables, and then we use the idea to develop spectral analysis of qualitative time series. In doing so, the spectral envelope and optimal scaling are introduced, and their properties are discussed. Spectral envelope and the corresponding optimal scaling are a population idea; consequently, efficient estimation is presented. Pertinent theoretical results are also summarized. Examples of using the methodology on a DNA sequence are given. The examples include a piecewise analysis of a gene in the Epstein–Barr virus (EBV). Often, one collects qualitative-valued time series in experimental designs. This problem is also explored as we discuss the analysis of replicated series that depend on covariates. The spectral envelope is intimately associated with the concept of principal component analysis of time series, and this relationship is explored in a separate section. Finally, we list an R script that can be used to calculate the spectral envelope and optimal scalings. The script can also be used to perform the aforementioned dynamic analysis.

13.2 Scaling Categorical Time Series

Our work on the spectral envelope was motivated by collaborations with researchers who collected categorical-valued time series with an interest in the cyclic behavior of the data. For example, Table 13.1 shows the per minute sleep state of an infant taken from a study on the effects of prenatal exposure to alcohol. Details can be found in Stoffer et al. (1988), but briefly, an electroencephalographic (EEG) sleep recording of approximately 2 hours (h) is obtained on a full-term infant 24–36 h after birth, and the recording is scored by a pediatric neurologist for sleep state. There are two main types of sleep, non-rapid eye movement (NON-REM), also known as *quiet sleep* and rapid eye movement (REM), also known as *active sleep*. In addition, there are four stages of NON-REM (NR1–NR4), with NR1 being the "most active" of the four states, and finally awake (AW), which naturally occurs briefly through the night. This particular infant was never awake during the study.

It is not too difficult to notice a pattern in the data if one concentrates on REM versus NON-REM sleep states. But it would be difficult to try to assess patterns in a longer sequence, or if there were more categories, without some graphical aid. One simple method would be to *scale* the data, that is, *assign numerical values to the categories* and then draw a time plot of the scales. Since the states have an order,* one obvious scaling is as follows:

$$\mathtt{NR4} = 1, \quad \mathtt{NR3} = 2, \quad \mathtt{NR2} = 3, \quad \mathtt{NR1} = 4, \quad \mathtt{REM} = 5, \quad \mathtt{AW} = 6, \tag{13.1}$$

and Figure 13.1 (often referred to as a *hypnogram*) shows the time plot using this scaling. Another interesting scaling might be to combine the quiet states and the active states:

$$\mathtt{NR4} = \mathtt{NR3} = \mathtt{NR2} = \mathtt{NR1} = 0, \quad \mathtt{REM} = 1, \quad \mathtt{AW} = 2. \tag{13.2}$$

TABLE 13.1

Per Minute Infant EEG Sleep States (Read Down and Across)

REM	NR2	NR4	NR2	NR1	NR2	NR3	NR4	NR1	NR1	REM
REM	REM	NR4	NR1	NR1	NR2	NR4	NR4	NR1	NR1	REM
REM	REM	NR4	NR1	NR1	REM	NR4	NR4	NR1	NR1	REM
REM	NR3	NR4	NR1	REM	REM	NR4	NR4	NR1	NR1	REM
REM	NR4	NR4	NR1	REM	REM	NR4	NR4	NR1	NR1	REM
REM	NR4	NR4	NR1	REM	REM	NR4	NR4	NR1	NR1	REM
REM	NR4	NR4	NR2	REM	NR2	NR4	NR4	NR1	NR1	NR2
REM	NR4	NR4	REM	REM	NR2	NR4	NR4	NR1	REM	
NR2	NR4	NR4	NR1	REM	NR2	NR4	NR4	NR1	REM	
REM	NR2	NR4	NR1	REM	NR3	NR4	NR2	NR1	REM	

* The so-called "ordering" of sleep states is somewhat tenuous. For example, sleep does not progress through these stages in sequence. For a typical normal healthy adult, sleep begins in stage NR1 and progresses into stages NR2, NR3, and NR4. Sleep moves through these stages repeatedly before entering REM sleep. Moreover, sleep typically transitions between REM and stage NR2. Sleep cycles through these stages approximately four or five times throughout the night. On average, adults enter the REM stage approximately 90 min after falling asleep. The first cycle of REM sleep might last only a short amount of time, but each cycle becomes longer.

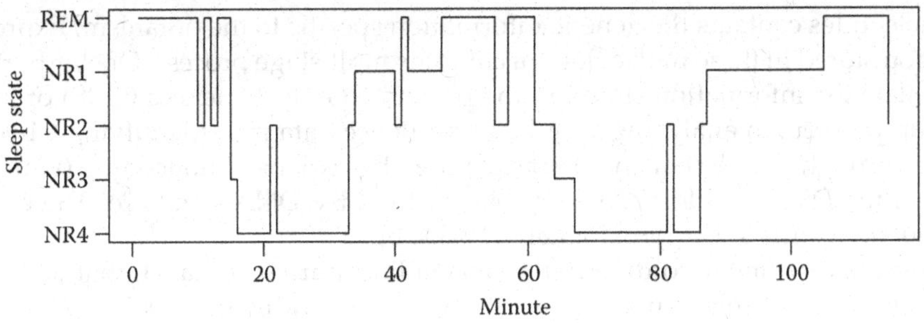

FIGURE 13.1
Time plot of the EEG sleep state data in Table 13.1 using the scaling in (13.1).

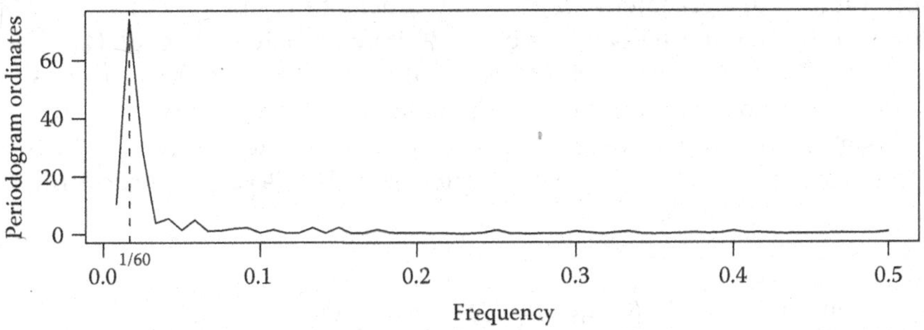

FIGURE 13.2
Periodogram of the EEG sleep state data in Table 13.1 based on the scaling in (13.1). The peak corresponds to a frequency of approximately one cycle every 60 min.

The time plot using (13.2) would be similar to Figure 13.1 as far as the cyclic (in and out of REM sleep) behavior of this infant's sleep pattern. Figure 13.2 shows the periodogram of the sleep data using the scaling in (13.1). Note that there is a large peak at the frequency corresponding to one cycle every 60 min. As one might imagine, the general appearance of the periodogram using the scaling (13.2) (not shown) is similar to Figure 13.2. Most of us would feel comfortable with this analysis even though we made an arbitrary and ad hoc choice about the particular scaling. It is evident from the data (without any scaling) that, if the interest is in infant's sleep cycling, this particular sleep study indicates that an infant cycles between REM and NON-REM sleep at a rate of about one cycle per hour.

The intuition used in the previous example is lost when one considers a long DNA sequence. Briefly, a DNA strand can be viewed as a long string of linked nucleotides. Each nucleotide is composed of a nitrogenous base, a five carbon sugar, and a phosphate group. There are four different bases that can be grouped by size, the pyrimidines, thymine (T) and cytosine (C), and the purines, adenine (A) and guanine (G). The nucleotides are linked together by a backbone of alternating sugar and phosphate groups with the 5′ carbon of one sugar linked to the 3′ carbon of the next, giving the string direction. DNA molecules occur naturally as a double helix composed of polynucleotide strands with the bases facing inward. The two strands are complementary, so it is sufficient to represent a DNA molecule by a sequence of bases on a single strand. Thus, a strand of DNA can be represented as a sequence of letters, termed base pairs (bp), from the finite alphabet {A, C, G, T}. The order of the nucleotides contains the genetic information specific to the organism. Expression of information stored in these molecules is a complex multistage process. One important task is to translate the information stored in the protein-coding sequences (CDS) of the DNA. A common problem in analyzing long DNA sequence data is in identifying CDS that are dispersed throughout the sequence and separated by regions of noncoding (which makes up most of the DNA). Table 13.2 shows part of the EBV DNA sequence. The entire EBV DNA sequence consists of approximately 172,000 bp.

One could try scaling according to the pyrimidine–purine alphabet, that is, $\mathtt{A} = \mathtt{G} = 0$ and $\mathtt{C} = \mathtt{T} = 1$, but this is not necessarily of interest for every CDS of EBV. There are numerous possible alphabets of interest, for example, one might focus on the strong–weak hydrogen bonding alphabet* $\mathtt{S} = \{\mathtt{C}, \mathtt{G}\} = 0$ and $\mathtt{W} = \{\mathtt{A}, \mathtt{T}\} = 1$. While model calculations as well as experimental data strongly agree that some kind of periodic signal exists in certain DNA sequences, there is a large disagreement about the exact type of periodicity. In addition, there is disagreement about which nucleotide alphabets are involved in the signals; for example, compare Ioshikhes et al. (1996) with Satchwell et al. (1986).

If we consider the naive approach of arbitrarily assigning numerical values (scales) to the categories and then proceeding with a spectral analysis, the result will depend on the particular assignment of numerical values. The obvious problem of being arbitrary is illustrated as follows: Suppose we observe the sequence ATCTACATG..., then setting $\mathtt{A} = \mathtt{G} = 0$

TABLE 13.2

Part of the Epstein–Barr Virus DNA Sequence (Read Across and Down)

AGAATTCGTC	TTGCTCTATT	CACCCTTACT	TTTCTTCTTG	CCCGTTCTCT	TTCTTAGTAT
GAATCCAGTA	TGCCTGCCTG	TAATTGTTGC	GCCCTACCTC	TTTTGGCTGG	CGGCTATTGC
CGCCTCGTGT	TTCACGGCCT	CAGTTAGTAC	CGTTGTGACC	GCCACCGGCT	TGGCCCTCTC
ACTTCTACTC	TTGGCAGCAG	TGGCCAGCTC	ATATGCCGCT	GCACAAAGGA	AACTGCTGAC
ACCGGTGACA	GTGCTTACTG	CGGTTGTCAC	TTGTGAGTAC	ACACGCACCA	TTTACAATGC
ATGATGTTCG	TGAGATTGAT	CTGTCTCTAA	CAGTTCACTT	CCTCTGCTTT	TCTCCTCAGT
CTTTGCAATT	TGCCTAACAT	GGAGGATTGA	GGACCCACCT	TTTAATTCTC	TTCTGTTTGC
ATTGCTGGCC	GCAGCTGGCG	GACTACAAGG	CATTTACGGT	TAGTGTGCCT	CTGTTATGAA
ATGCAGGTTT	GACTTCATAT	GTATGCCTTG	GCATGACGTC	AACTTTACTT	TTATTTCAGT
TCTGGTGATG	CTTGTGCTCC	TGATACTAGC	GTACAGAAGG	AGATGGCGCC	GTTTGACTGT
TTGTGGCGGC	ATCATGTTTT	TGGCATGTGT	ACTTGTCCTC	ATCGTCGACG	CTGTTTTGCA
GCTGAGTCCC	CTCCTTGGAG	CTGTAACTGT	GGTTTCCATG	ACGCTGCTGC	TACTGGCTTT
CGTCCTCTGG	CTCTCTTCGC	CAGGGGGCCT	AGGTACTCTT	GGTGCAGCCC	TTTTAACATT
GGCAGCAGGT	AAGCCACACG	TGTGACATTG	CTTGCCTTTT	TGCCACATGT	TTTCTGGACA
CAGGACTAAC	CATGCCATCT	CTGATTATAG	CTCTGGCACT	GCTAGCGTCA	CTGATTTTGG
GCACACTTAA	CTTGACTACA	ATGTTCCTTC	TCATGCTCCT	ATGGACACTT	GGTAAGTTTT
CCCTTCCTTT	AACTCATTAC	TTGTTCTTTT	GTAATCGCAG	CTCTAACTTG	GCATCTCTTT
TACAGTGGTT	CTCCTGATTT	GCTCTTCGTG	CTCTTCATGT	CCACTGAGCA	AGATCCTTCT
GGCACGACTG	TTCCTATATG	CTCTCGCACT	CTTGTTGCTA	GCCTCCGCGC	TAATCGCTGG
TGGCAGTATT	TTGCAAACAA	ACTTCAAGAG	TTTAAGCAGC	ACTGAATTTA	TACCCAGTGA

* S refers to guanine (G) or cytosine (C) for the *strong* hydrogen bonding interaction between the base pairs. W refers to adenine (A) or thymine (T) for the *weak* hydrogen bonding interaction between the base pairs.

and C = T = 1 yields the numerical sequence 011101010..., which is not very interesting. However, if we use the strong–weak bonding alphabet, W = {A, T} = 0 and S = {C, G} = 1, then the sequence becomes 001001001..., which is very interesting. It should be clear, then, that one does not want to focus on only one scaling. Instead, the focus should be on finding scalings that bring out all of the interesting features in the data. Rather than choose values arbitrarily, the spectral envelope approach selects scales that help emphasize any periodic feature that exists in a categorical time series of virtually any length in a quick and automated fashion. In addition, the technique can help in determining whether a sequence is merely a random assignment of categories.

13.3 Definition of Spectral Envelope

As a general description, the spectral envelope is a frequency-based, principal component technique applied to a multivariate time series. In this section, we will focus on the basic concept and its use in the analysis of categorical time series. Technical details can be found in Stoffer et al. (1993a).

Briefly, in establishing the spectral envelope for categorical time series, we addressed the basic question of how to efficiently discover periodic components in categorical time series. This was accomplished via nonparametric spectral analysis as follows. Let $\{X_t;\, t = 0, \pm 1, \pm 2, \ldots\}$ be a categorical-valued time series with finite state-space $\mathcal{C} = \{c_1, c_2, \ldots, c_{k+1}\}$. Assume that X_t is stationary and $p_j = \Pr\{X_t = c_j\} > 0$ for $j = 1, 2, \ldots, k+1$. For $\beta = (\beta_1, \beta_2, \ldots, \beta_{k+1})' \in \mathbb{R}^{k+1}$, denote by $X_t(\beta)$ the real-valued stationary time series corresponding to the scaling that assigns the category c_j the numerical value β_j, for $j = 1, 2, \ldots, k+1$. Our goal was to find scaling β so that the spectral density is in some sense interesting and to summarize the spectral information by what we called the spectral envelope.

We chose β to maximize the power (variance) at each frequency ω, across frequencies $\omega \in (-1/2, 1/2]$, relative to the total power $\sigma^2(\beta) = \mathrm{Var}\{X_t(\beta)\}$. That is, we chose $\beta(\omega)$, at each ω of interest, so that

$$\lambda(\omega) = \sup_{\beta} \left\{ \frac{f(\omega; \beta)}{\sigma^2(\beta)} \right\}, \tag{13.3}$$

over all β not proportional to $\mathbb{1}_{k+1}$, the $(k+1) \times 1$ vector of ones. Note that $\lambda(\omega)$ is not defined if $\beta = a\mathbb{1}_{k+1}$ for $a \in \mathbb{R}$ because such a scaling corresponds to assigning each category the same value a; in this case, $f(\omega; \beta) \equiv 0$ and $\sigma^2(\beta) = 0$. The optimality criterion $\lambda(\omega)$ possesses the desirable property of being invariant under location and scale changes of β.

As in most scaling problems for categorical data, it was useful to represent the categories in terms of the vectors $e_1, e_2, \ldots, e_{k+1}$, where e_j represents the $(k+1) \times 1$ vector with one in the jth row and zeros elsewhere. We then defined a $(k+1)$-dimensional stationary time series $\boldsymbol{Y}_t$ by $\boldsymbol{Y}_t = e_j$ when $X_t = c_j$. The time series $X_t(\beta)$ can be obtained from the $\boldsymbol{Y}_t$ time series by the relationship $X_t(\beta) = \beta' \boldsymbol{Y}_t$. Assume that the vector process $\boldsymbol{Y}_t$ has a continuous spectral density denoted by $f_Y(\omega)$. For each ω, $f_Y(\omega)$ is, of course, a $(k+1) \times (k+1)$ complex-valued Hermitian matrix. Note that the relationship $X_t(\beta) = \beta' \boldsymbol{Y}_t$ implies that

$f_Y(\omega;\beta) = \beta' f_Y(\omega)\beta = \beta' f_Y^{re}(\omega)\beta$, where $f_Y^{re}(\omega)$ denotes the real part of $f_Y(\omega)$. The optimality criterion can thus be expressed as

$$\lambda(\omega) = \sup_{\beta} \left\{ \frac{\beta' f_Y^{re}(\omega)\beta}{\beta' V \beta} \right\} \tag{13.4}$$

where V is the variance–covariance matrix of $\mathbf{Y}_t$. The resulting scaling $\beta(\omega)$ is called the optimal scaling.

The $\mathbf{Y}_t$ process is a multivariate point process, and any particular component of $\mathbf{Y}_t$ is the individual point process for the corresponding state (e.g., the first component of $\mathbf{Y}_t$ indicates whether or not the process is in state c_1 at time t). For any fixed t, $\mathbf{Y}_t$ represents a single observation from a simple multinomial sampling scheme. It readily follows that $V = D - p\,p'$, where $p = (p_1, \ldots, p_{k+1})'$, and D is the $(k+1) \times (k+1)$ diagonal matrix $D = \text{diag}\{p_1, \ldots, p_{k+1}\}$. Since, by assumption, $p_j > 0$ for $j = 1, 2, \ldots, k+1$, it follows that $\text{rank}(V) = k$ with the null space of V being spanned by $\mathbb{1}_{k+1}$. For any $(k+1) \times k$ full rank matrix Q whose columns are linearly independent of $\mathbb{1}_{k+1}$, $Q'VQ$ is a $k \times k$ positive definite symmetric matrix.

With the matrix Q as previously defined, and for $-1/2 < \omega \leq 1/2$, define $\lambda(\omega)$ to be the largest eigenvalue of the determinantal equation

$$|Q' f_Y^{re}(\omega) Q - \lambda Q' V Q| = 0,$$

and let $\boldsymbol{b}(\omega) \in \mathbb{R}^k$ be any corresponding eigenvector, that is,

$$Q' f_Y^{re}(\omega) Q \boldsymbol{b}(\omega) = \lambda(\omega) Q' V Q \boldsymbol{b}(\omega).$$

The eigenvalue $\lambda(\omega) \geq 0$ does not depend on the choice of Q. Although the eigenvector $\boldsymbol{b}(\omega)$ depends on the particular choice of Q, the equivalence class of scalings associated with $\beta(\omega) = Q\boldsymbol{b}(\omega)$ does not depend on Q. A convenient choice of Q is $Q = [I_k \mid \mathbf{0}\,]'$, where I_k is the $k \times k$ identity matrix and $\mathbf{0}$ is the $k \times 1$ vector of zeros. For this choice, $Q' f_Y^{re}(\omega) Q$ and $Q'VQ$ are the upper $k \times k$ blocks of $f_Y^{re}(\omega)$ and V, respectively. This choice corresponds to setting the last component of $\beta(\omega)$ to zero.

The value $\lambda(\omega)$ itself has a useful interpretation; specifically, $\lambda(\omega)d\omega$ represents the largest proportion of the total power that can be attributed to the frequencies $\omega d\omega$ for any particular scaled process $X_t(\beta)$, with the maximum being achieved by the scaling $\beta(\omega)$. This result is demonstrated in Figure 13.3. Because of its central role, $\lambda(\omega)$ was defined to be the *spectral envelope* of a stationary categorical time series.

The name spectral envelope is appropriate since $\lambda(\omega)$ envelopes the standardized spectrum of any scaled process. That is, given any β normalized so that $X_t(\beta)$ has total power one, $f(\omega;\beta) \leq \lambda(\omega)$ with equality if and only if β is proportional to $\beta(\omega)$.

Although the law of the process $X_t(\beta)$ for any one-to-one scaling β completely determines the law of the categorical process X_t, information is lost when one restricts attention to the spectrum of $X_t(\beta)$. Less information is lost when one considers the spectrum of $\mathbf{Y}_t$. Dealing directly with the spectral density $f_Y(\omega)$ itself is somewhat cumbersome since it is a function into the set of complex Hermitian matrices. Alternatively, one can view the spectral envelope as an easily understood, parsimonious tool for exploring the periodic nature of a categorical time series with a minimal loss of information.

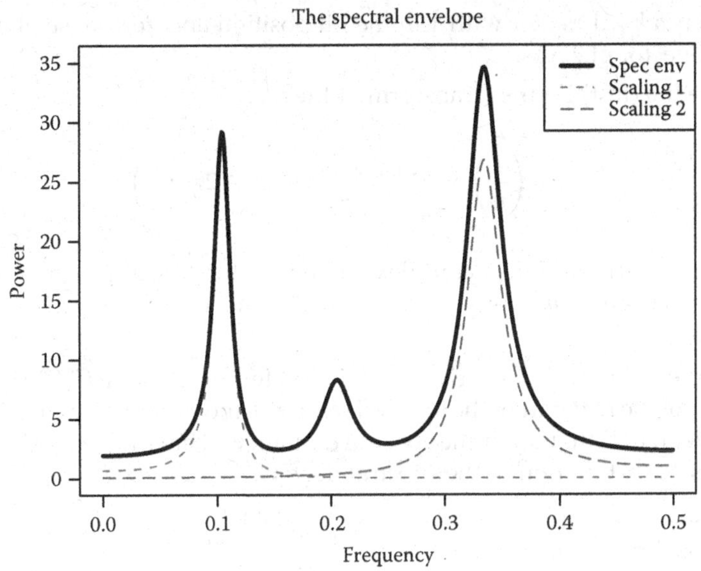

FIGURE 13.3
Demonstration of the spectral envelope. The short dashed line indicates a spectral density corresponding to some scaling. The long dashed line indicates a spectral density corresponding to a different scaling. The thick solid line is the spectral envelope, which can be thought of as throwing a blanket over all possible spectral densities corresponding to all possible scalings of the sequence. Because the exhibited spectral densities attain the value of the spectral envelope at the frequency near 0.1, the corresponding scaling is optimal at that frequency. The scaling at the frequency near 1/3 is close to optimal, but the spectral envelope indicates that there is a scaling that can get more power at frequency 1/3. In addition to finding interesting frequencies (e.g., there is something interesting near the frequency of 0.2 that neither scaling 1 nor 2 discovers), the spectral envelope reveals frequencies for which nothing is interesting (e.g., no matter which scaling is used, there is nothing interesting in this sequence in the frequency range above 0.4).

In view of (13.4), there is an apparent relationship of the spectral envelope and principal components. This relationship is discussed in Section 13.7

13.4 Estimation

In view of the dimension reduction mentioned in the previous section, the easiest way to estimate the spectral envelope is to fix the scale of the last state at 0, and then select the indicator vectors to be k-dimensional. More precisely, to estimate the spectral envelope and the optimal scalings given a stationary categorical sequence, $\{X_t;\, t = 1, \dots, n\}$, with state-space $C = \{c_1, \dots, c_{k+1}\}$, perform the following tasks.

(1) Form $k \times 1$ vectors $\{\boldsymbol{Y}_t,\, t = 1, \dots, n\}$ as follows:

$$\boldsymbol{Y}_t = \boldsymbol{e}_j \quad \text{if} \quad X_t = c_j, \quad j = 1, \dots, k;$$
$$\boldsymbol{Y}_t = \boldsymbol{0}_k \quad \text{if} \quad X_t = c_{k+1},$$

where e_j is a $k \times 1$ vector with 1 in the jth position and zeros elsewhere and $\mathbf{0}_k$ is the $k \times 1$ vector of zeros.

(2) Calculate the (fast) Fourier transform of the data,

$$d\left(\frac{j}{n}\right) = n^{-1/2} \sum_{t=1}^{n} Y_t \exp\left(-2\pi it\frac{j}{n}\right).$$

Note that $d(j/n)$ is a $k \times 1$ complex-valued vector. Calculate the periodogram, $I(j/n) = d(j/n)d^*(j/n)$, for $j = 1, \ldots, \lfloor n/2 \rfloor$, and retain only the real part, say $I^{re}(j/n)$.

(3) Smooth the real part of the periodogram as preferred to obtain $\widehat{f^{re}}(j/n)$, a consistent estimator of the real part of the spectral matrix. Time series texts such as Shumway and Stoffer (2011) that cover the spectral domain will have an extensive discussion on consistent estimation of the spectral density.

(4) Calculate the $k \times k$ covariance matrix of the data, $S = n^{-1} \sum_{t=1}^{n} (Y_t - \overline{Y})(Y_t - \overline{Y})'$, where $\overline{Y}$ is the sample mean of the data.

(5) For each $\omega_j = j/n$, $j = 1, \ldots, \lfloor n/2 \rfloor$, determine the largest eigenvalue and the corresponding eigenvector of the matrix $2n^{-1}S^{-1/2}\hat{f}^{re}(\omega_j)S^{-1/2}$. Note that $S^{-1/2}$ is the inverse of the unique square root matrix* of S.

(6) The sample spectral envelope $\widehat{\lambda}(\omega_j)$ is the eigenvalue obtained in the previous step. If $b(\omega_j)$ denotes the eigenvector obtained in the previous step, the optimal sample scaling is $\widehat{\beta}(\omega_j) = S^{-1/2}b(\omega_j)$; this will result in k values, the $(k+1)$-st value being held fixed at zero.

Any standard programming language can be used to do the calculations; basically, one only has to be able to compute fast Fourier transforms and eigenvalues and eigenvectors of real symmetric matrices. Some examples using the R Statistical Programming Language R Core Team (2013) may be found in Section 13.8; also, see Shumway and Stoffer (2011, Chapter 7). Inference for the sample spectral envelope and the sample optimal scalings are described in detail in Stoffer et al. (1993a). A few of the main results of that paper are as follows.

If X_t is an i.i.d. sequence and $\widetilde{\lambda}(\omega)$ is the largest eigenvalue of the periodogram matrix, $I(\omega_j)$, then the following large sample approximation based on the chi-square distribution is valid for $x > 0$:

$$\Pr\left\{n2^{-1}\widetilde{\lambda}\left(\omega_j\right) < x\right\} \doteq \Pr\left\{\chi^2_{2k} < 4x\right\} - \pi^{1/2}x^{(k-1)/2}e^{-x}\frac{\Pr\left\{\chi^2_{k+1} < 2x\right\}}{\Gamma\left(\frac{k}{2}\right)}, \tag{13.5}$$

where $k+1$ is the size of the alphabet being considered. Note that $I(\omega_j)$ has at most one positive eigenvalue and consequently, $\widetilde{\lambda}(\omega_j) = \operatorname{tr} I(\omega_j)$.

* If $S = P\Lambda P'$ is the spectral decomposition of S, then $S^{-1/2} = P\Lambda^{-1/2}P'$, where $\Lambda^{-1/2}$ is the diagonal matrix with the reciprocal of the root eigenvalues along the diagonal.

In the general case, if a smoothed estimator is used and $\lambda(\omega)$ is a distinct root (which implies that $\lambda(\omega) > 0$), then, independently, for any collection of Fourier frequencies $\{\omega_i; i = 1, \ldots, M\}$, M fixed, and for large n and m,

$$\nu_m \frac{\widehat{\lambda}(\omega_i) - \lambda(\omega_i)}{\lambda(\omega_i)} \sim \mathrm{AN}(0, 1) \tag{13.6}$$

and

$$\nu_m \left[\widehat{\beta}(\omega_i) - \beta(\omega_i)\right] \sim \mathrm{AN}(\mathbf{0}, \Sigma_i), \tag{13.7}$$

where $\Sigma_i = V^{-1/2}\Omega_i V^{-1/2}$ with

$$\Omega_i = \frac{\{\lambda(\omega_i)H(\omega_i)^+ f^{re}(\omega_i)H(\omega_i)^+ - \boldsymbol{a}(\omega_i)\boldsymbol{a}(\omega_i)'\}}{2},$$

and $H(\omega_i) = f^{re}(\omega_i) - \lambda(\omega_i)\mathrm{I}_{k-1}$, $\boldsymbol{a}(\omega_i) = H(\omega_i)^+ f^{im}(\omega_i)V^{1/2}\boldsymbol{u}(\omega_i)$, and $H(\omega_i)^+$ refers to the Moore–Penrose inverse of $H(\omega_i)$.

The term "ν_m" depends on the type of estimator being used. For example, in the case of estimation via weighted averaging of the periodogram, that is,

$$\widehat{f}(\omega) = \sum_{q=-m}^{m} h_q I(\omega_{j+q}),$$

where $\{\omega_{j+q}; q = 0, \pm 1, \ldots, \pm m\}$ is a band of frequencies where ω_j is the fundamental frequency closet to ω, and such that the weights $h_q = h_{-q}$ are positive and $\sum_{q=-m}^{m} h_q = 1$, then $\nu_m^{-2} = \sum_{q=-m}^{m} h_q^2$. If a simple average is used, $h_q = 1/(2m+1)$, then $\nu_m^2 = (2m+1)$. Based on these results, asymptotic normal confidence intervals and tests for $\lambda(\omega)$ can be readily constructed. Similarly, for $\beta(\omega)$, asymptotic confidence ellipsoids and chi-square tests can be constructed; details can be found in Stoffer et al. (1993a, Theorems 3.1–3.3). As a note, we mention that this technique is not restricted to the use of sinusoids. In Stoffer et al. (1993b), the use of the Walsh basis* of square-waves functions that take only the values ± 1 only is described.

A simple asymptotic test statistic for $\beta(\omega)$ can be obtained. Let $\widehat{H}(\omega) = \widehat{f}_Y^{re}(\omega) - \widehat{\lambda}(\omega)I_k$, and

$$\xi_m(\omega) = \frac{\sqrt{2}\,\nu_m \widehat{f}_Y^{re}(\omega)^{-1/2}\widehat{H}(\omega)\left(\widehat{\beta}(\omega) - \beta(\omega)\right)}{\widehat{\lambda}(\omega)^{1/2}}.$$

Then,

$$\xi_m(\omega)'\xi_m(\omega) \tag{13.8}$$

* The Walsh functions are a completion of the Haar functions; a summary of their use in statistics is given in Stoffer (1991).

converges ($m \to \infty$) in distribution to a distribution that is stochastically less than χ^2_k and stochastically greater than χ^2_{k-1}. Note that the test statistic (13.8) is zero if $\beta(\omega)$ is replaced by $\widehat{\beta}(\omega)$. One can check whether or not a particular element of $\widehat{\beta}(\omega)$ is zero by inserting $\widehat{\beta}(\omega)$ in for $\beta(\omega)$, but with the particular element zeroed out and the resulting vector rescaled to be of unit length, into (13.8).

Significance thresholds for a consistent spectral envelope estimate can easily be computed using the following approximations. Using a first-order Taylor expansion, we have

$$\log \widehat{\lambda}(\omega) \approx \log \lambda(\omega) + \frac{\widehat{\lambda}(\omega) - \lambda(\omega)}{\lambda(\omega)},$$

so that ($n, m \to \infty$)

$$\nu_m \left[\log \widehat{\lambda}(\omega) - \log \lambda(\omega)\right] \sim AN(0,1). \tag{13.9}$$

It also follows that $E[\log \widehat{\lambda}(\omega)] \approx \log \lambda(\omega)$ and $\mathrm{Var}[\log \widehat{\lambda}(\omega)] \approx \nu_m^{-2}$. If there is no signal present in a sequence of length n, we expect $\lambda(j/n) \approx 2/n$ for $1 < j < n/2$, and hence approximately $(1-\alpha)\times 100\%$ of the time, $\log \widehat{\lambda}(\omega)$ will be less than $\log(2/n)+(z_\alpha/\nu_m)$ where z_α is the $(1-\alpha)$ upper tail cutoff of the standard normal distribution. Exponentiating, the α critical value for $\widehat{\lambda}(\omega)$ becomes $(2/n)\exp(z_\alpha/\nu_m)$. Although this method is a bit crude, from our experience, thresholding at very small α-levels (say, $\alpha = 10^{-4}$ to 10^{-6}, depending on the size of n) works well. Some further insight into choosing α will be given in the numerical examples.

13.5 Numerical Examples

As a simple example of the kind of analysis that can be accomplished, we consider the gene BNRF1 (bp 1736–5689) of the EBV. Since we are considering the nucleotide sequence consisting of four bp, we use the following indicator vectors to represent the data:

$$Y_t = (1,0,0)' \text{ if } X_t = \mathtt{A}; \qquad Y_t = (0,1,0)' \text{ if } X_t = \mathtt{C};$$
$$Y_t = (0,0,1)' \text{ if } X_t = \mathtt{G}; \qquad Y_t = (0,0,0)' \text{ if } X_t = \mathtt{T},$$

so that the scale for the thymine nucleotide, T, is set to zero. Figure 13.4 shows the spectral envelope estimate of the entire coding sequence (3954 bp long). The figure also shows a strong signal at frequency 1/3; the corresponding optimal scaling was A $= 0.10$, C $=$ 0.61, G $= 0.78$, T $= 0$, which indicates that the signal is in the strong–weak bonding alphabet, $S = \{\mathtt{C},\mathtt{G}\}$ and $W = \{\mathtt{A},\mathtt{T}\}$.*

In Shumway and Stoffer (2011, Example 7.18), there is evidence that the gene is not homogeneous, and in fact, the last fourth of the gene is unlike the first three quarters of the gene. For example, Figure 13.5 shows a dynamic spectral envelope with a block size of 500. Precise details of the analysis can be found in the R code, Section 13.8. It is immediately

* W refers to adenine (A) or thymine (T) for the *weak* hydrogen bonding interaction between the base pairs. S refers to guanine (G) or cytosine (C) for the *strong* hydrogen bonding interaction between the base pairs.

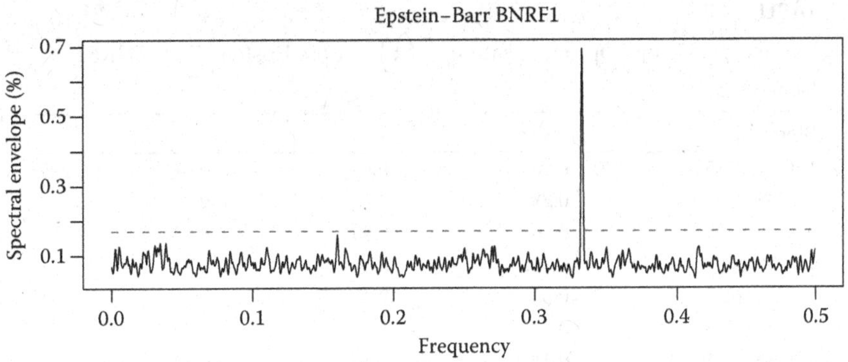

FIGURE 13.4
Smoothed sample spectral envelope of the BNRF1 gene from the Epstein–Barr virus.

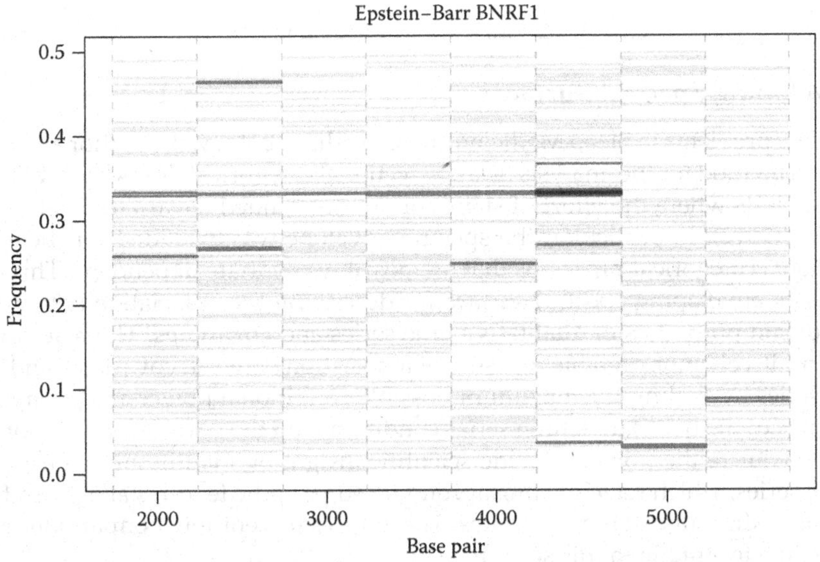

FIGURE 13.5
Dynamic spectral envelope estimates for the BNRF1 gene (bp 1736–5689) of the Epstein–Barr virus (EBV). The vertical dashed lines indicate the blocks, and values over the approximate 0.005 null significance threshold are indicated by darker regions.

evident from the figure that even within small segments of the gene, there is heterogeneity. There is, however, a basic cyclic pattern that exists through most of the gene as evidenced by the peak at $\omega = 1/3$, except at the end of the gene. Table 13.3 shows the optimal scalings at the one-third frequency and we note that the corresponding alphabets are somewhat consistent in the "significant" blocks, with each block in the beginning of the sequence indicating a weak–strong bonding alphabet (`A` = `T`, `C` = `G`). This alphabet starts to break down at block number five (bp 3736–4235). As previously indicated, there is a substantial difference in the final 1000 bp (blocks 7 and 8) of the gene.

TABLE 13.3

Blockwise (500 bp) Optimal Scaling, $\widehat{\beta}\left(\frac{1}{3}\right)$, for the Epstein–Barr BNRF1 Gene Example

Block (bp)	A	C	G	T
1. 1736–2235	0.26	0.69	0.68	0
2. 2236–2735	0.23	0.71	0.67	0
3. 2736–3235	0.16	0.56	0.82	0
4. 3236–3735	0.15	0.61	0.78	0
5. 3736–4235	0.30	0.35	0.89	0
6. 4236–4735	0.22	0.61	0.76	0
7. 4736–5235[a]	0.41	0.56	0.72	0
8. 5236–5689[a]	0.90	−0.43	−0.07	0

[a] $\widehat{\lambda}\left(\frac{1}{3}\right)$ is not significant in this block.

13.6 Enveloping Spectral Surfaces

Motivated by problems in Sleep Medicine and Circadian Biology, the author and colleagues developed a method for the analysis of cross-sectional categorical time series collected from multiple subjects where the effect of static continuous-valued covariates is of interest; see Krafty et al. (2012). In particular, the spectral envelope was extended for the analysis of cross-sectional categorical processes that are possibly covariate dependent. This extension introduces an enveloping spectral surface for describing the association between the frequency domain properties of qualitative time series and covariates. The resulting surface offers an intuitively interpretable measure of association between covariates and a qualitative time series by finding the maximum possible conditional power at a given frequency from scalings of the qualitative time series conditional on the covariates. The optimal scalings that maximize the power provide scientific insight by identifying the aspects of the qualitative series, which have the most pronounced periodic features at a given frequency conditional on the value of the covariates. The approach is entirely nonparametric, and we summarize the technique in this section.

In this section, we suppose we observe qualitative time series $\{X_{jt}; t = 0, \pm 1, \pm 2, \dots\}$ with finite state-space $\mathcal{C} = \{c_1, c_2, \dots, c_{p+1}\}$ and a covariate vector $\boldsymbol{S}_j = (S_{j1}, \dots, S_{jq})' \in \mathcal{S} \subset \mathbb{R}^q$ for $j = 1, \dots, N$ independent subjects. We assume that $\{X_{jt}; t = 0, \pm 1, \pm 2, \dots\}$ is stationary conditional on $\boldsymbol{S}_j$ such that

$$\inf_{s \in \mathcal{S}, k=1,\dots,p+1} \Pr\{X_{jt} = c_k \mid \boldsymbol{S}_j = s\} > 0, \tag{13.10}$$

so that there are no absorbing states. The covariates $\boldsymbol{S}_j$ are assumed to be independent and identically distributed second-order random variables with density function ϕ_s. Aside from making some smoothness assumptions about the conditional spectral distribution of X_{jt} given $\boldsymbol{S}_j$ to aid estimation, we will only assume a very general nonparametric model for X_{jt} and $\boldsymbol{S}_j$.

Analogous to the discussion in Section 13.3, we consider the quantitative time series $X_{jt}(\boldsymbol{\beta})$ obtained from scaling X_{jt} such that $X_{jt}(\boldsymbol{\beta}) = \beta_\ell$ when $X_{jt} = c_\ell$, for $1 \le \ell \le p$, and $X_{jt}(\beta) = 0$ when $X_{jt} = c_{p+1}$. Further, we suppose that the p-dimensional random vector process $\boldsymbol{Y}_{jt}$, which has one in the ℓth element if $X_{jt} = c_\ell$ for $\ell = 1, \ldots, p$ and zeros elsewhere, has a spectral density conditional on the value of the covariate. In this case, define the conditional spectral density and variance of $\boldsymbol{Y}_{jt}$ as

$$f(\omega, s) = \sum_{\tau=-\infty}^{\infty} \mathrm{Cov}\left(\boldsymbol{Y}_{jt}, \boldsymbol{Y}_{jt+\tau} \mid \boldsymbol{S}_j = s\right) e^{-2\pi i \omega \tau} \tag{13.11}$$

$$V(s) = \mathrm{Var}\left(\boldsymbol{Y}_{jt} \mid \boldsymbol{S}_j = s\right). \tag{13.12}$$

We will assume that $f(\omega, s)$ and $V(s)$ are nonsingular for all frequencies ω and $s \in \mathcal{S}$. Under this assumption, we have the existence of the spectral density of $X_{jt}(\beta)$ conditional on $\boldsymbol{S}_j = s$ for all $\beta \in \mathbb{R}^p \setminus \mathbf{0}_p$. Thus, define

$$f_x(\omega, s;\, \beta) = \sum_{\tau=-\infty}^{\infty} \mathrm{Cov}\left[X_{jt}(\beta), X_{jt+\tau}(\beta) \mid \boldsymbol{S}_j = s\right] e^{-2\pi i \omega \tau}.$$

As an extension of the spectral envelope, for every frequency ω and covariate $s \in \mathcal{S}$, we define the *enveloping spectral surface*, $\lambda(\omega, s)$, to be the maximal normalized power among all possible scalings at frequency ω, conditional on the covariate value $\boldsymbol{S}_j = s$.

Letting $V_x(s;\, \beta) = \mathrm{Var}\left[X_{jt}(\beta) \mid \boldsymbol{S}_j = s\right]$ be the conditional variance of the scaled time series, we formally defined the enveloping spectral surface as

$$\lambda(\omega, s) = \max_{\beta \neq \mathbf{0}_p} \frac{f_x(\omega, s;\, \beta)}{V_x(s;\, \beta)}. \tag{13.13}$$

In addition to the maximum value $\lambda(\omega, s)$, the scalings where this maximum is achieved can provide important information by locating the scalings of the qualitative time series for which cycles at a given frequency are most prominent conditional on the covariate vector. Equivalently, note that $\lambda(\omega, s)$ is the largest eigenvector associated with

$$g(\omega, s) = V^{-1/2}(s) f(\omega, s) V^{-1/2}(s),$$

and the optimal scaling is linearly related to the eigenvector associated with the largest eigenvalue.

An aspect of the enveloping spectral surface that can be of scientific interest are frequencies where the enveloping spectral surface changes based on covariate values. An interpretable measure of the dynamics of the enveloping spectral surface with respect to the covariates depends on the form of the covariates. Here, we consider the case where the covariates are continuous random variables such that the spectrum $g^{re}(\omega, \boldsymbol{S})$ is smooth.

For a metric space $\mathcal{D}$, let $\mathcal{C}^d(\mathcal{D})$ be the space of real-valued functions over $\mathcal{D}$ such that all dth-order partial derivatives exist and are continuous. We need the following two smoothness assumptions:

A(i): The support of the density function ϕ_s is $\mathcal{S}$ and $\phi_s \in \mathcal{C}^2(\mathcal{S})$.

A(ii): Each element of the $p \times p$ spectral density matrix g^{re} is an element of the space $\mathcal{C}^4[\mathbb{R} \times \mathcal{S}]$.

Under Assumptions A(i) and A(ii), the continuity of the eigenvalues of a matrix-valued function implies that the enveloping spectral surface λ is continuous in both the frequency and covariates. In this setting, the first-order partial derivatives of the enveloping spectral surface in the direction of the covariates provide an assessment of the dependence of the enveloping spectral surface on the covariates. Let $\boldsymbol{D}\boldsymbol{\lambda}(\omega, \boldsymbol{s})$ be the q-dimensional vector of partial derivatives of $\lambda(\omega, \boldsymbol{s})$ with respect to $\boldsymbol{s}$ which has jth element $D_j\lambda(\omega, \boldsymbol{s}) = \partial\lambda(\omega, \boldsymbol{s})/\partial s_j$. For a frequency $\omega \in \mathbb{R}$, if $\boldsymbol{D}\boldsymbol{\lambda}(\omega, \boldsymbol{s}) = \boldsymbol{0}$ for all $\boldsymbol{s} \in \mathcal{S}$, then there is no association between the the maximal amount of normalized power at frequency ω and the covariates. The following result, which follows from Magnus and Neudecker (1988, Chapter 8, Theorem 7) provides a computationally useful form for $D_j\lambda$ in terms of the derivatives of the real part of the spectral density of the indicator variables. In particular, under Assumptions A(i) and A(ii), if $\lambda(\omega, \boldsymbol{s})$ is a unique eigenvalue of $g^{re}(\omega, \boldsymbol{s})$, then

$$D_j\lambda(\omega, \boldsymbol{s}) = \gamma(\omega, \boldsymbol{s})' V^{1/2}(\boldsymbol{s}) D_j g^{re}(\omega, \boldsymbol{s}) V^{1/2}(\boldsymbol{s})\gamma(\omega, \boldsymbol{s}) \tag{13.14}$$

where $D_j g^{re}(\omega, \boldsymbol{s})$ is the $p \times p$ matrix with ℓmth element $D_j g^{re}_{\ell m}(\omega, \boldsymbol{s}) = \partial g^{re}_{\ell m}(\omega, \boldsymbol{s})/\partial s_j$, and $\gamma(\omega, \boldsymbol{s})$ is the corresponding eigenvector. Equation (13.14) provides a useful tool for developing an estimation procedure for the derivatives of the enveloping spectral surface directly from estimates of the eigenvectors and derivatives of g^{re}.

It is assumed that epochs of the qualitative time series of length T, $\{X_{jt};\, t = 1, \ldots, T\}$, are observed for $j = 1, \ldots, N$ subjects. Asymptotic properties are established as both the number of time points and the number of subjects are large so that $T, N \to \infty$. Since the spectral density is Hermitian and periodic with period 1, we restrict our attention to $\omega \in [0, 1/2]$.

Estimation procedures based on multivariate local quadratic regression are proposed. Let K_q and K_{q+1} be spherically symmetric q and $q+1$-dimensional compactly supported density functions that possess eighth-order marginal moments over $\mathcal{S}$ and $[0, 1/2] \times \mathcal{S}$, respectively. For $k = q$ or $k = (q + 1)$, the bandwidth will be parameterized by assuming that there exists a positive definite $k \times k$ symmetric real matrix H^* and a scaling bandwidth $h > 0$ such that the bandwidth is parameterized as $H = hH^*$ and the corresponding weight functions are $|H|^{-1}K_k[H^{-1}\boldsymbol{s}]$. Asymptotic properties will be established when $h \to 0$ as $N, T \to \infty$.

Local quadratic estimation will be used to estimate V, g^{re}, and $D_k g^{re}$. A comprehensive review of local polynomial regression is given in Fan and Gijbels (1996). Although V, g^{re}, and its derivatives can be of scientific interest in their own right, we are concerned with the estimation of these quantities exclusively for use in estimating and performing inference on the enveloping spectral surface and its derivatives.

To estimate g^{re} and its derivatives, first note that the conditional spectral density of $V^{-1/2}(\boldsymbol{S}_j)\boldsymbol{Y}_{jt}$ is $g(\omega, \boldsymbol{S}_j)$. Define the normalized periodogram I_{jk} for $j = 1, \ldots, N$ and $k = 1, \ldots, \lfloor T/2 \rfloor$ as $I_{jk} = \widehat{V}^{-1/2}(\boldsymbol{S}_j)\widetilde{\boldsymbol{Y}}_{jk}\widetilde{\boldsymbol{Y}}^*_{jk}\widehat{V}^{-1/2}(\boldsymbol{S}_j)$ where $\widetilde{\boldsymbol{Y}}_{jk} = T^{-1/2}\sum_{t=1}^{T}\left(\boldsymbol{Y}_{jt} - \overline{\boldsymbol{Y}}_j\right)e^{-2\pi i \omega_k t}$ is the finite Fourier transform of $\boldsymbol{Y}_{jt}$ at frequency $\omega_k = k/T$ and $\widetilde{\boldsymbol{Y}}^*_{jk}$ is the conjugate transpose of $\widetilde{\boldsymbol{Y}}_{jk}$. The components $I^{re}_{jk\ell m}$ provide asymptotically unbiased but inconsistent estimates of $g^{re}_{\ell m}(\omega_k, \boldsymbol{S}_j)$ and we apply a local quadratic regression to these real components of the periodograms to obtain a consistent estimate of $g^{re}(\omega, \boldsymbol{s})$. For ease of notation, define the $(q + 1)$-dimensional vector

$$\xi_{jk} = \begin{pmatrix} \omega_k - \omega \\ S_j - s \end{pmatrix}.$$

Then, for $\omega \in [0,1/2]$ and $s \in \mathcal{S}$, define the $p \times p$ matrices $\widehat{g^{re}}(\omega,s)$ and $\widehat{D_k g^{re}}(\omega,s)$ for $k = 1,\ldots,q$ with respective ℓmth elements $\widehat{g^{re}_{\ell m}}(\omega,s) = \widehat{\alpha}_0$ and $\widehat{D_k g^{re}_{\ell m}}(\omega,s) = \widehat{\alpha}_{k+1}$,

$$\begin{bmatrix} \widehat{\alpha}_0 \\ \widehat{\alpha} \\ \text{vec}(\widehat{Q}) \end{bmatrix} = \underset{\alpha_0 \in \mathbb{R}, \alpha \in \mathbb{R}^{q+1}, Q \in \mathcal{Q}_{q+1}}{\text{argmin}} \sum_{j=1}^{N} \sum_{k=0}^{\lfloor T/2 \rfloor} \left\{ I^{re}_{jk\ell m} - \alpha_0 - \alpha' \xi_{jk} - \xi'_{jk} Q \xi_{jk} \right\}^2$$
$$\times K_{q+1} H_g^{-1} \xi_{jk} \;/\; |H_g|,$$

$\widehat{\alpha} = (\widehat{\alpha}_1,\ldots,\widehat{\alpha}_{q+1})'$, and H_g is a symmetric positive definite $(q+1) \times (q+1)$ real matrix.

The following theorem provides the asymptotic consistency of $\widehat{V}$, $\widehat{g^{re}}$, and $\widehat{D_k g^{re}}$, which allows for the consistent estimation of the enveloping spectral surface and its derivatives.

Theorem 13.1 *Let $H_v = h_v H_v^*$ and $H_g = h_g H_g^*$ for positive definite symmetric matrices H_v^*, H_g^* and positive real numbers h_v, h_g. Under Assumptions A(i) and A(ii), the first-order optimal conditional mean squared error of $\widehat{V}(s)$, $\widehat{g^{re}}(\omega,s)$, and $\widehat{D_k g^{re}}(\omega,s)$ for $\omega \in [0,1/2]$ and $s \in \mathcal{S}$ are achieved when $h_v \sim (NT)^{-1/(q+6)}$ and $h_g \sim (NT)^{-1/(q+7)}$ as $N,T \to \infty$. If $h_v \sim (NT)^{-1/(q+6)}$, $h_g \sim (NT)^{-1/(q+7)}$, and $N \sim T^{(q/6)-\epsilon}$ for some $\epsilon \in (0,q/6)$, then*

$$\widehat{V}(s) = V(s) + O_p\left[(NT)^{-3/(q+6)}\right],$$
$$\widehat{g^{re}}(\omega,s) = g^{re}(\omega,s) + O_p\left[(NT)^{-3/(q+7)}\right],$$
$$\widehat{D_k g^{re}}(\omega,s) = D_k g^{re}(\omega,s) + O_p\left[(NT)^{-2/(q+7)}\right],$$

conditional on $S_1,\ldots,S_N$.

We can now estimate the enveloping spectral surface and optimal scalings. Define $\widehat{\lambda}(\omega,s)$ as the largest eigenvalue of $\widehat{g^{re}}(\omega,s)$. Let $\widehat{\gamma}(\omega,s) = \widehat{V}^{-1/2}(s)\widehat{\psi}(\omega,s)$ where $\widehat{\psi}(\omega,s)$ is the eigenvector of $\widehat{g^{re}}(\omega,s)$ associated with $\widehat{\lambda}(\omega,s)$ such that $\widehat{\gamma}(\omega,s)'\widehat{V}(s)\widehat{\gamma}(\omega,s) = 1$ and the first nonzero element of $\widehat{\gamma}(\omega,s)$ is positive. The next theorem establishes the consistency and asymptotic distribution of $\widehat{\lambda}(\omega,s)$ and the consistency of $\widehat{\gamma}(\omega,s)$.

Theorem 13.2 *If Assumptions A(i) and A(ii) hold, $h_v \sim (NT)^{-1/(q+6)}$, $h_g \sim (NT)^{-1/(q+7)}$ and $N \sim T^{(q/6)-\epsilon}$ for some $\epsilon \in (0,q/6)$ as $N,T \to \infty$, then for $\omega \in [0,1/2]$ and $s \in \mathcal{S}$*

$$\widehat{\lambda}(\omega,s) = \lambda(\omega,s) + O_p\left[(NT)^{-3/(q+7)}\right]$$
$$\widehat{\gamma}(\omega,s) = \gamma(\omega,s) + O_p\left[(NT)^{-3/(q+7)}\right]$$

conditional on $S_1, \ldots, S_N$. *In addition,* $\widehat{\lambda}(\omega, s)$ *is conditionally asymptotically normal with large sample variance*

$$\text{Var}\left[\, \widehat{\lambda}(\omega, s) \mid S_1, \ldots, S_N\right] \approx \frac{\lambda^2(\omega, s)\rho}{NT|H_g|\phi_s(s)},$$

where

$$\begin{aligned}\rho &= (\mu_{14} - \mu_{12})^{-2} \\ &\times \left[\mu_{20}\left(\mu_{14} + \mu_{12}^2\right)^2 - 2(q+1)\mu_{12}\mu_{22}\left(\mu_{14} + q\mu_{12}^2\right) + (q+1)\mu_{12}^2\left(\mu_{24} + q\mu_{22}^2\right)\right]\end{aligned}$$

for $\mu_{\ell m} = \int_{u \in \mathbb{R} \times \mathcal{S}} u_1^m K_{q+1}^{\ell}(u)du$.

Equation (13.14) can be used to obtain a consistent estimate of $D\lambda(\omega, s)$ from the estimates in Theorem 13.1. Define $\widehat{D\lambda}(\omega, s)$ as the q-vector with kth element $\widehat{D_k\lambda}(\omega, s)$ where

$$\widehat{D_k\lambda}(\omega, s) = \widehat{\gamma}(\omega, s)'\widehat{V}^{1/2}(s)\widehat{D_k g^{re}}(\omega, s)\widehat{V}^{1/2}(s)\widehat{\gamma}(\omega, s).$$

The consistency of $\widehat{D\lambda}(\omega, s)$ and the large sample distribution of its elements are established in the following theorem.

Theorem 13.3 *If Assumptions A(i) and A(ii) hold,* $h_v \sim (NT)^{-1/(q+6)}$, $h_g \sim (NT)^{-1/(q+7)}$ *and* $N \sim T^{(q/6)-\epsilon}$ *for some* $\epsilon \in (0, q/6)$ *as* $N, T \to \infty$, *then for* $\omega \in [0, 1/2]$ *and* $s \in \mathcal{S}$

$$\widehat{D\lambda}(\omega, s) = D\lambda(\omega, s) + O_p\left[(NT)^{-2/(q+7)}\right]$$

conditional on $S_1, \ldots, S_N$. *In addition,* $\widehat{D_k\lambda}(\omega, s)$ *is conditionally asymptotically normal with large sample variance*

$$\text{Var}\left[\widehat{D_k\lambda}(\omega, s) \mid S_1, \ldots, S_N\right] \approx \frac{\left(e_k' H_g^{-2} e_k\right)\mu_{22}}{NT|H_g|\phi_s(s)\mu_{12}^2}\Gamma(\omega, S)$$

where e_k *is the p-dimensional vector of zeros except for a one in k-th element,* $\mu_{\ell m} = \int_{u \in \mathbb{R} \times \mathcal{S}} u_1^m K_{q+1}^{\ell}(u)du$, *and* $\Gamma(\omega, s)$ *is uniquely determined by* $g(\omega, s)$.

The explicit form for $\Gamma(\omega, s)$ is given in Krafty et al. (2012, Appendix). Note that $\Gamma(\omega, s)$ depends on both $g^{re}(\omega, s)$ and $g^{im}(\omega, s)$. Subsequently, large sample confidence intervals for $D_k\lambda(\omega, s)$ require consistent estimates of both $g^{re}(\omega, s)$ and $g^{im}(\omega, s)$. An estimate for $g^{re}(\omega, s)$ has been obtained through $\widehat{g^{re}}(\omega, s)$. The local quadratic smoother applied to the real part of the periodograms can be applied to the imaginary part to obtain an estimate of $g^{im}(\omega, s)$. The ensuing pointwise confidence intervals can be used to assess if, for a given frequency, the first derivatives in the direction of the covariates are different from zero and that the enveloping spectral surface at that frequency depends on the covariate vector.

As an example of this extension to the spectral envelope, we consider data from the AgeWise study conducted at the University of Pittsburgh. The goal of the AgeWise study is to understand the connections between sleep and health, functioning, and well-being in older adults. In the study, data are collected from $N = 98$ adults between 60 and 89 years of age.

Each subject was monitored during a night of in-home sleep through ambulatory poly-somnography. Polysomnography is a comprehensive recording of the biophysiological changes that occur during sleep and includes the collection of electrocardiograph (ECG), electroencephalographic (EEG), electro-oculography (EOG), and electromyography (EMG) signals. The polysomnographic signals were used by a trained technician to score 20 seconds (s) epochs into stages of REM, NON-REM, or wakefulness. Our analysis considers 6 h immediately following the onset of sleep. The resulting qualitative time series for each subject consists of sleep stage as one of the $p = 3$ categories of REM, NON-REM, and wakefulness for each of the $T = 1080$ 20 s intervals during the first 6 h after sleep onset. Sleep stages for two subjects are plotted in Figure 13.6.

In addition to polysomnography, each patient completed the PSQI questionnaire from which a PSQI score was computed. The PSQI, which was introduced by Buysse et al. (1989), is a common instrument for measuring self-reported sleep quality. The PSQI questionnaire collects information about sleep quality and disturbance over a 1-month period. PSQI scores can take values between 0 and 21 with larger numbers representing poorer sleep quality. A score of 6 or larger has been shown to be clinically associated with poor sleep quality in a variety of settings. The observed PSQI scores in our sample of $N = 98$ subjects range from 1 to 16 with a mean of 8.88 and a standard deviation of 3.64. The goal of our investigation is to determine the association between the frequency domain properties of sleep stages and the single, that is, $q = 1$, quantitative variable of the PSQI score.

We choose NON-REM sleep as the referent group so that sleep stages for each subject are represented through indicator variables for wakefulness and for REM sleep. Local quadratic

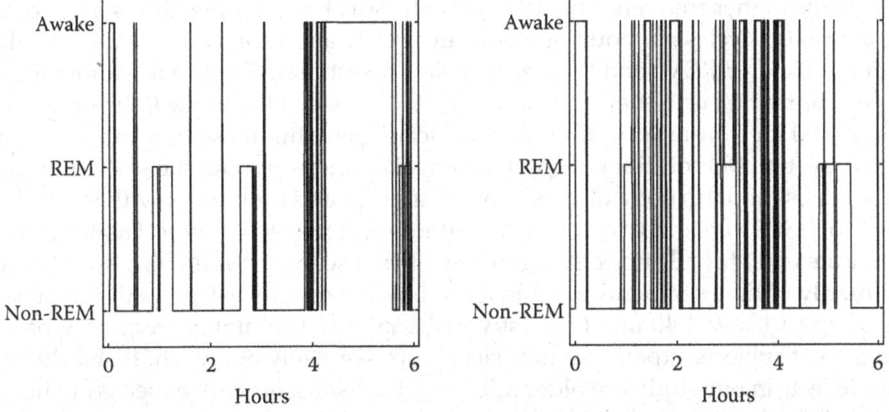

FIGURE 13.6
Sleep stages from two subjects for 6 h following sleep onset. The plot on the left is from a subject with a PSQI score of 1 while the plot on the right is from a subject with a PSQI score of 15.

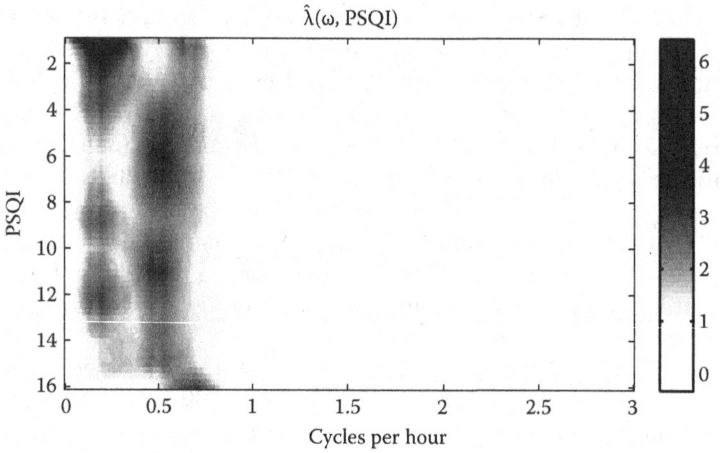

FIGURE 13.7
The estimated enveloping spectral surface for sleep staging conditional on PSQI score.

regression is implemented for the estimation of $\widehat{V}$ using the univariate Epanechnikov kernel and for the estimation of g^{re} using the spherical bivariate Epanechnikov kernel. The PSQI scores in our sample are approximately unimodal and symmetric; we estimate ϕ_s as the truncated normal distribution with support between 1 and 16, mean 8.88, and standard deviation 3.64.

The estimated enveloping spectral surface $\widehat{\lambda}(\omega, \text{PSQI})$ displayed in Figure 13.7 indicates that power exists primarily within two frequency bands: a band of low frequencies between 0.17 and 0.23 cycles per hour, or between one cycle every 4 h and 20 min and one cycle every 6 h, and a band of higher frequencies between 0.33 and 0.75 cycles per hour, or between one cycle every 1 h and 20 min and one cycle every 3 h. The estimated scalings $\widehat{\gamma}$ where the power in the enveloping spectral surface is achieved are approximately equivalent to an indicator variable for wakefulness for all frequencies in the low-frequency band and for all PSQI and approximately equivalent to an indicator variable for REM sleep for all frequencies in the higher-frequency band and for all PSQI . For example, the estimated optimal scaling at $\omega = 0.2$ cycles per hour, or 1 cycle every 5 h, and PSQI $= 2$ assigns the value 0.42 for awake, −0.09 for REM, and 0 for NON-REM. The estimated optimal scaling at $\omega = 0.5$ cycles per hour, or 1 cycle every 2 h, and PSQI $= 6$ assigns the value 0.01 for awake, 0.37 for REM, and 0 for NON-REM. To illustrate the scalings within these two bands, Figure 13.6 displays the estimated conditional spectra for wakefulness, or with the scaling that assigns 1 to wakefulness and 0 to both REM and NON-REM sleep, and for REM, or with the scaling that assigns 1 to REM, sleep and 0 to both wakefulness and NON-REM sleep. The low-frequency band indicates that for all subjects, regardless of PSQI score, 6 h following sleep onset consists primarily of time segments in which the subject is mostly asleep with the presence of epochs of up to 1 h and 40 min of mostly wakefulness. The higher-frequency band indicates that most subjects experience REM sleep approximately every 2 h. It should be noted that the subjects in our study are older adults and are subsequently expected to have more disturbed sleep than younger adults and children. The presented inference is specific to older adults and is not generalizable to other populations.

To investigate if the power of the enveloping spectral surface within the frequency bands for wakefulness and for REM sleep changes with PSQI score, we compute the estimated

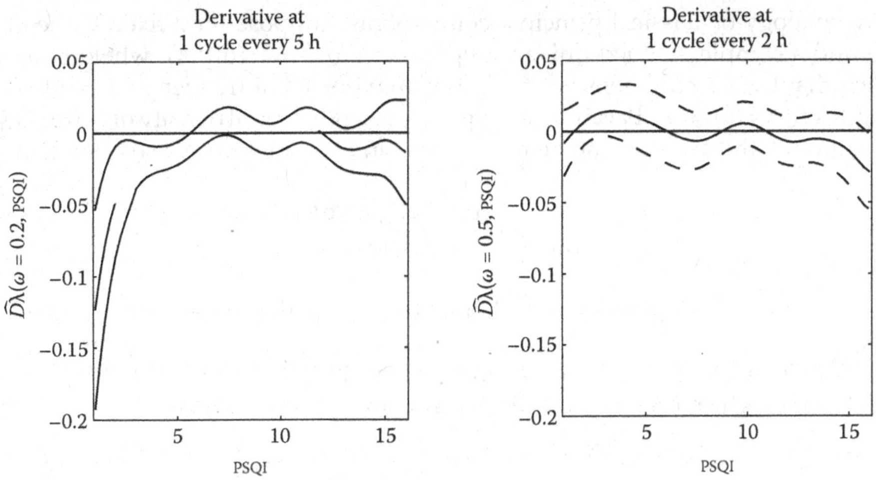

FIGURE 13.8
The estimated partial derivatives and pointwise 95% confidence intervals of the enveloping spectral surface for sleep staging with respect to the PSQI score for frequencies of one cycle every 5 h and one cycle every 2 h.

partial derivative of the enveloping spectral surface with respect to PSQI score at frequencies $\omega = 0.2$ cycles per hour, or one cycle every 5 h, and at $\omega = 0.5$, or one cycle every 2 h. Figure 13.8 displays $\widehat{D\lambda}(\omega = 0.2, \text{PSQI})$ and $\widehat{D\lambda}(\omega = 0.5, \text{PSQI})$ along with estimated pointwise 95% confidence intervals. The estimated derivative at one cycle every 5 h is negative for PSQI values less than 6 and not significantly different from zero for larger PSQI values. This indicates that, although significant power exists at periods wakefulness every 5 h for all PSQI values, this power decreases with an increase in PSQI for subjects with a PSQI less than 6 and is approximately the same for subjects with a PSQI greater than 6. Recall that the enveloping spectral surface is defined as the maximal normalized conditional power. From this result, we hypothesize that subjects experience approximately the same frequency of long periods of wakefulness regardless of their PSQI score but that subjects with a PSQI above 6 experience more short epochs of wakefulness. Consequently, the normalized conditional power for wakefulness once every 5 h is larger in subjects with smaller PSQI scores. The derivative of the enveloping spectral surface at the frequency of one cycle every 2 h is not significantly different from zero and we conclude that the amount of conditional power of REM sleep at one cycle every 2 h does not depend on PSQI score.

13.7 Principal Components

It may have been noticed that the theory associated with the spectral envelope is seemingly related to principal component analysis for time series. In this section, we summarize the technique so that the connection between the concepts is evident.

For the case of principal component analysis for time series, suppose we have a zero mean, $p \times 1$, stationary vector process X_t that has a $p \times p$ spectral density matrix given by $f_{xx}(\omega)$. Recall $f_{xx}(\omega)$ is a complex-valued, nonnegative definite, Hermitian matrix.

Using the analogy of classical principal components, suppose for a fixed value of ω, we want to find a complex-valued univariate process $Y_t(\omega) = c(\omega)^* X_t$, where $c(\omega)$ is complex, such that the spectral density of $Y_t(\omega)$ is maximized at frequency ω, and $c(\omega)$ is of unit length, $c(\omega)^* c(\omega) = 1$. Because, at frequency ω, the spectral density of $Y_t(\omega)$ is $f_y(\omega) = c(\omega)^* f_{xx}(\omega) c(\omega)$, the problem can be restated as, find complex vector $c(\omega)$ such that

$$\max_{c(\omega)\neq 0} \frac{c(\omega)^* f_{xx}(\omega) c(\omega)}{c(\omega)^* c(\omega)}. \tag{13.15}$$

Let $\{(\lambda_1(\omega), e_1(\omega)), \ldots, (\lambda_p(\omega), e_p(\omega))\}$ denote the eigenvalue–eigenvector pairs of $f_{xx}(\omega)$, where $\lambda_1(\omega) \geq \lambda_2(\omega) \geq \cdots \geq \lambda_p(\omega) \geq 0$, and the eigenvectors are of unit length. We note that the eigenvalues of a Hermitian matrix are real. The solution to (13.15) is to choose $c(\omega) = e_1(\omega)$, in which case the desired linear combination is $Y_t(\omega) = e_1(\omega)^* X_t$. For this choice,

$$\max_{c(\omega)\neq 0} \frac{c(\omega)^* f_{xx}(\omega) c(\omega)}{c(\omega)^* c(\omega)} = \frac{e_1(\omega)^* f_{xx}(\omega) e_1(\omega)}{e_1(\omega)^* e_1(\omega)} = \lambda_1(\omega). \tag{13.16}$$

This process may be repeated for any frequency ω, and the complex-valued process, $Y_{t1}(\omega) = e_1(\omega)^* X_t$, is called the first principal component at frequency ω. The kth principal component at frequency ω, for $k = 1, 2, \ldots, p$, is the complex-valued time series $Y_{tk}(\omega) = e_k(\omega)^* X_t$, in analogy to the classical case. In this case, the spectral density of $Y_{tk}(\omega)$ at frequency ω is $f_{y_k}(\omega) = e_k(\omega)^* f_{xx}(\omega) e_k(\omega) = \lambda_k(\omega)$.

The previous development of spectral domain principal components is related to the spectral envelope methodology as discussed in equation (13.4). In particular, the spectral envelope is a principal component analysis on the real part of $f_{xx}(\omega)$. Hence, the difference between spectral domain principal component analysis and the spectral envelope is that, for the spectral envelope, the $c(\omega)$ are restricted to be real. If, in the development of the spectral envelope, we allowed for complex scalings, the two methods would be identical.

Another way to motivate the use of principal components in the frequency domain was given in Brillinger (1981, Chapter 9). Although the technique appears to be different, it leads to the same analysis. In this case, we suppose we have a stationary, p-dimensional, vector-valued process X_t and we are only able to keep a univariate process Y_t such that, when needed, we may reconstruct the vector-valued process, X_t, according to an optimality criterion. Specifically, we suppose we want to approximate a mean-zero, stationary, vector-valued time series, X_t, with spectral matrix $f_{xx}(\omega)$, by a univariate process Y_t defined by

$$Y_t = \sum_{j=-\infty}^{\infty} c_{t-j}^* X_j, \tag{13.17}$$

where $\{c_j\}$ is a $p \times 1$ vector-valued filter, such that $\{c_j\}$ is absolutely summable; that is, $\sum_{j=-\infty}^{\infty} |c_j| < \infty$. The approximation is accomplished so the reconstruction of X_t from y_t, say,

$$\widehat{X}_t = \sum_{j=-\infty}^{\infty} b_{t-j} Y_j, \tag{13.18}$$

where $\{b_j\}$ is an absolutely summable $p\times 1$ filter, is such that the mean square approximation error

$$E\left\{\left(X_t - \widehat{X}_t\right)^* \left(X_t - \widehat{X}_t\right)\right\} \tag{13.19}$$

is minimized.

Let $b(\omega)$ and $c(\omega)$ be the transforms of $\{b_j\}$ and $\{c_j\}$, respectively. For example,

$$c(\omega) = \sum_{j=-\infty}^{\infty} c_j \exp(-2\pi i j \omega), \tag{13.20}$$

and, consequently,

$$c_j = \int_{-1/2}^{1/2} c(\omega) \exp(2\pi i j \omega) d\omega. \tag{13.21}$$

Brillinger (1981, Theorem 9.3.1) shows the solution to the problem is to choose $c(\omega)$ to satisfy (13.15) and to set $b(\omega) = \overline{c(\omega)}$. This is precisely the previous problem, with the solution given by (13.16). That is, we choose $c(\omega) = e_1(\omega)$ and $b(\omega) = \overline{e_1(\omega)}$; the filter values can be obtained via the inversion formula given by (13.21). Using these results, in view of (13.17), we may form the first principal component series, say Y_{t1}.

This technique may be extended by requesting another series, say, Y_{t2}, for approximating X_t with respect to minimum mean square error, but where the coherency between Y_{t2} and Y_{t1} is zero. In this case, we choose $c(\omega) = e_2(\omega)$. Continuing this way, we can obtain the first $q \le p$ principal component series, say, $Y_t = (Y_{t1}, \ldots, Y_{tq})'$, having spectral density $f_{yy}(\omega) = \text{diag}\{\lambda_1(\omega), \ldots, \lambda_q(\omega)\}$. The series Y_{tk} is the kth principal component series.

13.8 R Code

The following R script can be used to calculate the spectral envelope and optimal scalings. The script can also be used to perform dynamic analysis in a piecewise fashion. The scripts are used to produce Figures 13.4 and 13.5 and Table 13.3. The scripts require the use of the R package `astsa`, which must be downloaded and installed prior to running the script. The data are included in the package as `bnrf1ebv`, which lists the bp using the code `A` = 1, `C` = 2, `G` = 3, `T` = 4.

```
require(astsa)

u = factor(bnrf1ebv)                    # first, input the data as factors and then
x = model.matrix(~u-1)[,1:3]            # make an indicator matrix
Var = var(x)                            # var-cov matrix
xspec = mvspec(x, spans=c(7,7), plot=FALSE)  # spectral matrices are an array
                                             # called fxx
fxxr = Re(xspec$fxx)                    # fxxr is real(fxx)
```

```
#---  compute Q = Var^-1/2  ---#
ev = eigen(Var)
Q = ev$vectors%*%diag(1/sqrt(ev$values))%*%t(ev$vectors)

#--- compute spectral envelope and scale vectors ---#
num = xspec$n.used                # effective sample size
nfreq = length(xspec$freq)        # number of frequencies
specenv = matrix(0,nfreq,1)       # initialize the spectral envelope
beta = matrix(0,nfreq,3)          # initialize the scale vectors
 for (k in 1:nfreq){
  ev = eigen(2*Q%*%fxxr[,,k]%*%Q/num)  # get evalues of normalized spectral
                                        # matrix at freq k/n
  specenv[k] = ev$values[1]             # spec env at freq k/n is max evalue
  b = Q%*%ev$vectors[,1]                # beta at freq k/n
  beta[k,] = b/sqrt(sum(b^2))           # helps to normalize beta
 }

#---  output and graphics  ---#
 dev.new(height=3)
 par(mar=c(3,3,2,1), mgp=c(1.6,.6,0))
 frequency = (0:(nfreq-1))/num
plot(frequency, 100*specenv, type="l", ylab="Spectral Envelope (%)")
 title("Epstein-Barr  BNRF1")
## add significance threshold to plot ##
 m = xspec$kernel$m
 nuinv=sqrt(sum(xspec$kernel[-m:m]^2))
 thresh=100*(2/num)*exp(qnorm(.9999)*nuinv)*matrix(1,nfreq,1)
lines(frequency, thresh, lty="dashed", col="blue")

#--  details  --#
output = cbind(frequency, specenv, beta)  # results
colnames(output)=c("freq","specenv","A", "C", "G")

##--- dynamic part ---##
z = matrix(0,250,8)
output2 = array(0, dim=c(250,5,8))
colnames(output2) = c("freq","specenv","A", "C", "G")
for (j in 1:8){
   ind = (500*(j-1)+1):(500*j)
   if (j==8) ind=3501:length(bnrf1ebv)
xx = x[ind,]          # select subsequence -- the rest of the this part is the same
                      # as above
Var = var(xx)
xspec = mvspec(xx, spans=c(3,3), plot=FALSE)
fxxr = Re(xspec$fxx)
ev = eigen(Var)
Q = ev$vectors%*%diag(1/sqrt(ev$values))%*%t(ev$vectors)
num = xspec$n.used
nfreq = length(xspec$freq)
frequency = (0:(nfreq-1))/num
specenv = matrix(0, nfreq, 1)
beta = matrix(0, nfreq, 3)
 for (k in 1:nfreq){
  ev = eigen(2*Q%*%fxxr[,,k]%*%Q/num)
  specenv[k] = ev$values[1]
  b = Q%*%ev$vectors[,1]
  beta[k,] = b/sqrt(sum(b^2))
 }
```

```
if(j<8)  { z[,j] = specenv; output2[,,j] = cbind(frequency, specenv, beta) }
if(j==8) { z[1:240,8] = specenv; output2[1:240,,j] = cbind(frequency, specenv,
    beta) }
}

#--- output and graphics (results in output2)---#
zz = 100*t(z)          #  zz is 8x250
rowss = rep(1:8, each=2)
zz = zz[rowss,]        # now it's 16x250
# threshold
m = xspec$kernel$m
nuinv = sqrt(sum(xspec$kernel[-m:m]^2))
thresh=100*(2/num)*exp(qnorm(.995)*nuinv)

dev.new(height = 5)
par(mar=c(3,3,3,1), mgp=c(1.6,.6,0))
xa = 0:249/500
ya1 = 1736+500*0:8
rowss = c(1, rep(2:8, each=2), 9)
ya = ya1[rowss]
ya[seq(2,16,by=2)] = ya[seq(2,16,by=2)]-.5
levs = thresh*seq(0, 4.5, by=.5)
colr = gray(c(10,9,5,4.5,4,3,2,1,0)/10)
contour(ya, xa, zz, xlab="base pair", ylab="frequency", levels=levs, col=colr,
    main="Epstein-Barr  BNRF1", lwd=2, drawlabels=FALSE)
```

Acknowledgment

This work was supported, in part, by a grant from the U.S. National Science Foundation.

References

Brillinger, D. R. (1981). *Time Series: Data Analysis and Theory*, vol. 36. Society for Industrial and Applied Mathematics, Philadelphia.

Buysse, D. J., Reynolds, C. F., Monk, T. H., Berman, S. R., and Kupfer, D. J. (1989). The Pittsburgh sleep quality index: A new instrument for psychiatric practice and research. *Psychiatry Research*, 28(2):193–213.

Fan, J. and Gijbels, I. (1996). *Local Polynomial Modelling and Its Applications*, vol. 66. Chapman & Hall, London.

Ioshikhes, I., Bolshoy, A., Derenshteyn, K., Borodovsky, M., and Trifonov, E. N. (1996). Nucleosome DNA sequence pattern revealed by multiple alignment of experimentally mapped sequences. *Journal of Molecular Biology*, 262(2):129–139.

Krafty, R. T., Xiong, S., Stoffer, D. S., Buysse, D. J., and Hall, M. (2012). Enveloping spectral surfaces: Covariate dependent spectral analysis of categorical time series. *Journal of Time Series Analysis*, 33(5):797–806.

Magnus, J. and Neudecker, H. (1988). *Matrix Differential Calculus with Applications in Statistics and Econometrics*. Wiley & Sons, New York.

R Core Team (2013). *R: A Language and Environment for Statistical Computing*. R Foundation for Statistical Computing, Vienna, Austria.

Satchwell, S. C., Drew, H. R., and Travers, A. A. (1986). Sequence periodicities in chicken nucleosome core DNA. *Journal of Molecular Biology*, 191(4):659–675.

Shumway, R. and Stoffer, D. (2011). *Time Series Analysis and Its Applications*, 3rd edn. Springer, New York.

Stoffer, D. S. (1991). Walsh-Fourier analysis and its statistical applications. *Journal of the American Statistical Association*, 86(414):461–479.

Stoffer, D. S., Scher, M. S., Richardson, G. A., Day, N. L., and Coble, P. A. (1988). A Walsh-Fourier analysis of the effects of moderate maternal alcohol consumption on neonatal sleep-state cycling. *Journal of the American Statistical Association*, 83(404):954–963.

Stoffer, D. S., Tyler, D. E., and McDougall, A. J. (1993a). Spectral analysis for categorical time series: Scaling and the spectral envelope. *Biometrika*, 80(3):611–622.

Stoffer, D. S., Tyler, D. E., McDougall, A. J., and Schachtel, G. (1993b). Spectral analysis of DNA sequences (with discussion). *Bulletin of the International Statistical Institute*, 1:345–361.

14

Coherence Consideration in Binary Time Series Analysis

Benjamin Kedem

CONTENTS

14.1 Introduction

In a recent study of mortality forecasting in the United States, it has been found that quite often mortality patterns in a given state are influenced by mortality trends in neighboring states, and the inclusion of interaction terms from the latter in log-linear models can substantially improve mortality forecasting in the given state (Khan et al., 2004). This motivates the problem of identifying interaction terms expressed as products of covariates of the form $x_t x_{t-k}$ in other time series regression models including logistic regression for binary time series. This chapter discusses a spectral measure for interaction identification and its application in binary time series regression. The spectral measure for interaction identification, called *residual coherence*, depends on a certain nonlinear extension of the well-known measure of (squared) coherence. It is helpful, therefore, to provide first some background leading to the definition of residual coherence and illustrate its use. This is followed by an application to logistic regression for binary time series.

14.2 Coherence for Quadratic Systems

Let $x_t, t = \cdots, -1, 0, 1, \ldots$, be a zero mean stationary time series admitting a spectral representation in terms of a process of orthogonal increments $\xi_x(\lambda), \lambda \in (-\pi, \pi]$. Define a system of degree n with input x_t and output y_t by the nth degree polynomial functional

$$y_t = \int \cdots \int \exp[it(\lambda_1 + \cdots + \lambda_n)] H_n(\lambda_1, \ldots, \lambda_n) \xi_x(d\lambda_1) \cdots \xi_x(d\lambda_n)$$
$$+ \cdots + \int \exp(it\lambda_1) H_1(\lambda_1) \xi_x(d\lambda_1) + H_0 \tag{14.1}$$

where H_0 is a constant and $H_j, j = 1, \ldots, n$ are complex continuous and bounded kernels. H_n is said to be the leading kernel. Functionals x, y of the form (14.1) are said to be orthogonal if $E(x\bar{y}) = 0$. In addition, we shall assume that all relevant spectra and cross spectra are well defined.

Nonlinear systems that admit a representation of the form (14.1) have been studied in Tick (1961), Kimelfeld (1972, 1974), Nelson and Van Ness (1973a,b), and Priestley (1988) among others.

Consider now a quadratic system, that is, $n = 2$, with input x_t and output y_t. Then, following a procedure described in Lectures 1 through 4 of Wiener (1958), y_t can be expressed as a sum of orthogonal functionals $G_j(K_j, y_t)$ with leading kernels K_j,

$$y_t = \sum_{j=0}^{2} G_j(K_j, y_t) \tag{14.2}$$

where $G_0(K_0, y_t) = K_0$ is a constant, $G_1(K_1, y_t)$ is a first-order (linear) functional, and $G_2(K_2, y_t)$ is a quadratic functional with leading kernel K_2. Assume that $E(y_t) = 0$ for all t. Then $K_0 = 0$ with probability one, and by the orthogonality of G_1 and G_2, $K_1(\lambda) = f_{xy}(\lambda)/f_{xx}(\lambda)$ where f_{xx} is the spectral density of x_t, and f_{xy} is the cross-spectral density of x_t and y_t.

In general, it is difficult to determine K_2 without imposing conditions on x_t such as the Gaussian assumption (Priestley, 1988, Section 3.3). Tick (1961) determined the kernels of a quadratic functional when x_t is Gaussian employing the cross bi-spectrum between x_t and y_t. In Kimelfeld (1972, 1974), by the use of *lag processes*, it is shown how to bypass the Gaussian assumption by approximating G_2 itself without determining K_2, using a class of approximating functionals for which we can determine all the kernels as follows.

For integers $u_k, k = 1, \ldots, n$ define the lag processes $U_k(t)$ by the centered product,

$$U_k(t) = x_t x_{t+u_k} - R_{xx}(u_k) \tag{14.3}$$

where R_{xx} is the autocovariance of x_t. Under the assumption that the $U_k(t)$ are stationary, it can be shown that for sufficiently large n, y_t in (14.2) admits the mean-square representation,

$$y_t = G_1\left(\frac{f_{xy}(\lambda)}{f_{xx}(\lambda)}, y_t\right) + \sum_{k=1}^{n} \left[\int e^{it\lambda} B_k(\lambda) \xi_{U_k}(d\lambda) + \int e^{it\lambda} A_k(\lambda) \xi_x(d\lambda)\right], \tag{14.4}$$

where

$$A_k(\lambda) = -\frac{B_k(\lambda) f_{xu_k}(\lambda)}{f_{xx}(\lambda)}. \tag{14.5}$$

To get the $B_k(\lambda)$, define

$$\begin{aligned}\mathbf{f}_{uu}(\lambda) &= (f_{u_iu_j}(\lambda)),\\ \mathbf{f}_{ux}(\lambda) &= (f_{u_1x}(\lambda),\ldots,f_{u_nx}(\lambda))', \mathbf{f}_{uy}(\lambda) = (f_{u_1y}(\lambda),\ldots,f_{u_ny}(\lambda))',\\ \mathbf{B}(\lambda) &= (B_1(\lambda),\ldots,B_n(\lambda))'.\end{aligned}$$

Then, observing that $\mathbf{f}_{ux}(\lambda)$ is the conjugate transpose of $\mathbf{f}_{xu}(\lambda)$, we have

$$\mathbf{B}(\lambda) = \left(\mathbf{f}_{uu}(\lambda) - \frac{1}{f_{xx}(\lambda)}\mathbf{f}_{ux}(\lambda)\mathbf{f}_{xu}(\lambda)\right)^{-1}\left(\mathbf{f}_{uy}(\lambda) - \frac{f_{xy}(\lambda)}{f_{xx}(\lambda)}\mathbf{f}_{ux}(\lambda)\right). \tag{14.6}$$

It is important to note that the definition of A_k in (14.5) guarantees the orthogonality of the approximating sum in (14.4) and G_1. Thus, if we define

$$\begin{aligned}y_t &= G(t) + \epsilon_t\\ &\equiv G_1\left(\frac{f_{xy}(\lambda)}{f_{xx}(\lambda)}, y_t\right) + \sum_{k=1}^{n}\left[\int e^{it\lambda}B_k(\lambda)\xi_{U_k}(d\lambda) + \int e^{it\lambda}A_k(\lambda)\xi_x(d\lambda)\right] + \epsilon_t\end{aligned} \tag{14.7}$$

where ϵ_t is orthogonal to G_1 and to the quadratic sum, then clearly A_k and B_k do not change, and in addition,

$$0 \le S_2(\lambda; u_1,\ldots,u_n) \equiv \frac{f_{GG}(\lambda)}{f_{yy}(\lambda)} \le 1.$$

It is easy to see then that

$$\begin{aligned}S_2(\lambda; u_1,\ldots,u_n) &= \frac{|f_{xy}(\lambda)|^2}{f_{xx}(\lambda)f_{yy}(\lambda)}\\ &\quad + \frac{1}{f_{yy}(\lambda)}\mathbf{B}'(-\lambda)\left(\mathbf{f}_{uu}(\lambda) - \frac{1}{f_{xx}(\lambda)}\mathbf{f}_{ux}(\lambda)\mathbf{f}_{xu}(\lambda)\right)\mathbf{B}(\lambda)\end{aligned} \tag{14.8}$$

where we recognize that the first term on the right-hand side of (14.8) is the well-known (squared) coherence that measures the degree of linear relationship between x_t and y_t in (14.7); see Koopmans (1974, p. 137). The other term is due to the quadratic term in (14.7) corresponsing to the lag processes $U_k(t)$. We shall refer to $S_2(\lambda; u_1,\ldots,u_n)$ as *lagged coherence* (Kimelfeld, 1972). It measures the validity of models of the form (14.7) by observing that when $S_2(\lambda; u_1,\ldots,u_n)$ is close to 1 for all $\lambda \in [0,\pi]$ then the signal-to-noise ratio is high. Clearly, $S_2(\lambda; u_1,\ldots,u_n)$ may be close to one on all or part of $[0,\pi]$ due to the quadratic term in (14.7) represented by the sum and not as a result of the linear component G_1, in which case the system is substantially quadratic. Similarly, $S_2(\lambda; u_1,\ldots,u_n)$ could be large due to the linear component, in which case the system is substantially linear.

It is interesting to compare the lagged coherence (14.8) with the "quadratic coherency" of Tick (1961) which assumes that x_t is Gaussian,

$$\text{quad. coh}(\omega) = \frac{|f_{xy}(\lambda)|^2}{f_{xx}(\lambda)f_{yy}(\lambda)} + \frac{\frac{1}{2}}{f_{yy}(\lambda)} \int \frac{|\text{C.B.S.}(\omega-\lambda,\lambda)|^2}{f_{xx}(\omega-\lambda)f_{xx}(\lambda)} d\lambda \tag{14.9}$$

where C.B.S. stands for the cross bi-spectrum, the Fourier transform of $E[x_{t+t_2}x_{t+t_1}(y_t - E(y_t))]$ as a function of t_1, t_2. In both (14.8) and (14.9), the quadratic contribution is measured as an augmentation of the linear coherence, and both are between 0 and 1; for proofs see Kimelfeld (1972).

14.2.1 Residual Coherence

The lagged coherence $S_2(\lambda; u_1, \ldots, u_n)$ can also help in the selection of the lags themselves, where $u'_1, \ldots, u'_n$ are preferable to $u_1, \ldots, u_n$ if

$$S_2(\lambda; u'_1, \ldots, u'_n) \geq S_2(\lambda; u_1, \ldots, u_n), \quad \forall\lambda \in [0, \pi].$$

In this chapter, we make use of this idea in the case of a single lag u as follows. First, it is more convenient to define a lag process using the notation

$$X_u(t) = x_t x_{t-u} - R_{xx}(u), \quad u = 0, 1, 2, \ldots. \tag{14.10}$$

Consider the model

$$y_t = \sum_{k=-\infty}^{\infty} l_k x_{t-k} + \sum_{k=-\infty}^{\infty} b_k X_u(t-k) + \epsilon_t \tag{14.11}$$

where ϵ_t is independent noise. By adding and subtracting an appropriate linear term of the form $\sum_{k=-\infty}^{\infty} a_k x_{t-k}$ we can rewrite (14.11) as in (14.4), and in that case the lagged coherence reduces to something more palatable (Kedem-Kimelfeld, 1975),

$$S_2(\lambda; u) = S_1(\lambda) + \frac{|B(\lambda)|^2}{f_{yy}(\lambda)} \left[f_{x_u x_u}(\lambda) - \frac{|f_{xx_u}(\lambda)|^2}{f_{xx}(\lambda)} \right], \quad -\pi < \lambda \leq \pi \tag{14.12}$$

where $u = 0, 1, 2, \ldots$, $S_1(\lambda)$ is the linear coherence as in (14.8),

$$S_1(\lambda) = \frac{|f_{xy}(\lambda)|^2}{f_{xx}(\lambda)f_{yy}(\lambda)}, \tag{14.13}$$

and

$$B(\lambda) = \frac{f_{xx}(\lambda)f_{x_u y}(\lambda) - f_{x_u x}(\lambda)f_{xy}(\lambda)}{f_{xx}(\lambda)f_{x_u x_u}(\lambda) - |f_{xx_u}(\lambda)|^2} \quad -\pi < \lambda \leq \pi.$$

Clearly, $0 \le S_1(\lambda) \le 1$ for all $\lambda \in (-\pi, \pi]$, and similarly

$$0 \le S_2(\lambda; u) \le 1, \quad -\pi < \lambda \le \pi, u = 0, 1, 2, \ldots$$

For a given lag u, the influence of $X_u(t)$ on y_t can be measured by noting a significant increase in $S_2(\lambda; u)$ relative to the linear coherence $S_1(\lambda)$, for some or all $\lambda \in [0, \pi]$. Alternatively, as suggested recently in Khan et al. (2004), we can use the maximum *residual coherence* defined as

$$RS(u) = \max_{\lambda}\{S_2(\lambda; u) - S_1(\lambda)\}, \quad u = 0, 1, 2, \ldots \tag{14.14}$$

to measure the influence of the "interaction" $X_u(t)$ on y_t. This can be done graphically. As a graphical display, $RS(u)$ resembles the periodogram where the "lag is replaced by frequency," language related to me years ago by the late Melvin Hinich. In both measures, one tries to discern graphically conspicuous ordinates and identify by this important lags in the case of residual coherence and important frequencies in the case of the periodogram. And like in periodogram analysis, the residual coherence is elevated at secondary conspicuous lags provided the corresponding lag processes have significant coefficients in models of the form (14.4). This is illustrated in the analysis of model (14.16).

In this connection, Hinich (1979), assuming a stationary Gaussian input, presents a procedure for determining the values of multiple lags when there is a finite number of lag processes, taking advantage of the relationship between the weights of the lag processes in a quadratic system and the sample cross bi-spectrum between the input and output series. To determine the true lags, he too presents a graphical device called *lagstrum*, in which lag plays the role of frequency.

14.2.1.1 Examples: Residual Coherence Applied to Clipped Binary Series

Suppose x_t is a first-order autoregressive process $x_t = 0.3x_{t-1} + \epsilon_t$ where ϵ_t is standard logistic noise, and consider an autoregression plus a past interaction covariate $x_{t-1}x_{t-2}$,

$$z_t = 0.8z_{t-1} + 1.5x_{t-1}x_{t-2} + \eta_t, \quad t = 1, \ldots, 156 \tag{14.15}$$

where η is again standard logistic noise. Except for a constant, $x_{t-1}x_{t-2}$ is a lag process with $u = 1$, and we would expect the residual coherence obtained from (x_t, z_t) to peak at $u = 1$. This can be seen clearly in the bar plot at the top of Figure 14.1. Clipping z_t at level 5 we obtain a binary time series

$$y_t = \begin{cases} 1, & z_t \ge 5 \\ 0, & z_t < 5 \end{cases}, \quad t = 1, \ldots, 156.$$

The bar plot at the bottom of Figure 14.1 obtained from (x_t, y_t) again is maximized at $u = 1$ as expected, since in general clipping operations retain to a degree useful spectral information from the original baseline series which in the present case is z_t; see Kedem (1980). Very similar bar plots are obtained when ϵ_t and η_t are both Gaussian. Thus, the residual coherence $RS(u)$ points to a possible association between y_t and $x_{t-1}x_{t-2}$.

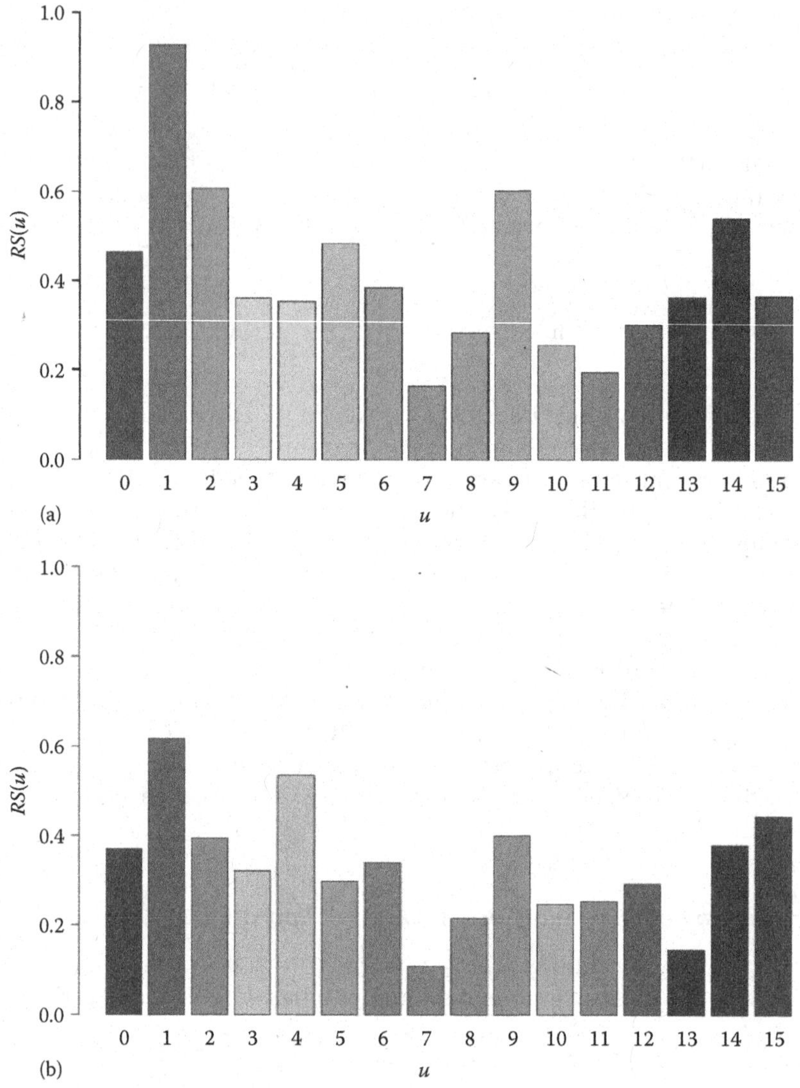

FIGURE 14.1
Maximum residual coherence $RS(u)$ obtained from (x_t, z_t) (a) and from (x_t, y_t) (b). $RS(u)$ peaks at $u = 1$.

Adding another past interaction term with lag $u = 2$, we have

$$z_t = 0.8z_{t-1} + 1.5x_{t-1}x_{t-2} - 1.3x_{t-1}x_{t-3} + \eta_t, \quad t = 1, \ldots, 156. \tag{14.16}$$

Repeating these same steps with the new z_t, we see from Figure 14.2 the residual coherence from both (x_t, z_t) and (x_t, y_t) peaks at $u = 1, 2$. We see that $RS(u)$ points to a possible association between y_t and the interaction terms $x_{t-1}x_{t-2}$ and $x_{t-1}x_{t-3}$.

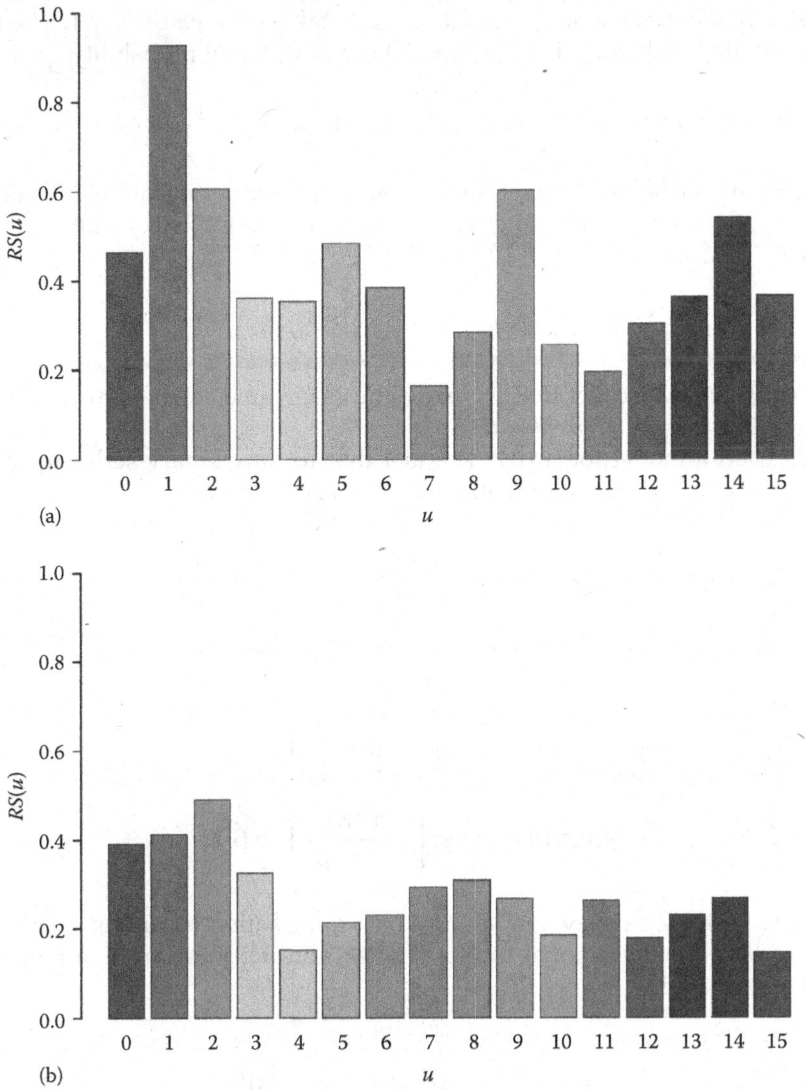

FIGURE 14.2
Maximum residual coherence $RS(u)$ obtained from (x_t, z_t) (a) and from (x_t, y_t) (b). $RS(u)$ peaks at $u = 1, 2$.

14.3 Logistic Regression for Binary Time Series

The following is a brief review of basic facts about logistic regression when the observations constitute a binary time series. To overcome the dependence in the data the inference is conditional, taking into account all that is known to the observer at time t. This may include past values of the time series of interest as well as past values of covariates; see Kedem and Fokianos (2002).

Consider a binary time series y_t, $t = 1, 2, \ldots, N$ taking the value 0 or 1. The regression problem associated with such data is to model the conditional probability

$$\pi_t(\beta) \equiv P_\beta(y_t = 1|\mathcal{F}_{t-1}) \tag{14.17}$$

where β is a p-dimensional parameter vector and $\mathcal{F}_{t-1}$ represents all that is known to the observer at time $t-1$ about the time series y_t itself and about its covariates. Clearly, with covariates $\mathbf{Z}_{t-1}, \mathbf{Z}_{t-2}, \ldots,$

$$\mathcal{F}_{t-1} = \sigma(y_{t-1}, y_{t-2}, \ldots, \mathbf{Z}_{t-1}, \mathbf{Z}_{t-2}, \ldots)$$

and $\mathcal{F}_{t-1} \subset \mathcal{F}_t$. This last fact that is very natural for time series is instrumental in the estimation of β using *partial likelihood*.

First, we need a model for $\pi_t(\beta)$. The fact that for any binary series the conditional log-odds is given by a linear combination of "interaction" terms

$$\log\left\{\frac{P(y_t = 1|y_{t-1}, \ldots, y_1)}{P(y_t = 0|y_{t-1}, \ldots, y_1)}\right\}$$
$$= \theta_t + \sum_{i<t} \theta_{it} y_i + \sum_{i<j<t} \theta_{ijt} y_i y_j + \cdots + \theta_{1\ldots t} y_1 \cdots y_{t-1} \tag{14.18}$$

motivates the logistic regression model which we shall use,

$$\text{logit}(\pi_t(\beta)) \equiv \log\left\{\frac{\pi_t(\beta)}{1 - \pi_t(\beta)}\right\} = \beta' \mathbf{Z}_{t-1}. \tag{14.19}$$

The covariate vector $\mathbf{Z}_{t-1}$ may contain any past information including past values of y_t itself. Using only the conditional model (14.19), and *not* joint information, the partial likelihood of β is simply

$$\text{PL}(\beta) = \prod_{t=1}^{N} [\pi_t(\beta)]^{y_t} [1 - \pi_t(\beta)]^{1-y_t}.$$

The solution $\hat{\beta}$ of

$$\mathbf{S}_N(\beta) \equiv \nabla \log \text{PL}(\beta) = \sum_{t=1}^{N} \mathbf{Z}_{t-1}(Y_t - \pi_t(\beta)) = \mathbf{0} \tag{14.20}$$

is the maximum partial likelihood estimator, and replacing N by t we obtain the score vector process $\{\mathbf{S}_t(\beta)\}$, $t = 1, \ldots, N$. It follows that $\{\mathbf{S}_t(\beta)\}$ is a martingale with respect to the filtration $\mathcal{F}_0 \subset \mathcal{F}_1 \subset \mathcal{F}_2 \subset \cdots$, a fact used in the derivation of the asymptotic distribution of $\hat{\beta}$.

To derive asymptotic results we need certain regularity conditions, including the requirement that the covariates are "well behaved." By this, we mean that there is a probability measure ν on R^p such that $\int_{R^p} \mathbf{z}\mathbf{z}'\nu(d\mathbf{z})$ is positive definite and such that for Borel sets $A \subset R^p$,

$$\frac{1}{N}\sum_{t=1}^{N} I_{[\mathbf{z}_{t-1}\in A]} \to \nu(A),$$

in probability as $N \to \infty$, at the true value of β. In particular, this implies that

$$\frac{\mathbf{H}_N(\beta)}{N} \to \mathbf{G}(\beta) = \int_{R^p} \frac{\exp(\beta'\mathbf{z})}{(1+\exp(\beta'\mathbf{z}))^2}\mathbf{z}\mathbf{z}'\nu(d\mathbf{z}),$$

in probability, where

$$\mathbf{H}_N(\beta) = \nabla\nabla'(-\log \mathrm{PL}(\beta)) = \sum_{t=1}^{N} \mathbf{Z}_{t-1}\mathbf{Z}'_{t-1}\frac{\exp(\beta'\mathbf{Z}_{t-1})}{[1+\exp(\beta'\mathbf{Z}_{t-1})]^2}$$

is the *observed information matrix*, and $\mathbf{G}(\beta)$ is the information matrix per single observation for estimating β. All this leads to the fact that β is asymptotically normal,

$$\sqrt{N}(\hat{\beta} - \beta) \to \mathcal{N}_p(\mathbf{0}, \mathbf{G}^{-1}(\beta)),$$

in distribution. From this fact, we derive the usual hypothesis tests including the log-partial likelihood ratio test. See Slud and Kedem (1994) and Kedem and Fokianos (2002) for many more results as well as proofs.

Since the score equation (14.20) is the same as the score equation that would have been obtained had the data been independent, we may use standard statistical packages for partial likelihood estimation and hypothesis testing. However, the results must be interpreted conditionally.

14.3.1 Interactions in Logistic Regression

The residual coherence introduced earlier can identify or point to potentially useful interactions, but it should be considered as a suggestive device only. Significance tests can be used to ascertain that an identified interaction term is indeed a useful covariate. Thus, the importance of the identified interactions is determined from within logistic regression and not by coherence testing. In this way, we avoid the inclusion of covariates judged useful on account of coherence but not significant on account of logistic regression analysis.

To illustrate this interplay between coherence analysis and logistic regression, consider the logistic regression model

$$\mathrm{logit}(\pi_t(\beta)) = \beta_0 + \beta_1 y_{t-1} + \beta_2 x_{t-1} + \beta_3 x_{t-2} + \beta_4 x_{t-1}x_{t-2} \tag{14.21}$$

where the binary time series y_t is obtained by clipping z_t in (14.15) at level 5. Recall that an application of residual coherence to (x_t, y_t) points to $x_{t-1}x_{t-2}$ as a potential covariate. This is supported by the R output in Table 14.1 which points to the great significance of this (shifted) interaction term $x_{t-1}x_{t-2}$. However, x_{t-1} and x_{t-2} are not significant.

TABLE 14.1

Logistic regression results for Model (14.21) showing the significance of the interaction $x_{t-1}x_{t-2}$. However, x_{t-1} and x_{t-2} are not significant.

	$\hat{\beta}$	SE	p-value
Intercept	−2.18220	0.38383	1.31e-08
y_{t-1}	3.66744	0.54258	1.39e-11
x_{t-1}	0.03879	0.15594	0.804
x_{t-2}	−0.22056	0.16708	0.187
$x_{t-1}x_{t-2}$	0.57688	0.12933	8.18e-06

TABLE 14.2

Logistic regression results for Model (14.22) showing the significance of the interaction covariates $x_{t-1}x_{t-2}$ and $x_{t-1}x_{t-3}$. However, x_{t-1}, x_{t-2}, and x_{t-3} are not significant.

	$\hat{\beta}$	SE	p-value
Intercept	−1.88572	0.38142	7.66e-07
y_{t-1}	2.90036	0.52766	3.87e-08
x_{t-1}	−0.06225	0.13068	0.634
x_{t-2}	−0.04295	0.15389	0.780
x_{t-3}	−0.14459	0.12116	0.233
$x_{t-1}x_{t-2}$	0.61570	0.12494	8.30e-07
$x_{t-1}x_{t-3}$	−0.45104	0.09702	3.33e-06

Repeating this for model (14.16) where the residual coherence identified two potential interaction covariates with lags $u = 1, 2$, the estimation results of the extended model

$$\text{logit}(\pi_t(\beta)) \\ = \beta_0 + \beta_1 y_{t-1} + \beta_2 x_{t-1} + \beta_3 x_{t-2} + \beta_4 x_{t-3} + \beta_5 x_{t-1}x_{t-2} + \beta_6 x_{t-1}x_{t-3} \quad (14.22)$$

in Table 14.2 show that these (shifted) interaction covariates $x_{t-1}x_{t-2}$ and $x_{t-1}x_{t-3}$ are indeed decisively significant. At the same time, x_{t-1}, x_{t-2}, and x_{t-3} themselves are not significant.

14.3.2 Application to LA Mortality

Shumway et al. (1988) analyzed filtered weekly mortality data in Los Angeles County from January 1, 1970 to December 31, 1979, using a regression model in terms of temperature (quadratic) and a log-pollution covariate, plus autoregressive noise. The data were reanalyzed in Kedem and Fokianos (2002) by Poisson regression, where it was found that total mortality z_t, $t = 1, \ldots, 508$, depends on itself, temperature T_t, and log carbon monoxide $C_t = \log(CO_t)$. It is interesting to see what further insight might be gained using logistic regression.

The filtered data range from 142.13 to 231.73 with mean 169.05, so that level 180 is above average. Define the clipped series,

$$y_t = \begin{cases} 1, & z_t \geq 180 \\ 0, & z_t < 180 \end{cases}, \quad t = 1, \ldots, 508.$$

This gives the binary time series

1111010010001010000100000000000010000001101011111111111111111000000000000000000010000100010
1010001111111111111000000000000001000000001000000000000010011011111111110100000000100000000
0000000000000000000000000100110000000010000000000000000000000000000000000000000000111111111
0000000000000000000000000000000000000000000000000011111000000000000000000000000000000000000
0000000000100000000000000000000000000000000000000000000000000001110110000000000000000000
0000000000000001000000000011111001000000000000000101000000000000000000000000000

Since we intend to use C_t with values ranging from 0.924 to 3.109, it is sensible to replace T_t by $x_t = T_t/10$. The residual coherence obtained from (x_t, y_t) is shown in Figure 14.3, suggesting (past of) $x_t x_{t-k}$, $k = 0, 2, 4$ as possible interaction covariates for logistic regression. Table 14.3 gives the Akaike information criterion (AIC) results for some models selected out of many more. The table shows that the AIC is minimized at the model

$$\text{logit}(\pi_t(\beta)) = \beta_0 + \beta_1 y_{t-1} + \beta_2 y_{t-2} + \beta_3 C_{t-1} + \beta_4 x_{t-1} x_{t-3} \tag{14.23}$$

containing the interaction $x_{t-1}x_{t-3}$ (past of $x_t x_{t-2}$) which appears more useful than its factors. The estimates are given in Table 14.4. Apparently, $x_{t-1}x_{t-3}$ is quite significant. Model (14.23) can be judged further from the plots of the estimated autocorrelation and cumulative periodogram of the residuals shown in Figure 14.4.

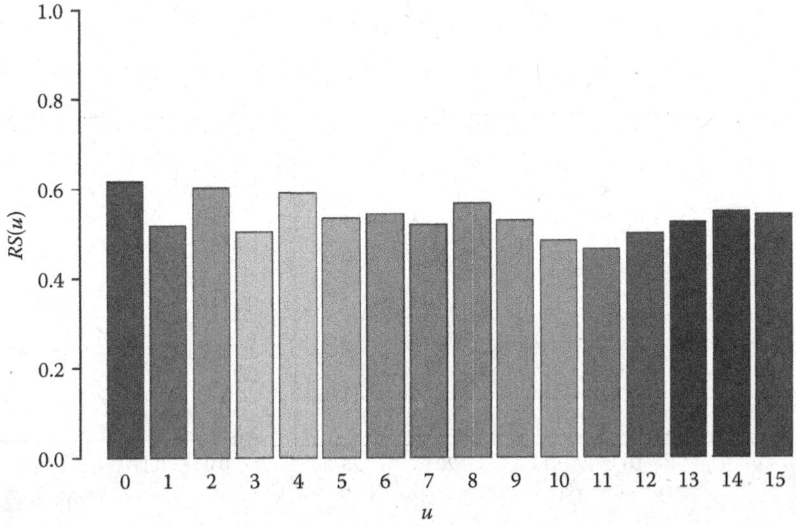

FIGURE 14.3
Maximum residual coherence $RS(u)$ obtained from scaled temperature and clipped mortality (x_t, y_t).

TABLE 14.3

Logistic regression models for clipped mortality data showing that the interaction $x_{t-1}x_{t-3}$ is a potentially useful covariate.

	Model	AIC
1.	$y_{t-1} + y_{t-2}$	343.97
2.	$y_{t-1} + C_{t-1}$	325.84
3.	$y_{t-1} + y_{t-2} + C_{t-1}$	306.95
4.	$y_{t-1} + C_{t-1} + x_{t-1}x_{t-5}$	302.71
5.	$y_{t-1} + y_{t-2} + C_{t-1} + x_{t-3}$	298.94
6.	$y_{t-1} + y_{t-2} + C_{t-1} + x_{t-1}^2$	298.47
7.	$y_{t-1} + y_{t-2} + C_{t-1} + x_{t-1}$	298.35
8.	$y_{t-1} + y_{t-2} + C_{t-1} + x_{t-1} + x_{t-2} + x_{t-3}$	297.18
9.	$y_{t-1} + y_{t-2} + C_{t-1} + x_{t-1} + x_{t-3}$	297.03
10.	$y_{t-1} + y_{t-2} + C_{t-1} + x_{t-1}x_{t-3} + x_{t-1}x_{t-5}$	295.85
11.	$y_{t-1} + y_{t-2} + C_{t-1} + x_{t-1}x_{t-3}$	295.08

TABLE 14.4

Logistic regression results for Model (14.23) showing the significance of the interaction covariate $x_{t-1}x_{t-3}$.

	$\hat{\beta}$	SE	p-value
Intercept	−3.544	1.315	7×10^{-3}
y_{t-1}	1.662	0.351	2.12×10^{-6}
y_{t-2}	1.148	0.361	10^{-3}
C_{t-1}	2.031	0.401	4.05×10^{-7}
$x_{t-1}x_{t-3}$	−0.059	0.017	5×10^{-4}

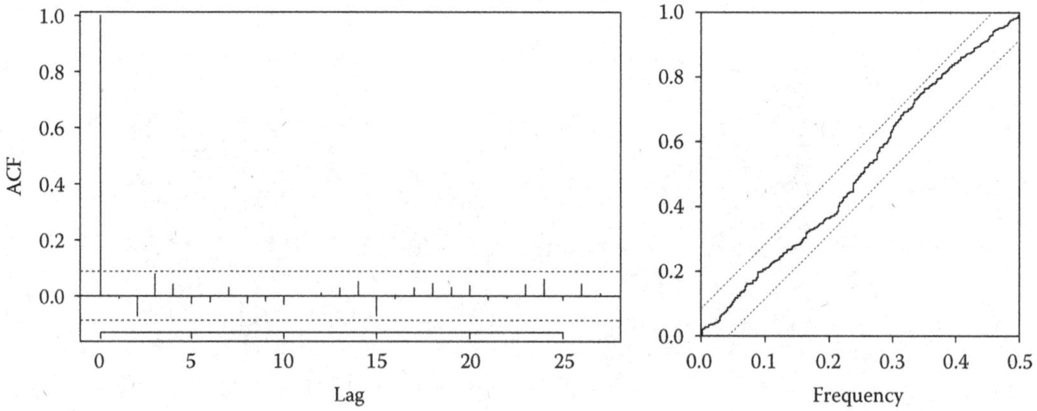

FIGURE 14.4
Autocorrelation and cumulative periodogram of the residuals from model (14.23).

14.4 Discussion

In this chapter, we explored the use of residual coherence as a graphical tool for identifying potentially useful interactions in logistic regression for binary time series. In many cases, in practice, the inclusion of interaction terms leads to improved models and, as was illustrated, an interaction covariate could be more significant than its factors. There are situations, however, when the difference between $S_2(\lambda; u)$ and $S_1(\lambda)$ is relatively large for a frequency band due to a local minimum in the linear coherence $S_1(\lambda)$ giving rise to a relatively large residual coherence $RS(u)$ for some u without a substantial contribution from $X_u(t)$. This is another reason why we need to couple the residual coherence with some further evidence as to the importance of an identified interaction covariate. This can be done, for example, by hypothesis testing and model selection criteria and by some sort of residual analysis.

Clearly, instead of prescreening the variables using residual coherence, we could simply include interaction terms in the model and apply model selection. The advantage of prescreening using residual coherence is that the search for useful interaction terms is facilitated using spectral information which accommodates model selection and hypothesis testing. Moreover, exploring potential general relationships among time series is an important time series problem where coherence measures, including residual coherence, as well as other measures have an important role before fitting parametric models.

References

Hinich, M. (1979). Estimating the lag structure of a nonlinear time series model. *Journal of the American Statistical Association*, 74(366):449–452.

Kedem, B. (1980). *Binary Time Series*. Dekker, New York.

Kedem, B. and Fokianos, K. (2002). *Regression Models for Time Series Analysis*. Wiley, Hoboken, NJ.

Kedem-Kimelfeld, B. (1975). Estimating the lags of lag processes. *Journal of the American Statistical Association*, 70(351):603–605.

Khan, D., Katzoff, M., and Kedem, B. (2014). Coherence structure and its application in mortality forecasting. *Journal of Statistcal Theory and Practice*, 8(4):578–590.

Kimelfeld, B. (1972). An orthogonal representation of nonlinear systems. Technical Report No. 66, Department of Statistics, Carnegie-Mellon University, Pittsburgh, PA.

Kimelfeld, B. (1974). Estimating the kernels of nonlinear orthogonal polynomial functionals. *Annals of Statistics*, 2(2):353–358.

Koopmans, L. (1974). *The Spectral Analysis of Time Series*. Academic Press, New York.

Nelson, J. and Van Ness, J. (1973a). Choosing a nonlinear predictor. *Technometrics*, 15(2):219–231.

Nelson, J. and Van Ness, J. (1973b). Formulation of a nonlinear predictor. *Technometrics*, 15(1):1–12.

Priestley, M. (1988). *Non-Linear and Non-Stationary Time Series Analysis*. Academic Press, San Diego, CA.

Shumway, R., Azari, A., and Pawitan, Y. (1988). Modeling mortality fluctuations in Los Angeles as functions of pollution and weather effects. *Environmental Research*, 45(2):224–241.

Slud, E. and Kedem, B. (1994). Partial likelihood analysis of logistic regression and autoregression. *Statistica Sinica*, 4(1):89–106.

Tick, L. (1961). The estimation of transfer functions of quadratic systems. *Technometrics*, 3(4):563–577.

Wiener, N. (1958). *Nonlinear Problems in Random Theory*. M.I.T. Press, Cambridge, MA.

Section IV

Discrete-Valued Spatio-Temporal Processes

15

Hierarchical Dynamic Generalized Linear Mixed Models for Discrete-Valued Spatio-Temporal Data

Scott H. Holan and Christopher K. Wikle

CONTENTS

15.1 Introduction

Discrete-valued spatio-temporal data arise frequently across a diverse range of subject-matter disciplines, including epidemiology, small area estimation in federal surveys, environmental science, and ecology, among others. In general, modeling this type of data can prove challenging due to the complexity of the observed data and underlying dynamical processes (e.g., see Cressie and Wikle, 2011, and the references therein). In this chapter, we focus primarily on modeling count data using spatio-temporal generalized linear models within a Bayesian hierarchical modeling (BHM) framework. In particular, we review some of the common methods in this context and describe some recent advances. For completeness, we provide brief discussion surrounding other types of discrete-valued spatio-temporal data, such as Bernoulli data and others. Finally, we provide a succinct real data illustration outlining the prediction of waterfowl migratory patterns across the north-central United States and Canada.

In the context of modeling count spatio-temporal data, several methods have emerged, including auto-Poisson models (Besag, 1974), generalized linear dynamical (spatio-temporal) mixed models (Wikle, 2002), and Bayesian nonparametric methods based on Dirichlet process mixtures (Kottas et al., 2008), among others. The direction pursued here focuses on generalized linear mixed models (GLMMs) (see McCulloch et al., 2001, for a brief overview of GLMMs). Specifically, we consider generalized linear models (GLMs) with a latent Gaussian process model (e.g., see the overview in Cressie and Wikle, 2011, and the references therein). In this context, we have a non-Gaussian data model along with a latent dynamic Gaussian model for the underlying unobserved process (Section 15.2) and, thus, the latent random effects cannot be integrated out analytically (Verbeke and Molenberghs, 2009). From this perspective, the models we describe are similar to the dynamic linear models framework in the time series (non-spatial) case. For further discussion surrounding Bayesian dynamic linear models see Gamerman et al. (2015; Chapter 8 in this volume) and the references therein.

To date, there have been many methodological contributions in the area of BHMs for count-valued spatio-temporal data. For example, Waller et al. (1997) consider a spatio-temporal count model for mapping disease rates, where the observations are assumed to come from a Poisson distribution. In the context of ecological modeling, Wikle (2003) introduces a Bayesian hierarchical spatio-temporal Poisson model to predict the relative population abundance of house finches over the eastern United States. Wikle and Anderson (2003) propose a spatio-temporal zero-inflated Poisson model that uses exogenous climate processes to model tornado counts.

Other diverse application areas include Wikle and Royle (2005) where the authors propose a dynamic spatio-temporal exponential family (Poisson) model for selecting sampling locations to estimate July brood counts in the Prairie Pothole Region of the United States. In contrast, Schrödle and Held (2011) describe spatio-temporal disease mapping models using integrated nested Laplace approximations (INLA) to facilitate fast computation in the context of space–time count data. Further, Lopes et al. (2011) introduce a class of spatio-temporal latent factor models for observations belonging to the exponential family of distributions. However, the models are illustrated using a Bernoulli data example to model rainfall. Finally, Wu et al. (2013) develop a class of Bayesian Conway-Maxwell Poisson (CMP) models with dynamic dispersion and illustrate the approach by estimating migratory waterfowl settling patterns.

The area of discrete-valued spatio-temporal modeling is expansive in terms of both methodological contributions and applications. The previous list of contributions is in no way meant to be exhaustive. Instead, it serves to illustrate the rich literature that exists on the subject. For further discussion, see Cressie and Wikle (2011) and the references therein.

This chapter proceeds as follows. Section 15.2 provides a general description of spatio-temporal modeling from a BHM perspective. Specifically, this section reviews the Bayesian hierarchical framework and details effective partitioning of the model hierarchy in terms of models for the observed data, latent processes, and parameters. Section 15.3 discusses various data models for discrete-valued spatio-temporal data, whereas modeling dynamics is pursued in Section 15.4. Section 15.5 provides an illustration of modeling spatio-temporal count data in the context of an application to forecasting migratory bird settling patterns. Specifically, this section methodologically illustrates the use of a spatio-temporal Poisson model, with a latent dynamic Gaussian process for the Poisson intensity parameter through a real data example. Finally, Section 15.6 provides concluding discussion.

15.2 Hierarchical Models

The hierarchical paradigm has experienced significant growth over the past two decades. The original ideas behind the process-based hierarchical modeling approach, as presented here, emerged largely out of the work of Berliner (1996) and have been further exposited by Wikle et al. (1998), Wikle (2003b), Cressie et al. (2009), and Wikle et al. (2013), among others. This approach is conceptually straightforward and it provides an extremely rich framework for modeling complex dependence structures in the context of discrete-valued spatio-temporal processes. Importantly, in addition to the process-based emphasis, the hierarchical framework presented here also emphasizes modeling parameters, which is often not the case in nested error regression-type hierarchical models.

Although the hierarchical modeling paradigm has become fairly well established (e.g., see Cressie and Wikle, 2011, and the references therein), we provide a brief description here for those readers less familiar with these ideas. The main idea underlying the BHMs presented here is to consider a joint probability model for the data, process, and parameters, which are generally specified through conditionally linked model components; that is, the data conditioned on the process and parameters and the process conditioned on the parameters. Several references focus on this type of hierarchical thinking including Royle and Dorazio (2008) and Cressie and Wikle (2011), among others, whereas more traditional presentations of hierarchal modeling can be found in Banerjee et al. (2003), Carlin and Louis (2011), Gelman et al. (2013), and the references therein.

Synthesis and effective utilization of information, both from direct and from indirect sources, are two paramount objectives in statistical modeling and data analysis. In fact, both direct and indirect sources of information play a key role in statistical modeling and often include expert opinion, physical laws, and previous empirical results. For specificity, consider the case where we have an underlying scientific process of interest, denoted by Y (a spatio-temporal process). Associated with this process we also have observed data, say Z. We assume that we have parameters θ_Z associated with the measurement process Z that might account for differences in the support and representativeness between the Z and the underlying true process Y defined at a given resolution of interest. Additionally, we assume that there are some parameters θ_Y, typically associated with the evolution operator and innovation covariances, that describe the dynamics of true underlying process of interest, Y.

Let $[Z|Y]$ and $[Y]$ denote the conditional distribution of Z given Y and the marginal distribution of Y, respectively. Then, assuming conditional independence of the parameters and using the law of total probability, the joint probability distribution of the data and process given the parameters can be decomposed as

$$[Z, Y|\theta_Z, \theta_Y] = [Z|Y, \theta_Z]\,[Y|\theta_Y], \tag{15.1}$$

where $[Z|Y, \theta_Z]$ is the data distribution (or "data model"—assuming conditional independence) and $[Y|\theta_Y]$ denotes the process distribution (or "process model").

In traditional statistics, typically, the data Z is given some specified distributional form along with associated parameters $\theta = (\theta_Z, \theta_Y)$ corresponding to the spatio-temporal mean, variances, and covariances. Although distributional assumptions for Z can be relaxed in the context of discrete-valued spatio-temporal data (e.g., Kottas et al., 2008), we limit our discussion to parametric models and focus primarily on count-valued spatio-temporal data. Integrating out the random process Y in (15.1) results in $[Z|\theta]$, in which case interest resides in estimating the parameters given the data. The disadvantage of such estimation is that it eliminates explicit estimation of the underlying true latent process Y. Instead, the distribution for Y is implicitly included through the first and second moments as a result of the integration.

Modeling spatio-temporal count data (or count time series for that matter) can proceed either from an observation-driven perspective or using a process-driven (parameter-driven) approach. By taking a process-based approach (i.e., explicitly modeling Y) several advantages arise. First, in many applications, one is actually interested in predicting the true underlying latent process Y, rather than just accounting for the co-variability. Second, given the complexity and high dimensionality of many real-world observed processes, it is often extremely difficult to specify the dependence structure associated with Z (e.g., due to non-Gaussianity, nonlinearity in time, and/or nonstationarity in space and/or time). Consequently, as a result of needing to specify a realistic dependence structure, likelihood-based inference in this context is challenging. In contrast, by placing emphasis on modeling the process Y instead, one can directly incorporate scientific insight into the model and more easily account for measurement (and/or sampling) and process uncertainty. For example, Markovian approximations and spatially and/or time-varying parameters can be readily incorporated in the model hierarchy. In other words, the hierarchical (conditional) specification allows extremely complicated marginal dependence structures to be replaced by a more scientific specification of the conditional mean as random process at a lower stage in the model hierarchy. This type of modeling is analogous to the traditional mixed model setting, where the practitioner must choose between the marginal model that arises from integrating out the random effects or the conditional model, where the random effects are predicted and the conditional covariance of the data model is less complicated (Demidenko, 2013). In contrast to the linear mixed model case, the generalized linear mixed model case, which is the focus here, is significantly more complicated. Importantly, for non-Gaussian data models, it is seldom possible to analytically integrate out the random effects. In other words, it is rarely the case that integrating out the random effect will result in a closed-form solution. Consequently, discrete-valued dynamic spatio-temporal generalized linear mixed models are typically quite computationally demanding, even after some form of dimension reduction.

In general, interest resides in estimating the posterior distribution of the process and parameters given the data. Using Bayes theorem, the fully BHM can be represented as

$$[Y, \theta | Z] \propto [Z | Y, \theta_Z] [Y | \theta_Y] [\theta_Z, \theta_Y], \tag{15.2}$$

where $\theta = (\theta_Y, \theta_Z)$ and it is necessary to specify a prior distribution for $[\theta_Z, \theta_Y]$. Note that in (15.2) the normalizing constant integrates over both the process Y and the parameters θ_Z and θ_Y.

Importantly, this representation facilitates a conditional way of thinking about complicated applications in a probabilistically consistent manner and naturally provides a means of quantifying uncertainty. In the context of spatio-temporal count-valued data, the *data model* (i.e., $[Z|Y, \theta_Z]$) will follow a count distribution such as a Poisson, negative binomial (NegBin), or CMP, among others (see Section 15.3).

An important aspect of (15.2) is that the right-hand side can be further decomposed into several submodels. For example, assuming conditional independence given the true underlying process, multiple data sets with different spatial and/or temporal supports could be accommodated through the following data model specification:

$$\left[Z^{(1)}, Z^{(2)} | Y, \theta_Z^{(1)} \theta_Z^{(2)}\right] = \left[Z^{(1)} | Y, \theta_Z^{(1)}\right] \left[Z^{(2)} | Y, \theta_Z^{(2)}\right], \tag{15.3}$$

where, for $j = 1, 2$, $Z^{(j)}$ and $\theta_Z^{(j)}$ correspond to the observations and parameters from the jth data set, respectively (e.g., see Wang et al., 2012). In this context, $Z^{(1)}$ and $Z^{(2)}$ need not have the same data distribution. Although, in practice, the assumption of conditional independence is often reasonable across a wide range of applications, when possible, this assumption should be validated.

For many applications, it is also natural to decompose the model for the process into subcomponents. In particular, in the context of discrete-valued spatio-temporal data, it is often natural to assume that the process has a Markov structure in time. Assuming a first-order Markov structure in time yields the following decomposition:

$$[\mathbf{Y}] = [\{Y_0, Y_1, \ldots, Y_T\}] = [Y_0] \prod_{t=1}^{T} [Y_t | Y_{t-1}] .$$

Alternatively, the process model could be further decomposed to accommodate multivariate structure. In this case, letting $[\mathbf{Y}] = \left[\mathbf{Y}^{(1)}, \mathbf{Y}^{(2)}\right]$, the process model can be expressed as $[\mathbf{Y}] = \left[\mathbf{Y}^{(2)} | \mathbf{Y}^{(1)}\right] \left[\mathbf{Y}^{(1)}\right]$, where the order of conditioning is usually suggested by the specific application and chosen by the practitioner (Royle and Berliner, 1999).

There is a vast literature on modeling non-Gaussian time series using a state-space approach (e.g., Carlin et al., 1992; Fahrmeir, 1992; Fahrmeir and Kaufmann, 1991; Gamerman, 1998; Kitagawa, 1987; West et al., 1985, among others). One major distinction between models in the time series case and the models described here is that in the spatio-temporal setting we now need to consider spatial dependence in addition to serial correlation, with these two dependence structures typically being nonseparable. Also, in contrast to the pure time series case, the spatio-temporal case often suffers from being extremely high dimensional. Specifically, consider a process that is measured at n locations and T times. Going from the pure time series case to the spatio-temporal setting results in an increase of $(n-1)T$ observations. Consequently, a necessary component of spatio-temporal modeling resides in effective dimension reduction.

In the context of discrete-valued spatio-temporal models, we assume that the *data model* comes from the exponential family and is non-Gaussian (e.g., Bernoulli, Poisson, NegBin, etc.); see Section 15.3. In particular, using similar notation to Cressie and Wikle (2011) we assume that $\mathbf{Z}_t$ denotes an m_t-dimensional vector of observations at time t from the exponential family of distributions. That is,

$$[\mathbf{Z}_t | \gamma_t] \propto \exp\left\{\gamma_t' \mathbf{Z}_t - b_t\left(\gamma_t\right) - c_t\left(\mathbf{Z}_t\right)\right\},$$

where γ_t denotes an m_t-dimensional set of natural parameters that depend on the process $\mathbf{Y}_t$ and $\mathrm{E}\left(\mathbf{Z}_t | \gamma_t\right) \equiv \mu_t$. Then, assuming the usual regularity conditions for the exponential family of distributions (McCulloch et al., 2001) we have that

$$g\left(\mu_t\right) = \mathbf{X}_t \beta + \mathbf{H}_t\left(\theta_h\right) \mathbf{Y}_t + \eta_t, \tag{15.4}$$

where $g(\cdot)$ is a known link function, $\mathbf{Y}_t$ is an n-dimensional spatial process vector of interest, $\mathbf{X}_t$ is a matrix of covariates (assumed known), β are the unknown "regression" coefficients associated with $\mathbf{X}$, $\mathbf{H}_t$ is the $m_t \times n$ observation matrix which is often assumed known but could also be specified in terms of the unknown hyperparameters θ_h, and η_t is an independent (across time) additive error term. It is important to note that, depending on the particular application, the additive error term, η_t, may not be warranted.

15.3 Data Models for Discrete-Valued Spatio-Temporal Data

As previously alluded to, discrete-valued spatio-temporal data arise across a broad range of subject-matter disciplines, with the specific distribution chosen to facilitate the analysis under consideration. Although, in principle, the framework presented here can accommodate virtually any discrete-valued distribution, we briefly describe a few of the more popular distributions that arise in practice. The distributions displayed here are not meant to constitute an exhaustive list of potential data model distributions that could be employed. Instead, the distributions we describe are merely meant to demonstrate the rich class of discrete-valued spatio-temporal models that can be constructed under the BHM (latent Gaussian process) framework described in Section 15.2. A specific example using the Poisson distribution is considered in Section 15.5.2.

Again, although we mainly focus on count-valued spatio-temporal data, many other discrete-valued data models could be considered. For example, when considering spatio-temporal binary data it is natural to use logistic regression such that the conditional distribution of $\mathbf{Z}_t$ given the n-dimensional vector of probabilities $\mathbf{p}_t$ is Bernoulli; that is,

$$\mathbf{Z}_t|\mathbf{p}_t \sim ind.\ Bern\left(\mathbf{H}_t\mathbf{p}_t\right).$$

In this case, it is natural to model $\mathbf{p}_t$ through the logit link (where through an abuse of notation we define $\text{logit}\left(\mathbf{p}_t\right) = \log\left\{\mathbf{p}_t/(1-\mathbf{p}_t)\right\}$ to be the logit transform applied to each element of $\mathbf{p}_t$). Alternatively, the probit link function could be considered in place of the logit link and in many cases when dealing with binary spatio-temporal data use of a probit link function along with data augmentation will facilitate computation (Albert and Chib, 1993). Although not considered here, the Bernoulli data model also arises in the context of spatio-temporal auto-logistic models (Zhu and Zheng [2015; Chapter 17 in this volume]; Zhu et al., 2008) and agent-based models (Hooten and Wikle, 2010; Wikle and Hooten [2015; Chapter 16 in this volume]).

In contrast to spatio-temporal binary data, one could consider a polychotomous outcome (i.e., outcomes with more than two ordered categories). In this case, a natural data model distribution is the multinomial. This gives rise to a spatio-temporal multinomial logistic regression. That is, assuming the usual conditions for the K category multinomial distribution, the data model is given by

$$\mathbf{Z}_t|\mathbf{n}_t, \mathbf{p}_{1,t}, \ldots, \mathbf{p}_{K,t} \sim ind.\ Mult\left(\mathbf{n}_t, \mathbf{H}_t\mathbf{p}_{1,t}, \ldots, \mathbf{H}_t\mathbf{p}_{K,t}\right). \tag{15.5}$$

In (15.5), one possible model for $\mathbf{p}_{k,t}$ $(k=1,\ldots,K)$ is the multinomial logit (e.g., see Congdon, 2007; Arab et al., 2012). Under this construction, it is natural to model the underlying process, the multinomial logit of $\mathbf{p}_{k,t}$ $(k=1,\ldots,K)$, as a latent Gaussian spatio-temporal process.

There are several popular choices for the *data model* when considering count-valued data (e.g., Poisson, NegBin, CMP, etc.), with the specific choice often based on the level of dispersion and/or computational considerations. In most cases, the spatial dispersion is seen to be overdispersed (i.e., the variance is greater than the mean). Although not as common, there are also several examples of underdispersion (i.e., the variance is less than the mean) (Ridout and Besbeas, 2004). The latter case of underdispersion typically arises in situation where the observations occur as "rare events." Finally, in practice, the case of equidispersion (i.e., the variance is equal to the mean) is rarely satisfied.

The Poisson distribution has become the *de facto* distribution when it comes to modeling spatio-temporal count-valued data and is the distribution we use for illustration (Section 15.5). Therefore, we defer detailed discussion of this distribution until Section 15.5. Although the model assumes equidispersion, the case of overdispersion is readily facilitated through a spatio-temporal random effect in a Gaussian latent process model for the logarithm of the Poisson intensity parameter. Assuming all locations are observed at each time point there is no need to include a mapping matrix from the observations to the process model for the intensity, unless interest resides in an aggregate or other (possibly weighted) function of the underlying process. Letting $\boldsymbol{\lambda}_t$ denote the spatial intensity process at time t, a typical model specification for a spatio-temporal Poisson model is given by

$$\mathbf{Z}_t|\boldsymbol{\lambda}_t \sim ind.\ Pois\left(\mathbf{H}_t\boldsymbol{\lambda}_t\right),$$

where $\log(\boldsymbol{\lambda}_t)$ can be specified similar to the right-hand side of (15.4).

Another popular distribution for modeling overdispersed count-valued spatio-temporal data is the NegBin (Greene, 2008). In contrast to the Poisson data model, this data model has an explicit parameter that controls the level of overdispersion. Assuming the "intensity" parameter, $\boldsymbol{\lambda}_t$, and dispersion parameter, ν, are greater than zero, the model can be specified as

$$\mathbf{Z}_t|\boldsymbol{\lambda}_t, \nu \sim ind.\ NegBin\left(\mathbf{H}_t\boldsymbol{\lambda}_t, \nu\right),$$

where $\log\left(\boldsymbol{\lambda}_t\right)$ can be specified similar to the right-hand side of (15.4) and ν (or $\log(\nu)$) can be given an appropriate hyperprior. For a random variable Z, it is well known that the expected value and variance of this distribution are given by $E(Z) = \lambda$ and $Var(Z) = \lambda + \nu\lambda^2$ (Greene, 2008), and thus, this distribution readily accommodates processes where the variance exceeds the mean. Finally, for this distribution, it is possible to let the dispersion parameter be space or time varying; however, only overdispersion can be accommodated.

A less common distribution used to model count data is the CMP distribution. As discussed in Wu et al. (2013), the CMP distribution can be used as a suitable *data model* distribution when considering count-valued spatio-temporal data. The advantage of this data model distribution is that it flexibly allows for both spatial (or temporal) overdispersion and underdispersion within the same model. Let $\boldsymbol{\lambda}_t$ and ν be positive and denote the CMP "intensity" and dispersion parameters, respectively. For this distribution, $\nu = 1$ corresponds to the Poisson distribution, whereas $\nu < 1$ and $\nu > 1$ correspond, respectively, to overdispersed and underdispersed distributions. Further, the CMP distribution generalizes to the geometric distribution (when $\nu = 0$ and $\lambda < 1$) and the Bernoulli distribution (as $\nu \longrightarrow \infty$) in the limiting cases (Shmueli et al., 2005). A spatio-temporal version of the CMP distribution is given by

$$\mathbf{Z}_t|\boldsymbol{\lambda}_t, \nu \sim ind.\ \mathrm{CMP}\left(\mathbf{H}_t\boldsymbol{\lambda}_t, \nu\right),$$

where $\log\left(\boldsymbol{\lambda}_t\right)$ can be specified similar to the right-hand side of (15.4) and $\log(\nu)$ is given a suitable hyperprior. Alternatively, as proposed by Wu et al. (2013), a dynamic model for the dispersion parameter could be imposed. Importantly, this distribution involves a normalizing constant that must be computed numerically since it involves the summation of an infinite series. For certain combinations of intensity and dispersion parameters, calculation of the normalizing constant can be computationally intensive. For these cases, Minka et al. (2003) derived an asymptotic approximation to the normalizing constant which is accurate for $\lambda > 10^{\nu}$. In contrast, Wu et al. (2013) proposed further improvements to computing the normalizing constant by taking advantage of parallel computing through Open Multiprocessing (OpenMP) and Compute Unified Device Architecture (CUDA), that is, graphics processing unit (GPU).

15.4 Modeling Dynamics

Dynamic models have long been considered in the non-Gaussian time series context (e.g., Carlin et al., 1992; Fahrmeir, 1992; Fahrmeir and Kaufmann, 1991; Gamerman, 1998; Kitagawa, 1987; West et al., 1985). Such models often take a more "econometrics" flavor, in which one seeks to accommodate multivariate temporal dependence through time-varying parameters. A major distinction between dynamical spatio-temporal models (DSTMs) and traditional multivariate time series models is that (1) there is spatial dependence (typically, nonstationary in space and nonseparable in space and time), (2) the spatial process changing through time is often of very high dimension, and (3) there is a scientific process that drives the way in which this spatial dependence changes through time. That is, there is a fundamental "process" that suggests modeling should be related to the evolution of a spatial process through time rather than simply modeling correlated time series or specifying marginal spatio-temporal dependence structures (Cressie and Wikle, 2011). We consider this process-driven approach here, recognizing that it fits naturally in the aforementioned hierarchical modeling framework. In this setting, we may be able to rely on fairly simple evolution models (e.g., autoregressive processes), but the modeling is complicated by high dimensionality and the need to reduce dimensionality either in terms of the process or in terms of parameter reduction. In addition, one must consider the possibility of modeling more complicated nonlinear scientific processes within the Markovian framework. Here, we focus on the case of discrete space and time, but note that the continuous space, discrete time case is closely related (e.g., Wikle and Cressie, 1999; Wikle and Holan, 2011; Wikle, 2002).

One can evolve the spatio-temporal process $\mathbf{Y}_t$ using the standard approaches from dynamical spatio-temporal models (e.g., Cressie and Wikle, 2011, Chapter 7) or through more traditional econometric-based dynamic linear models (e.g., Gamerman et al., 2007; Gelfand et al., 2005). However, as discussed in Cressie and Wikle (2011, Chapter 7), in the spatio-temporal context, realistic dynamical evolution of these processes requires transition (propagator) matrices that can accommodate real-world dynamical features (e.g., advection, diffusion, growth, etc.). Furthermore, the dimensionality of such processes often makes specification of the transition matrix a formidable challenge in terms of the number of parameters that must be estimated. Consequently, it is fairly typical to consider the evolution of so-called "spatial random effects," which are the projection coefficients of a basis function expansion of $\mathbf{Y}_t$ (e.g., Wikle and Cressie, 1999). Typically, the underlying dynamics of interest exist on a lower-dimensional manifold, allowing for a reduced rank representation, which also serves to reduce the parameter space associated with the process evolution. Thus, rather than model the spatio-temporal process $\mathbf{Y}_t$ directly, it is often convenient to consider the underlying spatio-temporal process to be decomposed into various components (e.g., Wikle et al., 2001; Wikle, 2003b; Wikle et al., 1998). For example, consider

$$\mathbf{Y}_t = \boldsymbol{\mu} + \boldsymbol{\Phi}^{(1)}\boldsymbol{\alpha}_t + \boldsymbol{\Phi}^{(2)}\boldsymbol{\beta}_t + \boldsymbol{\epsilon}_t,$$

where $\mathbf{Y}_t$ is an $n \times 1$ process vector defined at n spatial locations of interest, $\boldsymbol{\mu}$ is an $n \times 1$ spatial mean vector, $\boldsymbol{\Phi}^{(1)}$ is an $n \times p_1$ matrix, $\boldsymbol{\Phi}^{(2)}$ is an $n \times p_2$ matrix, $\boldsymbol{\alpha}_t$ and $\boldsymbol{\beta}_t$ are p_1-, p_2-dimensional vectors, respectively, and $\boldsymbol{\epsilon}_t$ is an $n \times 1$ mean zero spatial error process. In high-dimensional settings, $\boldsymbol{\Phi}^{(1)}$ is typically a "basis function" matrix, with $\boldsymbol{\alpha}_t$ denoting the corresponding expansion coefficients. The choice of the matrix $\boldsymbol{\Phi}^{(1)}$ in this context has been the source of considerable study in recent years, with many choices available, depending on whether these basis functions are specified (e.g., orthogonal polynomials, multiresolution wavelets or Wendland functions, splines, empirical orthogonal functions (EOFs), etc.), or whether they are in some sense estimated (e.g., discrete kernel convolutions, "predictive processes," dynamic factor models, etc.). Choices are typically made based on ideology, but should be made on more practical considerations such as whether the basis set is full rank ($p_1 = n$), rank reduced (i.e., $p_1 \ll n$), or over-complete ($p_1 \gg n$), or whether one wishes the $\boldsymbol{\alpha}_t$ coefficients to be spatially referenced (as in the discrete kernel convolution and "predictive process" approaches) or whether they live in "spectral" space. These issues are discussed in depth in Wikle (2010) and Cressie and Wikle (2011, Chapter 7). Our perspective is that these choices should consider the process dynamics, data, and computational demands of the problem at hand.

The choice of $\boldsymbol{\Phi}^{(2)}$ depends on the process $\mathbf{Y}_t$ and the choice of $\boldsymbol{\Phi}^{(1)}$ as well as the computational demands of the problem of interest. For example, if $\boldsymbol{\Phi}^{(1)}$ corresponds to a rank-reduced basis for a large-scale dynamical process, then one might consider $\boldsymbol{\Phi}^{(2)}$ to correspond to smaller scales, which may have different dynamics (e.g., Gladish and Wikle, 2014; Wikle et al., 2001). Alternatively, $\boldsymbol{\Phi}^{(2)}$ may correspond to covariates, or may be an identity matrix, in which case $\boldsymbol{\beta}_t$ are just "regression" coefficients or residual random effects (likely confounded with $\boldsymbol{\nu}_t$, the time-varying dispersion parameter, and $\boldsymbol{\epsilon}_t$), respectively. Clearly, not all of these components are required or useful in every spatio-temporal model–choices must be made relative to the process and data at hand. We will focus the discussion here on process-based dynamic models for $\boldsymbol{\alpha}_t$.

Let $\boldsymbol{\alpha}_t \equiv \left(\alpha_{1,t}, \ldots, \alpha_{p_1,t}\right)'$, where, depending on the choice of $\boldsymbol{\Phi}^{(1)}$, the index i in $\alpha_{i,t}$ may correspond to either physical space or "spectral" space. We are typically interested in a Markovian evolution model such as $\boldsymbol{\alpha}_t = \mathcal{M}\left(\boldsymbol{\alpha}_{t-1}; \boldsymbol{\eta}_t; \boldsymbol{\theta}\right), t = 1, 2, \ldots$, where $\mathcal{M}(\cdot)$ is an evolution operator, $\left\{\boldsymbol{\eta}_t\right\}$ an error process, and $\boldsymbol{\theta}$ parameters (that may, themselves, vary over space and/or time). Clearly, such a model is too general to be of much use beyond providing a conceptual framework. Rather, we consider the very general parametric class of models suggested by general quadratic nonlinearity (GQN) (Wikle and Holan, 2011; Wikle and Hooten, 2010):

$$\alpha_{i,t} = \sum_{j=1}^{p_1} m_{i,j,t}^{L} \alpha_{j,t-1} + \sum_{k=1}^{p_1} \sum_{\ell=1}^{p_1} m_{i,k\ell}^{Q} \alpha_{k,t-1} g\left(\alpha_{\ell,t-1}; \theta_g\right) + \eta_{i,t}, \tag{15.6}$$

for $i = 1, \ldots, p_1$, where $\left\{\eta_{i,t}\right\}$ is an error process (typically assumed to be a mean zero Gaussian process with some variance–covariance matrix given by $\mathbf{Q}_\alpha$), $m_{i,j,t}^{L}$ are linear interaction (transition) coefficients, $m^{Q}_{i,k\ell}$ are quadratic interaction coefficients, $g(\cdot)$ is some transformation of $\alpha_{\ell,t}$ that depends on parameters $\boldsymbol{\theta}_g$ and gives the process more generality than the simple dyadic interactions in the α coefficients alone. As described in Wikle and Hooten (2010), this framework is exceptionally flexible in that it can account for an extensive set of real-world mechanistic processes. Wikle and Holan (2011) show that this extends to higher-order interactions and the integro-difference continuous space case. However, even with quadratic interactions, the number of parameters that need to be estimated is on the order of p_1^3, which is a substantial curse of dimensionality. The efficient parameterization of (15.6) becomes the principle challenge in DSTM specification.

There are several simplifications and modeling approaches that can facilitate the specification of the parameter structure in (15.6). First, one can use the structure suggested by discretization of relevant mechanistic models (e.g., partial differential equations such as those given by reaction–diffusion and advection–diffusion processes) to simplify the parameters given by the linear and quadratic interaction coefficients in both physical and Galerkin (spectral) space (e.g., Hooten and Wikle, 2008; Wikle, 2003a; Wikle and Hooten, 2010; Wikle et al., 2001; Xu and Wikle, 2007). It is important to recognize that one uses these mathematical model representations to reduce the number of parameters, but this leaves many parameters that still must be estimated or modeled. Thus, such an approach is referred to as an mechanistically motivated model (e.g., see tutorial discussion in Cressie and Wikle, 2011, Chapters 6 and 7).

In situations where $\boldsymbol{\Phi}^{(1)}$ and $\boldsymbol{\Phi}^{(2)}$ correspond to large and medium/small-scale spatial basis functions, respectively, Wikle and Hooten (2010) make a case for dimension reduction based on arguments from turbulence theory. Wikle and Holan (2011) show that estimation and Bayesian inference can be substantially improved if one applies a stochastic search variable selection (e.g., George and McCulloch, 1993, 1997) approach to the linear and quadratic interaction parameters. Gladish and Wikle (2014) show that one can also effectively reduce the parameter space in this scenario if one assumes that medium scales influence the evolution of the large scales, but large scales do not influence the evolution of medium scales, which is also motivated by certain types of physical processes.

Critically, many real-world processes are very reasonably approximated by linear or quasi-linear dynamics, that is, the case where the quadratic interaction coefficients in (15.6) are zero. In this case, the model reduces to a vector autoregressive model with time-varying coefficients,

$$\boldsymbol{\alpha}_t = \mathbf{M}_t \boldsymbol{\alpha}_{t-1} + \boldsymbol{\eta}_t, \tag{15.7}$$

where $\boldsymbol{M}_t = \left[m^{L}_{i,j,t}\right]_{i,j}$ is the $p_1 \times p_1$ time-varying transition matrix and $\boldsymbol{\eta}_t \equiv \left(\eta_{1,t}, \ldots, \eta_{p_1,t}\right)' \sim Gau\,(\mathbf{0}, \mathbf{Q}_t)$ is the error process. Of course, the model given in (15.7) is just a multivariate autoregressive time series process. However, as stated previously, in dynamic spatio-temporal process modeling, the dimensionality of $\mathbf{M}_t$ may preclude direct estimation, and more critically, one should take into account the mechanistic nature of the dynamical evolution in the parameterization of this matrix. The approaches discussed earlier associated with mechanistically motivated specifications, spectral parameter reduction, and stochastic search variable selection can be used to facilitate this modeling. Importantly, it is often the case in real-world processes that a static transition matrix ($M_t \equiv M$) is adequate for modeling the spatio-temporal dependence.

15.5 Example: Forecasting Migratory Bird Settling Patterns

15.5.1 Breeding Population Survey Data

The migratory bird counts we consider come from the Breeding Population Survey (BPS) and constitute raw indicated pair count data for mallard ducks. The data are collected by the U.S. Fish and Wildlife Service in conjunction with the Canadian Wildlife Service and are publicly available at the FWS Division of Migratory Management website (https://migbirdapps.fws.gov/). The BPS is an ongoing annual survey that has been conducted since 1955 with the goal of providing detailed information about the spring population size of different duck species occurring across central Canada, Alaska, and the north-central United States.

The survey is conducted using two-person aerial crews flying fixed-wing aircraft at low altitudes along established 400 m wide transect lines that are divided into segments that are approximately 29 km long. Each aerial crew records mallard duck counts with the intent that subsequent comparisons will be made against a subset of available site counts collected by ground crews. Our analysis considers mallard duck counts for 2171 segments (falling between 85°–165°W longitude and 43°–69°N latitude) that are available during the period 1955–2010 and is limited to raw indicated pair counts, as measured by the presence of lone drakes or both paired ducks. For a complete description, see Wu et al. (2013) and the references therein.

15.5.2 Spatio-Temporal Poisson Models

For illustration, we consider a Bayesian hierarchical spatio-temporal Poisson model. In contrast, Wu et al. (2013) considered and highlighted a Bayesian hierarchical spatio-temporal CMP model with dynamic dispersion. In that context, the authors used the May values of the Palmer Drought Severity Index (PDSI) averaged over the Prairie Pothole Region in the United States as a regime switching climate covariate in a threshold autoregressive (TVAR) model. The model presented here is among the subset of models considered in Wu et al. (2013). However, in that context, the Poisson model was not the primary focus. Instead, modeling discussion was mainly concerned with the Bayesian hierarchical spatio-temporal CMP model with dynamic dispersion.

For $m_t \leq n$, let $\mathbf{Z}_t$ denote an m_t-dimensional vector with elements $\{Z_t(\mathbf{s}_i)\}$, where $t = 1,\ldots,T$ and $\mathbf{s}_i \in \{\mathbf{s}_1,\ldots,\mathbf{s}_n\} \in D \subset \mathbb{R}^2$. Here, $Z_t(\mathbf{s}_i)$ is a realization of an underlying count process at location $\mathbf{s}_i$ and time t. We note that for any given time there may be missing observations from the collection of n spatial locations. Additionally, let $\boldsymbol{\lambda}_t = (\lambda_t(\mathbf{s}_1),\ldots,\lambda_t(\mathbf{s}_n))'$ denote the spatially and temporally varying Poisson intensity parameter for each time t at all of the locations of interest, $\{\mathbf{s}_1,\ldots,\mathbf{s}_n\}$. Assuming the $Z_t(\mathbf{s}_i)$ given the true intensity $\lambda_t(\mathbf{s}_i)$ follows a Poisson distribution, the *data model* is given by

$$\mathbf{Z}_t|\boldsymbol{\lambda}_t \sim Pois(\mathbf{H}_t\boldsymbol{\lambda}_t), \quad t = 1,2,\ldots,T, \tag{15.8}$$

where $\mathbf{H}_t$ denotes an $m_t \times n$ incidence matrix that maps the observed data to the underlying latent process to account for missing data at each time t. The formulation in (15.8) involving $\mathbf{H}_t$ is fairly general and could also accommodate change of support (Wikle and Berliner, 2005).

The *process model* for the Poisson intensity is assumed to be governed by a dynamical process that resides on a low-dimensional manifold. For $p < n$, this p-dimensional process is denoted by $\boldsymbol{\alpha}_t$ and is obtained using the $n \times p$ basis function matrix $\boldsymbol{\Phi}^{(1)}$ to obtain a mapping from physical space. Consequently, the n-dimensional physical space representation of the low-dimensional process can be obtained as $\boldsymbol{\Phi}^{(1)}\boldsymbol{\alpha}_t$. Specifically, for $t = 1, \ldots, T$, the intensity process is given by

$$\log(\boldsymbol{\lambda}_t) = \boldsymbol{\mu} + \boldsymbol{\Phi}^{(1)}\boldsymbol{\alpha}_t + \boldsymbol{\epsilon}_t, \quad \boldsymbol{\epsilon}_t \overset{iid}{\sim} \text{Gau}\left(\mathbf{0}, \sigma_\epsilon^2 \mathbf{I}\right),$$

where $\boldsymbol{\mu}$ is a spatially referenced mean intensity and $\boldsymbol{\epsilon}_t$ is assumed uncorrelated and corresponds to small-scale spatio-temporal noise analogous to the "nugget" effect in spatial statistics.

The choice of basis function expansion matrix $\boldsymbol{\Phi}^{(1)}$ is typically problem specific and often chosen out of convenience or to match the desired goals of the analysis. In our case, it is important that $p \ll n$ and therefore, for this illustration, we work with EOFs (for details see Wu et al., 2013).

The evolution of the low-dimensional process $\boldsymbol{\alpha}_t$ remains to be specified. Although it is common for ecological and environmental processes to evolve nonlinearly, nonlinear dynamics can often be computationally expensive and, thus, a first-order Markov model of the form

$$\boldsymbol{\alpha}_t = \mathbf{M}_t \boldsymbol{\alpha}_{t-1} + \boldsymbol{\eta}_t, \quad \boldsymbol{\eta}_t \sim \text{Gau}\left(\mathbf{0}, \boldsymbol{\Sigma}_\eta\right) \tag{15.9}$$

is often used. In our application, (15.9) is simplified to consider three regimes, so $\mathbf{M}_t \in \{\mathbf{M}_1, \mathbf{M}_2, \mathbf{M}_3\}$ depending on a climate index covariate c_t. Here, this index is based on the PSDI and follows from the hypothesis that birds redistribute differently according to whether the landscape is wetter or dryer than normal. Specifically, assuming a TVAR model with three possible regimes for the latent process $\boldsymbol{\alpha}_t$, it follows that

$$\boldsymbol{\alpha}_t = \begin{cases} \mathbf{M}_1\boldsymbol{\alpha}_{t-1} + \boldsymbol{\eta}_t & \text{if } c_t < d_L, \\ \mathbf{M}_2\boldsymbol{\alpha}_{t-1} + \boldsymbol{\eta}_t & \text{if } d_L \le c_t \le d_U, \\ \mathbf{M}_3\boldsymbol{\alpha}_{t-1} + \boldsymbol{\eta}_t & \text{if } c_t > d_U, \end{cases}$$

where, for $t = 1, \ldots, T$, $\boldsymbol{\eta}_t \overset{iid}{\sim} \text{Gau}(\mathbf{0}, \boldsymbol{\Sigma}_\eta)$ and d_L, d_U are the threshold values that control the switching between climate regimes.

To complete the model, we must specify distributions for the *parameter models*. The overall spatial mean intensity, $\boldsymbol{\mu}$, is specified as

$$\boldsymbol{\mu} | \boldsymbol{\beta}, \sigma_\mu^2 \sim \text{Gau}\left(\mathbf{X}\boldsymbol{\beta}, \sigma_\mu^2 \mathbf{I}_n\right),$$

where $\mathbf{X}$ is a matrix of covariates, $\boldsymbol{\beta}$ is the associated regression coefficients, and σ_μ^2 is the associated error variance. In this example, we let $\mathbf{x}_i' = [1, \text{lon}_i, \text{lat}_i, \text{lon}_i \times \text{lat}_i, \text{lat}_i^2, \text{lon}_i^2]$, where $\mathbf{x}_i'$ is the ith row of matrix $\mathbf{X}$ and "lon," "lat" are longitude and latitude for location $\mathbf{s}_i$, $i = 1, \ldots, n$, respectively. The following prior distributions are specified for the parameters: $\text{vec}(\mathbf{M}_i) \sim \text{Gau}(\widetilde{\mathbf{m}}, \widetilde{\boldsymbol{\Sigma}}_m)$, for $i = 1, 2, 3$; $\boldsymbol{\Sigma}_\eta^{-1} \sim \text{Wishart}\left((v_\eta \mathbf{S}_\eta)^{-1}, v_\eta\right)$; $\boldsymbol{\beta} \sim \text{Gau}\left(\boldsymbol{\beta}_0, \boldsymbol{\Sigma}_\beta\right)$; $\boldsymbol{\alpha}_0 \sim \text{Gau}(\boldsymbol{\mu}_{\alpha_0}, \widetilde{\boldsymbol{\Sigma}}_{\alpha_0})$; $\sigma_\epsilon^2 \sim \text{IG}(q_\epsilon, r_\epsilon)$; and $\sigma_\mu^2 \sim \text{IG}(q_\mu, r_\mu)$.

15.5.3 Forecasting Application: Breeding Population Survey

The TVAR model presented here is well suited for investigating the hypothesis of drought-related "site philopatry"—the tendency for ducks to revisit their home site and consists of three redistribution matrices that are based on the May PDSI values $\mathbf{M}_1$, $\mathbf{M}_2$, and $\mathbf{M}_3$, defined according to dry, normal, and wet climate conditions, respectively. These three redistribution matrices allow the redistribution matrix to flexibly vary by climate regime. The threshold values that determine transition from one climate regime to another are given by $d_L = -0.183$ and $d_U = 1.701$ and are based upon the quantiles of the PDSI over the period of interest. Alternatively, though not pursued here, the threshold values could be estimated directly within the model hierarchy (e.g., see Wang and Holan, 2012).

Implementation of the Bayesian hierarchical model requires choices for the hyperparameters in our prior distributions. The exact choice of hyperparameters is given by $\widetilde{m} = \text{vec}(0.8\mathbf{I}_p)$, $\widetilde{\Sigma}_m = 100\mathbf{I}_{p^2}$, $v_\eta = p = 8$, $\mathbf{S}_\eta = 100\mathbf{I}_p$, $\boldsymbol{\beta}_0 = \mathbf{0}$, $\Sigma_\beta = 100\,\mathbf{I}_{n_\beta}$ where n_β is the length of $\boldsymbol{\beta}$. In terms of the variance components, $q_\epsilon = 2.18$ (shape parameter) and $r_\epsilon = 0.35$ (scale parameter) were chosen such that mean and variance of σ_ϵ^2 are 0.3 and 0.5, respectively. Similarly, $q_\mu = 3.28$ (shape parameter) and $r_\mu = 1.82$ (scale parameter) were chosen such that mean and variance of σ_μ^2 are 0.8 and 0.5, respectively. Lastly, $\boldsymbol{\mu}_{\alpha_0} = \boldsymbol{\Phi}^{(1)'} \boldsymbol{U}_0$, where $\boldsymbol{U}_0$ is the initial state in the lower-dimensional representation of the estimated latent process (see Wu et al., 2013, Appendix A) and $\Sigma_{\alpha_0} = \mathbf{10}\ \text{diag}(\zeta)$, where ζ are the eigenvalues related to the first $p = 8$ EOFs (which explained 45.6% of the variation and corresponded to the *elbow point* in the scree plot—not shown). Note that the specification of priors presented for this analysis are vague relative to the scale of the mallard data.

The model is estimated using Markov chain Monte Carlo (MCMC), with inference based on 6,000 MCMC posterior samples (i.e., 30,000 MCMC iterations after burn-in were thinned every fifth iteration). Convergence was assessed through a combination of visual inspection of the sample chains and through the Gelman–Rubin diagnostic (Brooks and Gelman, 1998), with no evidence of lack of convergence detected (for details, see Wu et al., 2013). To assess models' performance, we considered the Pearson correlation coefficient associated with the in-sample and out-of-sample predictions along with MSPE, where MSPE is calculated by taking the mean of the squared difference between the observations and the median of posterior predictive samples. Specifically, for the in-sample prediction the Pearson product-moment correlation coefficient between the median of posterior predictive samples ($\widehat{\mathbf{Y}}_{2009}$) and observed data ($\mathbf{Y}_{2009}$) was 98.89%, whereas for the out-of-sample prediction assessment (2010) the Pearson product-moment correlation coefficient was 79.64%.

Although we are primarily interested in forecasting, estimating a biologically motivated model with three $\mathbf{M}$ matrices also allows for potential assessment of the hypothesis of site philopatry. Figures 15.1 and 15.2 display the observed data for 2009 and 2010 along with the in-sample ($\widehat{\mathbf{Y}}_{2009}$) and out-of-sample ($\widetilde{\mathbf{Y}}_{2010}$) predictions and the median of posterior predictive samples. From looking at the observed data, the mallard counts clearly exhibit spatial structure. Further, both the in-sample and out-of-sample predictions appear to be doing a good job in terms of capturing the spatial pattern (relatively low bias), with relatively small uncertainty (as quantified using median absolute deviation—MAD). As expected, the in-sample predictions slightly outperform the out-of-sample predictions. Additionally, from a qualitative assessment of the three **M** matrices estimated in this analysis (Figure 15.3) the hypothesis of site philopatry appears to be supported. Nevertheless, it is important to note that with only 56 annual observations there is significant amount of uncertainty and, thus, diminished inferential capacity. Further, no attempt has been made to address any potential issues concerning multiple testing surrounding the entries of each **M** matrix.

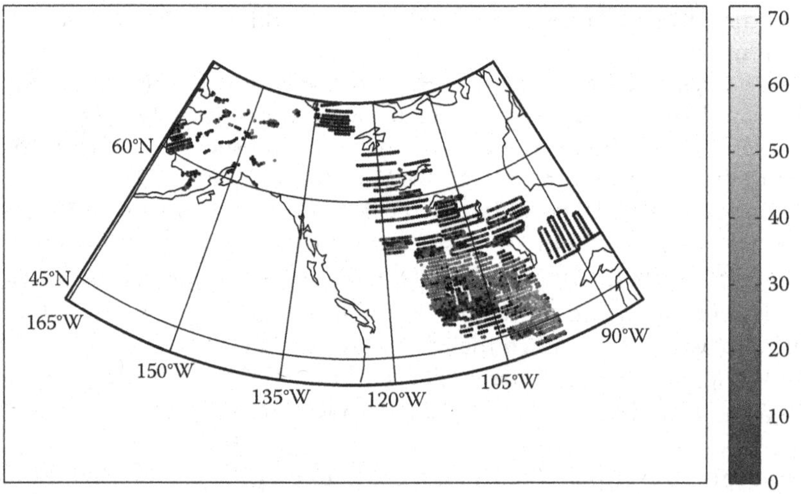

(a)

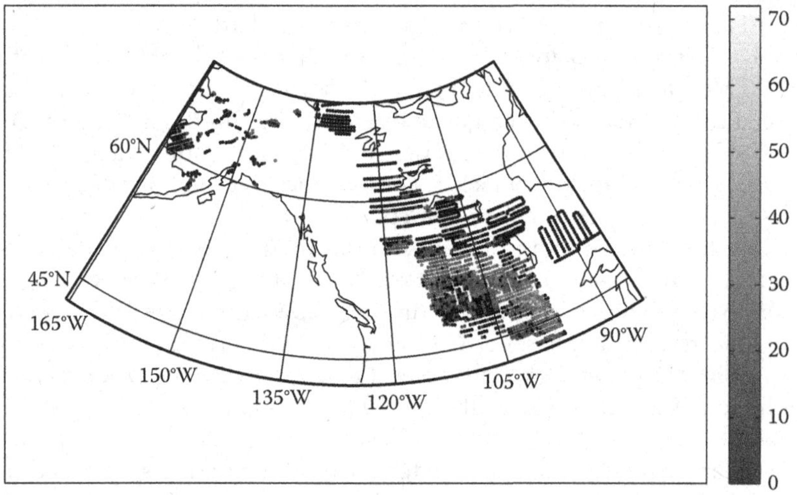

(b)

FIGURE 15.1
Observed and (in-sample) predicted counts for 2009 (using the median of the posterior predictive distribution) and their median absolute deviation (MAD) for the Poisson model (Section 15.5.3) using 8 EOFs. (a) Plot of Y_{2009}, (b) Plot of $\hat{Y}_{2009}$.

(Continued)

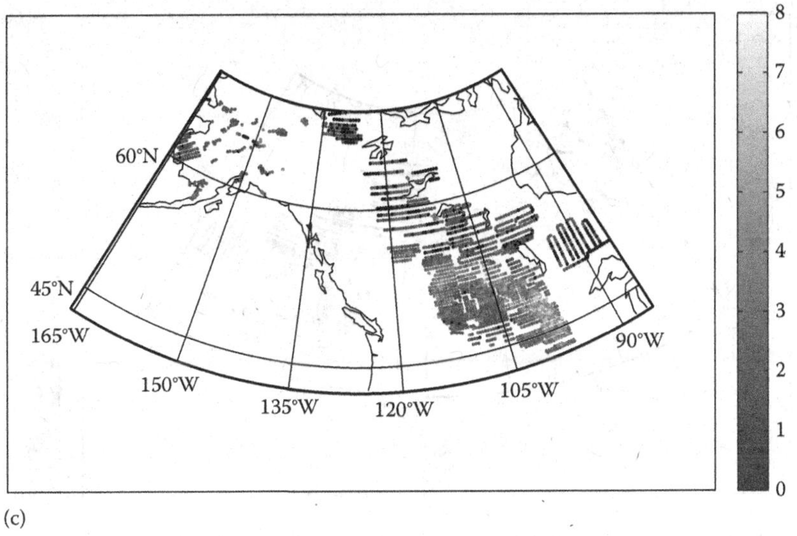

FIGURE 15.1 ***(Continued)***
Observed and (in-sample) predicted counts for 2009 (using the median of the posterior predictive distribution) and their median absolute deviation (MAD) for the Poisson model (Section 15.5.3) using 8 EOFs. (c) MAD of in-sample prediction.

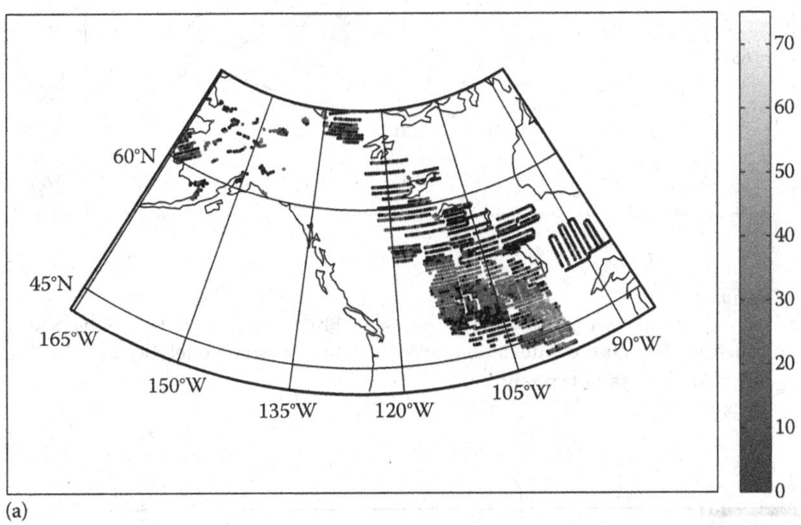

FIGURE 15.2
Observed and (one-step-ahead forecasts) predicted counts for 2010 (using the median of the posterior predictive distribution) and their median absolute deviation (MAD) for the Poisson model (Section 15.5.3) using 8 EOFs. (a) Plot of Y_{2010}.

(Continued)

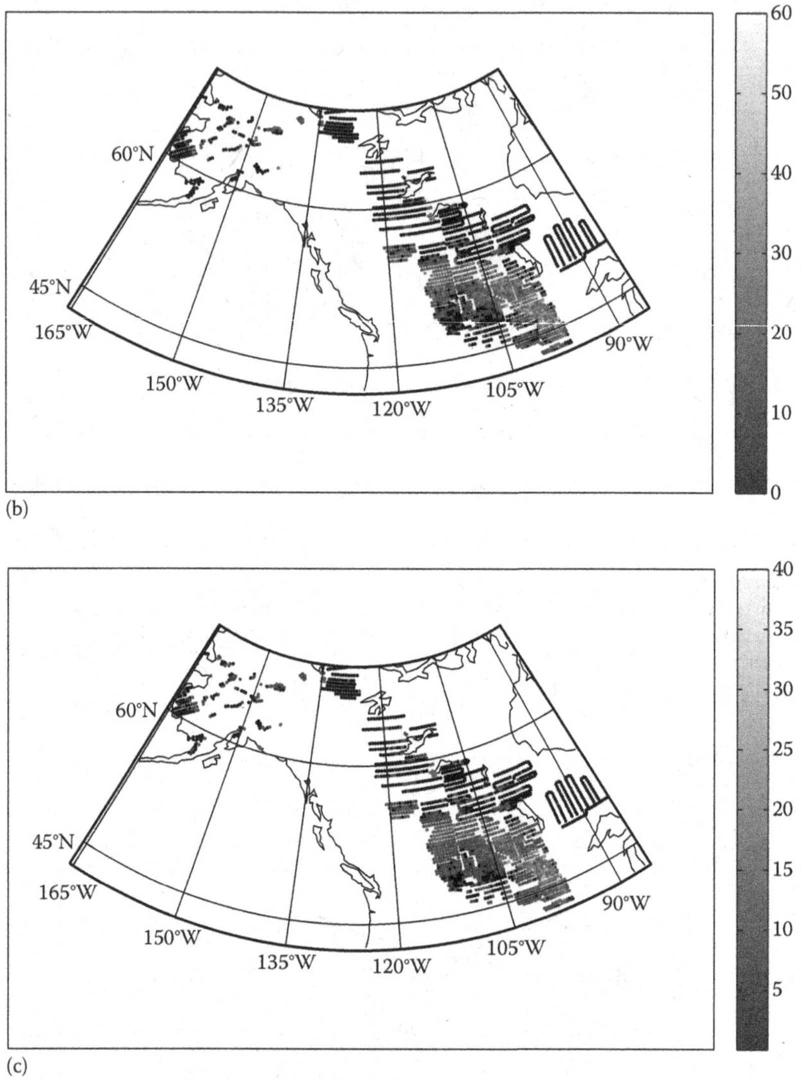

FIGURE 15.2 ***(Continued)***
Observed and (one-step-ahead forecasts) predicted counts for 2010 (using the median of the posterior predictive distribution) and their median absolute deviation (MAD) for the Poisson model (Section 15.5.3) using 8 EOFs. (b) Plot of $\hat{Y}_{2010}$, (c) MAD of in-sample prediction.

15.6 Conclusion

Modeling discrete-valued spatio-temporal data is often a challenging endeavor due to model complexity and high dimensionality. That is, specifying realistic dependence structure for a real-world observed spatio-temporal process can be extremely difficult due to inherent nonlinearities in time coupled with potential nonstationary behaviors in time

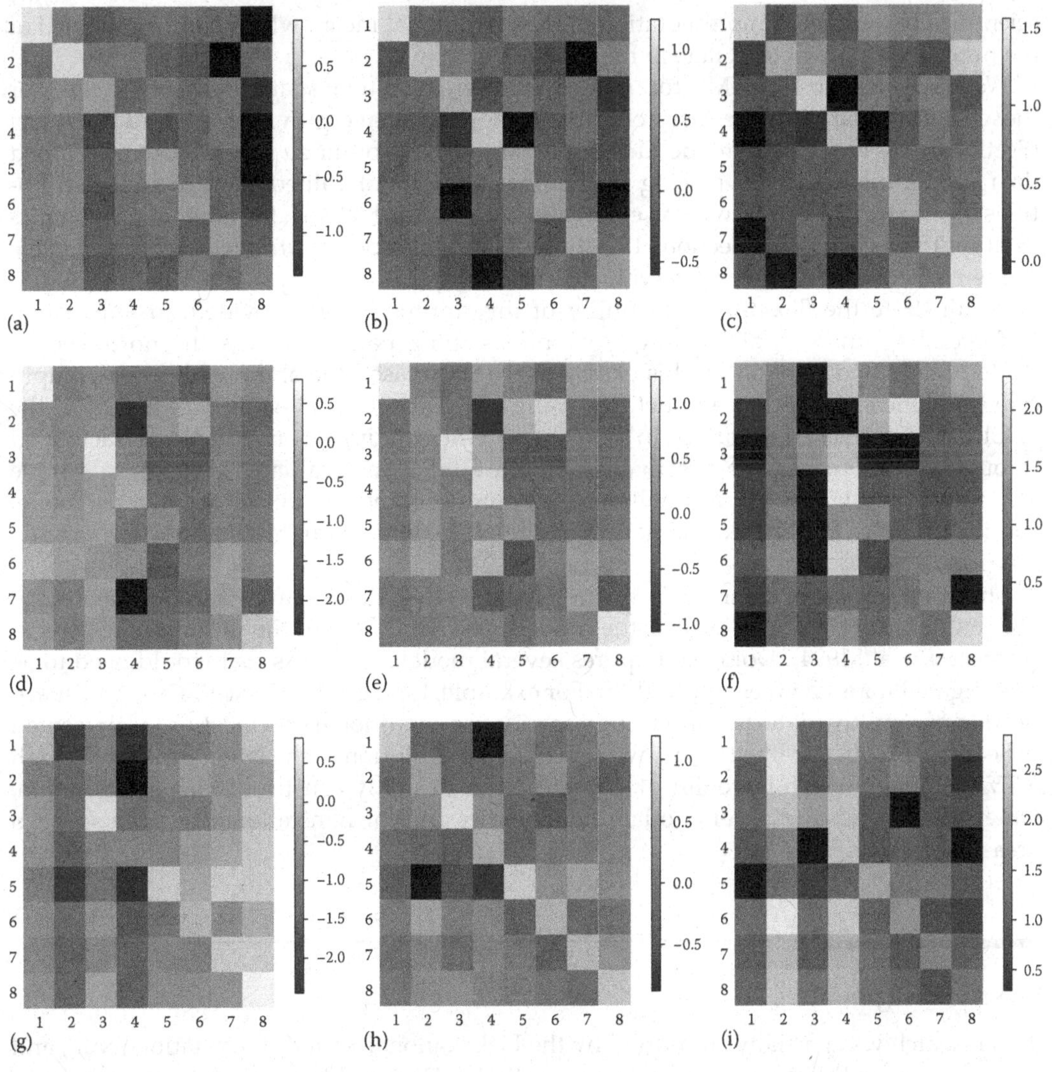

FIGURE 15.3
Image plots for element-wise posterior mean and 95% credible intervals of redistribution matrices $\mathbf{M}_i$, $i = 1, 2, 3$ for the Poisson model (Section 15.5.3) using 8 EOFs. Note that lower CI and upper CI denote the 2.5% and 97.5% quantiles of the posterior distributions, respectively. (a) M_1: Lower CI, (b) M_1: Posterior mean, (c) M_1: Upper CI, (d) M_2: Lower CI, (e) M_2: Posterior mean, (f) M_2: Upper CI, (g) M_3: Lower CI, (h) M_3: Posterior mean, and (i) M_3: Upper CI.

and/or space. Instead, utilizing a process-based approach and taking advantage of the Bayesian hierarchical paradigm often prove remarkably advantageous.

Although, in principle, there are many approaches to modeling discrete-valued spatio-temporal data, we presented a process-driven Bayesian hierarchical perspective. The key to this paradigm is that the joint probability model for the data, process, and parameters can be conditionally specified through linked model components. In particular, this formulation facilitates the specification of complicated marginal dependence through a more scientifically motivated representation of the conditional mean, which can be modeled at the process (or parameter) stage in the model hierarchy.

We have described BHMs for modeling discrete-valued spatio-temporal data. For those less familiar with the hierarchical paradigm, we have provided a broad overview (Section 15.2). To facilitate model development, we have outlined several discrete-valued data model distributions, placing an emphasis on count-valued data model distributions (Section 15.3). Lastly, we presented an expansive framework for modeling dynamics (Section 15.4). Combined, Sections 15.2 through 15.4 provide an extremely rich and flexible framework for modeling discrete-valued spatio-temporal data.

To illustrate the flexibility and utility of this approach, we provided an application to forecasting one-year-ahead migratory bird settling patterns across the north-central United States and Canada. In this example, we showcased the ability to specify complicated nonlinear dependence structures using a three-regime TVAR model based on the PDSI, a climate-related covariate. In this context, efficient dimension reduction is facilitated through EOFs. Further, the effectiveness of our model was evaluated through in-sample and out-of-sample prediction. Finally, as a by-product of the scientifically motivated model specification, we considered the notion of site philopatry through the Markov transition matrices.

There are several open areas of research for discrete-valued spatio-temporal modeling. One natural direction is to extend the models described here to the multivariate context (Fahrmeir et al., 1994). Doing so requires several modeling choices that would need to be investigated for each given application. For example, for a given application, would specification of a multivariate data model be superior to a conditional specification, as presented in Section 15.2? In our illustration, we specified the evolution of the latent process through a TVAR model, which is nonlinear. Nevertheless, this only constitutes one specification. Other specifications to accommodate nonlinearity and/or nonstationarity are also open areas of research.

Acknowledgments

This research was partially supported by the U.S. National Science Foundation (NSF) and the U.S. Census Bureau under NSF grant SES-1132031, funded through the NSF-Census Research Network (NCRN) program and through NSF grant DMS-1049093. The authors would like to thank Guohui Wu for his assistance with various aspects concerning the data analysis, including preparation of the figures.

References

Albert, J. H. and Chib, S. (1993). Bayesian analysis of binary and polychotomous response data. *Journal of the American Statistical Association*, 88(422):669–679.

Arab, A., Holan, S. H., Wikle, C. K., and Wildhaber, M. L. (2012). Semiparametric bivariate zero-inflated Poisson models with application to studies of abundance for multiple species. *Environmetrics*, 23(2):183–196.

Banerjee, S., Gelfand, A. E., and Carlin, B. P. (2003). *Hierarchical Modeling and Analysis for Spatial Data*. CRC Press.

Berliner, L. M. (1996). Hierarchical Bayesian time-series models. *Fundamental Theories of Physics*, 79:15–22.

Besag, J. (1974). Spatial interaction and the statistical analysis of lattice systems. *Journal of the Royal Statistical Society. Series B (Methodological)*, 36:192–236.

Brooks, S. and Gelman, A. (1998). General methods for monitoring convergence of iterative simulations. *Journal of Computational and Graphical Statistics*, 7:434–455.

Carlin, B. P. and Louis, T. A. (2011). *Bayesian Methods for Data Analysis*. Boca Raton, FL: CRC Press.

Carlin, B. P., Polson, N. G., and Stoffer, D. S. (1992). A Monte Carlo approach to nonnormal and nonlinear state-space modeling. *Journal of the American Statistical Association*, 87(418):493–500.

Congdon, P. (2007). *Bayesian Statistical Modelling*, vol. 704. Hoboken, NJ: John Wiley & Sons.

Cressie, N., Calder, C. A., Clark, J. S., Hoef, J. M. V., and Wikle, C. K. (2009). Accounting for uncertainty in ecological analysis: The strengths and limitations of hierarchical statistical modeling. *Ecological Applications*, 19(3):553–570.

Cressie, N. and Wikle, C. K. (2011). *Statistics for Spatio-Temporal Data*. Hoboken, NJ: John Wiley and Sons.

Demidenko, E. (2013). *Mixed Models: Theory and Applications with R*. Hoboken, NJ: John Wiley and Sons.

Fahrmeir, L. (1992). Posterior mode estimation by extended Kalman filtering for multivariate dynamic generalized linear models. *Journal of the American Statistical Association*, 87(418):501–509.

Fahrmeir, L. and Kaufmann, H. (1991). On Kalman filtering, posterior mode estimation and fisher scoring in dynamic exponential family regression. *Metrika*, 38(1):37–60.

Fahrmeir, L., Tutz, G., and Hennevogl, W. (1994). *Multivariate Statistical Modelling Based on Generalized Linear Models*. New York: Springer New York.

Gamerman, D. (1998). Markov chain Monte Carlo for dynamic generalised linear models. *Biometrika*, 85(1):215–227.

Gamerman, D., Abanto-Valle, C. A., Silva, R. S., and Martins, T. G. (2015). Dynamic Bayesian models for discrete-valued time series. In R. A. Davis, S. H. Holan, R. Lund and N. Ravishanker, eds., *Handbook of Discrete-Valued Time Series*, pp. 165–186. Chapman & Hall, Boca Raton, FL.

Gamerman, D., Salazar, E. and Reis, E. A. (2007). Dynamic Gaussian process priors with applications to the analysis of space-time data. In *Bayesian Statistics*, number 8 (eds Bernardo, J. M., Bayarri M., Berger J., Dawid A., Heckerman D., Smith A. and West M.). New York: Oxford University Press, pp. 149–74.

Gelfand, A. E., Banerjee, S., and Gamerman, D. (2005). Spatial process modelling for univariate and multivariate dynamic spatial data. *Environmetrics*, 16(5):465–479.

Gelman, A., Carlin, J. B., Stern, H. S., Dunson, D. B., Vehtari, A., and Rubin, D. B. (2013). *Bayesian Data Analysis*, 3rd edn. CRC press.

George, E. I. and McCulloch, R. E. (1993). Variable selection via Gibbs sampling. *Journal of the American Statistical Association*, 88(423):881–889.

George, E. I. and McCulloch, R. E. (1997). Approaches for Bayesian variable selection. *Statistica Sinica*, 7(2):339–373.

Gladish, D. W. and Wikle, C. K. (2014). Physically motivated scale interaction parameterization in reduced rank quadratic nonlinear dynamic spatio-temporal models. *Environmetrics*, 25(4):230–244.

Greene, W. (2008). Functional forms for the negative binomial model for count data. *Economics Letters*, 99(3):585–590.

Hooten, M. B. and Wikle, C. K. (2008). A hierarchical Bayesian non-linear spatio-temporal model for the spread of invasive species with application to the Eurasian collared-dove. *Environmental and Ecological Statistics*, 15(1):59–70.

Hooten, M. B. and Wikle, C. K. (2010). Statistical agent-based models for discrete spatio-temporal systems. *Journal of the American Statistical Association*, 105(489):236–248.

Kitagawa, G. (1987). Non-Gaussian state–space modeling of nonstationary time series. *Journal of the American statistical association*, 82(400):1032–1041.

Kottas, A., Duan, J. A., and Gelfand, A. E. (2008). Modeling disease incidence data with spatial and spatio temporal dirichlet process mixtures. *Biometrical Journal*, 50(1):29–42.

Lopes, H. F., Gamerman, D., and Salazar, E. (2011). Generalized spatial dynamic factor models. *Computational Statistics and Data Analysis*, 55(3):1319–1330.

McCulloch, C. E., Searle, S. R., and Neuhaus, J. M. (2001). *Generalized Linear, and Mixed Models*. New York: John Wiley and Sons.

Minka, T., Shmueli, G., Kadane, J., Borle, S., and Boatwright, P. (2003). Computing with the COM-Poisson distribution. Pittsburgh, PA: Department of Statistics, Carnegie Mellon University.

Ridout, M. S. and Besbeas, P. (2004). An empirical model for underdispersed count data. *Statistical Modelling*, 4(1):77–89.

Royle, J. A. and Berliner, L. M. (1999). A hierarchical approach to multivariate spatial modeling and prediction. *Journal of Agricultural, Biological, and Environmental Statistics*, 4(1):29–56.

Royle, J. A. and Dorazio, R. M. (2008). *Hierarchical Modeling and Inference in Ecology: The Analysis of Data from Populations, Metapopulations and Communities*. Burlington, MA: Academic Press.

Schrödle, B. and Held, L. (2011). Spatio-temporal disease mapping using INLA. *Environmetrics*, 22(6):725–734.

Shmueli, G., Minka, T., Kadane, J., Borle, S., and Boatwright, P. (2005). A useful distribution for fitting discrete data: Revival of the Conway-Maxwell Poisson distribution. *Journal of the Royal Statistical Society: Series C (Applied Statistics)*, 54(1):127–142.

Verbeke, G. and Molenberghs, G. (2009). *Linear Mixed Models for Longitudinal Data*. New York: Springer.

Waller, L., Carlin, B., Xia, H., and Gelfand, A. (1997). Hierarchical spatio-temporal mapping of disease rates. *Journal of the American Statistical Association*, 92:607–617.

Wang, J. and Holan, S. H. (2012). Bayesian multi-regime smooth transition regression with ordered categorical variables. *Computational Statistics and Data Analysis*, 56(2):4165–4179.

Wang, J. C., Holan, S. H., Nandram, B., Barboza, W., Toto, C., and Anderson, E. (2012). A Bayesian approach to estimating agricultural yield based on multiple repeated surveys. *Journal of Agricultural, Biological, and Environmental Statistics*, 17(1):84–106.

West, M., Harrison, P. J., and Migon, H. S. (1985). Dynamic generalized linear models and Bayesian forecasting. *Journal of the American Statistical Association*, 80(389):73–83.

Wikle, C. K. (2003a). Hierarchical Bayesian models for predicting the spread of ecological processes. *Ecology*, 84(6):1382–1394.

Wikle, C. K. (2010). Low rank representations as models for spatial processes, Chapter 8. In *Handbook of Spatial Statistics*. (eds Gelfand A. E., Fuentes M., Guttorp P. and Diggle P.). Boca Raton, FL: Chapman and Hall/CRC, pp. 107–18.

Wikle, C. K. and Anderson, C. (2003). Climatological analysis of tornado report counts using a hierarchical Bayesian spatio-temporal model. *Journal of Geophysical Research*, 108(D24):9005.

Wikle, C. K. and Berliner, L. M. (2005). Combining information across spatial scales. *Technometrics*, 47:80–91.

Wikle, C. K. and Cressie, N. (1999). A dimension-reduced approach to space-time Kalman filtering. *Biometrika*, 86(4):815.

Wikle, C. K. and Holan, S. H. (2011). Polynomial nonlinear spatio-temporal integro-difference equation models. *Journal of Time Series Analysis*, 32(4):339–350.

Wikle, C. K. and Hooten, M. B. (2010). A general science-based framework for dynamical spatio-temporal models. *Test*, 19(3):417–451.

Wikle, C. K. and Hooten, M. B. (2015). Hierarchical agent-based spatio-temporal dynamic models for discrete-valued data. In R. A. Davis, S. H. Holan, R. Lund and N. Ravishanker, eds., *Handbook of Discrete-Valued Time Series*, pp. 349–366. Chapman & Hall, Boca Raton, FL.

Wikle, C. K. Milliff, R., Nychka, D., and Berliner, L. M. (2001). Spatio-temporal hierarchical Bayesian modeling: Tropical ocean surface winds. *Journal of the American Statistical Association*, 96(454): 382–397.

Wikle, C. K. (2002). A kernel-based spectral model for non-Gaussian spatio-temporal processes. *Statistical Modelling*, 2(4):299–314.

Wikle, C. K. (2003b). Hierarchical models in environmental science. *International Statistical Review*, 71(2):181–199.

Wikle, C. K., Berliner, L. M., and Cressie, N. (1998). Hierarchical Bayesian space-time models. *Environmental and Ecological Statistics*, 5(2):117–154.

Wikle, C. K., Milliff, R. F., Herbei, R., and Leeds, W. B. (2013). Modern statistical methods in oceanography: A hierarchical perspective. *Statistical Science*, 28(4):466–486.

Wikle, C. K. and Royle, J. A. (2005). Dynamic design of ecological monitoring networks for non-Gaussian spatio-temporal data. *Environmetrics*, 16(5):507–522.

Wu, G., Holan, S. H., and Wikle, C. K. (2013). Hierarchical Bayesian Spatio-Temporal Conway–Maxwell Poisson Models with Dynamic Dispersion. *Journal of Agricultural, Biological, and Environmental Statistics*, 18:335–356.

Xu, K. and Wikle, C. K. (2007). Estimation of parameterized spatio-temporal dynamic models. *Journal of Statistical Planning and Inference*, 137(2):567–588.

Zhu, J. and Zheng, Y. (2015). Autologistic regression models for spatio-temporal binary data. In R. A. Davis, S. H. Holan, R. Lund and N. Ravishanker, eds., *Handbook of Discrete-Valued Time Series*, pp. 367–386. Chapman & Hall, Boca Raton, FL.

Zhu, J., Zheng, Y., Carroll, A. L., and Aukema, B. H. (2008). Autologistic regression analysis of spatial-temporal binary data via Monte Carlo maximum likelihood. *Journal of Agricultural, Biological, and Environmental Statistics*, 13(1):84–98.

16

Hierarchical Agent-Based Spatio-Temporal Dynamic Models for Discrete-Valued Data

Christopher K. Wikle and Mevin B. Hooten

CONTENTS

16.1 Introduction

In this chapter, we are concerned with discrete-valued spatio-temporal processes. To facilitate presentation, we will restrict our attention to such processes in discrete time, yet allow space to be continuous or discrete in principle. In spatio-temporal statistics, it is common to consider such models from a generalized linear mixed-model perspective (e.g., see Cressie and Wikle, 2011 and Holan and Wikle [2015; Chapter 15 in this volume]). This is a "top-down" approach whereby the spatio-temporal properties of the system are modeled in terms of a latent Gaussian spatio-temporal dynamical process (e.g., Wikle, 2002). Alternatively, one may consider such processes as Markov random fields (MRFs) using one of the classes of spatio-temporal "auto" models (e.g., spatio-temporal auto-logistic) as described in Zhu and Zheng (2015; Chapter 17 in this volume). The MRF approach is a local specification where relationships between neighbors are specified conditionally in a way to guarantee a valid joint distribution (see Section 16.2). In this chapter, we discuss an alternative "bottom-up" modeling strategy for discrete-valued spatio-temporal dynamical processes, which is agent based. Such *agent-based models* (ABMs) are prevalent in epidemiology and social sciences (e.g., Filatova et al., 2013; Gilbert, 2008; Keeling and Rohani, 2008; Sattenspiel, 2009) and also are called *individual-based models* in the ecological sciences (e.g., Grimm and Railsback, 2005), *multi-agent models* in engineering (e.g., Olfati-Saber, 2006), and *cellular automata* in the physical and mathematical sciences (e.g., Wolfram, 1984). All of these paradigms are characterized by autonomous agents (or individuals) that take on one of a discrete number of states (hence, discrete valued) that vary with time and space depending on a set of deterministic or probabilistic "rules." As discussed later, from a stochastic perspective, ABMs can sometimes be linked to MRF-based models through Markov network properties.

As a bottom-up modeling approach, ABMs have the property that the characteristics and actions of the agents ultimately define the properties and behavior of the system in which they exist. The most important part of such a modeling approach is that the autonomous agents interact with each other and with their environment. Although much of the literature associated with these models assumes that the agent interactions and evolution are governed by deterministic rules, from a statistical perspective, a crucial component of the interactions is stochasticity, and we focus on such probabilistic-based specifications in this chapter. In probabilistic ABMs, the evolution of the agent's state is defined through a parametric probability distribution, and thus, each spatio-temporal realization of the system is different (although, typically the properties of the system are the same in each realization).

Agents can be defined to represent various scales, for example, a virus, an individual animal, a node in a network, or a spatial unit (e.g., census tract, county). From this perspective, much of the distinction between the various ABMs considered across different disciplines (individual-based models, multi-agent models, and cellular automata) is lost, which makes it an ideal framework to consider from a statistical modeling perspective. As mentioned, ABMs are typically considered in discrete time. As such, one must decide if all agent states are updated at once during each time step (i.e., synchronous updating) or in a way so that as an agent is updated, its new state is available to the other agents (i.e., asynchronous updating). As discussed in Caron-Lormier et al. (2008), this can make a difference in the global system properties, but it is more common to consider synchronous updating, thus we only consider such updates here.

The primary statistical challenges associated with probabilistic ABMs are related to specification of the parametric probability distributions that govern agent behavior, as well as associated parameter (or rule) learning, and accounting realistically for uncertainty in observations and parameters. In addition, sensitivity of the local and global properties of the system to parameter variability is of interest, along with model selection. Of course, given the individual nature of such models and the fact that there are typically a very large number of autonomous agents, computational complexity can be daunting in a statistical framework. Note that computation can also be daunting in a deterministic ABM framework, especially if one is seeking to learn the rules of agent behavior. We have found that the principles of hierarchical statistical modeling can help mitigate many of the issues related to probabilistic ABMs (e.g., Hooten et al., 2010; Hooten and Wikle, 2010). In particular, the hierarchical framework is ideal for managing the uncertainty associated with observations, processes, and parameters. Perhaps the most critical advantage of the hierarchical paradigm in ABMs is that parameters describing individual interactions with other agents and the environment can themselves be stochastic processes with spatial and/or temporal variability that may include exogenous information about the environment. These models are still potentially computationally prohibitive, yet approximate inference methods such as statistical emulation and approximate Bayesian computation (ABC) can mitigate these computational issues in some cases.

In this chapter, we first present a somewhat general methodological framework for hierarchical ABMs in Section 16.2, followed by a specific hierarchical modeling example related to the spread of an animal disease in Section 16.3. This is followed by illustrations of the so-called "first-order emulator" in Section 16.4 and ABC in Section 16.5 to facilitate parameter estimation. We conclude in Section 16.6 with a brief discussion.

16.2 Hierarchical Dynamic Spatio-Temporal ABM Methodology

Let $y_{i,t}(\mathbf{s}_{i,t})$ correspond to a discrete-valued observation for the ith agent ($i = 1,\ldots,n$) at time $t = 1,2,\ldots,T$ and spatial location $\mathbf{s}_{i,t} \in D$, where D is some spatial domain that is a subset of R^d, typically with $d = 2$, where we assume that the agent locations in the spatial domain can, in principle, vary with time (i.e., $\mathbf{s}_{i,t}$). Here, we allow $\mathbf{s}_{i,t}$ to represent either a point in continuous space or a spatial region (e.g., a Census tract or grid cell) with positive support, depending on the application. We assume that $y_{i,t}(\mathbf{s}_{i,t})$ can take one of K possible values (i.e., states) at time t. As mentioned previously, there are many different types of ABMs, with a wide variety of notation and mathematical representations. However, one can consider most of these models to be some sort of network with each autonomous agent as a node in the network graph. As an example to introduce notation, consider first a static network in which only the value of the agent's state changes with time, not the location. We then define the graph $\mathcal{G} = (\mathcal{V},\mathcal{E})$ with a set of n vertices $\{v_i \in \mathcal{V}, i = 1,\ldots,n\}$ and associated edges, $\mathcal{E} \subset \{(i,j) : i,j \in \mathcal{V}, j \neq i\}$. We also define an adjacency matrix $\mathbf{A} = [a_{ij}]$ with nonzero elements, a_{ij}, specified if there exists an edge between the vertices v_i and v_j. Finally, we denote the set of neighbors of the ith node by $\mathcal{N}_i = \{j \in \mathcal{V} : a_{i,j} \neq 0\}$ or, equivalently, by $\mathcal{N}_i = \{j \in \mathcal{V} : (i,j) \in \mathcal{E}\}$. These neighbors may be specified *a priori* or based on some known proximity structure (e.g., county centroids or a regular grid). Dynamic graphs further allow the locations of these vertices to potentially move with time (i.e., so-called *dynamic agents*). Necessarily, the edges and adjacency matrices would then also be time dependent. For brevity of presentation in the following examples, we assume that only the discrete-valued states and adjacency matrix change with time, not the spatial locations of the network vertices.

We note that the graph structures associated with our ABM social network can have directed or undirected edges, depending on the application, or it can be a combination of both types of edges. An example of an undirected graphical structure is a Markov network whose probability distribution for a set of variables, say $\mathcal{X} = \{x_1,\ldots,x_n\}$, is given as the product

$$p(x_1,\ldots,x_n) = \frac{1}{Z}\prod_{c=1}^{C} \phi_c(\mathcal{X}_c),$$

where $\mathcal{X}_c \subset \mathcal{X}$ and $\phi_c(\mathcal{X}_c)$ is known as a *potential*, which is a nonnegative function of its argument. In this case, Z is a normalizing constant, called a *partition function*, that is a function of the potentials (e.g., Barber, 2012, Chapter 4). A special case of this Markov network is the MRF. The MRF is then defined by a set of conditional probability distributions, $p(x_i|\mathcal{N}_i), i = 1,\ldots,n$. That is, the model is defined such that each node is independent of all of the other nodes, given its neighbors on the undirected graph, $\mathcal{G}$. Note that one can then use the Hammersly–Clifford theorem to specify the functional conditions necessary for these conditional distributions to yield a valid joint distribution (e.g., Besag, 1974; Cressie, 1993; Guyon, 1995). Such models are useful in a variety of network analysis applications. In spatial statistics and image analysis, MRF models are typically constructed on a lattice. These models lead to the family of "auto" models (e.g., Besag, 1974; Cressie, 1993; Cressie and Wikle, 2011) and can be used to model discrete-valued spatio-temporal data as described in, for example, Besag (1972), Zhu et al. (2005), Besag and Tantrum (2003), Zheng and Zhu (2008; Chapter 17 in this volume).

We now consider the MRF perspective to denote a stochastic ABM. To simplify the exposition, assume that the agents exist on a fixed network, so that $\mathbf{s}_{i,t} = \mathbf{s}_i$ is fixed for each agent through time. We then denote the state of the random variable associated with the ith agent at time t by $Y_{i,t}$. We further consider our "neighborhood" in discrete time to be the current time t and one previous time lag $(t-1)$, although higher lagged neighborhoods could be considered, as could neighborhoods that consider the future (e.g., Zheng and Zhu, 2008). An alternative perspective that considers continuous time can be found in Rasmussen et al. (2007). In the discrete-time case, we then must specify the conditional distribution $p(Y_{i,t}|\{Y_{j,t}, Y_{j,t-1} : j \in \mathcal{N}_i\})$, for $i = 1, \ldots, n$. For example, if the agents can take only binary states, we might specify this distribution to be a Bernoulli distribution, in which case the model becomes a spatio-temporal auto-binomial model (e.g., Besag, 1972), depending on coefficients describing the weights of the current and previous neighborhood states, as well as potential interactions. In this case, the normalizing constant of the joint distribution implied by the Hammersly–Clifford theorem is intractable in general, complicating inference (e.g., Besag and Tantrum, 2003; Zhu et al., 2008). However, as discussed in Besag and Tantrum (2003), this distribution is substantially simplified if the time conditioning does not depend on the state of the neighbors at the current time t, but only the state of the neighbors at the past time, $t-1$. This is the typical situation in synchronous updating of ABMs and is one of the distinctions between MRF-based models and ABMs. Thus, we consider conditional distributions of this form, $p(Y_{i,t}|\{Y_{j,t-1} : j \in \mathcal{N}_i\})$.

Another major distinction between the spatio-temporal MRF approach and the hierarchical stochastic ABM approach considered here is that the parameters that control the conditional distributions, say given by the q-dimensional vectors $\boldsymbol{\theta}_i$, are modeled explicitly as random variables at the next stage of a model hierarchy. That is, if we had $p(Y_{i,t}|\{Y_{j,t-1} : j \in \mathcal{N}_i\}; \boldsymbol{\theta}_i)$ for $i = 1, \ldots, n$, these would be given prior distributions $\prod_{i=1}^{n} p(\boldsymbol{\theta}_i|\boldsymbol{\nu}_i)$ that might depend on the vector of hyperparameters, $\boldsymbol{\nu}_i$ (which could, in turn, depend on other parameters). These parameters $\{\boldsymbol{\theta}_i\}$ are typically modeled themselves as processes (in space, and possibly time) and can depend on exogenous variables that can allow the model to adapt to more complicated processes. That is, we are effectively modeling conditional ABMs, given parameters. Inference can then be obtained using Bayesian methods such as Markov chain Monte Carlo (MCMC) or ABC. These will be illustrated in the animal disease example in Section 16.3 and the epidemic model example in Section 16.5. We note that in nonhierarchical forms of these models, one can perform estimation through approximate likelihood methods (e.g., Besag and Tantrum, 2003; Zhu et al., 2005, 2008).

16.3 Statistical ABM for the Spread of Disease

During the past few decades, a raccoon (*Procyon lotor*) rabies epidemic has been occurring throughout the eastern United States This rabies epidemic has been the subject of many epidemiological research and was discussed in detail by Smith et al. (2002) and Wheeler and Waller (2008). Hooten and Wikle (2010) considered the data collected in Connecticut, USA, from years 1991–1995. These data were binary valued and arranged on a regular grid spanning the state of Connecticut (and a small portion of eastern New York, USA)

FIGURE 16.1
Presence (dark cells) or absence (light cells) of data for raccoon rabies in Connecticut, USA, during 1991–1995. Time for each image increases from top to bottom and left to right, in that order.

consisting of 109 approximately township-sized cells. The binary data represented presence or absence of rabies in each grid cell over a sequence of regularly spaced time periods spanning 1991–1995 (Figure 16.1). These data clearly indicate a spreading dynamic process from west to east throughout this time period. A unique aspect of the epidemic is that the spread is not as smooth as that which might arise from a partial differential equation (PDE) (e.g., Wikle, 2003) and it consists of several apparent long-distance dispersal events. In this case, long-distance dispersal is defined as a noncontiguous "jump" in disease status beyond that of the first- and second-order neighboring grid cells where the disease is present.

In what follows, we describe a hierarchical Bayesian ABM for those data that allow for heterogeneous spread of the disease similar to that presented by Hooten and Wikle (2010). We use the term "ABM" to describe this model because one could envision the areal units (i.e., grid cells or townships) themselves acting as "agents" which interact, effectively spreading the disease from one spatial region to another over time. In the network terminology, we might use the term "node" instead of agent, where a similar setting involving a virus could spread dynamically through a computer network, with servers acting as agents in that case.

Desirable features of an ABM that might be used to model the rabies epidemic are (1) the flexibility to allow for long-distance dispersal events as well as (2) a more typical diffusion-type process that accounts for the smoother, yet heterogeneous, spread of disease from neighbor to neighbor. Thus, consider the statistical data model for the rabies observations $y_{i,t}$ collected at grid cells $i = 1, \ldots, n$ for times $t = 1, \ldots, T$

$$y_{i,t} \sim \text{Bern}(\theta_{i,t}), \tag{16.1}$$

where the presence probabilities $\theta_{i,t}$ are assumed to control the observed process and are modeled as

$$\theta_{i,t} = \begin{cases} \phi, & y_{i,t-1} = 1 \\ \bar{p}_{i,t}, & (1 - y_{i,t-1})(\mathrm{I}_{\mathcal{N}_{i,t-1}}) = 1 \\ \psi, & (1 - \mathrm{I}_{\mathcal{N}_{i,t-1}}) = 1, \end{cases} \tag{16.2}$$

such that they depend on a set of indicator variables that identify when a given cell can be influenced by the neighboring cells (i.e., cells in the set $\mathcal{N}_{i,t-1}$, in this case the first- and second-order grid neighbors) that happen to be active at the previous time

$$\mathrm{I}_{\mathcal{N}_{i,t-1}} = \begin{cases} 1, & \sum_{j \in \mathcal{N}_i} y_{j,t-1} > 0 \\ 0, & \sum_{j \in \mathcal{N}_i} y_{j,t-1} = 0 \,. \end{cases} \tag{16.3}$$

Thus, the data depend on parameters $\theta_{i,t}$ which can take one of three forms (16.2): (1) persistence (ϕ), (2) short-distance dispersal ($\bar{p}_{i,t}$), or (3) long-distance dispersal (ψ). If the processes were fully persistent and not capable of long-distance dispersal, then only the middle term would be necessary in (16.2). It is this near-distance dispersal process ($\bar{p}_{i,t}$) that is the real workhorse of the model, allowing for "intelligent" spreading dynamics to occur. There are a number of ways one could specify the dynamics for the near-distance disperal probabilities $\bar{p}_{i,t}$. We take a similar approach as described by Hooten and Wikle (2010) where, given the disease status for the neighborhood at the previous time $\mathbf{y}_{\mathcal{N}_i,t-1} \equiv \{y_{j,t-1} : \mathbf{s}_j \in \mathcal{N}_i\}$ and an additional set of neighborhood interaction probabilities $\mathbf{p} \equiv (p_1, p_2, \ldots, p_{|\mathcal{N}_i|})$, we have the relationship

$$\bar{p}_{i,t} = 1 - \exp((\mathbf{y}_{\mathcal{N}_i,t-1})' \log(\mathbf{1} - \mathbf{p})). \tag{16.4}$$

This process model (16.4) essentially implies that the probability of disease presence in a given cell is a function of the incoming disease from neighboring cells at the previous time. Specifically, the probability of presence at time t in a previously unoccupied area is the union of the transition probabilities from its occupied neighbors. If we assume the interaction probabilities sum to 1 then a natural stochastic model for $\mathbf{p}$ is

$$\mathbf{p} \sim \text{Dirichlet}(\mathbf{a}), \tag{16.5}$$

where the hyperparameters $\mathbf{a}$ could be chosen to represent any prior knowledge pertaining to a bias in dispersal direction. Alternatively, both $\mathbf{p}$ and $\mathbf{a}$ could be allowed to vary either spatially or temporally. Hooten and Wikle (2010) describe a model that allows for spatial heterogeneity in the interaction probabilities $\mathbf{p}$ that correlate with the gradient of a potential field. To briefly describe how this generalization could be implemented, consider a potential field $\alpha(\mathbf{X})$ on the grid that depends on a set of variables $\mathbf{X}$ influencing the spread of disease. One possible hypothesis for disease spread might be that it responds to changes in the potential field α; for example, an abrupt change in a landscape feature such as a boundary between land and water would lead to an increase in the gradient perpendicular to the direction of the boundary. On a discrete spatial support, such as the one we are considering here for the rabies example, the first-order gradient could be summarized as a set of velocities $\mathbf{a}_i$, one in each direction from the cell of interest i to its neighboring cells in $\mathcal{N}_i$. These gradient values $\mathbf{a}_i$ (or some function of them) can then serve as hyperparameters in the Dirichlet model for the spatially varying interaction probabilities $\mathbf{p}_i$. Of course, the Dirichlet is only one way to model $\mathbf{p}_i$; other stochastic models or deterministic functions could be used to link $\mathbf{p}_i$ to $\mathbf{a}_i$, or $\mathbf{X}$.

Returning now to the model described in (16.1–16.5), we can express the posterior distribution for all unknowns as proportional to the conditionally factored joint distribution such that

$$[\mathbf{p}, \mathbf{a}, \phi, \psi | \{y_{i,t}, \forall i, t\}] \propto \prod_{t=1}^{T} \prod_{i=1}^{m} [y_{i,t} | \theta_{i,t}(\mathbf{p}, \phi, \psi)][\mathbf{p}|\mathbf{a}][\mathbf{a}][\phi][\psi] . \tag{16.6}$$

Also, recall that the presence probabilities $\theta_{i,t}$ are a function of the other model parameters; thus, inference can be obtained for them as derived quantities in the model. For example, the posterior mean and standard deviation of the space–time series for $\theta_{i,t}$ are shown in Figure 16.2. The left panel of Figure 16.2 shows the posterior mean for $\theta_{i,t}$ and gives us a quantitative understanding of presence probability along the front of the epidemic over time. Similarly, the posterior standard deviation for $\theta_{i,t}$ shown in the right panel of Figure 16.2 allows us to visualize the uncertainty pertaining to presence probability for all grid cells and times. Such statistical products could be used for short-term forecasting; for example, forecasting the presence for the next year. Such forecasts require a reasonable timescale to accommodate the required computation time.

Overall, the relatively simple statistical ABM presented in (16.1–16.5) represents a fundamentally different approach to modeling dynamics (i.e., bottom-up rather than top-down) that is quite general and capable of representing complicated dynamics based on only a simple set of rules describing the behavior among agents. As a reminder, in this example we let the grid cells act as agents, but there is no reason why individual raccoons could not serve as agents being explicitly modeled if sufficient individual-level data were collected to gain the desired statistical inference. Scale is a critical component of all models, but it seems especially important in statistical ABMs because we seek to invert the models and are thus limited by the scale on which the data were collected.

16.4 Hierarchical First-Order Emulators and ABMs

Parameter estimation, calibration, and validation can be difficult in deterministic and stochastic ABMs given the computational cost of simulation and, in most cases, the nonlinear relationships in the parameters. The use of statistical emulators (or surrogates)

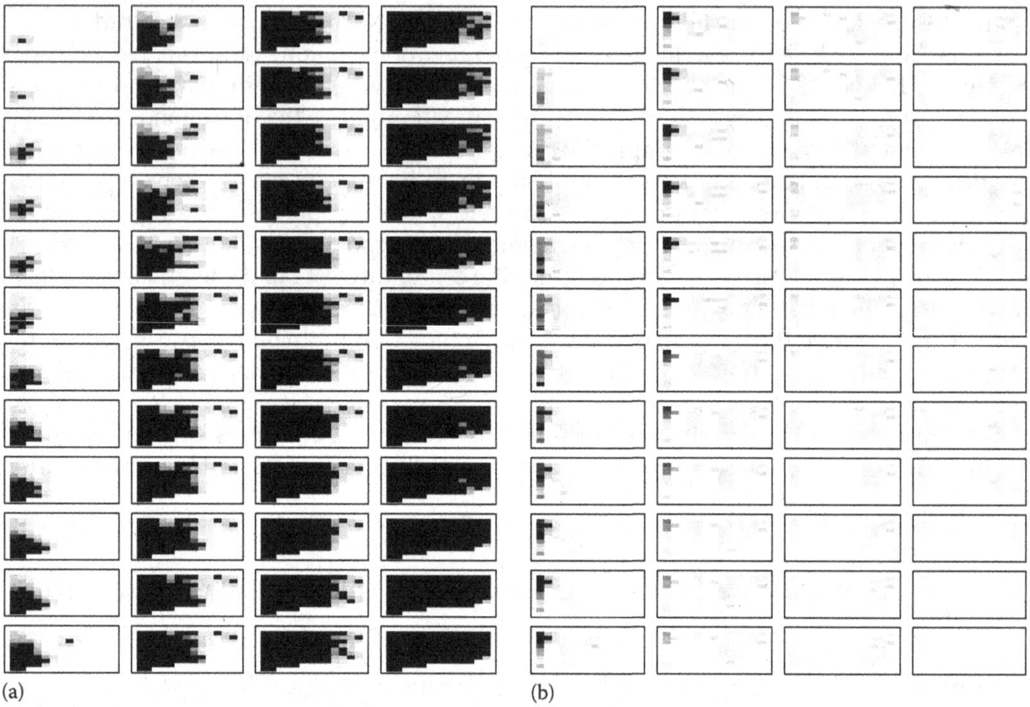

FIGURE 16.2
Posterior mean (a) and standard deviation (b) for the presence probabilities ($\theta_{i,t}$) for raccoon rabies in Connecticut, USA, during 1991–1995. Time for each image increases from top to bottom and left to right in that order. Left panel values range from zero (white) to one (black); right panel values range from zero (white) to $\sqrt{0.25}$ (black).

for complicated computer simulation models have, in recent years, proven useful to address these issues (e.g., Currin et al., 1991; Kennedy and O'Hagan, 2001; Sacks et al., 1989). Emulators act as a fast approximation to the computer simulation model, and because they are statistical models, are ideal to include within a Bayesian framework to perform uncertainty analysis, sensitivity analysis (OHagan, 2006), and model calibration and prediction (e.g., Higdon et al., 2008, 2004). That is, they allow a fairly high-fidelity representation of the simulation model but at a fraction of the computational cost. Statistical emulators have most often been implemented through the use of second-order (covariance) Gaussian process model specifications similar to geostatistical modeling (e.g., Kennedy and O'Hagan, 2001; OHagan, 2006; Sacks et al., 1989). Second-order emulators have recently been used to assess parameter sensitivity and estimation in ABMs (e.g., Dancik et al., 2010; Parry et al., 2013). As described in Hooten et al. (2011), it is often desirable to model the input–output relationship for a mechanistic model by using first-order characteristics (e.g., Frolov et al., 2009; Leeds et al., 2013, 2014; van der Merwe et al., 2007). In what follows, we provide a simple example to illustrate the use of first-order emulators to perform parameter estimation in the ABM context.

Assume we have a limited number of runs of the ABM simulation at input settings $\theta^{(1)}, \ldots, \theta^{(N)}$ and associated output $\mathbf{y}^{(1)}, \ldots, \mathbf{y}^{(N)}$, where $\mathbf{y}^{(i)}$ is a T-dimensional vector. Here, "input settings" may refer to model parameters, forcings, or even past values of the process state. We then seek a surrogate statistical model that predicts $\mathbf{y}^{*}$ at (untried) input settings θ^*. Following the first-order approach outlined in Hooten et al. (2011), we let $\mathbf{Y} = (\mathbf{y}^{(1)}, \ldots, \mathbf{y}^{(N)})$ and consider the singular value decomposition, $\mathbf{Y} = \mathbf{UDV}'$, where $\mathbf{U}$ is a $T \times T$ matrix of left singular vectors, $\mathbf{D}$ is a $T \times N$ matrix with the singular values along the principal diagonal, and $\mathbf{V}$ is an $N \times N$ matrix of right singular vectors. We then reduce the dimension by truncating the singular value decomposition matrices so that only the first q singular vectors and singular values are considered (e.g., selecting q to account for a significant portion of the variation in the output matrix $\mathbf{Y}$). We denote these truncated matrices by $\widetilde{\mathbf{U}}$, $\widetilde{\mathbf{D}}$, and $\widetilde{\mathbf{V}}$, so $\Psi \equiv \widetilde{\mathbf{U}}\widetilde{\mathbf{D}}$ is a $T \times q$ matrix and $\widetilde{\mathbf{V}}$ is an $N \times q$ matrix. We then seek to link the parameters to the model output through the right singular vectors, $\widetilde{\mathbf{V}}$. Specifically, note that the ith parameter vector $\theta^{(i)}$ corresponds to the ith row of $\widetilde{\mathbf{V}}$, which we denote $\mathbf{v}^{(i)}$. Thus, the ith output vector is approximated by $\mathbf{y}^{(i)} \approx \Psi \mathbf{v}^{(i)}$. We model these right singular vectors in terms of the parameters such that $\mathbf{v}^{(i)} = g(\theta^{(i)}, \boldsymbol{\alpha})$, where $g(\cdot)$ is some function of the input parameter vector $\theta^{(i)}$ and parameter vector $\boldsymbol{\alpha}$. We then model $\mathbf{y}^{(i)} = \Psi g(\theta^{(i)}, \boldsymbol{\alpha}) + \boldsymbol{\eta}_i$, where $\boldsymbol{\eta}_i$ is an error process. Typically, this error process, which accounts for truncation and emulator model representativeness, is assumed to be Gaussian with mean zero and variance–covariance matrix $\boldsymbol{\Sigma}_\eta$. Upon estimating the parameters $\boldsymbol{\alpha}$ from the model simulations, for a given input parameter vector θ^* we can predict the model output vector by $\mathbf{y}^*$. Hooten et al. (2011) consider both linear and nonlinear (random forest) functions for $g(\cdot)$, and Leeds et al. (2014) consider a parametric quadratic nonlinear model for $g(\cdot)$ as developed in Wikle and Hooten (2010) and Wikle and Holan (2011).

In a situation for which one has observation data corresponding to the simulation output vector, a Bayesian approach can be applied to efficiently obtain the posterior parameter distribution for θ by using the emulator in place of the simulation model. That is, given data vector $\mathbf{y}^D$, parameter estimates $\hat{\boldsymbol{\alpha}}$, prior distributions $[\theta]$, $[\boldsymbol{\tau}_\nu]$, and $[\boldsymbol{\Sigma}_\eta]$, a very simple Bayesian hierarchical model could be specified as

$$\mathbf{y}^D = \mathbf{y} + \boldsymbol{\nu}, \quad \boldsymbol{\nu} \sim (\boldsymbol{\tau}_\nu), \tag{16.7}$$

$$\mathbf{y} = \Psi g(\theta; \hat{\boldsymbol{\alpha}}) + \boldsymbol{\eta}, \quad \boldsymbol{\eta} \sim Gau(\mathbf{0}, \boldsymbol{\Sigma}_\eta),$$

$$[\theta][\boldsymbol{\tau}_\nu][\boldsymbol{\Sigma}_\eta],$$

where $\boldsymbol{\nu}$ is an error process with a distribution that depends on parameters $\boldsymbol{\tau}_\nu$. In addition, a non-Gaussian error distribution could be considered for $\boldsymbol{\eta}$ if warranted. As an alternative to the use of the simulation-derived parameter estimates for $\boldsymbol{\alpha}$, these parameters could be assigned a prior distribution with hyperparameters specified based on the model simulation estimates (e.g., prior mean given by $\hat{\boldsymbol{\alpha}}$) as in Leeds et al. (2014).

16.4.1 Simple Simulated Epidemic ABM Emulator Example

Consider a very simple cellular ABM to simulate an epidemic for susceptible, infected, and recovered (SIR) agents. This type of model is also referred to as a "compartment model." In particular, let the state of the ith agent at time t take one of the values

$$y_{i,t} \in \begin{cases} 0, & \text{susceptible} \\ 1, \ldots, K_I, & \text{infected} \\ K_I + 1, \ldots, K_I + K_R, & \text{recovered.} \end{cases}$$

Thus, there is one susceptible state, K_I infected states, and K_R recovered states. The ABM model is then specified such that susceptible agents (state 0) can be infected with some probability ($\theta_{i,t}$), with the odds of infection increasing depending on the number of neighbors infected. Once infected, the disease is assumed to follow a deterministic course, in which the infected individual goes through K_I infected states followed by K_R recovered states. The agent cannot be reinfected while it is in the recovered state. Specifically, this hybrid deterministic/stochastic SIR model is formulated so that the agent state evolves according to

$$y_{i,t} = \begin{cases} \text{Bern}(\theta_{i,t}), & \text{if } y_{i,t-1} = 0 \\ y_{i,t-1} + 1, & \text{if } 0 < y_{i,t-1} < K_I + K_R \\ 0, & \text{if } y_{i,t-1} = K_I + K_R, \end{cases}$$

where

$$\theta_{i,t} = \frac{J_{\mathcal{N}_i,t-1}\pi_i}{1 - \pi_i(1 - J_{\mathcal{N}_i,t-1})},$$

and

$$J_{\mathcal{N}_i,t-1} \equiv \sum_{j \in \mathcal{N}_i} 1_{(0<y_{j,t-1}<K_I)},$$

with π_i the prior probability of the ith agent being infected if exposed. In this case, $\theta_{i,t}$ reflects the probability of infection, for which the odds increase as the number of infected neighbors increases; in other words, for one neighbor the probability of infection is π_i, and the odds ratio is $J_{\mathcal{N}_i,t-1}$ (the number of infected neighbors), defined relative to the odds of infection if one neighbor is infected.

As an illustration, we consider a simulation with a regular grid of 32×32 agents, and simulate 150 times steps from a random initial SIR state (assuming a baseline probability of infection of 0.1), assuming 4 infection states ($N_I = 4$) and 4 recovered states ($N_R = 4$). Thus, $y_{i,t}$ can take discrete values $\{0, 1, \ldots, 8\}$. We allow the prior probability of infection to then be spatially varying according to an 8×8 grid of "regions" (each region contains a 4×4 grid of 16 agents). We generate the prior probability of infection for all cells in the jth region according to

$$p_j = \Phi^{-1}(\beta_1 x_{1,j} + \beta_2 x_{2,j}),$$

where Φ^{-1} is the inverse cumulative distribution function (CDF) of a standard normal distribution (although other link functions could be considered here) and $x_{1,j}$ and $x_{2,j}$ are assumed to be known covariates (in our case, corresponding to a sinusoidal pattern as shown in Figure 16.3a and b). Thus, the prior probability of infection for the ith agent, π_i, is equal to p_j if the ith agent is in region j. We simulate this epidemic assuming $\beta_1 = 0.3$ and $\beta_2 = -0.1$, which gives the probability of infection variation in space as shown in Figure 16.3c. The initial state and two realizations from this ABM simulation, separated by 10 time steps, are shown in Figure 16.3d through f. Note that the lower probability

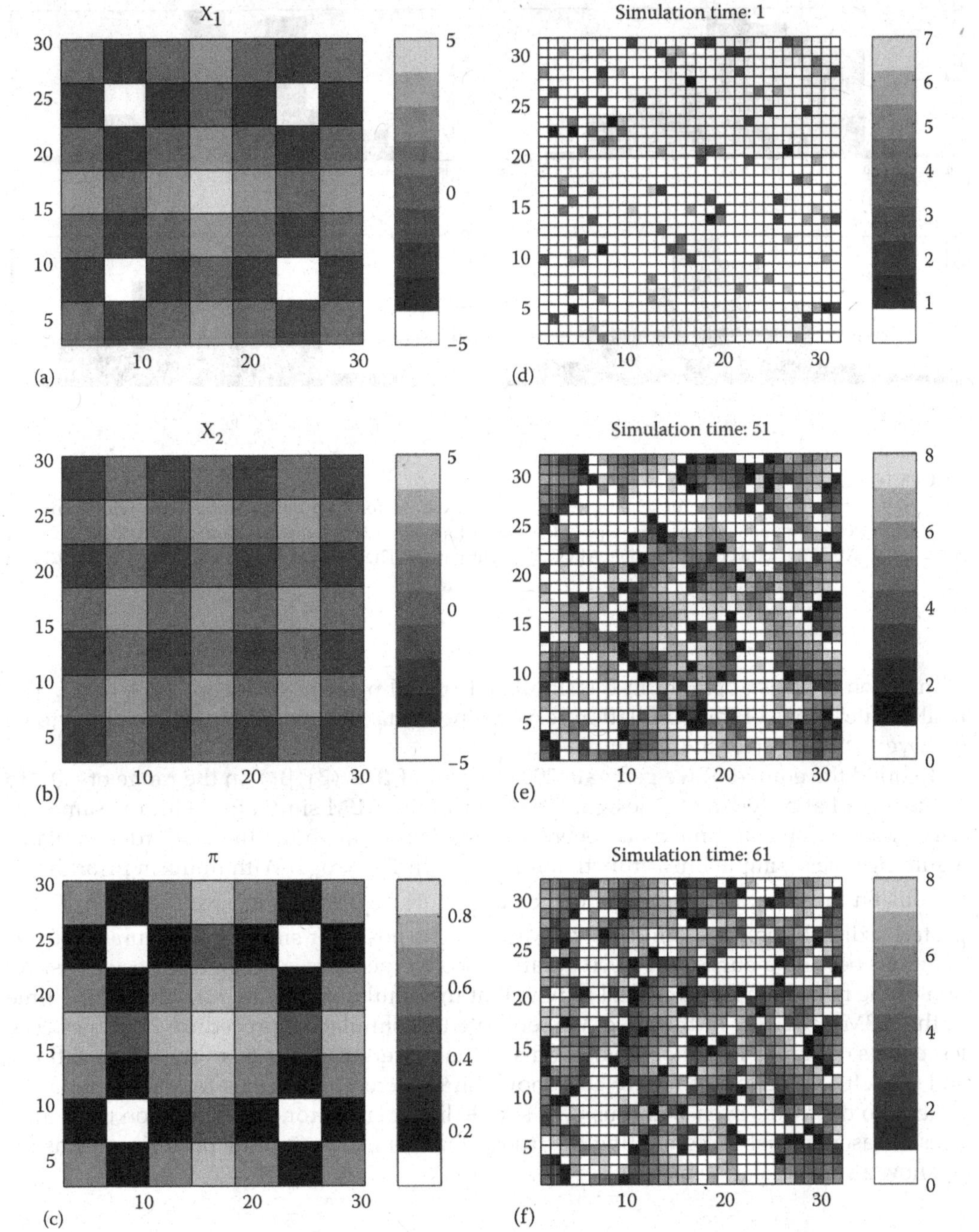

FIGURE 16.3
(a) X_1 covariate related to the probability of infection; (b) X_2 covariate related to the probability of infection; (c) prior probability of infection associated with the X_1 and X_2 spatial covariates; (d) initial state (time 1) for the SIR cellular ABM; (e) agent states at time $t = 51$ for the SIR cellular ABM; (f) agent state at time $t = 61$ for the SIR cellular ABM. In plots (d–f), state 0 (white) corresponds to susceptible agents, states (1–4) correspond to infected agents (black), and states (5–8) correspond to recovered agents (gray).

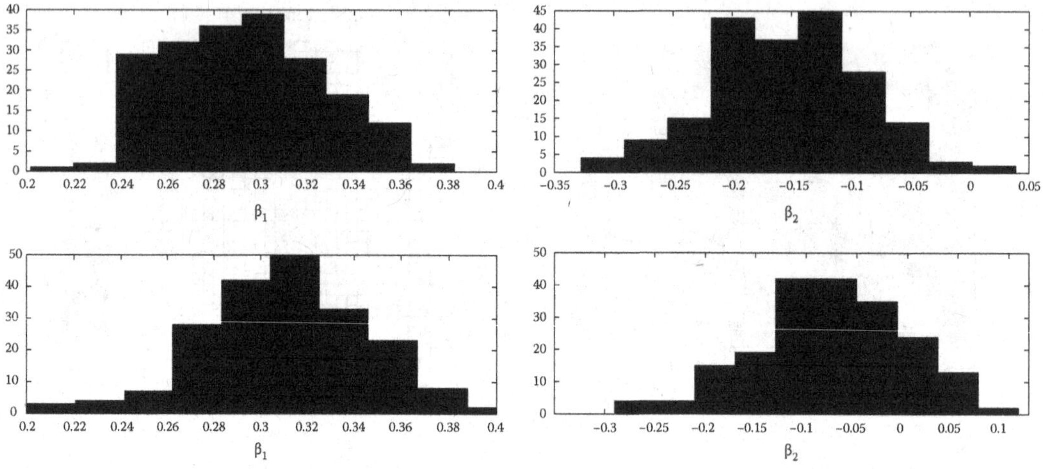

FIGURE 16.4
Histograms of posterior means for β parameters in the SIR ABM for each of 200 simulations with emulator- and ABC-based estimation approaches. Upper left panel: Emulator, β_1; Lower left panel: Emulator, β_2; Upper right panel: ABC, β_1; Lower right panel: ABC, β_2 (right panel). The true parameter values are $\beta_1 = 0.3$ and $\beta_2 = -0.1$.

of infection areas is apparent in the simulated spatial patterns in Figure 16.3e and f. The final 100 iterations of this simulation serve as the "data" (i.e., $\mathbf{y}^D$ in (16.7)), assumed to be observed without error in this simple example.

To build the emulator, we generate 200 samples of $\boldsymbol{\beta} \equiv (\beta_1, \beta_2)'$ in the range of -0.8 to 0.8, using a Latin-hypercube design. We then run the ABM simulator, with the same initial condition as the original data-generating simulation, and build the first-order emulator assuming a very simple linear function for $g(\cdot)$, with $\boldsymbol{\Sigma}_\eta = \sigma_\eta^2 \mathbf{I}$. With uniform prior distributions on the β parameters (-0.8 to 0.8) and a diffuse prior on σ_η^2, we sample from the posterior distribution for $\boldsymbol{\beta}$ and σ_η^2 via MCMC. The posterior simulation is run for 10,000 iterations beyond a burn-in of 1,000 iterations; convergence is evaluated by visual inspection of the two chains. It is critical to note that the emulator runs in a fraction of the time as the ABM simulator in this case. We replicate this simulation procedure 200 times. The left panels of Figure 16.4 show histograms of the posterior means from these simulations, and the left two panels of Figure 16.5 show the 95% credible regions for each simulation relative to the truth. Even with this very simple linear emulator, the estimation procedure does a reasonable job of estimating the true parameters, although the posterior means for β_2 show a small negative bias.

16.5 Approximate Bayesian Computation (ABC) and ABMs

There is an increasing interest in the statistical sciences for performing inference and calibration in situations where likelihoods are difficult to calculate or not available, or if simulation models can easily be used to generate "data." One of the most popular

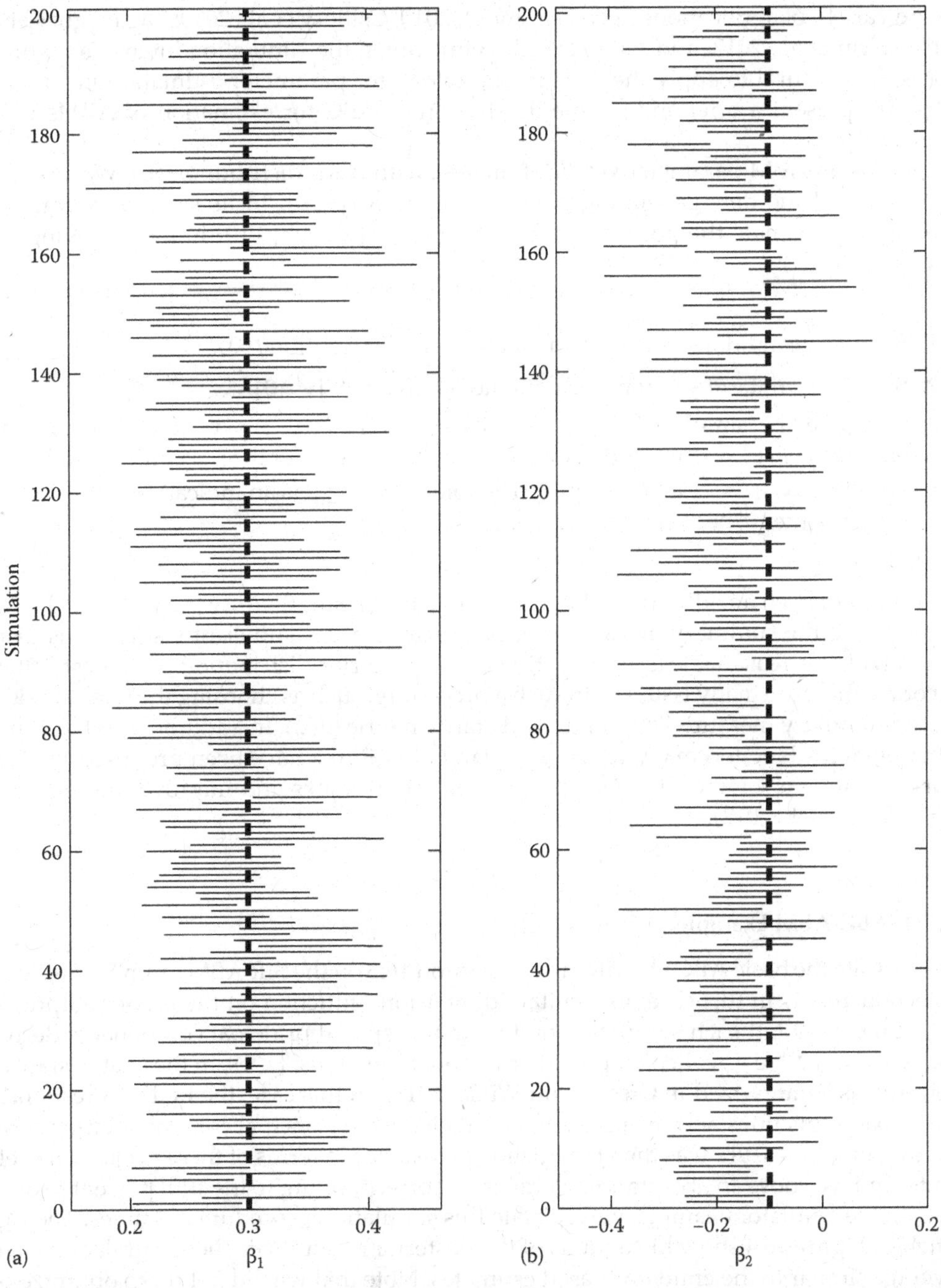

FIGURE 16.5
Posterior 95% credible intervals for β_1 (a) and β_2 (b) for each of 200 simulations for the SIR ABM. The true parameter values are shown by dashed lines.

methods for such situations is ABC, originally introduced in genetics by Tavaré et al. (1997) and popularized by Beaumont et al. (2002). This methodology has since been used across a wide variety of applications (e.g., Beaumont, 2010; Csilléry et al., 2010; Lagarrigues et al., 2014; Marin et al., 2012, and references therein). Given the simulation-intensive nature of ABMs, it is natural to apply the ABC methodology for parameter estimation in such settings. We present a brief and simple illustration of ABC for estimation of ABMs in the following.

Let θ be an ABM parameter vector of interest with prior distribution $[\theta]$. We also have data, $\mathbf{y}^D$, and the data-generating ABM distribution $[\mathbf{y}^D|\theta]$. In its simplest form, ABC seeks to approximate the posterior distribution $[\theta|\mathbf{y}^D] \propto [\mathbf{y}^D|\theta][\theta]$ in the following way (Toni et al., 2009):

1. Sample a candidate vector $\tilde{\theta}$ from a proposal distribution $f(\theta)$.
2. Simulate a dataset $\tilde{\mathbf{y}}^D$ from the data model given by $[\mathbf{y}^D|\tilde{\theta}]$.
3. Compare the simulated dataset (or its summary, $h(\tilde{\mathbf{y}}^D)$) with the data (or its summary, $h(\mathbf{y}^D)$) using a distance metric, $d(\cdot)$, and tolerance, ϵ: procedurally, if $d(h(\mathbf{y}^D), h(\tilde{\mathbf{y}}^D)) \le \epsilon$, then accept $\tilde{\theta}$. The tolerance $\epsilon \ge 0$ is an indication of the level of "agreement" between $\mathbf{y}^D$ and $\tilde{\mathbf{y}}^D$.

The output of such an ABC algorithm is a sample of θ from $[\theta|\ d(h(\mathbf{y}^D), h(\tilde{\mathbf{y}}^D)) \le \epsilon]$. Thus, in general, if the summary measure is appropriate and ϵ is sufficiently small, this distribution will be a reasonable approximation of the posterior distribution of interest, $[\theta|\mathbf{y}^D]$. Of course there are many issues with such a procedure, such as choosing the tolerance level (ϵ), the summary measure ($h(\cdot)$), and the distance metric ($d(\cdot)$). In addition, this basic algorithm often leads to too many rejections. Many algorithms have been proposed in recent years to mitigate some of these subjective calibration issues and improve the procedure (e.g., Marin et al., 2012).

16.5.1 ABC/ABM Example

We illustrate the basic ABC algorithm from Section 16.5 on the SIR ABM example presented in Section 16.4.1. In this case, to facilitate dimension reduction on the model output, we project the data and each simulation on the first 40 spatial principal components derived from the data, $\mathbf{y}^D$ (e.g., called empirical orthogonal functions, EOFs, in the spatial statistics literature as summarized in Cressie and Wikle, 2011). In this case, the EOFs correspond to spatial basis functions and the projection coefficients are principal component time series. Our summary measure was then the Euclidean distance in terms of these projection coefficients, and we chose an acceptance tolerance, ϵ, corresponding to the 10th percentile of $d()$. As with the emulator example, we replicate this simulation procedure 200 times. The right panels of Figure 16.4 show histograms of the posterior means from these simulations, with coverage similar to the emulator-based estimates. Note that we did not try to optimize our selection of distance metric nor tolerance region for this example.

Although the simulation ABM in this simple example is quite fast to compute, in many real-world situations this would not be the case. In such situations, the simulation model could also be approximated. For example, one could use an emulator approach as described in Section 16.4.

16.6 Conclusion

Rule-based ABMs can be an effective modeling paradigm to account for complex spatio-temporal dynamics for discrete state spaces. With the addition of stochastic terms, and the presence of data, this becomes a statistical modeling challenge. The Bayesian hierarchical modeling approach can facilitate such statistical modeling due to its facility with dealing with uncertainty in data, processes, and parameters. Estimation can be facilitated by ABC and statistical emulators of the ABM. In addition, these models can be considered from a network perspective, which also allows one to borrow many existing network modeling methodologies. Although the use of complicated network models in the ABM context is relatively new, we expect to see additional developments in this area. In addition, there are potentially important issues associated with model selection in ABMs (e.g., Piou et al., 2009), and this is likely to be an important area of future research.

Acknowledgments

Wikle acknowledges support from U.S. National Science Foundation (NSF) grant DMS-1049093 and the NSF and U.S. Census Bureau under NSF grant SES-1132031, funded through the NSF-Census Research Network (NCRN) program. We thank Scott Holan and an anonymous reviewer for helpful comments on an early version of this chapter.

References

Barber, D. (2012). *Bayesian Reasoning and Machine Learning*. Cambridge University Press, Cambridge, U.K.

Beaumont, M. A. (2010). Approximate Bayesian computation in evolution and ecology. *Annual Review of Ecology, Evolution, and Systematics*, 41:379–406.

Beaumont, M. A., Zhang, W., and Balding, D. J. (2002). Approximate Bayesian computation in population genetics. *Genetics*, 162(4):2025–2035.

Besag, J. (1972). Nearest-neighbour systems and the auto-logistic model for binary data. *Journal of the Royal Statistical Society. Series B (Methodological)*, 34:75–83.

Besag, J. (1974). Spatial interaction and the statistical analysis of lattice systems. *Journal of the Royal Statistical Society. Series B (Methodological)*, 36:192–236.

Besag, J. and Tantrum, J. (2003). Likelihood analysis of binary data in space and time. In P. J. Green, N. L. Hjort, and S. Richardson, (eds.), *Highly Structured Stochastic Systems*. Oxford University Press, Oxford, U.K.

Caron-Lormier, G., Humphry, R. W., Bohan, D. A., Hawes, C., and Thorbek, P. (2008). Asynchronous and synchronous updating in individual-based models. *Ecological Modelling*, 212(3): 522–527.

Cressie, N. (1993). *Statistics for Spatial Data*. John Wiley & Sons, New York.

Cressie, N. and Wikle, C. K. (2011). *Statistics for Spatio-Temporal Data*. John Wiley & Sons, Hoboken, NJ.

Csilléry, K., Blum, M. G., Gaggiotti, O. E., and François, O. (2010). Approximate Bayesian computation (abc) in practice. *Trends in Ecology & Evolution*, 25(7):410–418.

Currin, C., Mitchell, T., Morris, M., and Ylvisaker, D. (1991). Bayesian prediction of deterministic functions, with applications to the design and analysis of computer experiments. *Journal of the American Statistical Association*, 86(416):953–963.

Dancik, G. M., Jones, D. E., and Dorman, K. S. (2010). Parameter estimation and sensitivity analysis in an agent-based model of *Leishmania major* infection. *Journal of Theoretical Biology*, 262(3):398–412.

Filatova, T., Verburg, P. H., Parker, D. C., and Stannard, C. A. (2013). Spatial agent-based models for socio-ecological systems: Challenges and prospects. *Environmental Modelling & Software*, 45:1–7.

Frolov, S., Baptista, A. M., Leen, T. K., Lu, Z., and van der Merwe, R. (2009). Fast data assimilation using a nonlinear Kalman filter and a model surrogate: An application to the columbia river estuary. *Dynamics of Atmospheres and Oceans*, 48(1):16–45.

Gilbert, N. (2008). *Agent-Based Models*. Sage Publications, Los Angeles, CA.

Grimm, V. and Railsback, S. F. (2005). *Individual-Based Modeling and Ecology*. Princeton University Press, Princeton, NJ.

Guyon, X. (1995). *Random Fields on a Network: Modeling, Statistics, and Applications*. Springer-Verlag, New York.

Higdon, D., Gattiker, J., Williams, B., and Rightley, M. (2008). Computer model calibration using high-dimensional output. *Journal of the American Statistical Association*, 103(482), 570–583.

Higdon, D., Kennedy, M., Cavendish, J. C., Cafeo, J. A., and Ryne, R. D. (2004). Combining field data and computer simulations for calibration and prediction. *SIAM Journal on Scientific Computing*, 26(2):448–466.

Holan, S. H. and Wikle, C. K. (2015). Hierarchical dynamic generalized linear mixed models for discrete-valued spatio-temporal data. In R. A. Davis, S. H. Holan, R. Lund and N. Ravishanker, eds., pp. 327–348. *Handbook of Discrete-Valued Time Series*, Chapman & Hall, Boca Raton, FL.

Hooten, M. B., Johnson, D. S., Hanks, E. M., and Lowry, J. H. (2010). Agent-based inference for animal movement and selection. *Journal of Agricultural, Biological, and Environmental Statistics*, 15(4):523–538.

Hooten, M. B., Leeds, W. B., Fiechter, J., and Wikle, C. K. (2011). Assessing first-order emulator inference for physical parameters in nonlinear mechanistic models. *Journal of Agricultural, Biological, and Environmental Statistics*, 16(4):475–494.

Hooten, M. B. and Wikle, C. K. (2010). Statistical agent-based models for discrete spatio-temporal systems. *Journal of the American Statistical Association*, 105(489):236–248.

Keeling, M. J. and Rohani, P. (2008). *Modeling Infectious Diseases in Humans and Animals*. Princeton University Press, Princeton, NJ.

Kennedy, M. C. and O'Hagan, A. (2001). Bayesian calibration of computer models. *Journal of the Royal Statistical Society: Series B (Statistical Methodology)*, 63(3):425–464.

Lagarrigues, G., Jabot, F., Lafond, V., and Courbaud, B. (2014). Approximate Bayesian computation to recalibrate individual-based models with population data: Illustration with a forest simulation model. *Ecological Modelling*. doi:10.1016/j.ecolmodel.2014.09.023.

Leeds, W., Wikle, C., and Fiechter, J. (2014). Emulator-assisted reduced-rank ecological data assimilation for nonlinear multivariate dynamical spatio-temporal processes. *Statistical Methodology*, 17:126–138.

Leeds, W., Wikle, C., Fiechter, J., Brown, J., and Milliff, R. (2013). Modeling 3-d spatio-temporal biogeochemical processes with a forest of 1-d statistical emulators. *Environmetrics*, 24(1):1–12.

Marin, J.-M., Pudlo, P., Robert, C. P., and Ryder, R. J. (2012). Approximate Bayesian computational methods. *Statistics and Computing*, 22(6):1167–1180.

OHagan, A. (2006). Bayesian analysis of computer code outputs: A tutorial. *Reliability Engineering & System Safety*, 91(10):1290–1300.

Olfati-Saber, R. (2006). Flocking for multi-agent dynamic systems: Algorithms and theory. *IEEE Transactions on Automatic Control*, 51(3):401–420.

Parry, H. R., Topping, C. J., Kennedy, M. C., Boatman, N. D., and Murray, A. W. (2013). A Bayesian sensitivity analysis applied to an agent-based model of bird population response to landscape change. *Environmental Modelling & Software*, 45:104–115.

Piou, C., Berger, U., and Grimm, V. (2009). Proposing an information criterion for individual-based models developed in a pattern-oriented modelling framework. *Ecological Modelling*, 220(17):1957–1967.

Rasmussen, J. G., Møller, J., Aukema, B. H., Raffa, K. F., and Zhu, J. (2007). Continuous time modelling of dynamical spatial lattice data observed at sparsely distributed times. *Journal of the Royal Statistical Society: Series B (Statistical Methodology)*, 69(4):701–713.

Sacks, J., Welch, W. J., Mitchell, T. J., and Wynn, H. P. (1989). Design and analysis of computer experiments. *Statistical Science*, 4(4):409–423.

Sattenspiel, L. (2009). *The Geographic Spread of Infectious Diseases: Models and Applications*. Princeton University Press, Princeton, NJ.

Smith, D. L., Lucey, B., Waller, L. A., Childs, J. E., and Real, L. A. (2002). Predicting the spatial dynamics of rabies epidemics on heterogeneous landscapes. *Proceedings of the National Academy of Sciences*, 99(6):3668–3672.

Tavaré, S., Balding, D. J., Griffiths, R. C., and Donnelly, P. (1997). Inferring coalescence times from DNA sequence data. *Genetics*, 145(2):505–518.

Toni, T., Welch, D., Strelkowa, N., Ipsen, A., and Stumpf, M. P. (2009). Approximate Bayesian computation scheme for parameter inference and model selection in dynamical systems. *Journal of the Royal Society Interface*, 6(31):187–202.

van der Merwe, R., Leen, T. K., Lu, Z., Frolov, S., and Baptista, A. M. (2007). Fast neural network surrogates for very high dimensional physics-based models in computational oceanography. *Neural Networks*, 20(4):462–478.

Wheeler, D. C. and Waller, L. A. (2008). Mountains, valleys, and rivers: The transmission of raccoon rabies over a heterogeneous landscape. *Journal of Agricultural, Biological, and Environmental Statistics*, 13(4):388–406.

Wikle, C. K. (2002). A kernel-based spectral model for non-Gaussian spatio-temporal processes. *Statistical Modelling*, 2(4):299–314.

Wikle, C. K. (2003). Hierarchical Bayesian models for predicting the spread of ecological processes. *Ecology*, 84(6):1382–1394.

Wikle, C. K. and Holan, S. H. (2011). Polynomial nonlinear spatio-temporal integro-difference equation models. *Journal of Time Series Analysis*, 32(4):339–350.

Wikle, C. K. and Hooten, M. B. (2010). A general science-based framework for dynamical spatio-temporal models. *Test*, 19(3):417–451.

Wolfram, S. (1984). Cellular automata as models of complexity. *Nature*, 311:419–423.

Zheng, Y. and Zhu, J. (2008). Markov chain Monte Carlo for a spatial-temporal autologistic regression model. *Journal of Computational and Graphical Statistics*, 17(1):123–137.

Zhu, J., Huang, H.-C., and Wu, J. (2005). Modeling spatial-temporal binary data using Markov random fields. *Journal of Agricultural, Biological, and Environmental Statistics*, 10(2):212–225.

Zhu, J. and Zheng, Y. (2015). Autologistic regression models for spatio-temporal binary data. In R. A. Davis, S. H. Holan, R. Lund and N. Ravishanker, eds., *Handbook of Discrete-Valued Time Series*, pp. 367–386. Chapman & Hall, Boca Raton, FL.

Zhu, J., Zheng, Y., Carroll, A. L., and Aukema, B. H. (2008). Autologistic regression analysis of spatial-temporal binary data via Monte Carlo maximum likelihood. *Journal of Agricultural, Biological, and Environmental Statistics*, 13(1):84–98.

17

Autologistic Regression Models for Spatio-Temporal Binary Data

Jun Zhu and Yanbing Zheng

CONTENTS

17.1 Introduction

Binary data on a spatial lattice are often encountered in environmental and ecological studies. Spatial statistical methods have been developed for modeling spatial binary responses and their relations to covariates while properly accounting for spatial correlation. In this chapter, we review autologistic models in the class of Markov random fields that model spatial dependence via autoregression and consider extensions to autologistic regression models for spatio-temporal binary data. In particular, the introduction of autoregression in space and time results in an unknown normalizing constant in the likelihood function, which makes estimation and statistical inference challenging. We describe several approaches to the inference for spatio-temporal autologistic regression models and illustrate them by an ecological data example.

17.1.1 Markov Random Field and Autologistic Model

For site $i = 1, \ldots, n$, let Y_i denote the response variable at site i. Besag (1974) developed a Markov random field model that conditionally specifies the distribution of Y_i. Let $\boldsymbol{Y} = (Y_1, \ldots, Y_n)'$ and $\boldsymbol{Y}_{-i} = (Y_j : j \neq i)'$ denote the response variables with $\boldsymbol{y} = (y_1, \ldots, y_n)'$ and $\boldsymbol{y}_{-i} = (y_j : j \neq i)'$ denoting a corresponding realization at all n sites and all except site i, respectively. In a Markov random field model for $\boldsymbol{Y}$, the full conditional probability density of Y_i (conditional on all other sites) is assumed to depend only on the responses at neighboring sites; that is, $p(y_i|\boldsymbol{y}_{-i}) = p(y_i|y_j : j \in \mathcal{N}_i)$, where $\mathcal{N}_i$ denotes a prespecified neighborhood of site i. The conditional probability density is generally specified in an exponential form

$$p(y_i|\boldsymbol{y}_{-i}) = p(y_i|y_j : j \in \mathcal{N}_i) = \exp\left\{A_i(\boldsymbol{y}_{-i})y_i - B_i(\boldsymbol{y}_{-i}) + C_i(y_i)\right\} \tag{17.1}$$

where A_i is a natural parameter function, B_i is a function of the model parameters and $\boldsymbol{y}_{-i}$ but free of y_i, and C_i is a function of y_i but free of the model parameters.

To ensure that the resulting joint distribution of $\boldsymbol{Y}$ is valid, Besag (1974) defined a negpotential function

$$Q(\boldsymbol{y}) = \ln\left\{\frac{p(\boldsymbol{y})}{p(\boldsymbol{0})}\right\}, \tag{17.2}$$

which is essentially the logarithm of the joint probability density function $p(\boldsymbol{y})$ up to a normalizing constant since

$$p(\boldsymbol{y}) = \frac{\exp\{Q(\boldsymbol{y})\}}{\sum_{\boldsymbol{z}\in\Omega}\exp\{Q(\boldsymbol{z})\}}, \tag{17.3}$$

where Ω denotes a suitable space of responses. It has been shown that $Q(\boldsymbol{y})$ in (17.2) can be uniquely expanded on Ω and the expansion is made up of the conditional probabilities $p(y_i|\boldsymbol{y}_{-i})$ in (17.1) under a positivity condition (Besag, 1974; Cressie, 1993). The Hammersley–Clifford Theorem and its corollary establish the sparsity of the expansion and most importantly, the validity of the joint probability $p(\boldsymbol{y})$ through the negpotential function $Q(\boldsymbol{y})$.

For binary data on a spatial lattice, Besag (1972) developed an autologistic model in the framework of Markov random fields. In particular, the binary response variable $Y_i \in \{0, 1\}$ has a conditional Bernoulli distribution. The pairwise-only dependence is among neighboring sites according to the neighborhood $\mathcal{N}_i$. Thus, the natural parameter function is of the form

$$A_i(\boldsymbol{y}_{-i}) = \alpha_i + \sum_{j\in\mathcal{N}_i} \theta_{ij} y_j, \tag{17.4}$$

$B_i(y_{-i}) = \ln[1 + \exp\{A_i(y_{-i})\}]$, and $C_i(y_i) = 0$, where α_i is a constant, θ_{ij}s are spatial dependence parameters such that $\theta_{ij} = \theta_{ji}$ for $j \neq i$ and $\theta_{ii} = 0$ for site $i = 1, \ldots, n$. The corresponding negpotential function is

$$Q(y) = \sum_{i=1}^{n} y_i \left\{ \alpha_i + (1/2) \sum_{j \in \mathcal{N}_i} \theta_{ij} y_j \right\}. \tag{17.5}$$

By the Hammersley–Clifford Theorem, the joint probability density is

$$p(y) = \frac{\exp\left[\sum_{i=1}^{n} y_i \left\{\alpha_i + (1/2) \sum_{j \in \mathcal{N}_i} \theta_{ij} y_j\right\}\right]}{\sum_{z \in \Omega} \exp\left[\sum_{i=1}^{n} z_i \left\{\alpha_i + (1/2) \sum_{j \in \mathcal{N}_i} \theta_{ij} z_j\right\}\right]}. \tag{17.6}$$

In (17.6), the normalizing constant in the denominator involves the model parameters and generally does not have an analytical form, which makes it a challenge to directly maximize the likelihood function.

The traditional parameterization of autologistic models may not be intuitive when incorporating regression. Similar to the parametrization used for auto-Gaussian models, a centered parameterization of autologistic models was proposed recently (Caragea and Kaiser, 2009; Kaiser et al., 2012), which is perhaps more suitable for regression purposes. In the centered parameterization,

$$A_i(y_{-i}) = \ln\{\kappa_i/(1 - \kappa_i)\} + \sum_{j \in \mathcal{N}_i} \theta_{ij}(y_j - \kappa_j),$$

where $\kappa_i \in (0, 1), i = 1, \ldots, n$. A detailed description of the centered parameterization of autologistic regression models is given in Section 17.3.

A special case of the autologistic model is the Ising model. The Ising model was first developed by Ernst Ising in his doctoral thesis as an attempt to describe phase transitions in ferromagnets (Ising, 1924, 1925). The basic idea is that microscopic magnets are arranged on a square lattice such that there is one magnet at each lattice site. Each magnet is assumed to have two possible spin directions, generally labeled as up ($y_i = +1$) or down ($y_i = -1$), and is assumed to only interact with its four nearest neighbors. In the Ising model, the total energy, also known as the Hamiltonian, of the configuration is given by

$$H(y) = -\sum_{i=1}^{n} \sum_{j \in \mathcal{N}_i, j < i} \theta y_i y_j - \sum_{i=1}^{n} \alpha y_i, \tag{17.7}$$

where $\mathcal{N}_i$ denotes the neighborhood of site i comprising the four nearest neighbors, the coefficient θ represents the strength of interactions among the nearest neighbors, and the coefficient α represents an external magnetic field. The cases $\theta > 0$ and $\theta < 0$ correspond to ferromagnetism and antiferromagnetism, respectively. The joint probability density of a configuration is given by the so-called Boltzmann factor

$$Z_{\beta}^{-1} \exp\{-\beta H(y)\}, \tag{17.8}$$

where $\beta = (k_B T)^{-1} \geq 0$ with T denoting the Kelvin temperature and k_B denoting the Boltzmann constant, while $Z_\beta = \sum_y \exp\{-\beta H(y)\}$ is a partition function (or, normalizing constant). When the parameter β in (17.8) surpasses a threshold value, a phase transition from short-range to long-range interactions would occur, resulting in an ordered phase with nonzero limiting correlation (see, e.g., Pickard, 1976, 1977).

17.1.2 Spatio-Temporal Autologistic Model

For spatio-temporal binary data, let $Y_{i,t} \in \{\pm 1\}$ denote the binary response variable with $y_{i,t}$ denoting a realization at site $i = 1, \ldots, n$ and time t. Let $\mathbf{Y}_t = (Y_{1,t}, \ldots, Y_{n,t})'$ denote the vector of binary responses with realizations $\boldsymbol{y}_t = (y_{1,t}, \ldots, y_{n,t})'$ at all sites and a given time point t. Bartlett (1971, 1972) developed a Markov process with

$$P(Y_{i,t+\Delta t} = y_{i,t} | \boldsymbol{y}_t) = 1 - \lambda(\Delta t)\{1 - F_i(\boldsymbol{y}_t)\}, \tag{17.9}$$

where $\lambda \geq 0$, $\Delta t \geq 0$, and $F_i(\boldsymbol{y}_t)$ is a function of $\boldsymbol{y}_t$. The joint probability density of $\mathbf{Y}_t$, when the Markov process is at equilibrium, is

$$p(\boldsymbol{y}_t) = c(\alpha, \theta) \exp\left(-\alpha \sum_{i=1}^{n} y_{i,t} - \theta \sum_{i=1}^{n} y_{i,t} z_{i,t}\right), \tag{17.10}$$

where α and θ are two coefficients, under the condition that

$$\sum_{i=1}^{n} \{1 - F_i(\boldsymbol{y}_t)\} = \sum_{i=1}^{n} \{1 - F_i(\tilde{\boldsymbol{y}}_{i,t})\} \exp(2\alpha y_{i,t} + 4\theta y_{i,t} z^*_{i,t}), \tag{17.11}$$

where $z_{i,t}$ denotes a linear combination of $y_{j,t}$ for $j \neq i$, $z^*_{i,t}$ is a symmetrized form of $z_{i,t}$ (e.g., in the one-dimensional space, if $z_{i,t} = y_{i-1,t}$, then $2z^*_{i,t} = y_{i-1,t} + y_{i+1,t}$), and $\tilde{\boldsymbol{y}}_{i,t} = (y_{1,t}, \ldots, y_{i-1,t}, -y_{i,t}, y_{i+1,t}, \ldots, y_{n,t})'$.

A direct and symmetric solution to Equation (17.11) is

$$1 - F_i(\boldsymbol{y}_t) = \exp(\alpha y_{i,t} + 2\theta y_{i,t} z^*_{i,t}) f(\alpha y_{i,t}, \theta y_{i,t} z^*_{i,t}), \tag{17.12}$$

where f is a suitable, positive function, and even in both $y_{i,t}$ and $y_{i,t} z^*_{i,t}$.

Now, on a square lattice, let $Y_{i,i',t} \in \{0, 1\}$ denote the binary response with a realization $y_{i,i',t}$ at row i, column i', and time t. Besag (1972) proposed a Markov process of binary responses developing through time on the square lattice, which can be viewed as a special case of Bartlett (1971, 1972). In particular, for fixed α_y, $\theta_{y,1}$, and $\theta_{y,2}$,

$$\begin{aligned} &P\left(Y_{i,i',t+\Delta t} = y | Y_{i,i',t} = y, y_{\cdot,\cdot,t':t' \leq t+\Delta t}, \text{ excluding } y_{i,i',t+\Delta t}\right) \\ &= \left[1 + \Delta t \exp\left\{\alpha_y + \theta_{y,1}(y_{i-1,i',t} + y_{i+1,i',t}) + \theta_{y,2}(y_{i,i'-1,t} + y_{i,i'+1,t})\right\}\right]^{-1} + o(\Delta t) \\ &= 1 - \Delta t \exp\left\{\alpha_y + \theta_{y,1}(y_{i-1,i',t} + y_{i+1,i',t}) + \theta_{y,2}(y_{i,i'-1,t} + y_{i,i'+1,t})\right\} + o(\Delta t), \end{aligned} \tag{17.13}$$

gives the probability that $Y_{i,i',\cdot}$ remains unchanged in the time interval $(t, t + \Delta t]$, given all other values at or before time $t+\Delta t$. Further, it can be shown that its stationary distribution is an autologistic model with the full conditional distribution

$$p\left(y_{i,i',t} | y_{i-1,i',t}, y_{i+1,i',t}, y_{i,i'-1,t}, y_{i,i'+1,t}\right)$$
$$= \frac{\exp\left[y_{i,i',t}\left\{\alpha + \theta_1(y_{i-1,i',t} + y_{i+1,i',t}) + \theta_2(y_{i,i'-1,t} + y_{i,i'+1,t})\right\}\right]}{1 + \exp\left\{\alpha + \theta_1(y_{i-1,i',t} + y_{i+1,i',t}) + \theta_2(y_{i,i'-1,t} + y_{i,i'+1,t})\right\}}, \tag{17.14}$$

where $\alpha = \alpha_0 - \alpha_1$, $\theta_1 = \theta_{0,1} - \theta_{1,1}$, and $\theta_2 = \theta_{0,2} - \theta_{1,2}$.

17.2 Spatio-Temporal Autologistic Regression Model

17.2.1 Model

For the analysis of spatio-temporal binary data in practice, it is often of interest to account for possible effects of covariates. For example, Gumpertz et al. (1997) and Huffer and Wu (1998) incorporated covariates in an autologistic model by replacing the constant α_i in (17.6) with a linear regression term and the spatial lattice can be either regular or irregular. The resulting model is referred to as an autologistic regression model. Zhu et al. (2005) and Zheng and Zhu (2008) extended the autologistic regression model to a spatio-temporal autologistic regression model that accounts for covariates and spatio-temporal dependence simultaneously for binary responses measured repeatedly over discrete time points on a spatial lattice.

As earlier, let $i = 1, \ldots, n$ denote sites on a spatial lattice. Further, let $t \in \mathbb{Z}$ index discrete time points and $Y_{i,t} \in \{0, 1\}$ denote the binary response variable at site i and time t. Let $x_{0,i,t} \equiv 1$ and let $x_{k,i,t}$ denote the kth covariate at site i and time t, for $k = 1, \ldots, p$ and a total of p covariates. Zhu et al. (2005) developed a spatio-temporal autologistic regression model via the full conditional distributions:

$$p(y_{i,t} | y_{i',t'} : (i', t') \neq (i, t)) = p(y_{i,t} | y_{i',t'} : (i', t') \in \mathcal{N}_{i,t})$$
$$= \frac{\exp\left\{\sum_{k=0}^{p} \theta_k x_{k,i,t} y_{i,t} + \sum_{j \in \mathcal{N}_i} \theta_{p+1} y_{i,t} y_{j,t} + \theta_{p+2} y_{i,t}(y_{i,t-1} + y_{i,t+1})\right\}}{1 + \exp\left\{\sum_{k=0}^{p} \theta_k x_{k,i,t} + \sum_{j \in \mathcal{N}_i} \theta_{p+1} y_{j,t} + \theta_{p+2}(y_{i,t-1} + y_{i,t+1})\right\}}, \tag{17.15}$$

where $\mathcal{N}_{i,t} = \{(j, t) : j \in \mathcal{N}_i\} \cup \{(i, t-1), (i, t+1)\}$ denotes a spatio-temporal neighborhood for site i and time t and recall that $\mathcal{N}_i = \{j : \text{site } j \text{ is a neighbor of site } i\}$. The model parameters are the intercept θ_0, slope θ_k for covariate x_k with $k = 1, \ldots, p$, a spatial autoregressive coefficient θ_{p+1}, and a temporal autoregressive coefficient θ_{p+2}. Let $\theta = (\theta_0, \ldots, \theta_{p+2})'$ denote the vector of parameters in the model (17.15).

Let $\boldsymbol{Y}_t = (Y_{1,t},\ldots,Y_{n,t})'$ denote the binary responses at all sites and a given time point t for $t = 1,\ldots,T$ and a total of T sampling time points. Then, the joint distribution of $\boldsymbol{Y}_2,\ldots,\boldsymbol{Y}_{T-1}$ conditional on $\boldsymbol{Y}_1$ and $\boldsymbol{Y}_T$ is

$$p(\boldsymbol{y}_2,\ldots,\boldsymbol{y}_{T-1}|\boldsymbol{y}_1,\boldsymbol{y}_T;\boldsymbol{\theta})$$
$$= c(\boldsymbol{\theta})^{-1}\exp\left\{\sum_{t=2}^{T-1}\left(\sum_{i=1}^{n}\sum_{k=0}^{p}\theta_k x_{k,i,t}y_{i,t} + \frac{1}{2}\sum_{i=1}^{n}\sum_{j\in\mathcal{N}_i}\theta_{p+1}y_{i,t}y_{j,t}\right) + \sum_{t=2}^{T}\sum_{i=1}^{n}\theta_{p+2}y_{i,t}y_{i,t-1}\right\}, \tag{17.16}$$

where $c(\boldsymbol{\theta})$ is a normalizing constant and generally is intractable as it does not have an analytical form.

The full conditional distribution (17.15) is symmetric in time and thus depends on both past and future time points. For prediction at future time points, however, it would be more sensible to have the conditional distributions depend only on the past. For example, Zhu et al. (2008) proposed the following conditional distributions:

$$p(y_{i,t}|y_{j,t}: j \neq i, \boldsymbol{y}_{t'}: t' = t-1, t-2,\ldots)$$
$$= p(y_{i,t}|y_{j,t}: j \in \mathcal{N}_i, \boldsymbol{y}_{t'}: t' = t-1, t-2,\ldots,t-S)$$
$$= \frac{\exp\left(\sum_{k=0}^{p}\theta_k x_{k,i,t}y_{i,t} + \sum_{j\in\mathcal{N}_i}\theta_{p+1}y_{i,t}y_{j,t} + \sum_{s=1}^{S}\theta_{p+1+s}y_{i,t}y_{i,t-s}\right)}{1+\exp\left(\sum_{k=0}^{p}\theta_k x_{k,i,t} + \sum_{j\in\mathcal{N}_i}\theta_{p+1}y_{j,t} + \sum_{s=1}^{S}\theta_{p+1+s}y_{i,t-s}\right)}, \tag{17.17}$$

where $i = 1,\ldots,n$, $t = S+1,\ldots,T$, and S is the maximum temporal lag. The term in (17.17) is a full conditional distribution for a given time point t, even though it is not a full conditional distribution for all i and t. The spatial neighborhood $\mathcal{N}_i$ may be further partitioned into different orders of neighborhood. In particular, let $\mathcal{N}_i = \sum_{l=1}^{L}\mathcal{N}_i^{(l)}$, where $\mathcal{N}_i^{(l)}$ denotes the lth-order neighborhood that comprises the lth nearest neighbors for $l = 1,\ldots,L$. Similar to the model specified via (17.16), the transition probability $p(\boldsymbol{y}_t|\boldsymbol{y}_{t'}: t' = t-1,\ldots,t-S)$ and the subsequent joint distribution function can be obtained. For ease of presentation, we focus on (17.16).

17.2.2 Statistical Inference

The intractable normalizing constant in the joint distribution function poses challenges in the statistical inference for the autologistic model with or without regression, an area of active research in the last couple of decades. While Besag (1975) originally proposed maximum pseudo-likelihood estimates (MPLEs), Huffer and Wu (1998) used Markov chain Monte Carlo (MCMC) methods to approximate the unknown normalizing constant and developed Monte Carlo maximum likelihood estimates (MCMLE) for spatial autologistic models. Further, Huang and Ogata (2002) generalized the pseudo-likelihood function and showed better performance of the resulting estimates than MPLE in terms of standard errors and efficiency relative to maximum likelihood estimates (MLEs). Berthelsen and Møller (2003) developed path sampling to approximate the ratio of unknown normalizing constants in spatial point processes, which Zheng and Zhu (2008) used for computing the MCMLE. Friel et al. (2009) proposed a fast computation method for the estimation of the normalizing constant based on a reduced dependence approximation of the likelihood function. Later, we describe statistical inference based on MPLE, MCMLE, and Bayesian hierarchical modeling.

17.2.2.1 Maximum Pseudo-Likelihood Estimation

Maximum pseudo-likelihood, first introduced by Besag (1975) for autologistic models, is a popular approach to the statistical inference for autologistic regression models. The MPLE is the value of θ that maximizes the product of the full conditional distributions,

$$\tilde{\theta} = \arg\max_{\theta} L_{\mathrm{PL}}(Y;\theta),$$

where the pseudo-likelihood function for a spatio-temporal autologistic model is

$$L_{\mathrm{PL}}(Y;\theta) = \prod_{i,t} p(y_{i,t}|y_{i',t'} : (i',t') \neq (i,t))$$

$$= \prod_{i,t} \left[\frac{\exp\{\sum_{k=0}^{p} \theta_k x_{k,i,t} y_{i,t} + \sum_{j\in\mathcal{N}_i} \theta_{p+1} y_{i,t} y_{j,t} + \theta_{p+2} y_{i,t}(y_{i,t-1} + y_{i,t+1})\}}{1 + \exp\{\sum_{k=0}^{p} \theta_k x_{k,i,t} + \sum_{j\in\mathcal{N}_i} \theta_{p+1} y_{j,t} + \theta_{p+2}(y_{i,t-1} + y_{i,t+1})\}} \right]. \tag{17.18}$$

Although the pseudo-likelihood function (17.18) is not the true likelihood except in the trivial case of spatio-temporal independence, it can be shown that MPLEs are consistent and asymptotically normal under suitable regularity conditions (Guyon, 1995).

To maximize the pseudo-likelihood function and obtain the MPLE of θ, it is straightforward to apply the standard logistic regression that assumes independence, which can be implemented by, for example, `proc logistic` in `SAS` or the function `glm` in `R`. The corresponding standard errors and approximate confidence intervals can be obtained by a parametric bootstrap. Specifically, in the parametric bootstrap, M resamples of spatio-temporal binary responses are drawn according to the spatio-temporal autologistic regression model using Gibbs sampling or perfect sampling. For each resample, an MPLE is computed and the M resampled MPLEs are used to obtain an estimate of the variance of the MPLE based on the original data. In particular, perfect sampling uses coupling and upon coalescence of the coupled Markov chains, the resulting Monte Carlo samples are guaranteed to be from the target distribution (e.g., Propp and Wilson, 1996; Møller, 1999).

17.2.2.2 Monte Carlo Maximum Likelihood Estimation

The maximum pseudo-likelihood approach is computationally efficient, but is statistically less efficient than maximum likelihood (Gumpertz et al., 1997; Wu and Huffer, 1997; Zheng and Zhu, 2008). An alternative approach is Monte Carlo maximum likelihood (MCML), where the normalizing constant is approximated using MCMC and thus direct maximization of likelihood function can be obtained.

The likelihood function can be rewritten as

$$L(\boldsymbol{Y};\theta) = p(y_2,\ldots,y_{T-1}|y_1,y_T;\theta) = c(\theta)^{-1}\exp(\theta'\boldsymbol{z}),$$

where

$$\boldsymbol{z} = \left(\sum_{i,t} y_{i,t}, \sum_{i,t} x_{1,i,t}y_{i,t}, \ldots, \sum_{i,t} x_{p,i,t}y_{i,t}, \frac{1}{2}\sum_{i,t}\sum_{i'\in\mathcal{N}_i} y_{i,t}y_{i',t}, \sum_{i,t} \theta_{p+1}y_{i,t}y_{i,t-1}\right)'.$$

Based on a preselected parameter vector $\psi = (\psi_0,\ldots,\psi_{p+2})'$, approximate the ratio of two normalizing constants via importance sampling by

$$\frac{c(\theta)}{c(\psi)} = E_{\psi}\left\{\frac{\exp(\theta'\boldsymbol{z})}{\exp(\psi'\boldsymbol{z})}\right\} \approx M^{-1}\sum_{m=1}^{M}\frac{\exp(\theta'\boldsymbol{z}^m)}{\exp(\psi'\boldsymbol{z}^m)} = M^{-1}\sum_{m=1}^{M}\exp\{(\theta-\psi)'\boldsymbol{z}^m\},$$

where $\boldsymbol{z}^m$ is $\boldsymbol{z}$ evaluated at the mth Monte Carlo sample of $\boldsymbol{Y}$ for $m = 1,\ldots,M$. Monte Carlo samples of $\boldsymbol{Y}$ are generated from the joint distribution evaluated at ψ. Then the MLE can be approximated by maximizing a rescaled version of the likelihood function

$$c(\psi)L(\boldsymbol{Y};\theta) = \frac{c(\psi)}{c(\theta)}\exp(\theta'\boldsymbol{z}) = \left[M^{-1}\sum_{m=1}^{M}\exp\left\{(\theta-\psi)'\boldsymbol{z}^m\right\}\right]^{-1}\exp(\theta'\boldsymbol{z}).$$

The variances of the estimates can be estimated by using the diagonal elements of the inverse of the observed Fisher information matrix (Huffer and Wu, 1998; Geyer, 1994).

The MCMLE provides a good approximation of the MLE of the model parameters when the reference parameter ψ is close to the truth (Geyer and Thompson, 1992). The MPLE is a natural choice for the reference parameter. However, when the spatial or temporal dependence is strong, MPLE can be far away from the MLE, whereas MCMLE with MPLE as the reference parameter may not exist and the iteration may lead to a sequence of estimates that drift off to infinity. In this case, we select ψ to be an approximation obtained by a stochastic approximation algorithm. This is a two-stage MCMC stochastic approximation algorithm proposed by Gu and Zhu (2001) for computing the MLEs of model parameters for a class of spatial models. In the first stage, the estimates are moved into a feasible region quickly by using large gain constants in the stochastic approximation and in the second stage, an optimal procedure is implemented with a stopping criterion chosen so that a desired precision can be obtained. By the first stage of the algorithm, ψ can be obtained.

17.2.3 Bayesian Inference

Bayesian hierarchical modeling can be applied for the inference about spatio-temporal autologistic regression models. Møller et al. (2006) presented an auxiliary variable MCMC algorithm that allows the construction of a proposal distribution so that the normalizing constants cancel out in the Metropolis–Hastings (MH) ratio. Zheng and Zhu (2008) proposed a Bayesian approach for both model parameter inference and prediction at future time points using MCMC. They proposed an MH algorithm to generate Monte Carlo samples from the posterior distribution of the parameter θ, where the likelihood ratio in the acceptance probability is approximated by

$$\frac{p(y_2,\ldots,y_{T-1}|y_1,y_T;\theta^*)}{p(y_2,\ldots,y_{T-1}|y_1,y_T;\theta)} = \frac{\exp\{\theta^{*'}z\}}{\exp\{\theta'z\}} \times \frac{\frac{c(\theta)}{c(\psi)}}{\frac{c(\theta^*)}{c(\psi)}}$$

$$\approx \exp\{(\theta^* - \theta)'z\} \times \frac{\sum_{m=1}^{M} \exp\{(\theta - \psi)'z^m\}}{\sum_{m=1}^{M} \exp\{(\theta^* - \psi)'z^m\}}.$$

Here, M Monte Carlo samples of Y need to be generated from the joint distribution $p(y_2,\ldots, y_{T-1}|y_1, y_T, \psi)$ evaluated at ψ, but only once at the beginning of the MH algorithm, which makes the algorithm efficient. For the MH algorithm, a good choice of the parameter vector ψ helps to speed up the convergence process. The closer ψ is to the posterior mode of θ, the better the results are. Further, the variance of the proposal distribution needs to be adjusted to ensure a reasonable acceptance probability in the MH algorithm (Gelman et al., 2003).

Path sampling is an alternative way to calculate the ratio of two normalizing constants and is based on the following identity:

$$\ln\left\{\frac{c(\theta)}{c(\mathbf{0})}\right\} = \int_0^1 E_{\theta(s)}\left\{\frac{d}{ds}\theta(s)'z\right\} ds$$

where the expectation is with respect to the joint distribution evaluated at the parameter $\theta(s)$ along a path of $\theta(s) = s\theta$ for $s \in [0,1]$ from $\mathbf{0}$ to θ. However, the computation can be costly because multiple Monte Carlo samples of Y are required for computing the expectation.

For the spatio-temporal autologistic regression model, Zheng and Zhu (2008) compared the performance of MPL, MCML, and Bayesian inference. They demonstrated that parameter inference via MPL can be statistically inefficient when spatial and/or temporal dependence is strong, whereas the statistical properties of the MCML are comparable to the Bayesian approach and the computation of MCML estimates is faster. Further, using Bayesian inference, the posterior distribution of the model parameters can be obtained and it becomes straightforward to construct credible bands at desired levels.

17.2.4 Prediction

Let $\tilde{Y} = (Y_{T+1},\ldots,Y_{T+T^*})'$ denote the responses at future time points $T+1,\ldots,T+T^*$ with $T^* \geq 1$. For prediction of $\tilde{Y}$ based on model parameter estimates from MPL and MCML, a Gibbs sampler can be used to obtain the Monte Carlo samples of $\tilde{Y}$ from

$$p(\tilde{y}|y_T, y_{T+T^*+1};\theta)$$
$$\propto \exp\left\{\sum_{t=T+1}^{T+T^*}\left(\sum_{i=1}^{n}\sum_{k=0}^{p}\theta_k x_{k,i,t}y_{i,t} + \frac{1}{2}\sum_{i=1}^{n}\sum_{j\in\mathcal{N}_i}\theta_{p+1}y_{i,t}y_{j,t}\right) + \sum_{t=T+1}^{T+T^*+1}\sum_{i=1}^{n}\theta_{p+2}y_{i,t}y_{i,t-1}\right\}.$$

For prediction of $\tilde{Y}$ in the Bayesian framework, the posterior predictive distribution of $\tilde{Y}$ is

$$p(\tilde{y}|y, y_{T+T^*+1}) = \int p(\tilde{y}|y, y_{T+T^*+1}; \theta)p(\theta|y)d\theta.$$

To draw Monte Carlo samples of $\tilde{Y}$ from $p(\tilde{y}|y, y_{T+T^*+1})$, first draw θ from its posterior distribution $p(\theta|y)$ and then for each given θ, draw $\tilde{Y}$ from $p(\tilde{y}|y, y_{T+T^*+1}; \theta)$ using a Gibbs sampler (Zheng and Zhu, 2008).

17.3 Centered Autologistic Regression Model

In the aforementioned autologistic regression models, the interpretation of model parameters is not straightforward (Caragea and Kaiser, 2009; Kaiser and Caregea, 2009). In the presence of positive spatial and temporal dependence, under the uncentered parameterization, the conditional expectation of $Y_{i,t}$ given its neighbors is

$$E(Y_{i,t}|Y_{i',t'} = y_{i',t'} : (i', t') \in \mathcal{N}_{i,t}) \\ = \frac{\exp\left\{\sum_{k=0}^{p} \theta_k x_{k,i,t} + \sum_{j\in\mathcal{N}_i} \theta_{p+1} y_{j,t} + \theta_{p+2}(y_{i,t-1} + y_{i,t+1})\right\}}{1 + \exp\left\{\sum_{k=0}^{p} \theta_k x_{k,i,t} + \sum_{j\in\mathcal{N}_i} \theta_{p+1} y_{j,t} + \theta_{p+2}(y_{i,t-1} + y_{i,t+1})\right\}}. \quad (17.19)$$

The expectation (17.20) is larger than the expectation of $Y_{i,t}$ under independence,

$$\frac{\exp\left\{\sum_{k=0}^{p} \theta_k x_{k,i,t}\right\}}{1 + \exp\left\{\sum_{k=0}^{p} \theta_k x_{k,i,t}\right\}},$$

as long as $Y_{i,t}$ has nonzero spatial and/or temporal neighbors, but is never smaller. This may not be reasonable when most neighbors are zeros and thus can bias the realizations toward 1. Hence, the interpretation of dependence parameters is difficult. Further, the marginal expectation of $Y_{i,t}$ (i.e., $E(Y_{i,t}|x_{k,i,t}, k = 1, \ldots, p)$) is greater than the expectation of $Y_{i,t}$ under independence. A simulation study in Wang (2013) showed that $E(Y_{i,t}|x_{k,i,t}, k = 1, \ldots, p)$ varies across different levels of spatial and temporal dependence for fixed regression coefficients. These make the interpretation of regression coefficients unclear since these coefficients are to reflect the effects of covariates and should have a consistent interpretation across varying dependence levels.

For non-Gaussian Markov random field models of spatial lattice data, the idea of centered parameterization was first proposed by Kaiser and Cressie (1997) for a Winsorized Poisson conditional model. More recently, Kaiser and Caregea (2009) explored the centered parameterization for a general exponential family of Markov random field models. In particular, Caragea and Kaiser (2009) studied the centered parameterization for spatial autologistic regression models and showed that the centered parameterization overcomes the interpretation difficulties. Wang and Zheng (2013) extended this work to the case of spatio-temporal autologistic regression models.

17.3.1 Model with Centered Parameterization

For site $i = 1, \ldots, n$ and time t, let $\pi_{i,t}$ denote the probability of $Y_{i,t} = 1$ under spatio-temporal independence. That is,

$$\pi_{i,t} = \frac{\exp\left(\sum_{k=0}^{p} \theta_k x_{k,i,t}\right)}{1 + \exp\left(\sum_{k=0}^{p} \theta_k x_{k,i,t}\right)}.$$

Let $y_{i,t}^* = y_{i,t} - \pi_{i,t}$ denote a centered response at site i and time t, centering around $\pi_{i,t}$. For pairwise-only dependence, Wang and Zheng (2013) defined a centered spatio-temporal autologistic regression model via the following full conditional distributions:

$$\begin{aligned} &p(y_{i,t}|y_{i',t'} : (i',t') \neq (i,t)) = p(y_{i,t}|y_{i',t'} : (i',t') \in \mathcal{N}_{i,t}) \\ &= \frac{\exp\left\{\sum_{k=0}^{p} \theta_k x_{k,i,t} y_{i,t} + \sum_{j \in \mathcal{N}_i} \theta_{p+1} y_{i,t} y_{j,t}^* + \theta_{p+2} y_{i,t}(y_{i,t-1}^* + y_{i,t+1}^*)\right\}}{1 + \exp\left\{\sum_{k=0}^{p} \theta_k x_{k,i,t} + \sum_{j \in \mathcal{N}_i} \theta_{p+1} y_{j,t}^* + \theta_{p+2}(y_{i,t-1}^* + y_{i,t+1}^*)\right\}}. \end{aligned} \tag{17.20}$$

By the Hammersley–Clifford theorem and its corollary, the joint likelihood function of $\boldsymbol{Y}_2, \ldots, \boldsymbol{Y}_{T-1}$ conditioned on $\boldsymbol{Y}_1$ and $\boldsymbol{Y}_T$ is

$$\begin{aligned} &L(\boldsymbol{Y}; \theta) = p(\boldsymbol{y}_2, \ldots, \boldsymbol{y}_{T-1}|\boldsymbol{y}_1, \boldsymbol{y}_T; \theta) \\ &= c^*(\theta)^{-1} \exp\left\{\sum_{t=2}^{T-1}\left(\sum_{i=1}^{n}\sum_{k=0}^{p} \theta_k x_{k,i,t} y_{i,t}^* + \frac{1}{2}\sum_{i=1}^{n}\sum_{j \in \mathcal{N}_i} \theta_{p+1} y_{i,t}^* y_{j,t}^*\right) + \sum_{t=2}^{T}\sum_{i=1}^{n} \theta_{p+2} y_{i,t}^* y_{i,t-1}^*\right\}, \end{aligned} \tag{17.21}$$

where $c^*(\theta)$ is the normalizing constant. When the temporal autocorrelation coefficient is zero (i.e., $\theta_{p+2} = 0$), the model reduces to a spatio-only autologistic regression model (Caragea and Kaiser, 2009; Hughes et al., 2011).

Thus, the conditional expectation of $Y_{i,t}$ given its neighbors is

$$\begin{aligned} &E(Y_{i,t}|Y_{i',t'} = y_{i',t'} : (i',t') \in \mathcal{N}_{i,t}) \\ &= \frac{\exp\left\{\sum_{k=0}^{p} \theta_k x_{k,i,t} + \sum_{j \in \mathcal{N}_i} \theta_{p+1} y_{j,t}^* + \theta_{p+2}(y_{i,t-1}^* + y_{i,t+1}^*)\right\}}{1 + \exp\left\{\sum_{k=0}^{p} \theta_k x_{k,i,t} + \sum_{j \in \mathcal{N}_i} \theta_{p+1} y_{j,t}^* + \theta_{p+2}(y_{i,t-1}^* + y_{i,t+1}^*)\right\}}, \end{aligned}$$

which we denote as $\pi_{i,t}^*$. Suppose that the spatial autoregressive coefficient θ_{p+1} and the temporal autoregressive coefficient θ_{p+2} are positive. Then, $\pi_{i,t}^* > \pi_{i,t}$ when

$$\theta_{p+1} \sum_{j \in \mathcal{N}_i} y_{j,t} + \theta_{p+2}(y_{i,t-1} + y_{i,t+1}) > \theta_{p+1} \sum_{j \in \mathcal{N}_i} \pi_{j,t} + \theta_{p+2}(\pi_{i,t-1} + \pi_{i,t+1}),$$

where $\sum_{j \in \mathcal{N}_i} \pi_{j,t}$ and $\pi_{i,t-1} + \pi_{i,t+1}$ are the expected numbers of nonzero spatial and temporal neighbors under the independence model, respectively. Specifically, if $\theta_{p+2} = 0$, then

$\pi^*_{i,t} > \pi_{i,t}$ only when the observed number of nonzero spatial neighbors is greater than the expected number of nonzero spatial neighbors under independence. That is, $\sum_{j\in\mathcal{N}_i} y_{j,t} > \sum_{j\in\mathcal{N}_i} \pi_{j,t}$. If $\theta_{p+1} = 0$, then $\pi^*_{i,t} > \pi_{i,t}$ only when the observed number of nonzero temporal neighbors is greater than the expected number of nonzero temporal neighbors under independence. That is, $y_{i,t-1} + y_{i,t+1} > \pi_{i,t-1} + \pi_{i,t+1}$. Thus, the interpretation of θ_{p+1} and θ_{p+2} as local dependence parameters is more sensible. Further, the simulation study in Wang (2013) showed that the marginal expectation of $Y_{i,t}$ under the centered parameterization remains constant over moderate levels of spatial and temporal dependence (i.e., $E(Y_{i,t}|x_{k,i,t}, k=1,\ldots,p) \approx \pi_{i,t}$). The interpretation of regression coefficients as effects of covariates is more sensible as well.

17.3.2 Statistical Inference

For the model with centered parameterization, its statistical inference has been developed based on expectation–maximization pseudo-likelihood, Monte Carlo expectation–maximization likelihood, and Bayesian inference (Wang and Zheng, 2013).

17.3.2.1 Expectation–Maximization Pseudo-Likelihood Estimator

To obtain the maximum pseudo-likelihood estimates of the model parameters, the combination of an expectation–maximization (EM) algorithm and a Newton–Raphson algorithm, called the expectation–maximization pseudo-likelihood estimator (EMPLE), is considered. Specifically, update $\pi_{i,t}$, the expectation of $Y_{i,t}$ under the independent model, at the E step and then at the M step, update $\hat{\theta}^l$ by maximizing

$$\prod_{i,t} \frac{\exp\left\{\sum_{k=0}^{p} \theta_k x_{k,i,t} y_{i,t} + \sum_{j\in\mathcal{N}_i} \theta_{p+1} y_{i,t} y_{j,t}^{*(l-1)} + \theta_{p+2} y_{i,t}(y_{i,t-1}^{*(l-1)} + y_{i,t+1}^{*(l-1)})\right\}}{1+\exp\left\{\sum_{k=0}^{p} \theta_k x_{k,i,t} + \sum_{j\in\mathcal{N}_i} \theta_{p+1} y_{j,t}^{*(l-1)} + \theta_{p+2}(y_{i,t-1}^{*(l-1)} + y_{i,t+1}^{*(l-1)})\right\}},$$

where $y_{i,t}^{*(l)}$ is the centered response at the lth iteration. The M step can be carried out by a Newton–Raphson algorithm using the standard logistic regression and the E and M steps are repeated until convergence. A parametric bootstrap can be used to compute the standard error of the EMPLE. For the starting value θ_0 at the start of the algorithm, different starting points can impact how long it takes to convergence. The maximum MPLE from the uncentered autologistic regression model is a natural choice.

17.3.2.2 Monte Carlo Expectation–Maximization Likelihood Estimator

Let $z^*_\theta = (\sum_{i,t} x_{0,i,t} y^*_{i,t}, \ldots, \sum_{i,t} x_{p,i,t} y^*_{i,t}, \frac{1}{2}\sum_{i,t}\sum_{j\in\mathcal{N}_i} y^*_{i,t} y^*_{j,t}, \sum_{i,t} y^*_{i,t} y^*_{i,t-1})'$. We consider a rescaled version of the likelihood function

$$c^*(\psi)L(Y;\theta) = \frac{c^*(\psi)}{c^*(\theta)}\exp(\theta' z^*_\theta) = \left[E_\psi\left\{\frac{\exp(\theta' z^*_\theta)}{\exp(\psi' z^*_\psi)}\right\}\right]^{-1}\exp(\theta' z^*_\theta),$$

where ψ is a reference parameter and z^*_ψ is z^* with centers evaluated at ψ. Monte Carlo expectation–maximization likelihood (MCEML) estimator can be used by combining an EM algorithm and a Newton–Raphson algorithm. Specifically, first choose a reference parameter vector ψ and generate M Monte Carlo samples of $\mathbf{Y}$ from the likelihood function evaluated at ψ. Then for the lth iteration, at the E step, we update $\pi^{(l-1)}_{i,t}$ and set $y^{*(l-1)}_{i,t} = y_{i,t} - \pi^{(l-1)}_{i,t}$. At the M step, we maximize the rescaled version of the likelihood function

$$\exp(\theta' z^*_{\hat{\theta}^{l-1}}) \left[M^{-1} \sum_{m=1}^{M} \exp\left(\theta' z^{*(m)}_{\hat{\theta}^{l-1}} - \psi' z^{*(m)}_{\psi} \right) \right]^{-1},$$

where $z^*_{\hat{\theta}^{l-1}}$ is z^* with centered responses $y^{*(l-1)}_{i,t}$ and $z^{*(m)}_{\hat{\theta}^{l-1}}$ and $z^{*(m)}_{\psi}$ are z^* evaluated at the mth Monte Carlo sample of $\mathbf{Y}$ generated at the beginning of the algorithm with centers computed at $\hat{\theta}^{(l-1)}$ and ψ, respectively. The M step can be carried out using a Newton–Raphson algorithm. We compute the observed Fisher information matrix and obtain the standard errors of the MCEMLE as a by-product of the MCEML estimation.

17.3.2.3 Bayesian Inference

We consider an MH algorithm to generate Monte Carlo samples of θ from the posterior distribution $p(\theta|y)$ (Zheng and Zhu, 2008), where the likelihood ratio in $\alpha(\theta^*|\theta)$ in the acceptance probability is approximated as

$$\begin{aligned} \frac{p(y_2, \ldots, y_{T-1} | y_1, y_T, \theta^*)}{p(y_2, \ldots, y_{T-1} | y_1, y_T, \theta)} &= \frac{\exp(\theta^{*'} z^*_{\theta^*})}{\exp(\theta' z^*_{\theta})} \times \frac{\frac{c^*(\theta)}{c^*(\psi)}}{\frac{c^*(\theta^*)}{c^*(\psi)}} \\ &\approx \frac{\exp(\theta^{*'} z^*_{\theta^*})}{\exp(\theta' z^*_{\theta})} \times \frac{\sum_{m=1}^{M} \exp(\theta' z^{*(m)}_{\theta} - \psi' z^{*(m)}_{\psi})}{\sum_{m=1}^{M} \exp(\theta^{*'} z^{*(m)}_{\theta^*} - \psi' z^{*(m)}_{\psi})}, \end{aligned}$$

where $z^{*(m)}_{\theta}$, $z^{*(m)}_{\theta^*}$, and $z^{*(m)}_{\psi}$ are z^* evaluated at the mth Monte Carlo sample of $\mathbf{Y}$ with centers computed based on θ, θ^*, and ψ, respectively.

17.4 Data Example

For illustration, we consider the outbreak of southern pine beetle (SPB) in North Carolina. The data consist of indicators of outbreak or not (0 = no outbreak; 1 = outbreak) in the 100 counties of North Carolina from 1960 to 1996. Figure 17.1 gives a time series of the county-level outbreak maps. The average precipitation in the fall (in cm) will be the covariate and is mapped in Figure 17.2. We use the data from 1960 to 1991 for model fitting and set aside the data from 1992 to 1996 for model validation. Two counties are considered to be neighbors if the corresponding county seats are within 30 miles of each other.

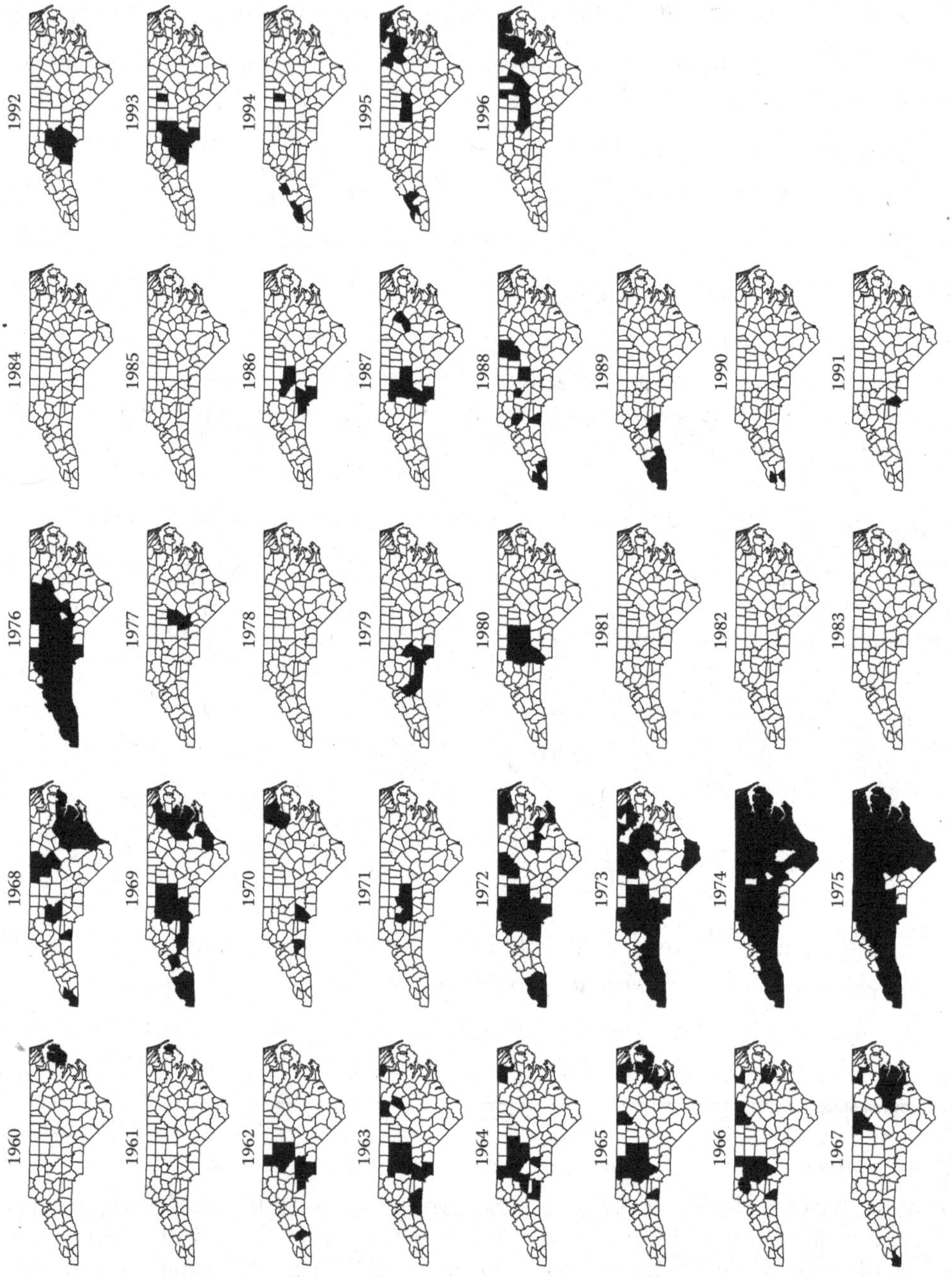

FIGURE 17.1
Maps of southern pine beetle outbreak from 1960 to 1996 in the counties of North Carolina. A county is filled black if there was an outbreak and is unfilled otherwise.

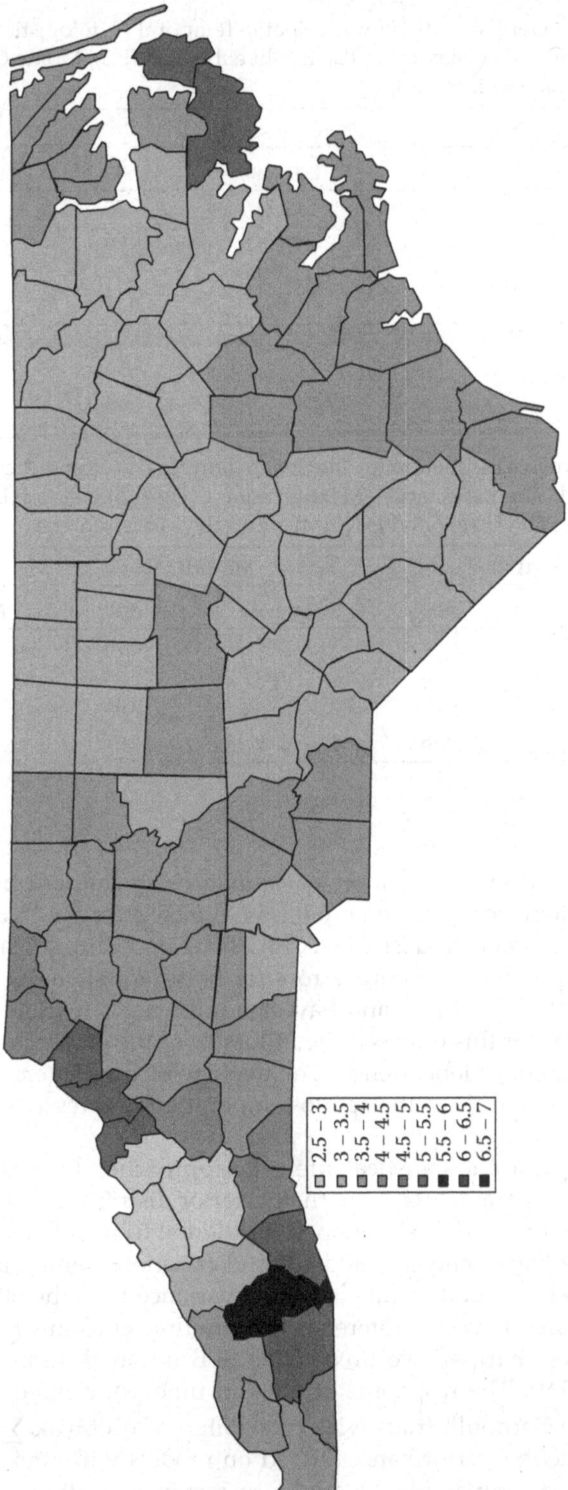

FIGURE 17.2
Map of mean fall precipitation in the counties of North Carolina.

TABLE 17.1

Comparison of Model Parameter Estimation for the Spatio-Temporal Autologistic Model with Uncentered Parameterization Using Maximum Pseudo-likelihood (MPL), Monte Carlo Maximum Likelihood (MCML), and Bayesian Inference

	MPL		MCML		Bayesian	
Parameters	**Estimate**	**SE**	**Estimate**	**SE**	**Mean**	**SD**
Intercept θ_0	−5.16	0.66	−2.71	0.20	−2.71	0.20
Slope θ_1	0.25	0.18	−0.14	0.05	−0.14	0.05
Spatial θ_2	1.45	0.14	0.91	0.05	0.91	0.06
Temporal θ_3	1.71	0.24	1.02	0.11	1.03	0.13

TABLE 17.2

Comparison of Model Parameter Estimation for the Spatio-Temporal Autologistic Model with Centered Parameterization Using Expectation–Maximization Pseudo-likelihood (EMPL), Monte Carlo Expectation–Maximization Likelihood (MCEML), and Bayesian Inference

	EMPL		MCEML		Bayesian	
Parameters	**Estimate**	**SE**	**Estimate**	**SE**	**Mean**	**SD**
Intercept θ_0	−4.96	0.32	−2.40	0.16	−2.86	0.29
Slope θ_1	0.21	0.08	−0.13	0.05	−0.13	0.08
Spatial θ_2	1.47	0.13	0.95	0.05	0.95	0.06
Temporal θ_2	1.75	0.18	0.89	0.07	0.89	0.10

Tables 17.1 and 17.2 give the model parameter estimates and standard errors from fitting the spatio-temporal autologistic regression models with uncentered and centered parameterization, respectively (see Wang and Zheng, 2013; Zheng and Zhu, 2008). The parameter estimates and the corresponding standard errors for both models using all of the three inference approaches, MPLE, MCMLE, and Bayesian inference, are quite close. One possible reason for this is that for this data set, the influence of the center is small relative to the strength of spatio-temporal dependence. The average of the centers $\pi_{i,t}$ evaluated at the MCEMLE is only 0.05, and thus, the spatio-temporal autoregressive terms dominate the outbreak probabilities.

For comparison among various statistical inference approaches, the results suggest that the inference for the model parameters using the posterior distribution matches well with MCML, but the inference from pseudo-likelihood is different from both Bayesian inference and MCML for both uncentered and centered parameterization models. In addition, estimation based on pseudo-likelihood results in higher variance than the other approaches. In terms of computing time, Bayesian inference is more time consuming compared with the other two approaches. Further, we predict the SPB outbreak from 1992 to 2001 in North Carolina (Table 17.3). The responses at the end time point (here, $y_{i,2002}$) are generated from independent Bernoulli trials with probability of outbreak $\sum_{t=1960}^{1991} y_{i,t}/31$ for $i = 1, \ldots, 100$. The prediction performances based on models with uncentered and centered parameterization are comparable. Overall, our recommendation is to use a model with uncentered parameterization if prediction is of primary interest, since the two

TABLE 17.3
Comparison of the Prediction Performance between Models with Centered and Uncentered Parameterization

	Centered			Uncentered		
Year	**EMPL**	**MCEML**	**Bayesian**	**MPL**	**MCML**	**Bayesian**
1992	0.65	0.18	0.14	0.66	0.09	0.09
1993	0.72	0.19	0.12	0.65	0.13	0.13
1994	0.70	0.20	0.14	0.74	0.06	0.08
1995	0.63	0.23	0.13	0.68	0.13	0.14
1996	0.62	0.24	0.09	0.61	0.17	0.16

Note: For centered parameterization, the prediction is based on statistical inference obtained using expectation–maximization pseudo-likelihood (EMPL), Monte Carlo expectation–maximization likelihood (MCEML); and Bayesian inference. For uncentered parameterization, the prediction is based on statistical inference obtained using maximum pseudo-likelihood (MPL), Monte Carlo maximum likelihood (MCML), and Bayesian inference. Reported are the prediction error rates for each year in 1992–1996.

parameterizations provide comparable performance in prediction but the centered parameterization is computationally more intensive. If the focus is on the interpretation of the regression coefficients, however, the centered parameterization is recommended.

17.5 Discussion

In this chapter, we have reviewed spatio-temporal autologistic regression models for spatio-temporal binary data. Alternatively, a generalized linear mixed model (GLMM) framework can be adopted for modeling such spatial data (Diggle and Ribeiro, 2007; Holan and Wikle [2015; Chapter 15 in this volume]). The response variable is modeled by a distribution in the exponential family and is related to covariates and spatial random effects in a link function. Thus, GLMM is flexible, as it is suitable for both Gaussian responses and non-Gaussian responses such as binomial and Poisson random variables. Statistical inference can be carried out using Bayesian hierarchical modeling, which is flexible as more complex structures can be readily placed on the model parameters. With suitable reduction of dimensionality for the spatio-temporal random effects, computation is generally feasible. In particular, faster computational algorithms are emerging such as integrated nested Laplace approximations (INLA) (Rue et al., 2009). Although likelihood-based approaches are suitable, it is sometimes a challenge to attain a full specification of the likelihood function, due to a lack of sufficient information and complex interactions among responses. In this case, an estimating equation approach may be attractive. For spatial binary data, Lin et al. (2008) developed a central limit theorem for a random field under various L_p metrics and derived the consistency and asymptotic normality of quasi-likelihood estimators. Lin (2010) further developed a generalized estimating equation (GEE) method for spatio-temporal binary data, but only a single binary covariate was considered and the spatio-temporal dependence is limited to be separable. Moreover, variable selection methods for identifying the suitable set of covariates are of interest. For example, Fu et al. (2013) developed adaptive Lasso for the selection of covariates in an autologistic regression model and extension to spatio-temporal autologistic regression model is discussed. Further research on this and other related topics will be worthwhile.

References

Bartlett, M. S. (1971). Physical nearest-neighbour models and non-linear time-series. *Journal of Applied Probability*, 8(2):222–232.

Bartlett, M. S. (1972). Physical nearest-neighbour models and non-linear time-series. II Further discussion of approximate solutions and exact equations. *Journal of Applied Probability*, 9:76–86.

Berthelsen, K. K. and Møller, J. (2003). Likelihood and nonparametric Bayesian MCMC inference for spatial point processes based on perfect simulation and path sampling. *Scandinavian Journal of Statistics*, 30:549–564.

Besag, J. (1972). Nearest-neighbour systems and the auto-logistic model for binary data. *Journal of the Royal Statistical Society Series B*, 34:75–83.

Besag, J. (1974). Spatial interaction and the statistical analysis of lattice systems. *Journal of the Royal Statistical Society Series B*, 36:192–225.

Besag, J. (1975). Statistical analysis of non-lattice data. *The Statistician*, 24:179–195.

Caragea, P. C. and Kaiser, M. S. (2009). Autologistic models with interpretable parameters. *Journal of Agricultural, Biological, and Environmental Statistics*, 14:281–300.

Cressie, N. (1993). *Statistics for Spatial Data, Revised Edition*. Wiley, New York.

Diggle, P. J. and Ribeiro, P. J. (2007). *Model-Based Geostatistics*. Springer, New York.

Friel, N., Pettitt, A. N., Reeves, R., and Wit, E. (2009). Bayesian inference in hidden Markov random fields for binary data defined on large lattices. *Journal of Computational and Graphical Statistics*, 18(2):243–261.

Fu, R., Thurman, A., Steen-Adams, M., and Zhu, J. (2013). On estimation and selection of autologistic regression models via penalized pseudolikelihood. *Journal of Agricultural, Biological, and Environmental Statistics*, 18:429–449.

Gelman, A., Carlin, J. B., Stern, H., and Rubin, D. (2003). *Bayesian Data Analysis*. Chapman & Hall, Boca Raton, FL.

Geyer, C. J. (1994). On the convergence of Monte Carlo maximum likelihood calculations. *Journal of the Royal Society of Statistics Series B*, 56:261–274.

Geyer, C. J. and Thompson, E. A. (1992). Constrained Monte Carlo maximum likelihood for dependent data (with discussion). *Journal of the Royal Statistical Society Series B*, 54:657–699.

Gu, M. G. and Zhu, H. T. (2001). Maximum likelihood estimation for spatial models by Markov chain Monte Carlo stochastic approximation. *Journal of the Royal Statistical Society Series B*, 63:339–355.

Gumpertz, M. L., Graham, J. M., and Ristaino, J. B. (1997). Autologistic models of spatial pattern of phytophthora epidemic in bell pepper: Effects of soil variables on disease presence. *Journal of Agricultural, Biological, and Environmental Statistics*, 2:131–156.

Guyon, X. (1995). *Random Fields on a Network: Modeling, Statistics, and Applications*. Springer, New York.

Holan, S. H. and Wikle, C. K. (2015). Hierarchical dynamic generalized linear mixed models for discrete-valued spatio-temporal data. In Davis, R., Holan, S., Lund, R., and Ravishanker, N., eds., *Handbook of Discrete-Valued Time Series*, pp. 327–348. Chapman & Hall, Boca Raton, FL.

Huang, F. and Ogata, Y. (2002). Generalized pseudo-likelihood estimates for Markov random fields on lattice. *Annals of Institutes of Statistical Mathematics*, 54:1–18.

Huffer, F. W. and Wu, H. (1998). Markov chain Monte Carlo for autologistic regression models with application to the distribution of plant speicies. *Biometrics*, 54:509–524.

Hughes, J. P., Haran, M., and Caragea, P. C. (2011). Autologistic models for binary data on a lattice. *Environmetrics*, 22:857–871.

Ising, E. (1924). Beitrag zur theorie des ferro- und paramagnetismus. PhD thesis, Mathematish-Naturewissenschaftliche Fakultät der Universität Hamburg.

Ising, E. (1925). Beitrag zur theorie des ferromagnetismus. *Zeitschrift für Physik A Hadrons and Nuclei*, 31:253–258.

Kaiser, M. S. and Caregea, P. C. (2009). Exploring dependence with data on spatial lattice. *Biometrics*, 65:857–865.

Kaiser, M. S., Caragea, P. C., and Furukawa, K. (2012). Centered parameterizations and dependence limitations in Markov random field models. *Journal of Statistical Planning and Inference*, 142:1855–1863.

Kaiser, M. S. and Cressie, N. (1997). Modeling Poisson variables with positive spatial dependence. *Statistics and Probability Letters*, 35:423–432.

Lin, P.-S. (2010). Estimating equations for separable spatial-temporal binary data. *Environmental and Ecological Statistics*, 17:543–557.

Lin, P.-S., Lee, H.-Y., and Clayton, M. (2008). Estimating equations for spatially correlated data in multi-dimensional space. *Biometrika*, 95:847–858.

Møller, J. (1999). Perfect simulation of conditionally specified models. *Journal of the Royal Statistical Society, Series B*, 61:251–264.

Møller, J., Pettitt, A. N., Reeves, R. W., and Berthelsen, K. K. (2006). An efficient Markov chain Monte Carlo method for distributions with intractable normalising constants. *Biometrika*, 93:451–458.

Pickard, D. K. (1976). Asymptotic inference for an Ising lattice. *Journal of Applied Probability*, 13:486–497.

Pickard, D. K. (1977). Asymptotic inference for an Ising lattice II. *Advances in Applied Probability*, 9:476–501.

Propp, J. G. and Wilson, D. B. (1996). Exact sampling with coupled Markov chains and applications to statistical mechanics. *Random Structures and Algorithms*, 9:223–252.

Rue, H., Martino, S., and Chopin, N. (2009). Exact sampling with coupled Markov chains and applications to statistical mechanics. *Journal of the Royal Statisical Society Series B*, 71:319–392.

Wang, Z. (2013). Analysis of binary data via spatial-temporal autologistic regression models. PhD thesis, Department of Statistics, University of Kentucky, Lexington, KY.

Wang, Z. and Zheng, Y. (2013). Analysis of binary data via a centered spatial-temporal autologistic regression model. *Environmental and Ecological Statistics*, 20:37–57.

Wu, H. and Huffer, F. W. (1997). Modeling the distribution of plant species using the autologistic regression model. *Environmental and Ecological Statistics*, 4:31–48.

Zheng, Y. and Zhu, J. (2008). Markov chain Monte Carlo for a spatial-temporal autologistic regression model. *Journal of Computational and Graphical Statistics*, 17:123–137.

Zhu, J., Huang, H.-C., and Wu, J.-P. (2005). Modeling spatial-temporal binary data using Markov random fields. *Journal of Agricultural, Biological, and Environmental Statistics*, 10:212–225.

Zhu, J., Zheng, Y., Carroll, A., and Aukema, B. H. (2008). Autologistic regression analysis of spatial-temporal binary data via Monte Carlo maximum likelihood. *Journal of Agricultural, Biological, and Environmental Statistics*, 13:84–98.

18

Spatio-Temporal Modeling for Small Area Health Analysis

Andrew B. Lawson and Ana Corberán-Vallet

CONTENTS

18.1 Introduction

Small area data arise in a variety of contexts. Usually, arbitrary geographic units (small areas) are the basic observation units in a study carried out within a predefined geographic study area (W). These could be administrative units such as zip codes, postal zones, census tracts, or larger units such as municipalities, counties, parishes, or even states. The study region W could be a predefined area such as a city, county, state, or country, or an arbitrarily defined group of units used for the specific study. It is common for health data to be collected within such units and that the resulting counts of disease are to be the focus of study. Health data usually consist of a particular disease incidence (new counts of disease in a fixed time period), or prevalence (counts within a longer time period). Diseases could range from noninfectious such as diabetes, asthma, or different types of cancers to infectious diseases such as HIV, influenza C, influenza A/H1N1, SARS, or corona virus. In the following, we will confine our attention to disease incidence within small areas and discrete time periods. Note that at a fine level of spatial and temporal resolution (residential location and date of diagnosis) the disease occurrence can form a spatio-temporal point process (Lawson, 2013, ch 12). We do not pursue this form here.

Assume that a chosen disease occurs within m spatial units and is also observed within T consecutive fixed time periods. The resulting observed count is $y_{it}, i = 1, \ldots, m; t = 1, \ldots, T$. The time evolution of the disease within each spatial unit can be considered an example of a discrete time series. Hence, the collection of spatial time series can be considered as an example of multivariate discrete time series.

Usually, for health data we assume that counts are described by a discrete probability model. For relatively rare diseases, a Poisson model is often assumed for y_{it} so that

$$y_{it} \sim Po(\lambda_{it}).$$

This assumption is in part justified theoretically from the aggregation of a Poisson process model for the underlying case events. The specification of the mean level (λ_{it}) requires some consideration. First, as disease occurs within a population that is "at risk" for the disease, the mean must be modulated by a population effect of some form. Usually, this modulation is considered via a multiplicative link to a modeled component such as

$$\lambda_{it} = e_{it}\theta_{it},$$

where e_{it} represents the population at risk and is usually computed as an *expected rate* or *count*. The estimator of e_{it} is usually based on a standard population rate (such as the whole study region or a larger area). Once estimated, the e_{it} is usually assumed fixed. It is important to note that some inferential sensitivity could arise in relation to the estimation method and population assumed for e_{it}. Second, the model component θ_{it}, which is known as *relative risk*, must be nonnegative. This is usually achieved by modeling θ_{it} on the log scale, that is, a linear parameterization is assumed for $\log(\theta_{it})$. The $\log(e_{it})$ is an offset.

Note that for finite populations within small areas we could assume a binomial likelihood as a variant, instead of the Poisson model. In that case, we assume that a (known) finite population n_{it} is found in the small area and out of this population a set of disease counts are observed. A classic example of this situation would be yearly births in counties of South Carolina (n_{it}) and births with abnormalities (y_{it}). In this case, we would assume a binomial model of the form

$$y_{it} \sim Bin(n_{it}, p_{it})$$

and the probability of abnormal birth in the ith area would be modeled over time as p_{it}. Often in this situation, the probability will be modeled with a suitable link to linear or nonlinear predictors and other terms. A logit, probit, or complimentary log–log link are commonly assumed.

18.2 Some Basic Space-Time Models

Space-time models can be roughly classified into two types. First, there are purely descriptive models that seek to provide a parsimonious description of the disease risk variation in space and time. Second, there are mechanistic models that seek to include some mechanism of disease occurrence within the model. These latter models are often assumed for infectious diseases where transmission from one time period to the next can be directly modeled (see Section 18.4.2). Descriptive models often use random effects to provide a parsimonious summary description of the risk variation. These are often most appropriate for noninfectious diseases. In what follows, we will discuss models for the relative risk under the Poisson model. Specification can be easily modified for a binomial likelihood.

18.2.1 Descriptive Models

A basic description of space-time variation would consist of a separate spatial and temporal effect model with a possible effect for the residual space-time interaction. Assume that

$$\log(\theta_{it}) = \alpha_0 + S_i + T_t + ST_{it} \tag{18.1}$$

where S_i, T_t, and ST_{it} represent the spatial, temporal, and space-time interaction terms, respectively. Here, $\exp(\alpha_0)$ represents the overall rate in space-time.

Some simple spatial models could consist of (1) spatial trend (e.g., $S_i = \alpha_1 s_{1i} + \alpha_2 s_{2i}$ where (s_{1i}, s_{2i}) is a coordinate pair for the geographic centroid of the ith small area), (2) uncorrelated heterogeneity (e.g., $S_i = v_i$ where v_i is an uncorrelated heterogeneity term), or (3) as for (2) but with correlated heterogeneity added (e.g., $S_i = v_i + u_i$ where u_i is a spatially correlated heterogeneity term). This latter model is sometimes called a *convolution* model. The temporal effect T_t can also take a variety of forms: (1) simple linear time trend, that is, $T_t = \beta\gamma_t$ where γ_t is the actual time of the tth period and (2) a random time effect such as an autoregressive lag 1 model (i.e., $T_t \sim N(\phi T_{t-1}, \tau_T^{-1})$) or a random walk (when $\phi = 1$). A simpler uncorrelated time effect could also be considered where $T_t \sim N(0, \tau_T^{-1})$, τ_* being the precision of the respective Gaussian distribution. Combinations of uncorrelated and correlated effects could also be considered for the time component. Finally, as a form of residual interaction, the space-time interaction term (ST_{it}) can also be included. Often, the specification of

$$\log(\theta_{it}) = \alpha_0 + v_i + u_i + \gamma_t + \psi_{it}, \tag{18.2}$$

where the interaction is assumed to be defined as $\psi_{it} \sim N(0, \tau_{\psi}^{-1})$ is found to be a robust and appropriate model for disease variation (see, e.g., Knorr-Held, 2000; Lawson, 2013, ch 12). More sophisticated models with nonseparable space-time variation are also possible (see, e.g., Cai et al., 2012, 2013). These models can sometimes be more effective in describing the space-time variation but are less immediately interpretable than separable models. In terms of inferential paradigms, it is commonly found that a Bayesian approach is adopted to the formulation of the hierarchical model structure and the ensuing estimation methods focus on posterior sampling via Markov chain Monte Carlo (MCMC). For the model specification in (18.2), the model hierarchy with suitable prior distributions could be

$$
\begin{aligned}
y_{it}|\lambda it &\sim Po(\lambda_{it} = e_{it}\theta_{it})\\
\log(\theta_{it}) &= \alpha_0 + v_i + u_i + \gamma_t + \psi_{it}\\
\alpha_0|\tau_0^{-1} &\sim N(0, \tau_0^{-1})\\
v_i|\tau_v^{-1} &\sim N(0, \tau_v^{-1})\\
u_i|\tau_u^{-1} &\sim ICAR\left(\frac{\tau_u^{-1}}{n_{\delta_i}}\right)\\
\gamma_t|\gamma_{t-1}, \tau_\gamma^{-1} &\sim N(\gamma_{t-1}, \tau_\gamma^{-1})\\
\psi_{it}|\tau_\psi^{-1} &\sim N(0, \tau_\psi^{-1}).
\end{aligned} \tag{18.3}
$$

The $ICAR(\tau_u^{-1}/n_{\delta_i})$ denotes an intrinsic conditional autoregressive spatial prior distribution and implies that the term u_i has a Markov random field specification: a conditional Gaussian distribution given its δ_i neighboring region set $(u_i|\cdots \sim N(\overline{u}_{\delta_i}, \tau_u^{-1}/n_{\delta_i}))$. The precisions (τ_*) could be assumed to have a gamma prior distribution, that is, $\tau_* \sim Ga(a, b)$, where a common choice is $a = 0.01$ and $b = 0.005$. Recently, the use of a noninformative uniform distribution has been recommended as a robust prior for standard deviation parameters (Gelman, 2006). The joint posterior distribution for the model parameters is analytically intractable but can be sampled using MCMC simulation techniques.

It has been found that the specification in (18.3) is a parsimonious and robust prescription for relative risk modeling in space and time (Ugarte et al., 2009). An example model fit using MCMC (WinBUGS; see Section 18.3) of this Bayesian hierarchical model to 10 years (1979–1988) of respiratory cancer mortality in Ohio (see Figure 18.1) led to a deviance information criterion (DIC) of 5751.4 with the effective number of parameters pD = 129. Whereas a model without the interaction term yielded a DIC = 5759.0 with pD = 80. This suggests that the space-time interaction model provides an improved fit to these data over a simple separable model.

18.2.2 Mechanistic Models

While descriptive models can perform well in describing space-time variation of noninfectious diseases, it is often more appropriate to consider transmission mechanisms when modeling infectious diseases. This is especially true when considering the prediction of the infection process. Transmission mechanisms usually require the specification of a transmission rate related to a pool of potential cases. The standard model that is usually proposed is a compartment model where a reservoir of people (susceptibles: S) can become infected cases (infected: I) and then be removed from the process (removed: R). A fundamental feature of these models is that they resolve to linked count models within discrete time periods. These Susceptible-Infected-Recovered (SIR) models can be formulated (and extended) in a variety of ways. In Section 18.4.2, we discuss the application of these to infectious diseases in space and time.

18.2.3 Kalman Filtering

Another relatively mechanistic modeling approach is to consider a linked two-component system of equations. These two components represent a *system equation* and a *measurement*

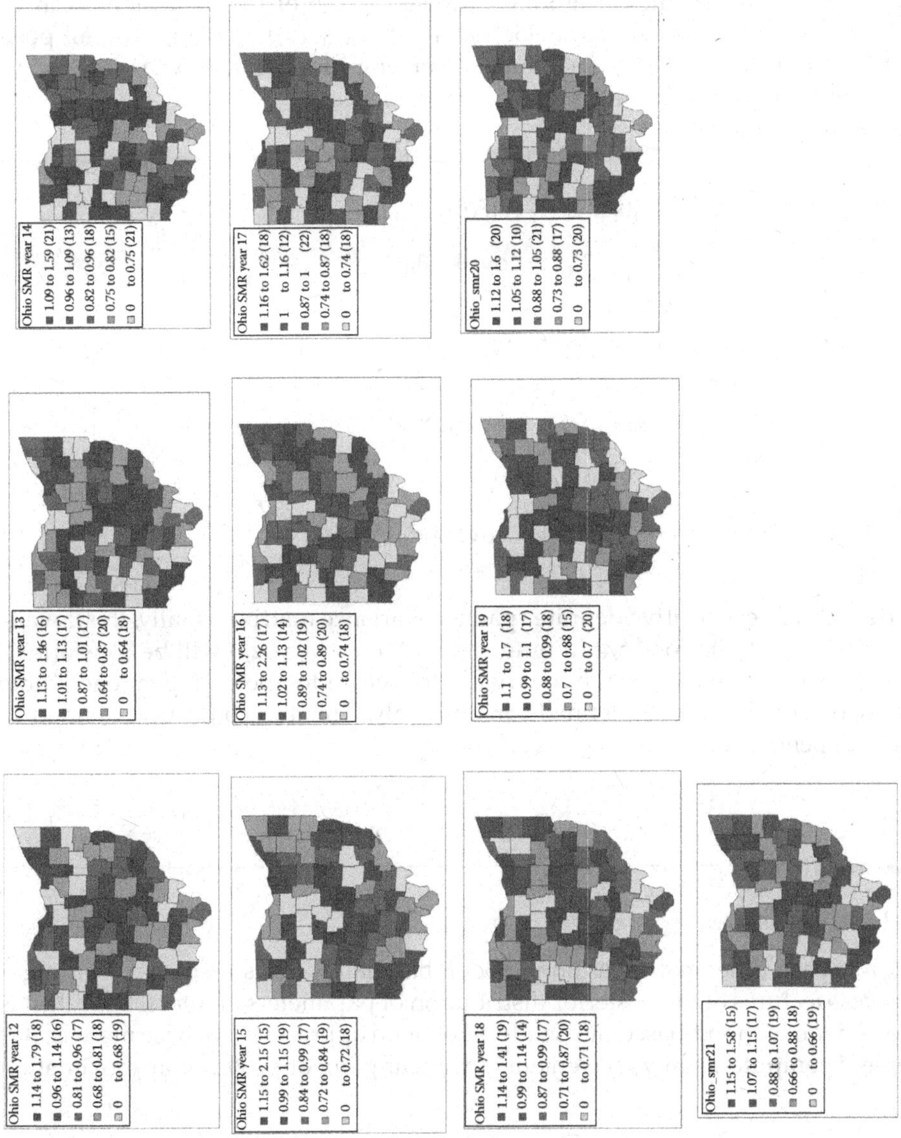

FIGURE 18.1
County-level Ohio respiratory cancer mortality: 10 years (1979–1988) displayed as standardized mortality ratios with expected rate computed from the state × 21 year average rate (1968–1988).

equation. It is natural to consider these components as a mechanistic description of the system behavior, and as an observational process. Cressie and Wikle (2011) have presented a thorough exposition of such a modeling paradigm where their *data* model is the measurement equation and their *process* model is their system equation. From the standpoint of space-time count data modeling in geographic health studies, it is convenient to first reparameterize the process model to focus on a transformed variable, see Holan and Wikle (2015; Chapter 15 in this volume). The non-Gaussian nature of the data model can be avoided by a transformation where $z_{it} = \log[(y_{it} + e_p)/(e_{it} + e_p)]$, e_p being a small positive constant. This transforms the observations into an empirical log relative risk. This is a close-to-Gaussian form for most small area disease incidence.

The system/process model is now

$$\theta_{it}|\theta_{t-1}, \Gamma \sim N(f(\theta_{i,t-1}), \Gamma)$$
$$\Gamma_{tk} = \mathrm{cov}(\theta_{it}, \theta_{ik})$$

with the observational/data process specified as

$$z_{it}|\mu_{it}, \Sigma^t \sim N(\mu_{it}, \Sigma^t)$$
$$\exp(\mu_{it}) = \theta_{it}$$
$$\Sigma^t_{il} = \mathrm{cov}(z_{it}, z_{lt})$$

where Σ^t is defined to be a positive definite spatial covariance matrix. Usually, this covariance would be thought to be constant in time and so $\Sigma_{il} = \mathrm{cov}(z_{it}, z_{lt})$ will be constant $\forall t$. However, it is possible to generalize the covariance to include temporal dependence. Note that the covariance of the risks is defined for time only. This too could be extended to include spatial dependence.

18.3 Model Fitting Issues

Model fitting for space-time small area count models has mainly focussed on Bayesian algorithms that access features of the posterior distribution of parameters of interest. While it is feasible to consider likelihood-based or pseudo-likelihood approaches to these models, it is now simpler and computationally convenient to use sampling or posterior approximation approaches.

18.3.1 Posterior Sampling

Once a model is specified, it is usually convenient to consider a hierarchical framework within which parameter conditioning occurs. Conditional distributions of parameters within a hierarchy can lead naturally to a Bayesian approach. In that case, we specify the posterior distribution of parameters given data as $p(\theta|\mathbf{y}) \propto l(\mathbf{y}|\theta)\mathbf{p}(\theta)$. Often in spatio-temporal models, it is difficult to obtain summaries of quantities from $p(\theta|\mathbf{y})$. The usual approach then is to employ a posterior sampling algorithm. This sampling algorithm will generate samples from the distribution in question and we can then use the samples to approximate posterior quantities, such as means, medians, quartiles, or quantiles. MCMC is often employed to generate such samples (Robert and Casella, 2005; Brooks et al., 2011). This consists of an iterative algorithm whereby new parameter values are generated from previously sampled values and which, after sufficient run time, approximates samples from the correct posterior distribution. The software package WinBUGS and more recent OpenBUGS have been developed to accommodate a range of MCMC sampling techniques. For the spatio-temporal examples discussed earlier, a wide range of code is available. The site http://academicdepartments.musc.edu/phs/research/lawson/data.htm/, (accessed April 22, 2015.) contains a variety of examples of spatio-temporal models which can be fitted using WinBUGS or OpenBUGS.

18.3.2 INLA

A recent development in the use of approximations to Bayesian models has been proposed by Rue et al. (2009). The basic idea is that a wide range of models that have a latent Gaussian structure can be approximated via integrated nested Laplace approximation (INLA). These approximations can be seen as successive approximations of functions within integrals. The integrals are then approximated by fixed integration schemes. This approximation approach is now available in R (package R-inla: www.r-inla.org). The INLA website contains many examples of the use of this approximation package, including spatial analyses. INLA provides a fast and reasonably accurate alternative approach to MCMC for posterior parameter estimation. It is particularly useful for large datasets ($m > 10{,}000$, say) where conventional sampling programs would be extremely slow. The main advantages of INLA in its current form are as follows: fast computation, flexible model specification, and application to log-linear Gaussian models. The main disadvantages are (currently) that it cannot handle certain types of missing data, certain types of measurement error or mixtures, certain models not expressible in log-linear form, and has a limited range of prior distributions. For applications to spatio-temporal health data, refer to Schrödle and Held (2011), Blangiardo et al. (2013), and Lawson (2013), Appendix D.

18.4 Advanced Modeling for Special Topics

18.4.1 Latent Components

It is possible to extend space-time models to consider the inclusion of latent components in either space or time dimensions. While the random effect models of Section 18.2 allow for some random variation, they do not allow for unobserved latent structure.

For example, we could conceive that a range of temporal (latent) profiles underlie the incidence in any area. These latent profiles are unobserved but we would like to estimate them if possible. This type of model can be thought of as spatial clustering of temporal profiles, so that some areas have different temporal profiles from others. In essence, this is a form of disaggregation of risk by categorizing groups of areas with similar temporal variation of risk. One such model could be defined as

$$y_{it} \sim Po(e_{it}\theta_{it}) \tag{18.4}$$

$$log(\theta_{it})|\psi_{it} = \alpha_0 + \sum_l w_{il}\psi_{lt}, \tag{18.5}$$

where, for each small area i, the weights satisfy two conditions, $0 < w_{il} \leq 1$ and $\sum_l w_{il} = 1$. The latent components ψ_l are indexed in time and there are $l = 1, \ldots, L$ unobserved components. In this formulation, each area has a set of probabilistic weights assigned to any given temporal component and so can be regarded as "voting" for a component in an area. Prior distributions for the components in this model are important for identifiability, and usually, a correlated prior distribution is assumed for ψ_l. For example, a first-order random walk prior distribution is often assumed

$$\psi_{lt}|\psi_{l,t-1}, \tau_{\psi_l}^{-1} \sim N(\psi_{l,t-1}, \tau_{\psi_l}^{-1}), \quad \forall l = 1, 2, \ldots, L$$

Alternatively, an AR(1) prior distribution could be assumed. A variety of choices are available for prior distributions for the weights. These could be spatially correlated or uncorrelated. A common choice is to assume that the vector $w_i = (w_{i1}, w_{i2}, \ldots, w_{iL})'$ has a singular multinomial distribution of the form

$$p_{il}^* \sim Ga(1, 1)$$

$$p_{il} = \frac{p_{il}^*}{\sum_k p_{ik}^*}$$

$$w_i|p_{i1}, p_{i2}, \ldots, p_{iL} \sim \text{Mult}(1, (p_{i1}, p_{i2}, \ldots, p_{iL})).$$

This leads to a hard classification of the area weight. A soft classification can also be defined using

$$\alpha_i|\Sigma_\alpha \sim MCAR(\Sigma_\alpha)$$

$$w_{il}^*|\alpha_{il}, \tau_\alpha \sim LN(\alpha_{il}, \tau_\alpha)$$

$$w_{il} = \frac{w_{il}^*}{\sum_k w_{ik}^*}$$

where MCAR denotes a multivariate CAR prior distribution, which admits correlation between spatially correlated fields (Gelfand and Vounatsou, 2003). The covariance matrix Σ_α can have a Wishart prior distribution. A fuller discussion and evaluation of these models can be found in Lawson et al. (2010) and Choi and Lawson (2011).

18.4.2 Infectious Diseases

In recent years, there has been rapid progress in developing statistical models for understanding and controlling the spread of infectious diseases, which remain a leading cause of morbidity and mortality worldwide. Unlike the analysis of noninfectious diseases, models describing infectious disease dynamics must take into account the transmissible nature of infections. The traditional approach to model the progress of an epidemic include the so-called compartmental models (Keeling and Rohani, 2008; Vynnycky and White, 2010). Within this class of models, the SIR model stratifies the population into three subgroups: those who are susceptible to being infected, those who are infected, and those who are immune. The discrete-time model describes the progression of the infection through the number of individuals in each compartment at discrete time steps. The following difference equations determine the number of individuals in different categories at a particular time period t

$$S_t = S_{t-1} - \beta I_{t-1} S_{t-1},$$

$$I_t = I_{t-1} + \beta I_{t-1} S_{t-1} - r I_{t-1},$$

$$R_t = R_{t-1} + r I_{t-1},$$

where the disease transmission rate β represents the rate at which two individuals come into effective contact (a contact that will lead to infection). Here, the transmission rate is assumed to be constant, but it can be allowed to vary in time. The parameter r represents the proportion of infected who recover and become immune. Based on the nature of the infection, alternative compartmental models, such as the Susceptible-Infected-Susceptible (SIS), Susceptible-Infected-Recovered-Susceptible (SIRS), Susceptible-Exposed-Infected-Recovered (SEIR), or Susceptible-Exposed-Infected-Recovered-Susceptible (SEIRS) models, can also be used.

Morton and Finkenstädt (2005) proposed a stochastic version of the discrete-time SIR model and showed its Bayesian analysis. An extension of that model to the spatial domain was proposed by Lawson and Song (2010), where a neighborhood infection effect was incorporated into the model specification to account for spatial transmission. Hooten et al. (2010) showed the application of an SIRS model to state-level influenza-like illness (ILI) data.

Ideally, spatio-temporal modeling of infectious diseases would be done at individual level (Lawson and Leimich, 2000; Deardon et al., 2010). By tracking the status of every individual in a population, these models provide an accurate description of the spread of epidemics through time and space. In addition, they allow for heterogeneity in the population via individual-level covariates. However, information about individual movement and contact behavior is scarcely ever available. In practice, only partial information about the total number of infected individuals in each small area and time period is available.

For aggregated counts within small areas and time periods, it is also common to assume a Poisson data-level model. Hierarchical Poisson models may be appropriate when the number of susceptibles is unknown and disease counts are small relative to the population size. One approach within this scenario is to assume that counts of disease y_{it} are Poisson distributed with mean $\lambda_{it} = e_{it}\theta_{it}$, where e_{it} is the number of cases expected during nonepidemic conditions and θ_{it} is the relative risk in area i and time period t, $i = 1, \ldots, m$ and $t = 1, \ldots, T$. Mugglin et al. (2002) described the evolution of epidemics through changes in the relative risks of disease, which are defined by a vector autoregressive model. Once the change points have been chosen, stability, growth, and recession of infection are described by modifying the mean of the innovation term in the autoregressive process. Knorr-Held and Richardson (2003) modeled the log of the relative risks through latent spatial and temporal components. An extra term that is a function of the previous number of cases is incorporated into the relative risk model during epidemic periods, which are differentiated through latent binary indicators, to explain the increase in incidence. An alternative approach, which is motivated from a branching process model with immigration, was proposed by Held et al. (2005). In that model, disease incidence is separated into two components as follows:

$$\lambda_{it} = \nu_{it} + \gamma y_{i,t-1} + \phi \sum_{j \sim i} y_{j,t-1}$$

where the endemic component ν_{it} relates disease incidence to latent parameters describing endemic seasonal patterns and the notation $i \sim j$ denotes that i is a neighbor of j. The epidemic component, which is modeled with an autoregression on the previous numbers of cases, captures occasional outbreaks beyond seasonal epidemics. Extensions of this model can be found in Held et al. (2006) and Paul et al. (2008).

18.5 Prospective Analysis and Disease Surveillance

Most of the models described in the previous sections have been developed for retrospective analyses of disease maps. However, there are situations where real-time modeling and prediction play a crucial part. This is the case, for instance, of public health surveillance, which is defined as (Thacker and Berkelman, 1992)

> the ongoing, systematic collection, analysis, and interpretation of health data essential to the planning, implementation, and evaluation of public health practice, closely integrated with the timely dissemination of these data to those who need to know. The final link of the surveillance chain is the application of these data to prevention and control.

Hence, sequential analyses of all the data collected so far are a key concept to early detection of changes in disease incidence and, consequently, to facilitate timely public health response.

Most work on surveillance methodology has evolved in temporal applications, and so a wide range of methods including process control charts, temporal scan statistics, regression techniques, and time series methods have been proposed to monitor univariate time series of counts of disease (Sonesson and Bock, 2003; Unkel et al., 2012). Regression models have been widely used for outbreak detection. For instance, the log-linear regression model of Farrington et al. (1996) is used by the Health Protection Agency to detect aberrations in laboratory-based surveillance data in England and Wales. At each time point, the observed count of disease is declared aberrant if it lies above a threshold, which is computed from the estimated model using a set of recent observations with similar conditions. Within the time series scenario, hidden Markov models have proved to be successful in monitoring epidemiological data. The basic idea of these models is to segment the time series of disease counts into epidemic and nonepidemic phases (Le Strat and Carrat, 1999). Martínez-Beneito et al. (2008) used a hidden Markov model to detect the onset of influenza epidemics. Unlike previous hidden Markov models, the authors modeled the series of differenced rates rather than the series of incidence rates. More recently, Conesa et al. (2015) have proposed an enhanced modeling framework that incorporates the magnitude of the incidence to better distinguish between epidemic and nonepidemic phases.

Increasingly, surveillance systems are capturing data on both the time and location of events. The use of spatial information enhances the ability to detect small localized outbreaks of disease relative to the surveillance of the overall count of disease cases across the entire study region, where increases in a relatively small number of regional counts may be diluted by the natural variation associated with overall counts. In addition, spatio-temporal surveillance facilitates public health interventions once an increased regional count has been identified. Consequently, practical statistical surveillance usually implies analyzing simultaneously multiple time series that are spatially correlated.

Unlike testing methods (Kulldorff, 2001; Rogerson, 2005), modeling for spatio-temporal disease surveillance is relatively recent, and this is a very active area of statistical research (Robertson et al., 2010). Models describing the behavior of disease in space and time allow covariate effects to be estimated and provide better insight into etiology, spread, prediction, and control of disease.

Kleinman et al. (2004) proposed a method based on generalized linear mixed models to evaluate whether observed counts of disease are larger than would be expected on the basis of a history of naturally occurring disease. In that model, the number of cases in area i and time t (y_{it}) is assumed to follow a binomial distribution with parameters n_{it} and p_{it}, n_{it} being the population and p_{it} the probability of an individual being a case, which is modeled as a function of covariate and spatial random effects. Once the model is fitted using historical data observed under endemic conditions, the probability of seeing more cases than the current observed count of disease is calculated for each small area and time period to detect unusually high counts of disease.

An alternative approach to prospective disease surveillance is the use of hidden Markov models. Watkins et al. (2009) provided an extension of a purely temporal hidden Markov model to incorporate spatially referenced data. More recently, Heaton et al. (2012) have proposed a spatio-temporal conditional autoregressive hidden Markov model with an absorbing state. By considering the epidemic state to be absorbing, the authors avoid undesirable behavior such as day-to-day switching between the epidemic and nonepidemic states. This feature, however, limits the application of the model to a single outbreak of disease at each location.

Bayesian hierarchical Poisson models, which are extensively used in disease mapping, have also proved to perform well in the prospective surveillance context. In Vidal Rodeiro and Lawson (2006), a Poisson distribution with a mean which is a function of the expected count of disease and the unknown area-specific relative risk was assumed as a data-level model, that is, $y_{it}||\lambda_{it} \sim Po(\lambda_{it} = e_{it}\theta_{it})$. The logarithm of the relative risk was then modeled as

$$\log(\theta_{it}) = v_i + u_i + \gamma_t + \psi_{it},$$

where v_i and u_i represent, respectively, spatially uncorrelated and correlated heterogeneity; γ_t is a smooth temporal trend, and ψ_{it} is the space-time interaction effect. To detect changes in the relative risk pattern of disease, the authors proposed to monitor at each time $t = 2, 3, \ldots, T$ the surveillance residuals defined as

$$r^s_{it} = y_{it} - \frac{1}{J}\sum_{j=1}^{J} e_{it}\theta^{(j)}_{i,t-1},$$

$\left\{\theta^{(j)}_{i,t-1}\right\}$ being a set of relative risks sampled from the posterior distribution that corresponds to the previous time period.

In an effort to overcome the estimation problem arising when Bayesian hierarchical Poisson models are used in a spatio-temporal surveillance context, Zhou and Lawson (2008) presented an approximated procedure where a spatial convolution model is fitted to the data observed at each time period t. Changes in the risk pattern of disease can then be detected by comparing the estimated relative risk for each small area $\hat{\theta}_{it}$ with a baseline level $\tilde{\theta}_{it}$, $i = 1, 2, \ldots, m$, calculated as an exponentially weighted moving average (EWMA) of historical estimates

$$\tilde{\theta}_{it} = \kappa\hat{\theta}_{i,t-1} + (1 - \kappa)\tilde{\theta}_{i,t-1}.$$

In particular, the authors defined a sample-based Monte Carlo p-value as

$$\text{MCP}_{it} = \frac{1}{J}\sum_{j=1}^{J} I(\hat{\theta}_{it}^{(j)} < \tilde{\theta}_{it}^{(j)}),$$

extremely small p-values indicating that an increase in disease risk might have occurred. The estimated percentage of increase in disease relative risk, which is defined as

$$\text{PIR}_{it} = \frac{\frac{1}{J}\sum_{j=1}^{J}(\hat{\theta}_{it}^{(j)} - \tilde{\theta}_{it}^{(j)})}{\frac{1}{J}\sum_{j=1}^{J}\tilde{\theta}_{it}^{(j)}} \times 100,$$

can also be calculated to assess the magnitude of change in disease risk. This is a computationally quick technique that has shown to have a good performance in outbreak detection. However, a sliding window of length one cannot guarantee accurate model estimates. Also, an EWMA approach is only justified under Gaussian model assumptions.

More recently, Corberán-Vallet and Lawson (2011) have introduced the surveillance conditional predictive ordinate (SCPO) as a general Bayesian model–based surveillance technique to detect small areas of unusual disease aggregation. The SCPO is based on the conditional predictive ordinate (CPO), which was introduced by Geisser (1980) as a Bayesian diagnostic to detect observations discrepant from a given model. The CPO was further discussed in Gelfand et al. (1992), where a cross-validation approach based on conditional predictive distributions arising from single observation deletion was proposed to address model determination. In particular, the SCPO is calculated for each small area and time period as

$$\text{SCPO}_{it} = f(y_{it}|y_{1:t-1}) = \int\int f(y_{it}|\theta_{it})\, p(\theta_{it}|\theta_{i,t-1}, y_{1:t-1})\, p(\theta_{i,t-1}|y_{1:t-1})\, d\theta_{i,t-1}\, d\theta_{it},$$

where $y_{1:t-1}$ means all the data up to time $t-1$; $f(y_{it}|\theta_{it})$ is Poisson, $p(\theta_{it}|\theta_{i,t-1}, y_{1:t-1})$ can be derived from the model describing the relative risk surface, and $p(\theta_{i,t-1}|y_{1:t-1})$ represents the marginal posterior distribution of parameter $\theta_{i,t-1}$ at time $t-1$.

In that paper, the convolution model was used to model the behavior of nonseasonal disease data, since the inclusion of adaptive time components may hinder detection of changes in risk. In that case, the SCPO simplifies to

$$\text{SCPO}_{it} = f(y_{it}|y_{1:t-1}) = \int f(y_{it}|\theta_i)\, p(\theta_i|y_{1:t-1})\, d\theta_i.$$

A Monte Carlo estimate for the SCPO, which does not have a closed form, can be obtained from a posterior sampling algorithm as

$$\text{SCPO}_{it} \approx \frac{1}{J}\sum_{j=1}^{J} Po(y_{it}|\, e_{it}\theta_i^{(j)}),$$

where $\{\theta_i^{(j)}\}_{j=1}^{J}$ is a set of relative risks sampled from the posterior distribution that corresponds to the previous time period. Hence, if there is no change in risk, y_{it} will be representative of the data expected under the previously fitted model. Otherwise, SCPO values close to zero will be obtained.

Corberán-Vallet and Lawson (2011) showed an application of the SCPO to Salmonellosis cases in South Carolina from January 1995 to December 2003 (see Figure 18.2). In order to detect occasional outbreaks beyond seasonal patterns, a generalization of the convolution model allowing for seasonal effects was used to model the regular behavior of disease. Figure 18.3 displays the spatial distribution of the SCPO for a selection of 6 months periods: September–October 1996, February–March 2001, and October–November 2002. As can be seen, values of the SCPO close to zero alert us to counties presenting unusually high counts of disease.

An important feature of the SCPO is that it can be easily extended to incorporate information from the spatial neighborhood, which facilitates outbreak detection capability when changes in risk affect neighboring areas simultaneously. Also, a multivariate SCPO (MSCPO) integrating information from $K \geq 2$ diseases can be computed to improve both detection time and recovery of the true outbreak behavior when changes in disease incidence happen simultaneously for two or more diseases. For each area i and time t, let $y_{it} = (y_{it1}, y_{it2}, \ldots, y_{itK})$ be the vector of observed counts of disease, $e_{it} = (e_{it1}, e_{it2}, \ldots, e_{itK})$ the vector of expected counts, $\hat{\theta}_i = (\hat{\theta}_{i1}, \hat{\theta}_{i2}, \ldots, \hat{\theta}_{iK})$ the vector of posterior relative risk estimates at the previous time point using a spatial-only shared-component model, and

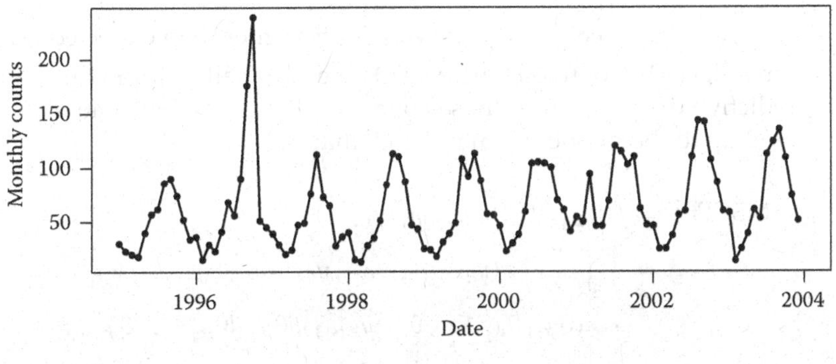

FIGURE 18.2
Monthly counts of reported Salmonellosis cases in South Carolina for the period 1995–2003.

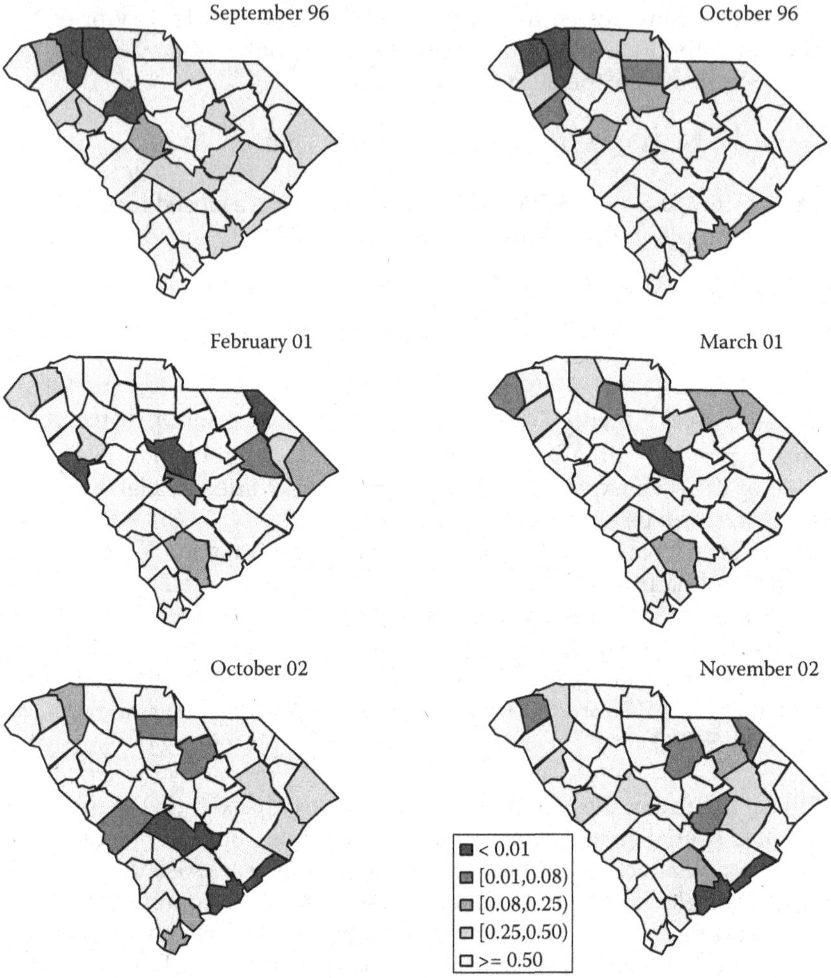

FIGURE 18.3
Spatial distribution of the scaled SCPO for the Salmonellosis data at those months undergoing a possible outbreak of disease.

$y_{it}^h = (y_{itk_1}, y_{itk_2}, \ldots, y_{itk_n})$ the vector of observed counts higher than expected, that is $y_{itk} > e_{itk}\hat{\theta}_{ik}$. Corberán-Vallet (2012) defined the MSCPO for each small area and time period as the conditional predictive distribution of those counts of disease higher than expected given the data observed up to the previous time period, that is,

$$\begin{aligned}\text{MSCPO}_{it} &= f(y_{itk_1}, y_{itk_2}, \ldots, y_{itk_n} | y_{1:t-1}) \\ &= \int\int \cdots \int f(y_{itk_1}, y_{itk_2}, \ldots, y_{itk_n} | \theta_{ik_1}, \theta_{ik_2}, \ldots, \theta_{ik_n}) \\ &\quad \times \pi(\theta_{ik_1}, \theta_{ik_2}, \ldots, \theta_{ik_n} | y_{1:t-1}) d\theta_{ik_1}\, d\theta_{ik_2} \ldots d\theta_{ik_n}.\end{aligned}$$

Values of the MSCPO close to zero alert to both small areas of increased disease incidence and the diseases causing the alarm within each area.

This line of research is particularly useful, since surveillance systems are often focused on more than one disease within a predefined study region. On those occasions when outbreaks of disease are likely to be correlated, the use of multivariate surveillance techniques enhances sensitivity and timeliness of outbreak detection for events that are present in more than one data set. Yet, little work has been conducted within this scenario (Kulldorff et al., 2007; Banks et al., 2012; Corberán-Vallet, 2012).

Much of the new literature in the area of prospective disease surveillance relates to syndromic surveillance, which has been introduced as an efficient tool to detect the outbreaks of disease at the earliest possible time, possibly even before definitive disease diagnosis is obtained. The idea is to monitor syndromes associated with disease such as school and work absenteeism, over-the-counter medication sales, medical consultations, etc. For instance, Kavanagh et al. (2012) adapted the Farrington et al. (1996) algorithm to monitor calls received by the National Health Service telephone helpline in Scotland. As Chan et al. (2012) emphasized, syndromic surveillance is effective to provide early awareness. However, alerts based on syndrome aberrations surely contain uncertainty, and so they should be evaluated with a proper probabilistic measure. Chan et al. (2012) used a space-time Bayesian hierarchical model incorporating information from meteorological factors to model influenza-like illness visits. The risk of an outbreak was assessed using $Pr(y_{it} >$ threshold $|y_{1:t-1})$. A very fruitful area for further research in the context of syndromic surveillance would be the development of a multivariate model to describe the disease of interest and the syndromic data jointly. This approach would allow us to quantify the effect that changes in the behavior of the syndromes have on the incidence of the disease under study.

18.6 Conclusions

In this chapter, we have attempted to survey the varied approaches to spatio-temporal series found in small area health studies. Our focus has been on descriptive models of space-time disease incidence data and also, in later sections, on prospective surveillance. Both areas provide a ready source of challenging problems for the modeler and much work is still needed to develop flexible and relevant models that can be used more widely by the Public Health community. While there are many testing approaches available to spatio-temporal incidence data, we have taken a Bayesian model–based approach. We firmly believe that this provides much greater flexibility for handling the complexities of spatio-temporal variation and will provide the greatest insight into spatial health dynamics in the future.

Acknowledgments

This work was supported by Grant Number R03CA162029 from the National Cancer Institute. The content is solely the responsibility of the authors and does not necessarily represent the official views of the National Cancer Institute or the National Institutes of Health.

References

Banks, D., Datta, G., Karr, A., Lynch, J., Niemi, J., and Vera, F. (2012). Bayesian CAR models for syndromic surveillance on multiple data streams: Theory and practice. *Information Fusion*, 13:105–116.

Blangiardo, M., Cameletti, M., Baio, G., and Rue, H. (2013). Spatial and spatio-temporal models with R-INLA. *Spatial and Spatio-temporal Epidemiology*, 4:33–49.

Brooks, S., Gelman, A., Jones, G., and Meng, X., editors (2011). *Handbook of Markov Chain Monte Carlo*. CRC Press, New York.

Cai, B., Lawson, A. B., Hossain, M. M., and Choi, J. (2012). Bayesian latent structure models with space-time dependent covariates. *Statistical Modelling*, 12:145–164.

Cai, B., Lawson, A. B., Hossain, M. M., Choi, J., Kirby, R., and Liu, J. (2013). Bayesian semi-parametric model with spatially-temporally varying coefficient selection. *Statistics in Medicine*, 23:3670–3685.

Chan, T.-C., King, C.-C., Yen, M.-Y., Chiang, P.-H., Huang, C.-S., and Hsiao, C. (2012). Probabilistic daily ILI syndromic surveillance with a spatio-temporal Bayesian hierarchical model. *PLoS ONE*, 5:e11626.

Choi, J. and Lawson, A. B. (2011). Evaluation of Bayesian spatial-temporal latent models in small area health data. *Environmetrics*, 22(8):1008–1022.

Conesa, D., Martínez-Beneito, M., Amorós, R., and López-Quílez, A. (2015). Bayesian hierarchical Poisson models with a hidden Markov structure for the detection of influenza epidemic outbreaks. *Statistical Methods in Medical Research*, 24:206–223.

Corberán-Vallet, A. (2012). Prospective surveillance of multivariate spatial disease data. *Statistical Methods in Medical Research*, 21:457–477.

Corberán-Vallet, A. and Lawson, A. (2011). Conditional predictive inference for online surveillance of spatial disease incidence. *Statistics in Medicine*, 30:3095–3116.

Cressie, N. and Wikle, C. (2011). *Statistics for Spatio-Temporal Data*. Wiley, New York.

Deardon, R., Brooks, S., Grenfell, B., Keeling, M., Tildesley, M., Savill, N., Shaw, D., and Woolhouse, M. (2010). Inference for individual-level models of infectious diseases in large populations. *Statistica Sinica*, 20:239–261.

Farrington, C., Andrews, N., Beale, A., and Catchpole, M. (1996). A statistical algorithm for the early detection of outbreaks of infectious disease. *Journal of the Royal Statistical Society, Series A*, 159:547–563.

Geisser, S. (1980). Discussion on sampling and Bayes' inference in scientific modelling and robustness (by g.e.p. box). *Journal of the Royal Statistical Society, Series A*, 143:416–417.

Gelfand, A., Dey, D., and Chang, H. (1992). *Bayesian Statistics 4*, chapter Model determination using predictive distributions with implementation via sampling-based methods. Oxford University Press, Oxford, U.K.

Gelfand, A. and Vounatsou, P. (2003). Proper multivariate conditional autoregressive models for spatial data. *Biostatistics*, 4:11–25.

Gelman, A. (2006). Prior distributions for variance parameters in hierarchical models. *Bayesian Analysis*, 1:515–534.

Heaton, M., Banks, D., Zou, J., Karr, A., Datta, G., Lynch, J., and Vera, F. (2012). A spatio-temporal absorbing state model for disease and syndromic surveillance. *Statistics in Medicine*, 31:2123–2136.

Held, L., Hofmann, M., and Höhle, M. (2006). A two-component model for counts of infectious diseases. *Biostatistics*, 7:422–437.

Held, L., Höhle, M., and Hofmann, M. (2005). A statistical framework for the analysis of multivariate infectious disease surveillance counts. *Statistical Modelling*, 5:187–199.

Holan, S. H. and Wikle, C. K. (2015). Hierarchical dynamic generalized linear mixed models for discrete-valued spatio-temporal data. In R. A. Davis, S. H. Holan, R. Lund and N. Ravishanker, eds., *Handbook of Discrete-Valued Time Series*, pp. 327–348. Chapman & Hall, Boca Raton, FL.

Hooten, M., Anderson, J., and Waller, L. (2010). Assessing North American influenza dynamics with a statistical SIRS model. *Spatial and Spatio-temporal Epidemiology*, 1:177–185.

Kavanagh, K., Robertson, C., Murdoch, H., Crooks, G., and McMenamin, J. (2012). Syndromic surveillance of influenza-like illness in Scotland during the influenza A H1N1v pandemic and beyond. *Journal of the Royal Statistical Society, Series A*, 175:939–958.

Keeling, M. and Rohani, P. (2008). *Modeling Infectious Diseases in Humans and Animals*. Princeton University Press, Princeton, NJ.

Kleinman, K., Lazarus, R., and Platt, R. (2004). A generalized linear mixed models approach for detecting incident clusters of disease in small areas, with an application to biological terrorism. *American Journal of Epidemiology*, 159:217–224.

Knorr-Held, L. (2000). Bayesian modelling of inseparable space-time variation in disease risk. *Statistics in Medicine*, 19:2555–2567.

Knorr-Held, L. and Richardson, S. (2003). A hierarchical model for space-time surveillance data on meningococcal disease incidence. *Journal of the Royal Statistical Society, Series C*, 52:169–183.

Kulldorff, M. (2001). Prospective time periodic geographical disease surveillance using a scan statistic. *Journal of the Royal Statistical Society, Series A*, 164:61–72.

Kulldorff, M., Mostashari, F., Duczmal, L., Yih, W., Kleinman, K., and Patt, R. (2007). Multivariate scan statistics for disease surveillance. *Statistics in Medicine*, 26:1824–1833.

Lawson, A. and Leimich, P. (2000). Approaches to the space-time modelling of infectious disease behaviour. *IMA Journal of Mathematics Applied in Medicine and Biology*, 17:1–13.

Lawson, A. and Song, H.-R. (2010). Bayesian hierarchical modeling of the dynamics of spatio-temporal influenza season outbreaks. *Spatial and Spatio-temporal Epidemiology*, 1:187–195.

Lawson, A. B. (2013). *Bayesian Disease Mapping: Hierarchical Modeling in Spatial Epidemiology*, 2nd edn. CRC Press, New York.

Lawson, A. B., Song, H.-R., Cai, B., Hossain, M. M., and Huang, K. (2010). Space-time latent component modeling of geo-referenced health data. *Statistics in Medicine*, 29:2012–2017.

Le Strat, Y. and Carrat, F. (1999). Monitoring epidemiologic surveillance data using hidden Markov models. *Statistics in Medicine*, 18:3463–3478.

Martínez-Beneito, M., Conesa, D., López-Quílez, A., and López-Maside, A. (2008). Bayesian Markov switching models for the early detection of influenza epidemics. *Statistics in Medicine*, 27:4455–4468.

Morton, A. and Finkenstädt, B. (2005). Discrete time modelling of disease incidence time series by using Markov chain Monte Carlo methods. *Journal of the Royal Statistical Society, Series C*, 54:575–594.

Mugglin, A., Cressie, N., and Gemmell, I. (2002). Hierarchical statistical modelling of influenza epidemic dynamics in space and time. *Statistics in Medicine*, 21:2703–2721.

Paul, M., Held, L., and Toschke, A. (2008). Multivariate modeling of infectious disease surveillance data. *Statistics in Medicine*, 27:6250–6267.

Robert, C. and Casella, G. (2005). *Monte Carlo Statistical Methods*, 2nd edn. Springer, New York.

Robertson, C., Nelson, T., MacNab, Y., and Lawson, A. (2010). Review of methods for space-time disease surveillance. *Spatial and Spatio-temporal Epidemiology*, 1:105–116.

Rogerson, P. (2005). *Spatial and Syndromic Surveillance for Public Health*, chapter Spatial surveillance and cumulative sum methods. Wiley, Chichester.

Rue, H., Martino, S., and Chopin, N. (2009). Approximate Bayesian inference for latent Gaussian models by using integrated nested Laplace approximations. *Journal of the Royal Statistical Society B*, 71:319–392.

Schrödle, B. and Held, L. (2011). Spatio-temporal disease mapping using INLA. *Environmetrics*, 22:725–734.

Sonesson, C. and Bock, D. (2003). A review and discussion of prospective statistical surveillance in public health. *Journal of the Royal Statistical Society, Series A*, 166:5–21.

Thacker, S. and Berkelman, R. (1992). *Public Health Surveillance*, chapter History of public health surveillance. John Wiley & Sons, New York.

Ugarte, M., Goicoa, T., Ibanez, B., and Militano, A. F. (2009). Evaluating the performance of spatio-temporal Bayesian models in disease mapping. *Environmetrics*. DOI: 10.1002/env.969.

Unkel, S., Farrington, C., Garthwaite, P., Robertson, C., and Andrews, N. (2012). Statistical methods for the prospective detection of infectious diseases outbreaks: A review. *Journal of the Royal Statistical Society, Series A*, 175:49–82.

Vidal Rodeiro, C. and Lawson, A. (2006). Monitoring changes in spatio-temporal maps of disease. *Biometrical Journal*, 48:463–480.

Vynnycky, E. and White, R. (2010). *An Introduction to Infectious Disease Modelling*. Oxford University Press, Oxford, U.K.

Watkins, R., Eagleson, S., Veenendaal, B., Wright, G., and Plant, A. (2009). Disease surveillance using a hidden Markov model. *BMC Medical Informatics and Decision Making*, 9:39.

Zhou, H. and Lawson, A. (2008). Ewma smoothing and Bayesian spatial modeling for health surveillance. *Statistics in Medicine*, 27:5907–5928.

Section V

Multivariate and Long Memory Discrete-Valued Processes

19

Models for Multivariate Count Time Series

Dimitris Karlis

CONTENTS

19.1 Introduction

We have seen a tremendous increase in models for discrete-valued time series over the past few decades. Although there is a flourishing literature on models and methods for univariate integer-valued time series, the literature is rather sparse for the multivariate case, especially for multivariate count time series. Multivariate count data occur in several different disciplines like epidemiology, marketing, criminology, and engineering, just to name a few. For example, in syndromic surveillance systems, we record the number of patients with a given symptom. An abrupt change in this number could indicate a threat to public health, and our goal would be to discover such a change as early as possible. In practice, a large number of symptoms are counted creating possibly associated multiple time series of counts. An adequate analysis of such multiple time series requires models that can take into account the correlation across time as well as the correlations between the different symptoms.

Another example comes from geophysical research, where data are collected over time on the number of earthquakes whose magnitudes are above a certain threshold (Boudreault and Charpentier, 2011). Different series can be generated from adjacent areas, making an important scientific question the correlation between the two areas. In criminology, one counts the number of occurrences of one type of crime in successive time periods (say, weeks). Analyzing together more than one type of crime generates many count time series that may be correlated. In finance, an analyst might wish to model the number of *bids and asks* for a stock, or the number of trades of different stocks in a portfolio. Similar examples may be seen for the number of purchases of different but related products in marketing, the number of claims of different policies in actuarial science, etc.

The underlying similarity in all the earlier mentioned examples is that the collected data are correlated counts observed at different time points. Hence, we have two sources of correlation, serial correlation since the data are time series and cross-correlation since the component time series are correlated at each time point. The need to account for both serial and cross-correlation complicates model specification, estimation, and inference. The literature on statistical models for multivariate time series of counts is rather sparse, perhaps because the analytical and computational issues are complicated. In recent years, new models have been developed to facilitate the modeling approach, which we discuss in this chapter.

We start in Section 19.2 with a brief review of some models for multivariate count data and a discussion of the problems that arise. These models form the basis for the time series models that will be discussed in the following sections along three main avenues: models based on thinning (Section 19.3), parameter-driven models (Section 19.4), and observation-driven models (Section 19.5). Additional models will be mentioned in Section 19.6. Concluding remarks are given in Section 19.7.

19.2 Models for Multivariate Count Data

Even ignoring the time correlation there are not many models for multivariate counts in the literature. Inference for multivariate counts is analytically and computationally demanding. Perhaps the case is easier and more developed in the bivariate case but there are several bivariate models that cannot easily generalize to the multivariate. This is a major obstacle for the development of flexible models to be used also in the time series context. We briefly explore some of the issues.

19.2.1 Use of Multivariate Extensions of Simple Models

Consider, for example, the simplest extension of the univariate Poisson distribution to the bivariate case. As in Kocherlakota and Kocherlakota (1992), the bivariate Poisson has probability mass function (*pmf*)

$$\begin{aligned} P(y_1,y_2) &= P(Y_1 = y_1, Y_2 = y_2; \boldsymbol{\Theta}) \\ &= e^{-(\theta_1+\theta_2+\theta_0)} \frac{\theta_1^{y_1}}{y_1!} \frac{\theta_2^{y_2}}{y_2!} \sum_{s=0}^{\min(y_1,y_2)} \binom{y_1}{s} \binom{y_2}{s} s! \left(\frac{\theta_0}{\theta_1\theta_2}\right)^s, \end{aligned} \tag{19.1}$$

where $\theta_1, \theta_2, \theta_0 \geq 0$, $y_1, y_2 = 0, 1, \ldots$, $\boldsymbol{\Theta} = (\theta_1, \theta_2, \theta_0)$. θ_0 is the covariance while the marginal means and variances are equal to $\theta_1 + \theta_0$ and $\theta_2 + \theta_0$, respectively. The marginal distributions are Poisson. One can easily see that this *pmf* involves a finite summation that can be computationally intensive for large counts. This bivariate Poisson distribution only allows positive correlation. We denote this by $BP(\theta_1, \theta_2, \theta_0)$. For $\theta_0 = 0$, we get two independent Poisson distributions. We may generalize this model by considering mixtures of the bivariate Poisson. Although there are a few schemes, two ways to do this have been studied in detail. Most of the literature assumes a $BP(\alpha\theta_1, \alpha\theta_2, \alpha\theta_0)$ distribution and places a mixing distribution on α. Depending on the choice of the distribution of α, such a model produces overdispersed marginal distributions but with always positive correlation. The correlation comes from two sources, the first is the intrinsic one from θ_0 and the second is due to the use of a common α.

A more refined model can be produced by assuming a $BP(\theta_1, \theta_2, 0)$ and letting θ_1, θ_2 jointly vary according to some bivariate continuous distribution, as, for example, in Chib and Winkelmann (2001) where a bivariate lognormal distribution is assumed. Here, the correlation comes from the correlation of the joint mixing distribution, and thus, it can be negative as well. The obstacle is that we do not have flexible bivariate distributions to use for the mixing, or some of them may lead to computational problems. The bivariate Poisson lognormal distribution in Chib and Winkelmann (2001) does not have closed-form *pmf* and bivariate integration is needed.

It is interesting to point out that generalization to higher dimensions is not straightforward even for simple models. For example, generalizing the bivariate Poisson to the multivariate Poisson with one correlation parameter for every pair of variables leads to multiple summation, see the details in Karlis and Meligkotsidou (2005). We will see later some ideas on how to overcome these problems.

19.2.2 Models Based on Copulas

A different avenue to build multivariate models is to apply the copula approach. Copulas (see Nelsen, 2006) have found a remarkably large number of applications in finance, hydrology, biostatistics, etc., since they allow the derivation and application of flexible multivariate models with given marginal distributions. The key idea is that the marginal properties can be separated from the association properties, thus leading to a wealth of potential models. For the case of discrete data, copula-based modeling is less developed. Genest and Nešlehová (2007) provided an excellent review on the topic. It is important to keep in mind that some of the desirable properties of copulas are not valid when dealing with count data. For example, dependence properties cannot be fully separated from marginal properties. To see this, consider the Kendall's tau correlation coefficient. The probability for a tie is not zero for discrete data and depends on the marginal distribution, hence the value of Kendall's tau is also dependent on the marginal distributions. Furthermore, the *pmf* cannot be derived through derivatives but via finite differences which can be cumbersome in larger dimensions. For a recent review on copulas for discrete data, see Nikoloulopoulos (2013b).

To help the exposition we first discuss bivariate copulas.

Definition (Nelsen, 2006). A bivariate copula is a function C from $[0,1]^2$ to $[0,1]$ with the following properties: (a) for every $\{u, v\} \in [0,1]$, $C(u,0) = 0 = C(0,v)$ and $C(u,1) = u$,

$C(1, v) = v$ and (b) for every $\{u_1, u_2, v_1, v_2\} \in [0, 1]$ such that $u_1 \leq u_2$ and $v_1 \leq v_2$, $C(u_2, v_2) - C(u_2, v_1) - C(u_1, v_2) + C(u_1, v_1) \geq 0$.

That is, copulas are bivariate distributions with uniform marginals. Recall the inversion theorem, central in simulation, where starting from a uniform random variable and applying the inverse transform of a distribution function we can generate any desired distribution. Copulas extend this idea in the sense that we start from two correlated uniforms and hence we end up with variables from whatever distribution we like which are still correlated.

If $F(x)$ and $G(y)$ are the *cdf*s of the univariate random variables X and Y, then $C(F(x), G(y))$ is a bivariate distribution for (X, Y) with marginal distributions F and G, respectively. Conversely, if H is a bivariate *cdf* with univariate marginal *cdf*'s F and G, then, according to Sklar's theorem (Sklar, 1959) there exists a bivariate copula C such that for all (X, Y), $H(x, y) = C(F(x), G(y))$. If F and G are continuous, then C is unique, otherwise, C is uniquely determined on *range* $F \times$ *range* G. This lack of uniqueness is not a problem in practical applications as it implies that there may exist two copulas with identical properties.

Copulas provide the joint cumulative function. In order to derive the joint density (for continuous data) or the joint probability function (for discrete data) we need to take the derivatives or the finite differences of the copula. For bivariate discrete data, the *pmf* is obtained by finite differences of the *cdf* through its copula representation (Genest and Nešlehová, 2007), that is,

$$h(x, y; \alpha_1, \alpha_2, \theta) = C(F(x; \alpha_1), G(y; \alpha_2); \theta) - C(F(x-1; \alpha_1), G(y; \alpha_2); \theta) - C(F(x; \alpha_1), G(y-1; \alpha_2); \theta) + C(F(x-1; \alpha_1), G(y-1; \alpha_2); \theta),$$

where $F(\cdot)$ and $G(\cdot)$ are the marginal *cdf*s and α_1 and α_2 are the parameters associated with the respective marginal distributions and θ denotes the parameter(s) of the copula. This poses a big problem in higher dimensions. In order to take differences, we need to evaluate the copula eight times in the trivariate case and 2^d times for d dimensions.

Copulas are *cdf*s and thus in many cases they are given as multidimensional integrals and not as simple formulas. A simple example is the bivariate Gaussian copula which is defined as a bivariate integral. In this case, one needs to evaluate multidimensional integrals many times in order to evaluate the *pmf*. To avoid this extensive integration, one can switch to copulas that are given in simple form without the need to integrate (e.g., the Frank copula). But even in this case, one needs to add and subtract several numbers (which are usually very close to 0) leading to possible truncation errors.

Another problem relates to the shortage of copulas that can allow for flexible correlation structure. For example, the multivariate Archimedean copulas assign the same correlation to all pairs of variables, which is too restrictive in practice. Also if one needs to specify both positive and negative correlations, more restrictions apply. To sum up, a big issue in working with models defined via copulas is the lack of a framework that allows flexible structure while maintaining computational simplicity.

19.2.3 Other Multivariate Models for Counts

There are other strategies to build flexible models for multivariate counts such as models based on conditional distributions (Berkhout and Plug, 2004), or finite mixtures (Karlis and Meligkotsidou, 2007). In most cases, things are more complicated than continuous models where the multivariate normal distributions is a cornerstone allowing for great flexibility and feasible calculations.

19.3 Models Based on Thinning

19.3.1 The Standard INAR Model

Among the most successful integer-valued time series models proposed in the literature are the INteger-valued AutoRegressive (INAR) models, introduced by McKenzie (1985) and Al-Osh and Alzaid (1987). Since then, several articles have been published to extend and generalize these models. The reader is referred to McKenzie (2003) and Jung and Tremayne (2006) for a comprehensive review of such models. The extension of the simple INAR(1) process to the multidimensional case is interesting as it provides a general framework for multivariate count time series modeling. The model had been considered in Franke and Rao (1995) and Latour (1997) but since then, there has been a long hiatus since this topic was addressed again by Pedeli and Karlis (2011) and Boudreault and Charpentier (2011).

Definition A sequence of random variables $\{Y_t : t = 0, 1, \ldots\}$ is an INAR(1) process if it satisfies a difference equation of the form

$$Y_t = \alpha \circ Y_{t-1} + R_t; \quad t = 1, 2, \ldots,$$

where $\alpha \in [0, 1]$, R_t is a sequence of uncorrelated nonnegative integer-valued random variables with mean μ and finite variance σ^2 (called hereafter as the innovations), and Y_0 represents an initial value of the process.

The operator "$\circ$" is defined as

$$\alpha \circ Y = \sum_{i=1}^{Y} Z_i = Z, \tag{19.2}$$

where Z_i are independently and identically distributed Bernoulli random variables with $P(Z_i = 1) = 1 - P(Z_i = 0) = \alpha$. This operator, known as the binomial thinning operator, is due to Steutel and van Harn (1979) and mimics the scalar multiplication used for normal time series models so as to ensure that only integer values will occur.

Note that by assuming any distribution other than Bernoulli for Z_is in (19.2), we get a generalized Steutel and van Harn operator. Other operators can also be similarly defined. A review on thinning operators can be found in Weiß (2008).

The model can be generalized to have p terms (i.e., INAR(p)), but there is no unique way to do this. Moving average (MA) terms can be added to the model leading to INARMA models. Covariates can also be introduced to model the mean of the innovation term to allow measuring the effect of additional information leading to INAR regression models. We next present extensions to the multivariate case by first extending the thinning operator to a matrix-valued form and then presenting the multivariate INAR model.

Let $\mathbf{A}$ be an $r \times r$ matrix with elements α_{ij}, $i,j=1,\ldots,r$ and $\mathbf{Y}$ be a nonnegative integer-valued r-dimensional vector. The matrix-valued operator "$\circ$" is defined as

$$\mathbf{A} \circ \mathbf{Y} = \begin{pmatrix} \sum_{j=1}^{r} \alpha_{1j} \circ Y_j \\ \vdots \\ \sum_{j=1}^{r} \alpha_{rj} \circ Y_j \end{pmatrix}.$$

The univariate operations $\alpha \circ X$ and $\beta \circ Y$ are independent if and only if the counting processes in their definitions are independent. Hence, the matrix-valued operator implies independence between the univariate operators. Properties of this operator can be found in Latour (1997).

Using this operator, Latour (1997) defined a multivariate generalized INAR process of order p (MGINAR(p)) by assuming that

$$\mathbf{Y}_t = \sum_{j=1}^{p} \mathbf{A}_j \circ \mathbf{Y}_{t-j} + \epsilon_t,$$

where $\mathbf{Y}_t$ and ϵ_t are r-vectors and $\mathbf{A}_j$, $j = 1,\ldots,p$ are $r \times r$ matrices and gave conditions for existence and stationarity. A more focused presentation of the model follows.

19.3.2 Multivariate INAR Model

Let $\mathbf{Y}$ and $\mathbf{R}$ be nonnegative integer-valued random r-vectors and let $\mathbf{A}$ be an $r \times r$ matrix with elements $\{\alpha_{ij}\}_{i,j=1,\ldots,r}$. The MINAR(1) process can be defined as

$$\mathbf{Y}_t = \mathbf{A} \circ \mathbf{Y}_{t-1} + \mathbf{R}_t = \begin{pmatrix} \alpha_{11} & \alpha_{12} & \ldots & \alpha_{1r} \\ \alpha_{21} & \alpha_{22} & \ldots & \alpha_{2r} \\ \vdots & & \ddots & \vdots \\ \alpha_{r1} & \alpha_{r2} & \ldots & \alpha_{rr} \end{pmatrix} \circ \begin{pmatrix} Y_{1,t-1} \\ \vdots \\ \vdots \\ Y_{r,t-1} \end{pmatrix} + \begin{pmatrix} R_{1t} \\ \vdots \\ \vdots \\ R_{rt} \end{pmatrix}, \quad t = 1,2\ldots \tag{19.3}$$

The vector of innovations $\mathbf{R}_t$ follows an r-variate discrete distribution, which characterizes the marginal distribution of the $\mathbf{Y}_t$ as well. More on this will follow. When $\mathbf{A}$ is diagonal, we will call the model diagonal MINAR. Clearly, this has less structure.

The nonnegative integer-valued random process $\{\mathbf{Y}_t\}_{t\in\mathbb{Z}}$ is the unique strictly stationary solution of (19.3), if the largest eigenvalue of the matrix $\mathbf{A}$ is less than 1 and $E\,\|\mathbf{R}_t\| < \infty$ (see also Franke and Rao, 1995; Latour, 1997).

To help the exposition consider the case with $r = 2$. The two series can be written as

$$Y_{1t} = \alpha_{11} \circ Y_{1,t-1} + \alpha_{12} \circ Y_{2,t-1} + R_{1t},$$
$$Y_{2t} = \alpha_{22} \circ Y_{2,t-1} + \alpha_{21} \circ Y_{1,t-1} + R_{2t}.$$

This helps to understand the dynamics. The cross correlation between the two series comes from sharing common elements as well as from the joint distribution of (R_{1t}, R_{2t}). If $\mathbf{A}$ is a diagonal matrix in this bivariate example, so that $\alpha_{12} = \alpha_{21} = 0$, then the two series are univariate INAR models but are still correlated due to the joint *pmf* of (R_{1t}, R_{2t}).

Taking expectations on both sides of (19.3), it is straightforward to obtain

$$\boldsymbol{\mu} = E(\mathbf{Y}_t) = [\mathbf{I} - \mathbf{A}]^{-1} E(\mathbf{R}_t). \tag{19.4}$$

The variance–covariance matrix $\boldsymbol{\gamma}(0) = E\left[(\mathbf{Y}_t - \boldsymbol{\mu})(\mathbf{Y}_t - \boldsymbol{\mu})'\right]$ satisfies a difference equation of the form

$$\boldsymbol{\gamma}(0) = \mathbf{A}\boldsymbol{\gamma}(\mathbf{0})\mathbf{A}' + \mathrm{diag}(\mathbf{B}\boldsymbol{\mu}) + \mathrm{Var}(\mathbf{R}_t). \tag{19.5}$$

The innovation series $\mathbf{R}_t$ consists of identically distributed sequences $\{R_{it}\}_{i=1}^{r}$ and has mean $E(\mathbf{R}_t) = \boldsymbol{\lambda} = (\lambda_1, \ldots, \lambda_r)'$ and variance

$$\mathrm{Var}(\mathbf{R}_t) = \begin{bmatrix} \upsilon_1\lambda_1 & \phi_{12} & \cdots & \phi_{1r} \\ \phi_{12} & \upsilon_2\lambda_2 & \cdots & \phi_{2r} \\ \vdots & \vdots & \ddots & \vdots \\ \phi_{1r} & \phi_{2r} & \cdots & \upsilon_r\lambda_r \end{bmatrix},$$

where $\upsilon_i > 0, i = 1, \ldots, r$. Depending on the value of the parameter υ_i, the assumptions of equidispersion ($\upsilon_i = 1$), overdispersion ($\upsilon_i > 1$), and underdispersion ($\upsilon_i \in (0,1)$) can be obtained.

In the bivariate case, that is, when $r = 2$, it can be proved that the vector of expectations (19.5) has elements

$$\mu_1 = \frac{(1-\alpha_{22})\lambda_1 + \alpha_{12}\lambda_2}{(1-\alpha_{11})(1-\alpha_{22}) - \alpha_{12}\alpha_{21}},$$
$$\mu_2 = \frac{(1-\alpha_{11})\lambda_2 + \alpha_{21}\lambda_1}{(1-\alpha_{11})(1-\alpha_{22}) - \alpha_{12}\alpha_{21}},$$

while the elements of $\gamma(0)$ are

$$\begin{aligned}
\gamma_{11}(0) &= \mathrm{Var}(Y_{1t}) \\
&= \frac{1}{(1-\alpha_{11}^2)}\Big\{\alpha_{12}^2\mathrm{Var}(Y_{2t}) + 2\alpha_{11}\alpha_{12}\mathrm{Cov}(Y_{1t}, Y_{2t}) \\
&\quad + \alpha_{11}(1-\alpha_{11})\mu_1 + \alpha_{12}(1-\alpha_{12})\mu_2 + \upsilon_1\lambda_1\Big\}, \\
\gamma_{22}(0) &= \mathrm{Var}(Y_{2t}) \\
&= \frac{1}{(1-\alpha_{22}^2)}\Big\{\alpha_{21}^2\mathrm{Var}(Y_{1t}) + 2\alpha_{22}\alpha_{21}\mathrm{Cov}(Y_{1t}, Y_{2t}) \\
&\quad + \alpha_{22}(1-\alpha_{22})\mu_2 + \alpha_{21}(1-\alpha_{21})\mu_1 + \upsilon_2\lambda_2\Big\}, \\
\gamma_{12}(0) &= \gamma_{21}(0) = \mathrm{Cov}(Y_{1t}, Y_{2t}) \\
&= \frac{\alpha_{11}\alpha_{21}\mathrm{Var}(Y_{1t}) + \alpha_{22}\alpha_{12}\mathrm{Var}(Y_{2t}) + \phi}{1-\alpha_{11}\alpha_{22}-\alpha_{12}\alpha_{21}},
\end{aligned}$$

where ϕ is the covariance between the innovations.

Note that $\mathrm{Cov}(Y_{it}, R_{jt}) = \mathrm{Cov}(R_{it}, R_{jt})$, $i, j = 1, \ldots, r$, $i \neq j$ (Pedeli and Karlis, 2011). That is, the covariance between the current value of one process and the innovations of the other process at time t is equal to the covariance of the innovations of the two series at the same time t.

Regarding the covariance function $\gamma(h) = E\left[(\mathbf{Y}_{t+h} - \mu)(\mathbf{Y}_t - \mu)'\right]$ for $h > 0$, iterative calculations provide us with an expression of the form

$$\gamma(h) = \mathbf{A}\gamma(h-1) = \mathbf{A}^h\gamma(0), \quad h \geq 1, \tag{19.6}$$

where $\gamma(0)$ is given by (19.5).

Applying the well-known Cayley–Hamilton theorem to (19.6), it is straightforward to show that the marginal processes will have an ARMA $(r, r-1)$ correlation structure. Since $\mathbf{A}$ is an $r \times r$ matrix, the Cayley–Hamilton theorem ensures that there exist constants $\xi_1, \ldots, \xi_r$, such that $\mathbf{A}^r - \xi_1\mathbf{A}^{r-1} - \cdots - \xi_r\mathbf{I} = \mathbf{0}$. Thus, $\gamma(h)$ satisfies

$$\gamma(h) - \xi_1\gamma(h-1) - \cdots - \xi_r\gamma(h-r) = 0, \quad h \geq r. \tag{19.7}$$

Equations (19.6) and (19.7) hold for every element in $\gamma(h)$, and hence, the autocorrelation function of $\{Y_{jt}\}, j = 1, \ldots, r$ satisfies

$$\rho_{jj}(h) - \sum_{i=1}^{r}\xi_i\rho_{jj}(h-i) = 0, \quad h \geq r.$$

Thus, each component has an ARMA(r,$r-1$) correlation structure (see also McKenzie, 1988; Dewald et al., 1989). In the simplest case of a BINAR(1) model, the marginal processes have ARMA(2,1) correlations with $\xi_1 = \alpha_{11} + \alpha_{22}$ and $\xi_2 = \alpha_{12}\alpha_{21} - \alpha_{11}\alpha_{22}$. For the diagonal MINAR(p) case, the marginal process is the simple univariate INAR(p) process.

Al-Osh and Alzaid (1987) expressed the marginal distribution of the INAR(1) model in terms of the innovation sequence $\{R_t\}$, that is, $Y_t \stackrel{d}{=} \sum_{i=0}^{\infty} \alpha^i \circ R_{t-i}$. This result was easily extended to the case of a diagonal MINAR(1) process (Pedeli and Karlis, 2013c) where

$$Y_{jt} \stackrel{d}{=} \sum_{i=0}^{\infty} \alpha_j^i \circ R_{j,t-i}.$$

For the general MINAR(1) process, the distribution of such a process can also be expressed in terms of the multivariate innovation sequence $\mathbf{R}_t$ as

$$\mathbf{Y}_t \stackrel{d}{=} \sum_{i=0}^{\infty} \mathbf{A}^i \circ \mathbf{R}_{t-i},$$

where $A^i = \mathbf{P}\mathbf{D}^i\mathbf{P}^{-1}$. Here, $\mathbf{P}$ is the matrix of the eigenvectors of $\mathbf{A}$ and $\mathbf{D}$ is the diagonal matrix of the eigenvalues of $\mathbf{A}$. Since all the eigenvalues should be smaller than 1 in order for stationarity to hold, the matrix $\mathbf{D}^i$ tends to a zero matrix as $i \to \infty$ and hence $\mathbf{A}^i$ tends to zero as well.

The usefulness of such expressions is that they facilitate the derivation of the (joint) probability generating function (pgf) of the (multivariate) process, thus revealing its distribution. Assuming stationarity, the joint pgf $G_{\mathbf{Y}}(\mathbf{s})$ satisfies the difference equation

$$G_{\mathbf{Y}}(\mathbf{s}) = G_{\mathbf{Y}}(\mathbf{A}^T\mathbf{s})G_{\mathbf{R}}(\mathbf{s}).$$

More details can be found in Pedeli and Karlis (2013c).

Extensions of the model mentioned earlier are possible. One can add covariates to the mean of the innovations using a log link function. This allows us to fit the effect of some other covariates to the observed multivariate time series, see Pedeli and Karlis (2013b) for such an application. Also, extensions to higher order are straightforward but lead to rather complicated models.

19.3.3 Estimation

The least squares approach for estimation was discussed in Latour (1997). However, based on parametric assumptions for the innovations, other estimation methods are available. Parametric models also offer more flexibility for predictions.

For the estimation of the BINAR(1) model, the method of conditional maximum likelihood can be used. The conditional density of the BINAR(1) model can be constructed as the convolution of

$$f_1(k) = \sum_{j_1=0}^{k} \binom{y_{1,t-1}}{j_1} \binom{y_{2,t-1}}{k-j_1} \alpha_{11}^{j_1}(1-\alpha_{11})^{y_{1,t-1}-j_1} \alpha_{12}^{k-j_1}(1-\alpha_{12})^{y_{2,t-1}-k+j_1},$$

$$f_2(s) = \sum_{j_2=0}^{s} \binom{y_{2,t-1}}{j_2} \binom{y_{1,t-1}}{s-j_2} \alpha_{22}^{j_2}(1-\alpha_{22})^{y_{2,t-1}-j_2} \alpha_{21}^{s-j_2}(1-\alpha_{21})^{y_{1,t-1}-s+j_2},$$

and a bivariate distribution of the form $f_3(r_1, r_2) = P(R_{1t} = r_1, R_{2t} = r_2)$. The functions $f_1(\cdot)$ and $f_2(\cdot)$ are the *pmfs* of a convolution of two binomial variates. Thus, the conditional density takes the form

$$f(\mathbf{y}_t|\mathbf{y}_{t-1}, \theta) = \sum_{k=0}^{g_1}\sum_{s=0}^{g_2} f_1(k) f_2(s) f_3(y_{1t}-k, y_{2t}-s),$$

where $g_1 = \min(y_{1t}, y_{1,t-1})$ and $g_2 = \min(y_{2t}, y_{2,t-1})$. Maximum likelihood estimates of the vector of unknown parameters θ can be obtained by maximization of the conditional likelihood function

$$L(\theta|\mathbf{y}) = \prod_{t=1}^{T} f(\mathbf{y}_t|\mathbf{y}_{t-1}, \theta) \tag{19.8}$$

for some initial value $\mathbf{y}_0$. The asymptotic normality of the conditional maximum likelihood estimate $\hat{\theta}$ has been shown in Franke and Rao (1995) after imposing a set of regularity conditions and applying the results of Billingsley (1961) for the estimation of Markov processes.

Numerical maximization of (19.8) is straightforward with standard statistical packages. The binomial convolution implies finite summation and hence it is feasible. Note also that since the pgf of a binomial distribution is a polynomial, one can derive the *pmf* of the convolution easily via polynomial multiplication using packages in R. Depending on the choice for the innovation distribution, the conditional maximum likelihood (CML) approach can be applied. In Pedeli and Karlis (2013c), a bivariate Poisson and a bivariate negative binomial distribution were used. For the parametric models prediction was discussed. An interesting result is that for the bivariate Poisson innovations the univariate series have a Hermite marginal distribution. In Karlis and Pedeli (2013), a copula-based bivariate innovation distribution was used allowing negative cross-correlation.

When moving to the multivariate case things become more demanding. First of all, a multivariate discrete distribution is needed for the innovations. As discussed in Section 19.2, such models can be complicated. In Pedeli and Karlis (2013a), a multivariate Poisson distribution is assumed with a diagonal matrix $\mathbf{A}$. Even in this case, the *pmf* of the multivariate Poisson distribution is demanding since multiple summation is needed. The conditional likelihood can be derived as in the bivariate case but now this is a convolution of several binomials and a multivariate discrete distribution. Alternatively, a composite likelihood approach can be used. Composite likelihood methods are based on the idea of constructing lower-dimensional score functions that still contain enough information about the structure considered but they are computationally more tractable (Varin, 2008). See also Davis and Yau (2011) for asymptotic properties of composite likelihood methods applied to linear time series models.

Application of composite likelihood approach implies the usage, for example, of bivariate marginal log-likelihood functions over all pairs instead of the usage of the multivariate likelihood. As an illustration, consider a trivariate probability function $P(x, y, z; \theta)$, where θ is a vector of parameters to estimate. The log-likelihood to be maximized is of the form $\ell(\theta) = \sum_{i=1}^{n} \log P(x_i, y_i, z_i; \theta)$ while the composite log-likelihood is

$$\ell_c(\theta) = \sum_{i=1}^{n} \left[\log P(x_i, y_i; \theta) + \log P(x_i, z_i; \theta) + \log P(y_i, z_i; \theta)\right],$$

that is, we replace the trivariate distribution by the product of the bivariate ones. The price to be paid is some loss of efficiency which can be very large for some models. On the other hand, the optimization problem is usually easier. Simulations for the diagonal trivariate MINAR(1) model have shown small efficiency loss (Pedeli and Karlis, 2013a). Therefore, the composite likelihood method makes feasible the application of some multivariate models for time series and this is worth further exploration. Alternatively, one may employ an expectation-maximization (EM) algorithm making use of the latent structure imposed by the convolution (Pedeli and Karlis, 2013a).

Finally, Bayesian estimation of the BINAR model with bivariate Poisson innovations is described in Sofronas (2012).

19.3.4 Other Models in This Category

Ristic et al. (2012) developed a simple bivariate integer-valued time series model with positively correlated geometric marginals based on the negative binomial thinning mechanism. Bulla et al. (2011) described a model based on another operator called the signed binomial operator allowing fitting integer-valued data in $\mathbb{Z}$. Recently, Scotto et al. (2014) derived a model for correlated binomial data and Nastic et al. (2014) discussed a model with Poisson marginals with same means.

Quoreshi (2006, 2008) described properties of a bivariate moving-average model (BINMA). The BINMA(q_1, q_2) model takes the form

$$\begin{aligned} y_{1t} &= u_{1t} + a_{11} \circ u_{1,t-1} + \cdots + a_{1q_1} u_{1,t-q_1} \\ y_{2t} &= u_{2t} + a_{21} \circ u_{2,t-1} + \cdots + a_{2q_2} u_{2,t-q_2}, \end{aligned}$$

where u_{it} are innovation terms following some positive discrete distribution. Estimation using conditional least squares, feasible least squares, and generalized method of moments is described. Cross-correlation between the series is implied by assuming dependent innovations terms. Extensions to the multivariate case are described in Quoreshi (2008). Similar to the MINAR model, the multivariate vector integer MA model can allow for mixing between the series. In both the bivariate and the multivariate models, no parametric assumptions for the innovations were given. A BINMA model has been studied in Brännäs and Nordström (2000).

19.4 Parameter-Driven Models

Parameter-driven models have also been considered for count time series. The main idea is that the serial correlation imposed is due to some correlated latent processes on the parameter space. They offer useful properties since the serial correlation of the latent process drives the serial correlation properties for the count model. On the other hand, estimation is usually much harder. We describe such multivariate models in the following.

19.4.1 Latent Factor Model

Jung et al. (2011) presented a factor model which in fact belongs to the family of parameter-driven models. The model was applied to the number of trades of five stocks belonging to two different sectors within a 5 min interval for a period of 61 days. The model assumed that the number of trades y_{it} for the ith stock at time t follows a Poisson distribution with mean θ_{it}, while

$$\log \theta_{it} = \mu_i + \gamma\lambda_t + \delta_i \tau_{s_i t} + \phi_i \omega_{it},$$

where μ_i is a mean specific to the ith stock, which perhaps may relate to some covariates specific to the stock as well, λ_t is a latent common market factor, $\tau_t = (\tau_{1t}, \tau_{s_i t})'$ is a latent vector of industry-specific factors and $\omega_t = (\omega_{1t}, \ldots, \omega_{Jt})$ are latent stock-specific factors.

The model assumes an AR(1) specification for the latent factors, namely,

$$\lambda_t \mid \lambda_{t-1} \sim N\left(\kappa_\lambda + \nu_\lambda \lambda_{t-1}, \sigma^2_\lambda\right)$$

$$\tau_{st} \mid \tau_{s,t-1} \sim N\left(\kappa_{\tau_s} + \nu_{\tau_s} \nu_{s,t-1}, \sigma^2_{\tau_s}\right)$$

$$\omega_{it} \mid \omega_{i,t-1} \sim N\left(\kappa_{\omega_i} + \nu_{\omega_i} \omega_{i,t-1}, \sigma^2_{\omega_i}\right).$$

Clearly, cross-correlation enters the model by assuming common factors, while serial correlation from the latent AR processes. The likelihood is complicated, and the authors developed an efficient importance sampling algorithm to evaluate the log-likelihood and apply the simulated likelihood approach.

19.4.2 State Space Model

Jorgensen et al. (1999) proposed a multivariate Poisson state space model with a common factor that can be analyzed by a standard Kalman filter. The model assumes that the multiple counts y_{it}, $i = 1, \ldots, d$ at time t follow conditionally independent Poisson distributions, namely,

$$Y_{it} \sim \text{Poisson}(\alpha_{it}\theta_t)$$

with $\alpha_{it} = \exp(\mathbf{x}_t'\alpha_i)$, where $\mathbf{x}_t$ is a vector of time-varying covariates including a constant and further

$$\theta_t|\theta_{t-1} \sim \text{Gamma}\left(b_t\theta_{t-1}, \frac{\sigma^2}{\theta_{t-1}}\right),$$

where Gamma(a, b^2) is a gamma distribution with mean a and coefficient of variation b. Kalman filtering is used for the latent process.

Finally, Lee et al. (2005) proposed a model starting from a bivariate zero-inflated Poisson regression with covariates, adding random effects that are correlated in time.

19.5 Observation-Driven Models

In observation-driven models, serial correlation comes from the fact that current observations relate directly to the previous ones. The Poisson autoregression model defined in Fokianos et al. (2009) constitutes an important member of this class for univariate series. The model has a feedback mechanism and is defined as

$$Y_t|\mathcal{F}_{t-1} \sim \text{Poisson}(\lambda_t), \;\; \lambda_t = d + a\lambda_{t-1} + bY_{t-1},$$

for $t \geq 1$, where the parameters d, a, b are assumed to be positive. In addition, assume that Y_0 and λ_0 are fixed and $\mathcal{F}_{t-1}$ is the information up to time $t-1$. The model is called INGARCH, but the name perhaps is very ambitious. Properties of the model can be seen in Fokianos et al. (2009). The model has found a lot of work after this (Trostheim, 2012; Davis and Liu, 2015). The Poisson INGARCH is incapable of modeling negative serial dependence in the observations which is, however, possible by the self-excited threshold Poisson autoregression model (see Wang et al., 2014). Extensions to higher-order INGARCH(p,q) (see Weiß, 2009), nonlinear relationships, and other distributional assumptions have been also proposed in the literature. This type of model offers a much richer autocorrelation structure than models based on thinning, like long memory properties, for example. Their estimation is more computational demanding, the same is true for deriving their properties.

Extension to higher dimensions has also been proposed. Liu (2012) proposed a bivariate Poisson integer-valued GARCH (BINGARCH) model. This model is capable of modeling the time dependence between two time series of counts. Consider two time series Y_{1t} and Y_{2t}. We assume that

$$\mathbf{Y}_t = (Y_{1t}, Y_{2t})|\mathcal{F}_{t-1} \sim BP2(\lambda_{1t}, \lambda_{2t}, \phi),$$

where $BP2(\lambda_1, \lambda_2, \phi)$ denotes a bivariate Poisson distribution with marginal means λ_1 and λ_2, respectively, and covariance equal to ϕ. This is a reparameterized version of the distribution in (19.1). Furthermore, we assume for the general BIV. INGARCH(m,q) model that

$$\lambda_t = \delta + \sum_{i=1}^{m} \mathbf{A}_i\lambda_{t-i} + \sum_{j=1}^{q} \mathbf{B}_j\mathbf{Y}_{t-j},$$

$\lambda_t = (\lambda_{1t}, \lambda_{2t})'$, $\delta > 0$ is 2-vector and $\mathbf{A}_i$ and $\mathbf{B}_j$ are 2×2 matrices with nonnegative entries. Conditions to ensure the positivity of λ are given.

For example, the BIV.INGARCH(1,1) model takes the form

$$\lambda_t = \delta + \mathbf{A}\lambda_{t-1} + \mathbf{B}\mathbf{Y}_{t-1}$$

or equivalently

$$\begin{pmatrix} \lambda_{1t} \\ \lambda_{2t} \end{pmatrix} = \begin{pmatrix} \delta_1 \\ \delta_2 \end{pmatrix} + \begin{bmatrix} a_{11} & a_{12} \\ a_{21} & a_{22} \end{bmatrix} \begin{pmatrix} \lambda_{1,t-1} \\ \lambda_{2,t-1} \end{pmatrix} + \begin{bmatrix} b_{11} & b_{12} \\ b_{21} & b_{22} \end{bmatrix} \begin{pmatrix} Y_{1,t-1} \\ Y_{2,t-1} \end{pmatrix}.$$

Stability properties for this model are not simple. Liu (2012) used the iterated random functions approach that allows us to derive the stability properties under a contracting constraint on the coefficient matrices. Inference procedures are also presented and applied to real data in the area of traffic accident analysis.

Heinen and Rengifo (2007) developed a model called the Autoregressive Conditional double Poisson model for a d-dimensional vector of counts $\mathbf{y}_t$. The model relates to the one given earlier. Their model was based on the double Poisson distribution defined in Efron (1986):

$$Y_{it} \mid \mathcal{F}_{t-1} \sim DP(\lambda_{it}, \phi_i), \quad i = 1, \ldots, d,$$

where λ_{it} is the mean of the double Poisson distribution of the ith count at time t and ϕ_i is an overdispersion parameter. For $\phi_i = 1$, we get the Poisson distribution. Then the mean vector $\lambda_t = (\lambda_{1t}, \ldots, \lambda_{dt})$ is defined as a VARMA process

$$\lambda_t = \omega + \mathbf{A}\mu_{t-1} + \mathbf{B}\mathbf{Y}_{t-1},$$

where $\mathbf{A}$ and $\mathbf{B}$ are appropriate matrices. Stationarity is ensured as long as the eigenvalues of $(\mathbf{I} - \mathbf{A} - \mathbf{B})$ lie within the unit circle. The VARMA order specification can be modified. The cross-correlation between the counts is imposed via a Gaussian copula. The model uses a trick to avoid problems when working with discrete-valued data and copulas, by using a continued extension argument, that is, by adding some noise to the counts to make them continuous and working with the continuous versions. The latter adjustment may create some noise around the model. A recent paper (Nikoloulopoulos, 2013a) discusses the problems with such a method.

Bien et al. (2011) used copulas to create a multivariate time series model defined on $\mathbb{Z}$ and not only on positive integers. They modeled a bivariate time series of bid and ask quote changes sampled at a high frequency. The marginal models used were assumed to follow a dynamic integer count hurdle (ICH) process, tied together with a copula, which was constructed by properly taking into account the discreteness of the data.

Finally, Held et al. (2005) described a model with multivariate counts where at time t counts of the other series at time $t-1$ enter as covariates for the mean of each series. Also, Brandt and Sandler (2012) used the model of Chib and Winkelmann (2001) and made it dynamic by adding autoregressive terms in the mean of the mixing distribution.

19.6 More Models

The earlier mentioned list of models does not limit other families of models to be derived. Such an example is the creation of hidden markov models (HMMs) for multivariate count time series. In the univariate case, Poisson HMMs have been used to model integer-valued time series data. Extensions to the multivariate case are possible but multivariate discrete distributions are needed to model the state distribution. In Orfanogiannaki and Karlis (2013), multivariate Poisson distributions based on copulas have been considered for this purpose.

Another approach to create multivariate time series models for counts could be through the discretization of standard continuous models. Consider, for example, the vector autoregressive model based on a bivariate normal distribution. Discretizing the output can lead to the desired time series. However, such a discretization is not unique and perhaps problems may occur while estimating the parameters, as multivariate integrals need to be calculated.

Joe (1996) described an approach to create time series models based on additively closed families. Working with bivariate distributions with this property, as, for example, the bivariate Poisson distribution, one can derive such a time series model. Such models share common elements with models based on thinning operations; see Joe (1996) for a methodology to create an appropriate operator.

19.7 Discussion

In this chapter, we have pulled together the existing literature on multivariate integer-valued time series modeling. Table 19.1 summarizes the models. An obstacle for such models is the lack of, or at least the lack of familiarity with, multivariate discrete distributions which are basic tools for their construction. Given greater availability of such basic tools, we expect that more models will become available in the near future. Also, ideas for tackling estimation problems like the composite likelihood approach can help a lot to this direction.

Such models can have also some other interesting potential. For example, taking the difference of the two time series from a bivariate model, we end up with a time series

TABLE 19.1

Models for multivariate count time series

Type of Model	References
Models based on thinning	Franke and Rao (1995); Latour (1997); Pedeli and Karlis (2011); Pedeli and Karlis (2013a,c); Pedeli and Karlis (2013b); Karlis and Pedeli (2013); Boudreault and Charpentier (2011); Quoreshi (2006, 2008); Ristic et al. (2012); Brännäs and Nordström (2000); Bulla et al. (2011)
Observation-driven models	Liu (2012); Heinen and Rengifo (2007); Bien et al. (2011); Held et al. (2005); Brandt and Sandler (2012)
Parameter-driven models	Jung et al. (2011); Jorgensen et al. (1999); Lee et al. (2005)

defined on $\mathbb{Z}$. Models for time series on $\mathbb{Z}$ are becoming popular in several disciplines and some of them can be derived by higher-dimension models. Multivariate time series in $\mathbb{Z}^d$ would also be of interest. For example, in finance one needs to model the number of ticks that a stock is going up or down during consecutive time points, and this can create a large number of time series, when managing a portfolio.

Acknowledgment

The author would like to thank Dr. Pedeli for helpful comments during the preparation of this chapter.

References

Al-Osh, M. and Alzaid, A. (1987). First–order integer–valued autoregressive process. *Journal of Time Series Analysis*, 8(3):261–275.

Berkhout, P. and Plug, E. (2004). A bivariate Poisson count data model using conditional probabilities. *Statistica Neerlandica*, 58(3):349–364.

Bien, K., Nolte, I., and Pohlmeier, W. (2011). An inflated multivariate integer count hurdle model: An application to bid and ask quote dynamics. *Journal of Applied Econometrics*, 26(4):669–707.

Billingsley, P. (1961). *Statistical Inference for Markov Processes*. University of Chicago Press, Chicago, IL.

Boudreault, M. and Charpentier, A. (2011). Multivariate integer-valued autoregressive models applied to earthquake counts. Available at http://arxiv.org/abs/1112.0929.

Brandt, P. and Sandler, T. (2012). A Bayesian Poisson vector autoregression model. *Political Analysis*, 20(3):292–315.

Brännäs, K. and Nordström, J. (2000). A bivariate integer valued allocation model for guest nights in hotels and cottages. *Umeå Economic Studies, 547*, Department of Economics, Umea University, Umea, Sweden.

Bulla, J., Chesneau, C., and Kachour, M. (2011). A bivariate first-order signed integer-valued autoregressive process. Available at http://hal.archives-ouvertes.fr/hal-00655102.

Chib, S. and Winkelmann, R. (2001). Markov chain Monte Carlo analysis of correlated count data. *Journal of Business & Economic Statistics*, 19(4):428–435.

Davis, R. A. and Liu, H. (2015). Theory and inference for a class of observation-driven models with application to time series of counts. *Statistica Sinica*. doi:10.5705/ss.2014.145t (to appear).

Davis, R. A. and Yau, C. Y. (2011). Comments on pairwise likelihood in time series models. *Statistica Sinica*, 21(1):255–277.

Dewald, L., Lewis, P., and McKenzie, E. (1989). A bivariate first–order autoregressive time series model in exponential variables BEAR(1). *Management Science*, 35(10):1236–1246.

Efron, B. (1986). Double exponential families and their use in generalized linear regression. *Journal of the American Statistical Association*, 81(395):709–721.

Fokianos, K., Rahbek, A., and Tjøstheim, D. (2009). Poisson autoregression. *Journal of the American Statistical Association*, 104:1430–1439.

Frank, M. J. (1979). On the simultaneous associativity of $F(x, y)$ and $x + y - F(x, y)$. *Aequationes Mathematicae*, 19:194–226.

Franke, J. and Rao, T. (1995). Multivariate first–order integer–valued autoregressions. Technical Report, Math. Dep., UMIST.

Genest, C. and Nešlehová, J. (2007). A primer on copulas for count data. *The ASTIN Bulletin*, 37:475–515.

Heinen, A. and Rengifo, E. (2007). Multivariate autoregressive modeling of time series count data using copulas. *Journal of Empirical Finance*, 14(4):564–583.

Held, L., Hohle, M., and Hofmann, M. (2005). A statistical framework for the analysis of multivariate infectious disease surveillance counts. *Statistical Modelling*, 5(3):187–199.

Joe, H. (1996). Time series models with univariate margins in the convolution-closed infinitely divisible class. *Journal of Applied Probability*, 33(3):664–677.

Jorgensen, B., Lundbye-Christensen, S., Song, P.-K., and Sun, L. (1999). A state space model for multivariate longitudinal count data. *Biometrika*, 86(1):169–181.

Jung, R., Liesenfeld, R., and Richard, J. (2011). Dynamic factor models for multivariate count data: An application to stock-market trading activity. *Journal of Business and Economic Statistics*, 29(1):73–85.

Jung, R. and Tremayne, A. (2006). Binomial thinning models for integer time series. *Statistical Modelling*, 6(2):81–96.

Karlis, D. and Meligkotsidou, L. (2005). Multivariate Poisson regression with covariance structure. *Statistics and Computing*, 15(4):255–265.

Karlis, D. and Meligkotsidou, L. (2007). Finite multivariate Poisson mixtures with applications. *Journal of Statistical Planning and Inference*, 137:1942–1960.

Karlis, D. and Pedeli, X. (2013). Flexible bivariate INAR(1) processes using copulas. *Communications in Statistics—Theory and Methods*, 42(4):723–740.

Kocherlakota, S. and Kocherlakota, K. (1992). *Bivariate Discrete Distributions, Statistics: Textbooks and Monographs*, vol. 132. Markel Dekker, New York.

Latour, A. (1997). The multivariate GINAR(p) process. *Advances in Applied Probability*, 29(1): 228–248.

Lee, A. H., Wang, K., Yau, K. K., Carrivick, P. J., and Stevenson, M. R. (2005). Modelling bivariate count series with excess zeros. *Mathematical Biosciences*, 196(2):226–237.

Liu, H. (2012). Some models for time series of counts. PhD thesis, Columbia University, New York.

McKenzie, E. (1985). Some simple models for discrete variate time series. *Water Resources Bulletin*, 21(4):645–650.

McKenzie, E. (1988). Some ARMA models for dependent sequences of Poisson counts. *Advances in Applied Probability*, 20(4):822–835.

McKenzie, E. (2003). *Discrete Variate Time Series*, vol. 21. In C. Rap. and D. Shanbhag, (eds.), *Stochastic Processes: Modelling and Simulation, Handbook of Statistics*, vol. 21, pp. 573606. Elsevier Science B.V., North-Holland, Amsterdam, the Netherlands.

Nastic, A., Ristic, M., and Popovic, P. (2014). Estimation in a bivariate integer-valued autoregressive process. In *Communications in Statistics, Theory and Methods*, page (forthcoming).

Nelsen, R. B. (2006). *An Introduction to Copulas*, 2nd edn. Springer–Verlag, New York.

Nikoloulopoulos, A. (2013a). On the estimation of normal copula discrete regression models using the continuous extension and simulated likelihood. *Journal of Statistical Planning and Inference*, 143(11):1923–1937.

Nikoloulopoulos, A. K. (2013b). Copula-based models for multivariate discrete response data. In *Copulae in Mathematical and Quantitative Finance*, pp. 231–249. Lecture Notes in Statistics, Springer.

Orfanogiannaki, K. and Karlis, D. (2013). Hidden Markov models in modeling time series of earthquakes. In *Proceedings of the 18th European Young Statisticians Meeting*, Osijek, Croatia, 2013, pp. 95–100. Available at http://www.mathos.unios.hr/eysm18/Proceedings_web.pdf.

Pedeli, X. and Karlis, D. (2011). A bivariate INAR(1) process with application. *Statistical Modelling*, 11(4):325–349.

Pedeli, X. and Karlis, D. (2013a). On composite likelihood estimation of a multivariate INAR(1) model. *Journal of Time Series Analysis*, 34(2):206–220.

Pedeli, X. and Karlis, D. (2013b). On estimation of the bivariate Poisson INAR process. *Communications in Statistics: Simulation and Computation*, 42(3):514–533.

Pedeli, X. and Karlis, D. (2013c). Some properties of multivariate INAR(1) processes. *Computational Statistics and Data Analysis*, 67:213–225.

Quoreshi, A. (2006). Bivariate time series modeling of financial count data. *Communications in Statistics—Theory and Methods*, 35(7):1343–1358.

Quoreshi, A. (2008). A vector integer-valued moving average model for high frequency financial count data. *Economics Letters*, 101(3):258–261.

Ristic, M., Nastic, A., Jayakumar, K., and Bakouch, H. (2012). A bivariate INAR(1) time series model with geometric marginals. *Applied Mathematics Letters*, 25(3):481–485.

Scotto, M. G., Weiß, C. H., Silva, M. E., and Pereira, I. (2014). Bivariate binomial autoregressive models. *Journal of Multivariate Analysis*, 125:233–251.

Sklar, M. (1959). Fonctions de répartition à *n* dimensions et leurs marges. l'Institut de Statistique de Paris., 8:229–231.

Sofronas, G. (2012). Bayesian estimation for bivariate integer autoregressive model. Unpublished Master thesis, Department of Statistics, Athens University of Economics, Athens, Greece.

Steutel, F. and van Harn, K. (1979). Discrete analogues of self–decomposability and stability. *The Annals of Probability*, 7:893–899.

Trostheim, D. (2012). Some recent theory for autoregressive count time series. *Test*, 21(3):413–438.

Varin, C. (2008). On composite marginal likelihoods. *Advances in Statistical Analysis*, 92(1):1–28.

Wang, C., Liu, H., Yao, J.-F., Davis, R. A., and Li, W. K. (2014). Self-excited threshold Poisson autoregression. *Journal of the American Statistical Association*, 109:777–787.

Weiß, C. H. (2008). Thinning operations for modeling time series of counts—a survey. *AStA Advances in Statistical Analysis*, 92(3):319–341.

Weiß, C. H. (2009). Modelling time series of counts with overdispersion. *Statistical Methods and Applications*, 18(4):507–519.

20

Dynamic Models for Time Series of Counts with a Marketing Application

Nalini Ravishanker, Rajkumar Venkatesan, and Shan Hu

CONTENTS

20.1 Introduction

In many applications, including marketing, we observe at different times and for different subjects counts of some event of interest. Accurate modeling of such time series of counts (responses) for N subjects over T time periods as functions of relevant covariates (subject-specific and time-varying), and incorporating dependence over time, is becoming increasingly important in several applications. In situations where we observe a vector of counts for each subject at each time, we are also interested in incorporating the association between the components of the count vectors. In this chapter, we describe the modeling of univariate and multivariate time series of counts in the context of a marketing application that involves modeling/predicting product sales.

While count data regression is a widely used applied statistical tool today (Kedem and Fokianos, 2002), models for count time series are less common. The main approaches include a regression-type approach using quasi-likelihood as discussed in Zeger (1988), a Poisson–Gamma mixture modeling approach described by Harvey and Fernandes (1989), the generalized linear autoregressive moving average (GLARMA) model discussed in

Davis et al. (2003), and the dynamic generalized linear model (GLM) discussed in Gamerman (1998) and Landim and Gamerman (2000). We propose to employ hierarchical dynamic models and illustrate on a marketing example. For Gaussian dynamic linear models (DLMs), also often referred to as Gaussian state space models, Kalman (1960) and Kalman and Bucy (1961) popularized a recursive algorithm for optimal estimation and prediction of the state vector, which then enables the prediction of the observation vector; see West (1989) for details. Carlin et al. (1992) described the use of Markov chain Monte Carlo (MCMC) methods for non-Gaussian and nonlinear state space models. Chen et al. (2000) is an excellent reference text for MCMC methods.

Hierarchical dynamic linear models (HDLMs) combine the stratified parametric linear models (Lindley and Smith, 1972) and the DLMs into a general framework, and have been particularly useful in econometric, education, and health care applications (Gamerman and Migon, 1993). The Gaussian HDLM includes a set of one or more dimensions reducing structural equations along with the observation equation and state (or system) equation of the DLM. Landim and Gamerman (2000) further extended the Gaussian HDLM to a more general class of models where the response vector has a matrix-valued normal distribution. For situations where the time series of responses consists of counts, DLMs have been generalized to dynamic generalized linear models (DGLMs) or exponential family state space models, which assume that the sampling distribution is a member of the exponential family of distributions, such as the Poisson or negative binomial distributions. The DGLMs may be viewed as dynamic versions of the GLMs (McCullagh and Nelder, 1989). For univariate time series, Fahrmeir and Kaufmann (1991) discussed Bayesian inference via an extended Kalman filter approach, while Gamerman (1998) described the use of the Metropolis–Hastings algorithm combined with the Gibbs sampler in repeated use of an adjusted version of Gaussian DLM. Applications of state space models of counts include Weinberg et al. (2007) in operations management and Aktekin et al. (2014) in finance, for instance. Wikle and Anderson (2003) described a dynamic zero-inflated Poisson (ZIP) model framework for tornado report counts, incorporating spatial and temporal effects. Gamerman et al. (2015; Chapter 8 in this volume) gives an excellent discussion of Bayesian DGLMs, with illustrations.

In many applications, the response consists of a vector-valued time series of counts, and there is a need to develop statistical modeling approaches for estimation and prediction. Fahrmeir (1992) described posterior inference via extended Kalman filtering for multivariate DGLMs. In this chapter, we describe hierarchical dynamic models for univariate and multivariate count times series. Specifically, we discuss a ZIP sampling distribution for the univariate case and a multivariate Poisson (MVP) sampling distribution for the multivariate case, incorporating covariates that may vary over location and/or time. The use of the MVP distribution enables us to model associations between the components of the count response vector, while the dynamic framework allows us to model the temporal behavior. The hierarchical structure enables us to capture the location (or subject)-specific effects over time. We also propose a multivariate dynamic finite mixture (MDFM) model framework to reduce the dimension of the state parameter and also to include the possibilities of negative correlations between the component of the multivariate time series.

The format of the chapter is as follows. Section 20.2 gives a description of the marketing application, including a description of the data. Section 20.3 describes a dynamic ZIP model for univariate count time series. Section 20.4 first reviews the MVP distribution and finite mixtures of MVP distributions and then describes Bayesian inference for a hierarchical dynamic model fit to multivariate time series of counts, where the coefficients are customer specific and also vary over time. Section 20.5 shows details and modeling results from a parsimonious MDFM model. Section 20.6 provides a brief summary.

20.2 Application to Marketing Actions of a Pharmaceutical Firm

We describe statistical analyses pertaining to marketing data from a large multinational pharmaceutical firm. Analysis of the drivers of new prescriptions written by physicians is of interest to marketing researchers. However, most of the existing research focuses on physician-level sales for a single drug within a category and do not consider the association over time between the sales of a drug and its competitors within a category. We are also interested in the effect of a firm's detailing on the sales of its own drug and competitors and would like to decompose the association in sales among competing drugs between marketing activities of a drug in the category and coincidence induced by general industry trends.

We carry out an empirical analysis using monthly data over a three-year period from a multinational pharmaceutical company, pertaining to physician prescriptions, and sales calls directed toward the physicians. Sales calls denote visits made to physician offices by the firm's representatives. Similar to most other research studies in this context, we treat physicians as customers of the pharmaceutical firm. The *behavioral data* collected monthly by the firm over a period of 3 years consist of the number of new prescriptions from a physician (sales) and the number of sales calls directed toward the physicians (detailing). The number of sales is the primary customer behavior on which we focus in this study. We focus on one of the newer drugs launched by the firm in a large therapeutic drug category (1 of the 10 largest therapeutic categories in the United States) as the *own drug*. The therapeutic category contains more than three major drugs, and the own drug possesses an intermediate market share. Due to confidentiality concerns, we are unable to reveal any other information about the drug category or the pharmaceutical firm. We are interested in modeling patterns in the number of prescriptions written by a physician on the *own drug*, a *leader drug*, and a *challenger drug*. We classify those drugs with the highest market shares in this specific therapeutic category as leaders and the other competing drugs as challengers. The sales calls directed toward the customer by the firm constitute *customer relationship management* (CRM) actions.

Existing research shows that detailing and sampling (giving the physician drug samples) influence new prescriptions from physicians (Mizik and Jacobson, 2004). Montoya et al. (2010) state that after accounting for dynamics in physician prescription writing behavior, detailing seems to be most effective in acquiring new physicians, whereas sampling is most effective in obtaining recurring prescriptions from existing physicians. The database consists of monthly prescription histories on the own drug for 45 continuous months within the last decade from a sample of physicians from the American Medical Association (AMA) database. The time window of our data starts 1 year after introduction of the own drug. Exploratory data analysis shows that while the firm obtains on average three new prescriptions per month from a single physician, and salespeople call on a physician on average about twice a month, there is considerable variation in both the monthly level of sales per physician and the number of sales calls directed toward the physician each month.

During these 45 months, the pharmaceutical firm also collected *attitudinal data*, that is, monthly information on customer (physician) attitudes regarding all the drugs in the therapeutic category and their corresponding salespeople. The sampling frame for the survey was obtained by combining the list of physicians available in the firm database and the AMA database and was expected to cover over 95% of the physicians in the United States. The information obtained in the survey that is relevant to our study include (1) physician ratings (on a seven-point scale) of the salesperson for each drug in terms of overall performance, credibility, knowledge of the disease, and knowledge of medications; (2) physician ratings (on a seven-point scale) of each drug on its overall performance; (3) demographic information such as the physician zip code and specialty; and (4) estimates of the number of times a salesperson from the drug company visited the physician in the last month. While the overall response rate was about 15% of all contacted physicians, the response rate among physicians who prescribed the own drug at least once before the time frame of the data was 35%. These statistics are similar to the response rates obtained in other studies in the pharmaceutical industry (Ahearne et al., 2007). Overall, 6249 physicians had responded at least once to the survey regarding at least one drug in the therapeutic category. An exploratory analysis of variance (ANOVA) analysis excluded the presence of selection bias. Although customer attitudes are known to influence customer reactions to marketing communications of a firm, they are rarely included in models that determine customer value. Exploratory analysis shows that sales calls and attitudes toward the firm correlate positively with sales, and attitudes toward competition correlate negatively with sales of the own drug.

The own drug represents a significantly different chemical formulation and further targets a different function of the human body to cure the disease condition than the drugs available at the time of introduction in the therapeutic category. It is, therefore, reasonable to expect that physicians will learn about the efficacy of the drug over time, resulting in variation (either increase or decrease) in sales and attitudes over time. This expectation is supported by multiple exploratory tests of the customer sales histories. We observe that the average level of sales (across all customers) ranges from about one in the first month to four in the last month. An ANOVA test rejected the null hypothesis that the mean level of sales was the same across the months. The variation in sales over time motivated us to develop a dynamic model framework where the coefficients in the customer-level sales response model could vary across customers and over time. Venkatesan et al. (2014) discussed a dynamic ZIP framework that combines sparse survey–based physician attitude data that are not available at regular intervals, with physician-level transaction and marketing histories that are available at regular time intervals. Univariate (own or competitor) prescription counts are modeled in order to discuss retention and sales; this model is discussed in Section 20.3 of this chapter.

An important step of the marketing research is to jointly model the number of prescriptions of different drugs written by the physicians over time, taking into account possible associations between them. Almost all the current research focuses on physician-level sales for a single drug within a category and does not consider the association between the sales over time of a drug and its competitions within a category. The effect of a firm's detailing on the sales of its own drug and competitors is also of interest. We describe an MDFM model, which provides a useful framework for studying the evolution of sales of a set of competing drugs within a category. Using this parsimonious model for multivariate counts, association in sales among competing drugs may be decomposed between marketing activities of a drug in the category and coincidence induced by general industry trends.

The model employs a mixture of multivariate Poisson distributions to model the vector of counts and induces parsimony by allowing some model coefficients to dynamically evolve over time and others to be subject specific and static over time. A richer and more general model formulation involves a hierarchical setup where the sampling distribution is a mixture of multivariate Poisson and all the coefficients in the model formulation are allowed to be dynamic and subject specific. The general hierarchical multivariate dynamic model (HMDM) framework is outlined in Section 20.4, followed by a description of the MDFM model in Section 20.5.

20.3 Dynamic ZIP Models for Univariate Prescription Counts

Let $Y_{i,t}$ denote the observed new prescription counts of the own drug from physician i in month t, for $i = 1, \ldots, N$ and $t = 1, \ldots, T$, and let $D_{i,t}$ be the level of detailing (sales calls) directed at the ith physician in month t. We assume that $Y_{i,t}$ follows a ZIP model (Lambert, 1992), that is, the ith physician at time t can belong to either of two latent (unobserved) states, the inactive state corresponding to $B_{i,t} = 1$, or the active state with $B_{i,t} = 0$. The states have the interpretation that zero new prescriptions will be observed with high probability from physicians in the inactive state. When the physician is in the active state, the number of new prescriptions can assume values $k = 0, 1, 2, \ldots$. Due to market forces, marketing and other influences, a physician is likely to move from the active to the inactive state and vice versa. In our context, we interpret a physician in the active state as being retained by a firm and a physician in the inactive state to be dormant, and we assume that a physician never quits a relationship, and there is always a finite probability that the physician will return to prescribing the own drug. This formulation is a special case of the hidden Markov model with two states, active and inactive (Netzer et al., 2008; Zucchini and MacDonald, 2010). Let $\lambda_{i,t} > 0$ be the mean prescription count for physician i at time t, and let $0 < \Pi_{i,t} < 1$ denote $P(B_{i,t} = 1)$. The ZIP model formulation is

$$
\begin{aligned}
P(Y_{i,t} = 0 | \lambda_{i,t}, \Pi_{i,t}) &= \Pi_{i,t} + (1 - \Pi_{i,t}) \exp(-\lambda_{i,t}), \\
P(Y_{i,t} = k | \lambda_{i,t}, \Pi_{i,t}) &= (1 - \Pi_{i,t}) \exp(-\lambda_{i,t}) \lambda_{i,t}^{k} / k!, \quad k = 1, 2, \ldots
\end{aligned} \tag{20.1}
$$

The distribution for $Y_{i,t}$ can be written as a mixture distribution, that is, $Y_{i,t} = V_{i,t}(1 - B_{i,t})$, where $B_{i,t} \sim \text{Bernoulli}(\Pi_{i,t})$, $V_{i,t} \sim \text{Poisson}(\lambda_{i,t})$, and $B_{i,t}$ and $V_{i,t}$ are independent, latent physician-specific dynamic parameters that are modeled as functions of observed covariates. It is reasonable to model the natural logarithm of $\lambda_{i,t}$ and the logit of $\Pi_{i,t}$ as

$$
\begin{aligned}
\ln(\lambda_{i,t}) &= \beta_{0,i,t}^{\lambda} + \beta_{1,i,t}^{\lambda} \ln(D_{i,t} + 1) + \beta_{2,i,t}^{\lambda} R_{i,t}, \\
\text{logit}(\Pi_{i,t}) &= \beta_{0,i,t}^{\Pi} + \beta_{1,i,t}^{\Pi} \ln(D_{i,t} + 1) + \beta_{2,i,t}^{\Pi} R_{i,t},
\end{aligned} \tag{20.2}
$$

where $R_{i,t}$ denotes the RFM variable (see later) which is widely used to predict customer response to offers in direct marketing (Blattberg et al., 2008), and captures behavioral loyalty of customers. In general, recency (R) refers to the time since a customer's last purchase (with a ceiling at 3 months), frequency (F) is the number of times a customer made a purchase in the last 3 months, and monetary value (M) is the amount spent in the last 3 months.

RFM is then a weighted average of these variables, and a common rule of thumb for the weights is 60%, 30%, and 10% for R, F, and M respectively. The three component variables (R, F, and M) are calculated at time t for each physician i as moving averages over 3 months prior to time t, wih respective weights 0.6, 0.3, and 0.1. To verify the validity of these weights in our study, we ran a logistic regression on the calibration data, where the response variable is a binary variable taking the value 1 if physician i wrote a new prescription at time t and taking value 0 otherwise, and R, F, and M are predictors in this model. Suppose the estimated regression coefficients are denoted by $\widehat{\theta}_R$, $\widehat{\theta}_F$, and $\widehat{\theta}_M$, the weight for recency (R) was computed as $\widehat{\theta}_R/(\widehat{\theta}_R+\widehat{\theta}_F+\widehat{\theta}_M)$. Weights for F and M may be obtained similarly.

Let $\boldsymbol{\beta}^{\lambda}_{i,t}=\left(\beta^{\lambda}_{0,i,t},\beta^{\lambda}_{1,i,t},\beta^{\lambda}_{2,i,t}\right)'$ and $\boldsymbol{\beta}^{\Pi}_{i,t}=\left(\beta^{\Pi}_{0,i,t},\beta^{\Pi}_{1,i,t},\beta^{\Pi}_{2,i,t}\right)'$, so that $\boldsymbol{\beta}_{i,t}=\left(\boldsymbol{\beta}^{\lambda\prime}_{i,t},\boldsymbol{\beta}^{\Pi\prime}_{i,t}\right)'$ is a $p=6$-dimensional vector. We assume the hierarchical (or structural) equation

$$\boldsymbol{\beta}_{i,t}=\boldsymbol{\gamma}_t+\mathbf{AO}_{i,t}\boldsymbol{\Gamma}_1+\mathbf{AC}_{i,t}\boldsymbol{\Gamma}_2+\mathbf{CD}_{i,t}\boldsymbol{\Gamma}_3+\mathbf{Z}_i\boldsymbol{\Delta}+\mathbf{v}_{i,t}, \tag{20.3}$$

where $\mathbf{AO}_{i,t}=\text{diag}(ao_{i,t},\ldots,ao_{i,t})$ denote attitudes towards the own drug, $\mathbf{AC}_{i,t}=\text{diag}(ac_{i,t},\ldots,ac_{i,t})$ denote attitudes towards the competitive drug, $\mathbf{CD}_{i,t}$ denote the estimates made by physician i of the competitive detailing at time t, $\mathbf{Z}_i$ represents physician demographics, $\mathbf{v}_{i,t}\sim N_p(\mathbf{0},\mathbf{V}_i)$ denote the errors, and $\boldsymbol{\gamma}_t$ is the p-dimensional state vector whose dynamic evolution is described by the state (or system) equation

$$\boldsymbol{\gamma}_t=\mathbf{G}\boldsymbol{\gamma}_{t-1}+\mathbf{w}_t, \tag{20.4}$$

where $\mathbf{G}$ is an identity matrix since a random walk evolution is assumed, and $\mathbf{w}_t\sim N_p(\mathbf{0},\mathbf{W})$ are the state errors. The model structure assumes that customer attitudes form in all time periods, but are observed only when customers respond to the survey. If customer attitudes are observed at time t, they affect the dynamic response coefficients in the hierarchical equation, that is, $\boldsymbol{\beta}_{i,t}$ and thus affect $\boldsymbol{\gamma}_l$ for $\ell=t,t+1,\ldots$ as well. Note that in (20.3), the predictor $R_{i,t}$ may be replaced by $\ln(Y_{i,t-1}+1)$.

Venkatesan et al. (2014) also included a model for handling the endogeneity of sales calls by modeling $D_{i,t}$ as a Poisson distribution conditional on its mean $\eta_{i,t}$, and modeling $\ln(\eta_{i,t})=\zeta_0+\sum_{k=1}^{p}\zeta_k\beta_{i,t,k}$. The ζ coefficients enable us to infer whether the firm considers customer sales potential and responsiveness to sales calls in its detailing plans. In general, endogeneity between sales and sales calls may be handled in two ways. One approach consists of including lagged detailing as well as $D_{i,t}$ in (20.2). We, however, use another approach that accommodates the endogeneity by explicitly modeling the process that generates detailing D_{it}, so that including only $D_{i,t}$ in (20.2), and not its lagged values, is sufficient. Note the similarity to the incidental parameter issue raised by Lancaster (2000).

Fairly standard, conditionally conjugate prior distributions, as usually adopted in HDLMs (Landim and Gamerman, 2000), are assumed: $\pi(\mathbf{V}_i)$ is an inverse-Wishart, $IW(n_v,\mathbf{S}_v)$ and $\pi(\mathbf{W})$ is $IW(n_w,\mathbf{S}_w)$, with $n_v=n_w=2p+1$, and $\mathbf{S}_v=\mathbf{S}_w=(2p+1)\mathbf{I}_{2p+1}$; $\pi(\boldsymbol{\Gamma}_1)$, $\pi(\boldsymbol{\Gamma}_2)$, and $\pi(\boldsymbol{\Gamma}_3)$ are each $MVN(\mathbf{0},100\mathbf{I}_p)$; $\pi(\boldsymbol{\Delta})$ is $MVN(\mathbf{0},100\mathbf{I}_{Kp})$ where K denotes the number of customer demographic predictors; $\pi(\boldsymbol{\zeta})$ is $MVN(\boldsymbol{\zeta}_0,\mathbf{V}_{\zeta})$; and $\pi(\boldsymbol{\gamma}_0)$ is

$MVN(\mathbf{0}, 100\mathbf{I}_p)$. For details on the choice of hyperparameters, see Venkatesan et al. (2014). A Gibbs sampling algorithm is employed to estimate the posterior distribution of the model parameters. The coefficients $\mathbf{\Gamma}_1, \mathbf{\Gamma}_2, \mathbf{\Gamma}_3$, and $\mathbf{\Delta}$ are obtained through suitable multivariate normal draws, the variances are routine draws from inverse Wishart distributions, the Forward-Filtering-Backward-Sampling (FFBS) algorithm enables sampling γ_t (see Carter and Kohn 1994; Fruhwirth-Schnatter 1994), and the Metropolis–Hastings algorithm is used to generate samples from other parameters. Modeling details as well as detailed results and comparisons with several other models are given in Venkatesan et al. (2014). In particular, the deviance information criterion (DIC) was the smallest for the hierarchical dynamic ZIP model that included attitudes in (20.3), followed by the corresponding model without attitudes. The dynamic models performed better than the corresponding static models. The hierarchical dynamic ZIP model also showed the best in-sample and hold-out predictive performance, giving the smallest mean absolute deviation (MAD) both for 1-month-ahead and 12-month-ahead predictions. Physician attitudes, when available, affected $\beta_{i,t}$ and γ_t. Information provided by posterior and predictive distributions from convergent MCMC samples for the model parameters of the hierarchical dynamic ZIP model enables the firm to make decisions about customer selection and resource allocation by analyzing the customer lifetime value (CLV) metric. CLV was computed over 35 months, because the firm revealed that it did not plan its sales force allocations over 3 years ahead, and is

$$CLV_i = \sum_{i=T^*+1}^{T^*+36} \frac{(1-\widehat{\Pi}_{i,t})\widehat{Y}_{i,t} - c_{i,t}\widehat{D}_{i,t}}{(1+d^*)^{t-T^*}}, \tag{20.5}$$

where $T^* = 10$, d^* is the discount coefficient, $c_{i,t}$ is the unit cost of a sales call, and $\widehat{Y}_{i,t}$ and $\widehat{D}_{i,t}$ denote the predicted means of the sales and detailing, respectively.

Ongoing collection of physician attitudes via surveys requires an annual investment of over $1 million from the firm, which would wish to evaluate whether the financial returns from collecting and using these attitudes in modeling exceeds the investment. Venkatesan et al. (2014) used customer selection and customer-level resource allocation based on a hold-out sample of 1000 physicians. The objective of the customer selection process is to identify the physicians who would be profitable in the future so that they can be prioritized for targeting. Physician-level sales and retention were predicted from months 10 to 45, and these predictions were used to compute the physician's CLV using (20.5). Missing attitudes in the hold-out sample were imputed using an ordered probit model (Albert and Chib, 1993).

Predictive results from a hierarchical dynamic ZIP model that includes physican attitude information in (20.3) were compared to results from a model that does not include data on attitudes, in order to quantify the implications to the firm and discuss selection of profitable physicians. Physicians can be classified into quintiles based on the actual CLV, the CLV predicted from the hierarchical dynamic ZIP model that includes customer attitudes, and the CLV predicted from a hierarchical dynamic ZIP model that did not include customer attitudes. The incremental profit from including customer attitudes was equivalent to 0.93% of the total CLV obtained from physicians identified to be in the top quintile based on their observed profits. This implies that if the firm was targeting the top quintile of its customer base, the returns from including customer attitudes to select the most likely physicans to target will be 0.93% higher than not including customer attitudes. Similarly, the returns from including customer attitudes would be higher by 3.57%, 29.62%, 79.33%, and 24.12% relative to not including customer attitudes, if the firm targets the second, third, fourth, and the fifth quintiles, respectively. The incremental profits from including attitudes were highest for the mid-tier groups, that is, third and fourth quintiles.

20.4 Hierarchical Multivariate Dynamic Models for Prescription Counts

Let $\boldsymbol{Y}_{it} = (Y_{1,it}, \ldots, Y_{m,it})$, for $t = 1, \ldots, T$, denote the m-dimensional time series of new prescription counts from physician i, where $i = 1, \ldots, N$. The components of the vector correspond to counts of the firm's own drug and the competing drugs. We propose a finite mixture of multivariate Poisson distributions as a sampling distribution of the m-dimensional vector, which allows negative as well as positive associations between counts of the own drug and the competing drugs. We start with a review of mixtures of multivariate Poisson distributions in Section 20.4.1 and then show a general hierarchical dynamic modeling framework in Section 20.4.2.

20.4.1 Finite Mixtures of Multivariate Poisson Distributions

Following Mahamunulu (1967) and Johnson et al. (1997), the definition of an m-variate Poisson distribution for a random vector of counts $\mathbf{Y}$ is based on a mapping $g : \mathbb{N}^q \rightarrow \mathbb{N}^m$, $q \geq m$, such that $\boldsymbol{Y} = g(\boldsymbol{X}) = \mathbf{AX}$. Here, $\boldsymbol{X} = (X_1, \ldots, X_q)'$ is a vector of unobserved independent Poisson random variables, that is, $X_r \sim \text{Poisson}(\lambda_r)$ for $r = 1, \ldots, q$; and $\mathbf{A}$ is an arbitrary $m \times q$ matrix which determines the properties of the multivariate Poisson distribution. The m-dimensional vector $\boldsymbol{Y} = (Y_1, \ldots, Y_m)' = \mathbf{AX}$ follows a multivariate Poisson distribution with parameters $\boldsymbol{\lambda} = (\lambda_1, \ldots, \lambda_q)'$ and pmf $MP_m(\boldsymbol{y}|\boldsymbol{\lambda})$ given by

$$P(\boldsymbol{Y} = \boldsymbol{y}|\boldsymbol{\lambda}) = \sum_{\boldsymbol{x} \in g^{-1}(\boldsymbol{y})} P(\boldsymbol{X} = \boldsymbol{x}) = \sum_{\boldsymbol{x} \in g^{-1}(\boldsymbol{y})} \prod_{r=1}^{q} P(X_r = x_r|\lambda_r), \tag{20.6}$$

where $g^{-1}(\boldsymbol{Y})$ denotes the inverse image of $\boldsymbol{Y} \in \mathbb{N}^m$ and for $r = 1, \ldots, q$, the pmf of the univariate Poisson distribution is $P(X_r = x_r|\lambda_r) = \exp(-\lambda_r)\lambda_r^{x_r}/x_r!$. The mean vector and variance–covariance matrix of $\boldsymbol{Y}$ conditional on $\boldsymbol{\lambda}$ are given by

$$E(\boldsymbol{Y}|\boldsymbol{\lambda}) = \boldsymbol{A\lambda}; \quad \text{Cov}(\boldsymbol{Y}|\boldsymbol{\lambda}) = \boldsymbol{A\Sigma A}', \tag{20.7}$$

where $\boldsymbol{\Sigma} = \text{diag}(\lambda_1, \ldots, \lambda_q)$. When $m = 1$, $MP_m(\boldsymbol{y}|\boldsymbol{\lambda})$ in (20.6) reduces to the univariate Poisson pmf $P(Y = y|\lambda) = \exp(-\lambda)\lambda^y/y!$. Use of the multivariate Poisson distribution for modeling applications has been sparse, possibly due to the complicated form of the pmf (20.6) which does not lend itself to easy computation.

Karlis and Meligkotsidou (2005) proposed a two-way covariance structured multivariate Poisson distribution, which permits more realistic modeling of multivariate counts in practical applications. This distribution is constructed by setting $\boldsymbol{A} = [\boldsymbol{A}_1\ \boldsymbol{A}_2]$, where

$A_1 = I_m$ captures the main effects; A_2 captures the two-way covariance effects; A_2 is an $m \times [m(m-1)]/2$ binary matrix; each column of A_2 has exactly two ones and $(m-2)$ zeros and no duplicate columns exist; and $q = m + [m(m-1)]/2$. Correspondingly, split the parameter $\boldsymbol{\lambda}$ into two parts, that is, $\boldsymbol{\lambda}^{(1)} = (\lambda_1, \ldots, \lambda_m)'$, which corresponds to the m main effects, and $\boldsymbol{\lambda}^{(2)} = (\lambda_{m+1}, \ldots, \lambda_q)'$ which corresponds to the $m(m-1)/2$ pairwise covariance effects. When $m = 2, q = 3$, let $\boldsymbol{Y} = (Y_1, Y_2)'$, and let $Y_1 = X_1 + X_3$ and $Y_2 = X_2 + X_3$, where $X_i \sim \text{Poisson}(\lambda_i)$, $i = 1, 2, 3$. The two-way covariance structured bivariate Poisson pmf is

$$MP_2(\mathbf{y}|\boldsymbol{\lambda}) = \exp\{-(\lambda_1 + \lambda_2 + \lambda_3)\}\frac{\lambda_1^{y_1}}{y_1!}\frac{\lambda_2^{y_2}}{y_2!}\sum_{i=0}^{s}\binom{y_1}{i}\binom{y_2}{i}i!\left(\frac{\lambda_3}{\lambda_1\lambda_2}\right)^i, \tag{20.8}$$

where $s = \min(y_1, y_2)$. When $m = 3, q = 6$, let $\boldsymbol{Y} = (Y_1, Y_2, Y_3)'$, and let $Y_1 = X_1 + X_4 + X_5$, $Y_2 = X_2 + X_4 + X_6$, and $Y_3 = X_3 + X_5 + X_6$, where $X_i \sim \text{Poisson}(\lambda_i)$ for $i = 1, \ldots, 6$. The two-way covariance structured trivariate Poisson pmf is

$$MP_3(\mathbf{y}|\boldsymbol{\lambda}) = \exp\left\{-\sum_{i=1}^{6}\lambda_i\right\}\sum_{(X_4, X_5, X_6)\in C}\frac{\lambda_1^{y_1 - X_4 - X_5}\lambda_2^{y_2 - X_4 - X_6}}{(y_1 - X_4 - X_5)!(y_2 - X_4 - X_6)!} \times \frac{\lambda_3^{y_3 - X_5 - X_6}\lambda_4^{X_4}\lambda_5^{X_5}\lambda_6^{X_6}}{(y_3 - X_5 - X_6)!X_4!X_5!X_6!}, \tag{20.9}$$

where the summation is over the set C such that $C = [(X_4, X_5, X_6) \in \mathbb{N}^3 : (X_4 + X_5 \leq y_1) \cap (X_4 + X_6 \leq y_2) \cap (X_5 + X_6 \leq y_3)] \neq \emptyset]$. For $m = 2$ and $m = 3$, the matrix $\mathbf{A}$ has the respective forms

$$\begin{pmatrix} 1 & 0 & 1 \\ 0 & 1 & 1 \end{pmatrix} \quad \text{and} \quad \begin{pmatrix} 1 & 0 & 0 & 1 & 1 & 0 \\ 0 & 1 & 0 & 1 & 0 & 1 \\ 0 & 0 & 1 & 0 & 1 & 1 \end{pmatrix}.$$

Under this structure, the variance–covariance matrix of $\boldsymbol{Y}$ given in (20.7) does not accommodate negative associations among the components of $\boldsymbol{Y}$ (Karlis and Meligkotsidou, 2005).

We proposed an approach for calculating the multivariate Poisson pmf which is faster than the recursive scheme proposed by Tsiamyrtzis and Karlis (2004). When $m = 2$, let y_1 and y_2 denote the observed counts, and without loss of generality, assume that $y_1 \leq y_2$, so that $\min(y_1, y_2) = y_1$. Since X_3 is the common term in the definitions of Y_1 and Y_2, it is straightforward to obtain the set of possible values that X_3 can assume, that is, $x_3 = 0, \ldots, \min(y_1, y_2)$, and obtain the corresponding values assumed by X_1 and X_2 to be, respectively, $X_1 = y_1 - x_3$ and $X_2 = y_2 - x_3$. We have solved for all possible sets of values for the inverse image of $\boldsymbol{y}$, that is, $\boldsymbol{x} \in g^{-1}(\boldsymbol{y})$. The pmf for the bivariate Poisson distribution can be calculated using (20.8). When $m = 3$, without loss of generality, we assume that $y_1 \leq y_2 \leq y_3$. The possible values for x_4 and x_5 are in the set $C_1 = (0, \ldots, y_1)$, and the possible values for x_6 are in the set $C_2 = (0, \ldots, y_2)$. We have in total L different combinations for (x_4, x_5, x_6), where $L = (\text{length of set } C_1)^2 \times (\text{length of set } C_2) = (y_1 + 1)^2(y_2 + 1)$. The corresponding values for X_1, X_2, X_3 can be calculated from (20.9). Let C^* denote the set of L different combinations of possible values for all $q = 6$ independent Poisson variables.

Since it is possible that in the set C^*, X_1, X_2, or X_3 may assume negative values, a subset of C^* which only contains nonnegative values of X_1, X_2, and X_3 is the inverse image of y. The pmf of the trivariate Poisson distribution is then obtained using (20.9). Computing times for evaluating the multivariate Poisson pmfs is discussed in Hu (2012).

Karlis and Meligkotsidou (2007) proposed finite mixtures of multivariate Poisson distributions, which allow for overdispersion in both the marginal distributions and negative correlations, and thus offer a wide range of models for real data applications. The pmf of a finite mixture of H multivariate Poisson distributions with mixing proportions $\pi_1, \ldots, \pi_H$ is given by

$$p(\mathbf{y}|\mathbf{\Phi}) = \sum_{h=1}^{H} \pi_h MP_m(\mathbf{y}|\boldsymbol{\lambda}_h),$$

where $\mathbf{\Phi}$ denotes the set of parameters $(\boldsymbol{\lambda}_1, \ldots, \boldsymbol{\lambda}_H, \pi_1, \ldots, \pi_{H-1})$. The expectation and covariance of $\mathbf{Y}$ conditional on $\boldsymbol{\lambda}$ are

$$E(\mathbf{Y}|\boldsymbol{\lambda}) = \sum_{h=1}^{H} \pi_h \mathbf{A}\boldsymbol{\lambda}_h; \quad \mathrm{Cov}(\mathbf{Y}) = \mathbf{A}\left[\sum_{h=1}^{H} \pi_h(\mathbf{\Sigma}_h + \boldsymbol{\lambda}_h\boldsymbol{\lambda}_h') - \left(\sum_{h=1}^{H} \pi_h\boldsymbol{\lambda}_h\right)\left(\sum_{h=1}^{H} \pi_h\boldsymbol{\lambda}_h\right)'\right]\mathbf{A}',$$

where $\mathbf{\Sigma}_h = \mathrm{diag}(\lambda_{1,h}, \ldots, \lambda_{q,h})$.

20.4.2 HMDM Model Description

A general framework for an HMDM allowing only for positive associations between components of $\mathbf{Y}_{i,t}$ is discussed in Ravishanker et al. (2014), by assuming a multivariate Poisson sampling distribution. Here, we extend this general formulation to a mixture of multivariate Poisson sampling distribution. The observation equation and a model for the latent process $\lambda_{j,i,t,h}$ of the extended HMDM are given in the following:

$$p(\mathbf{y}_{i,t}|\boldsymbol{\lambda}_{i,t,h}) = \sum_{h=1}^{H} \pi_h MP_m(\mathbf{y}_{i,t}|\boldsymbol{\lambda}_{i,t,h}),$$

$$\ln \lambda_{j,i,t,h} = \boldsymbol{B}_{j,i,t}'\boldsymbol{\delta}_{j,i,t,h} + \boldsymbol{S}_{j,i,t}'\boldsymbol{\eta}_{j,h}, \quad j = 1, \ldots, q, \tag{20.10}$$

where $\boldsymbol{B}_{j,i,t} = (B_{j,i,t,1}, \ldots, B_{j,i,t,a_j})'$ is an a_j-dimensional vector of exogenous predictors with location-time-varying (dynamic) coefficients $\boldsymbol{\delta}_{j,i,t,h} = (\delta_{j,i,t,h,1}, \ldots, \delta_{j,i,t,h,a_j})$ and $\boldsymbol{S}_{j,i,t} = (S_{j,i,t,1}, \ldots, S_{j,i,t,b_j})'$ is a b_j-dimensional vector of exogenous predictors with static coefficients $\boldsymbol{\eta}_{j,h} = (\eta_{j,h,1}, \ldots, \eta_{j,h,b_j})'$. We assume that the model either includes $\delta_{j,i,t,h,1}$ which represents the location-time-varying intercept, or includes $\eta_{j,h,1}$ which represents the static intercept, that is, either $D_{j,i,t,1} = 1$ or $S_{j,i,t,1} = 1$. A simple formulation of (20.10) could set $a_j = 1$ for $j = 1, \ldots, q$, set $b_j = b > 1$ for $j = 1, \ldots, m$, and $b_j = 0$ for $j = m+1, \ldots, q$, which implies using only the location-specific and time-dependent intercept to model the Poisson means corresponding to the association portion, and the location–time intercept together with an equal number of static coefficients for exogenous predictors corresponding to the main effects portion of the multivariate Poisson specification.

Let $p_d = \sum_{j=1}^{q} a_j$ and $p_s = \sum_{j=1}^{q} b_j$. Let $\boldsymbol{\beta}_{i,t}$ be a p_d-dimensional vector constructed by stacking the a_j coefficients $\delta_{j,i,t}$ for $j = 1, \ldots, q$. The structural equation relates the location-time-varying parameter $\boldsymbol{\beta}_{i,t}$ to an aggregate (pooled) state parameter $\boldsymbol{\gamma}_t$

$$\boldsymbol{\beta}_{i,t} = \boldsymbol{\gamma}_t + \mathbf{A}_{i,t}\boldsymbol{\Gamma}_1 + \mathbf{Z}_i\boldsymbol{\Gamma}_2 + \boldsymbol{v}_{i,t}, \tag{20.11}$$

where the errors $\boldsymbol{v}_{i,t}$ are assumed to be i.i.d. $N_{p_d}(\mathbf{0}, \boldsymbol{V}_i)$, and $\mathbf{A}_{i,t}$, $\mathbf{Z}_i$, $\boldsymbol{\Gamma}_1$, and $\boldsymbol{\Gamma}_2$ were defined below (20.3). The state (or system) equation of the HMDM is

$$\boldsymbol{\gamma}_t = \boldsymbol{G}\boldsymbol{\gamma}_{t-1} + \boldsymbol{w}_t, \tag{20.12}$$

where $\boldsymbol{G}$ is the p_d-dimensional state transition matrix and the state errors $\boldsymbol{w}_t$ are assumed to be i.i.d. $N_{p_d}(\mathbf{0}, \boldsymbol{W})$. If $H = 1$ and $m = 1$, the HMDM in (20.10)–(20.12) simplifies to the hierarchical DGLM (HDGLM), with the univariate Poisson pmf as the sampling distribution.

20.4.3 Bayesian Inference for the HMDM Model

Let $\mathbf{Y}$, $\mathbf{B}$, and $\mathbf{S}$, respectively, denote the responses y_{it}, the dynamic predictors, and the static predictors, for $t = 1, \ldots, T$ and $i = 1, \ldots, N$. Let $\boldsymbol{\eta}$ and $\boldsymbol{\beta}$ denote all the coefficients $\boldsymbol{\eta}_j$ and $\boldsymbol{\beta}_{it}$ for $j = 1, \ldots, q$, $t = 1, \ldots, T$ and $i = 1, \ldots, N$, and let $\boldsymbol{\gamma}$ denotes all the coefficients $\boldsymbol{\gamma}_t$ for $t = 1, \ldots, T$. The likelihood function under the model described by (20.10)–(20.12) is

$$L(\boldsymbol{\eta}, \boldsymbol{\beta}, \boldsymbol{\gamma}; \mathbf{Y}, \mathbf{D}, \mathbf{S}) = \prod_{i=1}^{N}\prod_{t=1}^{T} MP_m(\boldsymbol{y}_{i,t}|\boldsymbol{\beta}_{i,t}, \boldsymbol{\gamma}_t) p_{\text{normal}}(\boldsymbol{\eta}) \times p_{\text{normal}}(\boldsymbol{\beta}_{i,t}|\boldsymbol{\gamma}_t) \times p_{\text{normal}}(\boldsymbol{\gamma}_t|\boldsymbol{\gamma}_{t-1}),$$

where we have suppressed the terms $\mathbf{B}$ and $\mathbf{S}$ on the right side for brevity. We assume multivariate normal priors for the initial state vector and the static coefficients, that is, $\boldsymbol{\gamma}_0 \sim N_{p_d}(\boldsymbol{m}_0, \boldsymbol{C}_0)$ and $\boldsymbol{\eta} \sim N_{p_s}(\boldsymbol{\mu}_\eta, \boldsymbol{\Sigma}_\eta)$. We assume inverse Wishart priors for the variance terms $\boldsymbol{V}_i$ and $\boldsymbol{W}$, that is, $\boldsymbol{V}_i \sim IW(n_v, \boldsymbol{S}_v)$ and $\boldsymbol{W} \sim IW(n_w, \boldsymbol{S}_w)$. We assume a product prior specification, and the hyperparameters are selected to correspond to a vague prior specification.

The joint posterior of the unknown parameters is proportional to the product of the likelihood and the prior

$$\begin{aligned}
&\pi(\boldsymbol{\beta}_{it}, \boldsymbol{\gamma}_t, \boldsymbol{\eta}, \boldsymbol{V}_i, \boldsymbol{W}|\mathbf{Y}, \mathbf{D}, \mathbf{S}) \\
&\propto \left[\prod_{t=1}^{T}\prod_{i=1}^{N} MP_m(\boldsymbol{y}_{i,t}|\boldsymbol{\lambda}_{i,t})|\boldsymbol{V}_i|^{-1/2} \exp\left\{-\frac{1}{2}(\boldsymbol{\beta}_{i,t} - \boldsymbol{\gamma}_t)'\boldsymbol{V}_i^{-1}(\boldsymbol{\beta}_{i,t} - \boldsymbol{\gamma}_t)\right\}\right] \\
&\quad \times |\boldsymbol{\Sigma}_\eta|^{-1/2} \exp\left\{-\frac{1}{2}(\boldsymbol{\eta} - \boldsymbol{\mu}_\eta)'\boldsymbol{\Sigma}_\eta^{-1}(\boldsymbol{\eta} - \boldsymbol{\mu}_\eta)\right\}
\end{aligned}$$

$$\times \left[\prod_{t=1}^{T} |W|^{-1/2} \exp\left\{ -\frac{1}{2}(\gamma_t - G\gamma_{t-1})' W^{-1} (\gamma_t - G\gamma_{t-1}) \right\} \right]$$
$$\times |W|^{-n_w/2} \exp\left\{ -\frac{1}{2} tr(W^{-1} S_w) \right\}$$
$$\times |C_0|^{-1/2} \exp\left\{ -\frac{1}{2}(\gamma_0 - m_0)' C_0^{-1} (\gamma_0 - m_0) \right\} \left[\prod_{i=1}^{N} |V_i|^{-n_v/2} \exp\left\{ -\frac{1}{2} tr(V_i^{-1} S_v) \right\} \right].$$

The Gibbs sampler proceeds by sequentially sampling from the complete conditional distributions of the parameters, which are proportional to the joint posterior. Details on these distributions are provided in the Appendix. We use the FFBS algorithm described in Carter and Kohn (1994) and Fruhwirth-Schnatter (1994) to generate a random sample from the complete conditional distribution of γ_t, for $t = 1, \ldots, T$. We make inverse Wishart draws for V_i and W, Dirichlet draws for the vector of mixing proportions, and a Metropolis–Hastings algorithm is used for sampling $\beta_{i,t}$ and η. For details, and an ecology illustration, see Ravishanker et al. (2014).

For the marketing data on prescription counts, we explore a more parsimonious model, where some coefficients have a dynamic evolution over time but are not physician specific, while other coefficients are static over time, but are physician specific. We refer to this as the MDFM model and describe it in the next section.

20.5 Multivariate Dynamic Finite Mixture Model

For parsimony, we fit the MDFM model to the prescriptions data.

20.5.1 MDFM Model Description

We study patterns in the dynamic evolution of sales of the own drug along with a challenger drug and a leader drug. Here, $m = 3$, $q = 6$, $\mathbf{y}_{i,t} = (y_{i,t,1}, y_{i,t,2}, y_{i,t,3})'$, and we assume a mixture of H trivariate Poisson distributions with two-way covariate structure as the sampling distribution. The observation equation of the MDFM model is

$$p(\mathbf{y}_{i,t}|\boldsymbol{\lambda}_{i,t,h}) = \sum_{h=1}^{H} \pi_h MP_3(\mathbf{y}_{i,t}|\boldsymbol{\lambda}_{i,t,h}),$$

where $\boldsymbol{\lambda}_{i,t,h} = (\lambda_{i,t,1,h}, \ldots, \lambda_{i,t,q,h})'$, for $h = 1, \ldots, H$ and $MP_3(.)$ was defined in (20.9). For $h = 1, \ldots, H$, the underlying independent Poisson means are modeled as

$$\ln(\lambda_{i,t,k,h}) = \beta_{0,t,k,h} + \beta_{1,i,k,h} \ln(D_{i,t} + 1) + \beta_{2,i,k,h} \ln(y_{i,(t-1),k} + 1) \quad \text{for } k = 1, \ldots, m,$$
$$\ln(\lambda_{i,t,\ell,h}) = \beta_{0,t,\ell,h} \quad \text{for } \ell = m+1, \ldots, q.$$

Note that the q intercepts are not physician specific, but are allowed to evolve dynamically via the system equation. The partial regression coefficients corresponding to detailing and lagged prescription counts are not time evolving, although they are physician specific. This is a parsimonious model that we fit to the vector of prescription counts.

Let $\boldsymbol{\beta}_{0,t,h} = (\beta_{0,t,1,h}, \ldots, \beta_{0,t,q,h})'$. In the system equation, we assume a random walk evolution of the state parameter vector, so that for $h = 1, \ldots, H$,

$$\boldsymbol{\beta}_{0,t,h} = \mathbf{G}\boldsymbol{\beta}_{0,t-1,h} + \mathbf{w}_{t,h},$$

where $\mathbf{G}$ is an identity matrix, and $\mathbf{w}_{t,h} \sim N_p(\mathbf{0}, \mathbf{W}_h)$.

20.5.2 Bayesian Inference for the MDFM Model

We write the mixture model in terms of missing (or incomplete) data; see Dempster et al. (1977) and Diebolt and Robert (1994). For $i = 1, \ldots, N$ and $t = 1, \ldots, T$, recall that $p(\mathbf{y}_{i,t}|\boldsymbol{\lambda}_{i,t,h}) = \sum_{h=1}^{H} \pi_h MP_3(\mathbf{y}_{i,t}|\boldsymbol{\lambda}_{i,t,h})$. Let $\mathbf{z}_{i,t} = (z_{i,t,1}, \ldots, z_{i,t,H})'$ be an H-dimensional vector indicating the component to which $\mathbf{y}_{i,t}$ belongs, so that $z_{i,t,h} \in \{0, 1\}$ and $\sum_{h=1}^{H} z_{i,t,h} = 1$. The pmf of the complete data $(\mathbf{y}_{i,t}, \mathbf{z}_{i,t})$ is

$$f(\mathbf{y}_{i,t}, \mathbf{z}_{i,t}|\boldsymbol{\lambda}_{i,t,h}) = \prod_{h=1}^{H} \pi_h^{z_{i,t,h}} [p(\mathbf{y}_{i,t}|\boldsymbol{\lambda}_{i,t,h})]^{z_{i,t,h}}. \tag{20.13}$$

For $h = 1, \ldots, H$, we make the following prior assumptions. We assume that the p-dimensional initial state parameter vector $\boldsymbol{\beta}_{0,1,h} \sim MVN(\mathbf{a}_{0,h}, \mathbf{R}_{0,h})$, the m-dimensional subject-specific coefficient $\boldsymbol{\beta}_{1,i,h} \sim MVN(\mathbf{a}_{1,i,h}, \mathbf{R}_{1,i,h})$, and the m-dimensional parameter $\boldsymbol{\beta}_{2,i,h} \sim MVN(\mathbf{a}_{2,i,h}, \mathbf{R}_{2,i,h})$. We assume inverse Wishart priors, $IW(n_h, \mathbf{S}_h)$, for the variance terms $\mathbf{W}_h$, and assume the conjugate Dirichlet$(d_1, \ldots, d_H)$ prior for $\boldsymbol{\pi} = (\pi_1, \ldots, \pi_H)$, which simplifies to the Beta(d_1, d_2) prior for π_1 when $H = 2$.

The joint posterior density of the unknown parameters is proportional to the product of the complete data likelihood and the product of the priors discussed earlier. We denote the set of unknown parameters by $\boldsymbol{\Theta} = (\boldsymbol{\beta}_0, \boldsymbol{\beta}_1, \boldsymbol{\beta}_2, \mathbf{W}, \boldsymbol{\pi})$. The Gibbs sampler enables posterior inference by drawing samples using suitable techniques such as direct draws when the complete conditionals have known forms and sampling algorithms such as the Metropolis–Hastings algorithm otherwise. The complete conditional densities are proportional to the joint posterior and are shown in the Appendix, along with details of the sampling algorithms.

We fit the MDFM model with $H = 2$. The posterior mean and standard deviation of the mixing proportion π_1 are, respectively, 0.758 and 0.008, and the 95% highest posterior density (HPD) interval is (0.742, 0.770). Table 20.1 shows the posterior summary of the state variances $\mathbf{W}_h$ for $h = 1, 2$.

We calculate the independent Poisson means $\lambda_{i,t,h,1}, \ldots, \lambda_{i,t,h,q}$ for $h = 1, 2$ and $q = 6$. Through the unconditional expectation formula for the finite mixture of multivariate Poisson distributions, we obtain the predicted means of the own drug and the competing drugs. We then make Poisson draws with the corresponding predicted means. Figure 20.1 shows time series plots of the observed prescription counts for the three drugs from four randomly selected physicians, and Figure 20.2 shows time series plots of the corresponding predicted means. Figure 20.3 shows one-month-ahead predictions for all physicians. The absolute differences between observed counts and predicted counts are 1.869, 1.980, and 2.996, respectively, for the own, challenger, and leader drugs.

TABLE 20.1

Posterior summaries for $\mathbf{W}_h$

	Posterior Mean	Posterior SD	95 % HPD Interval
$W_{1,1,1}$	0.0086	0.0011	(0.0067, 0.0111)
$W_{1,2,2}$	0.0089	0.0012	(0.0067, 0.0109)
$W_{1,3,3}$	0.0085	0.0011	(0.0065, 0.0105)
$W_{1,4,4}$	0.0087	0.0011	(0.0069, 0.0111)
$W_{1,5,5}$	0.0090	0.0013	(0.0068, 0.0115)
$W_{1,6,6}$	0.0090	0.0013	(0.0068, 0.0116)
$W_{2,1,1}$	0.0089	0.0014	(0.0066, 0.0121)
$W_{2,2,2}$	0.0088	0.0012	(0.0068, 0.0112)
$W_{2,3,3}$	0.0088	0.0012	(0.0070, 0.0113)
$W_{2,4,4}$	0.0085	0.0011	(0.0066, 0.0105)
$W_{2,5,5}$	0.0087	0.0011	(0.0066, 0.0112)
$W_{2,6,6}$	0.0092	0.0013	(0.0068, 0.0117)

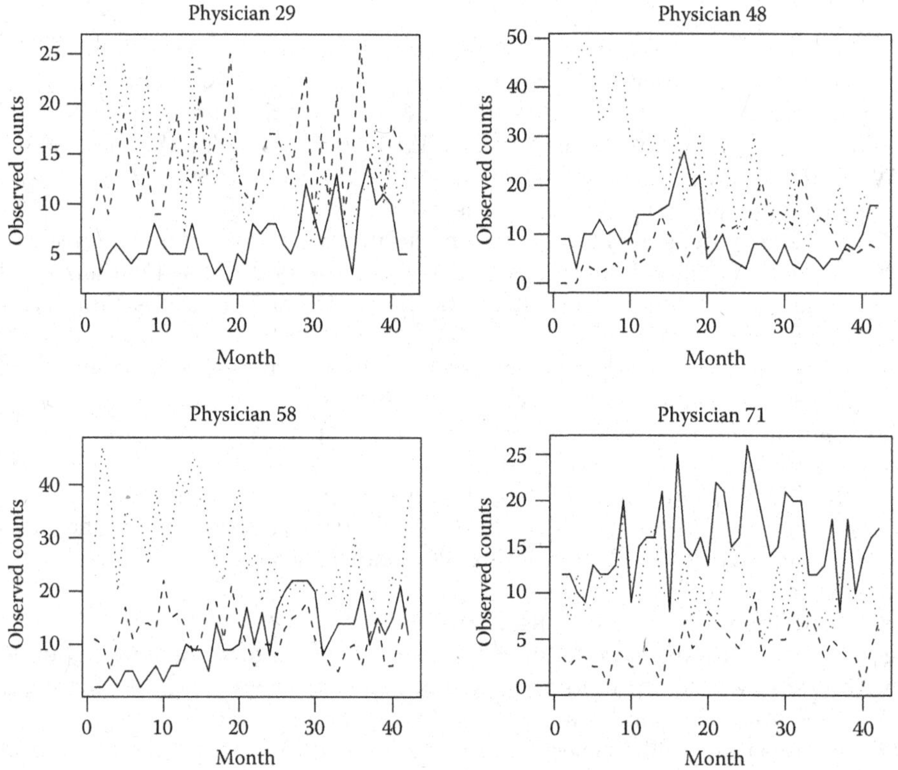

FIGURE 20.1
Time series plots of observed counts of own drug (solid line), challenger drug (dashed line), and leader drug (dotted line).

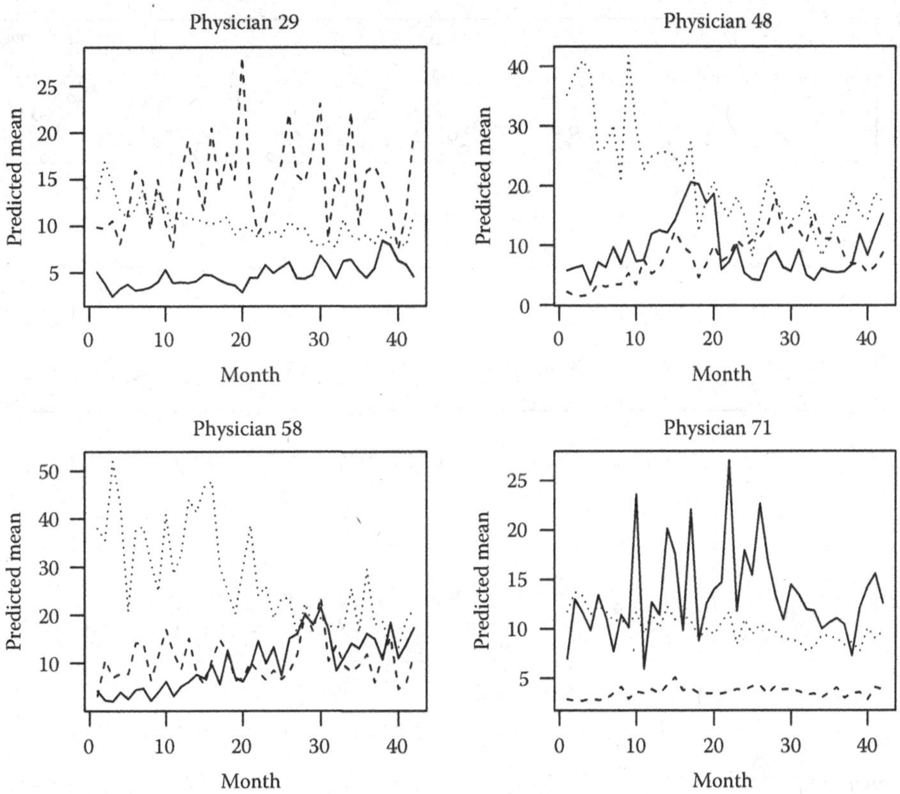

FIGURE 20.2
Time series plots of predictions of own drug (solid line), challenger drug (dashed line), and leader drug (dotted line).

20.6 Summary

This chapter describes a hierarchical dynamic modeling framework for univariate and multivariate time series of counts and investigates their utility for a marketing data set on new prescriptions written by physicians. We discuss a dynamic ZIP framework that combines sparse survey–based customer attitude data that are not available at regular intervals, with customer-level transaction and marketing histories that are available at regular time intervals. Univariate count time series of new prescriptions is modeled in order to discuss retention and sales. We also describe the use of the multivariate Poisson distribution as the sampling distribution in a fully Bayesian framework for inference in the context of multivariate count time series. This enables us to address a useful aspect in marketing research, which to jointly model the number of prescriptions of different drugs written by the physicians over time, taking into account possible associations between them, and studying the effect of a firm's detailing on the sales of its own drug and competitors. This is done using the parsimonious MDFM model, which enables us to study the evolution of sales of a set of competing drugs within a category. Work on this topic is ongoing, and it is also

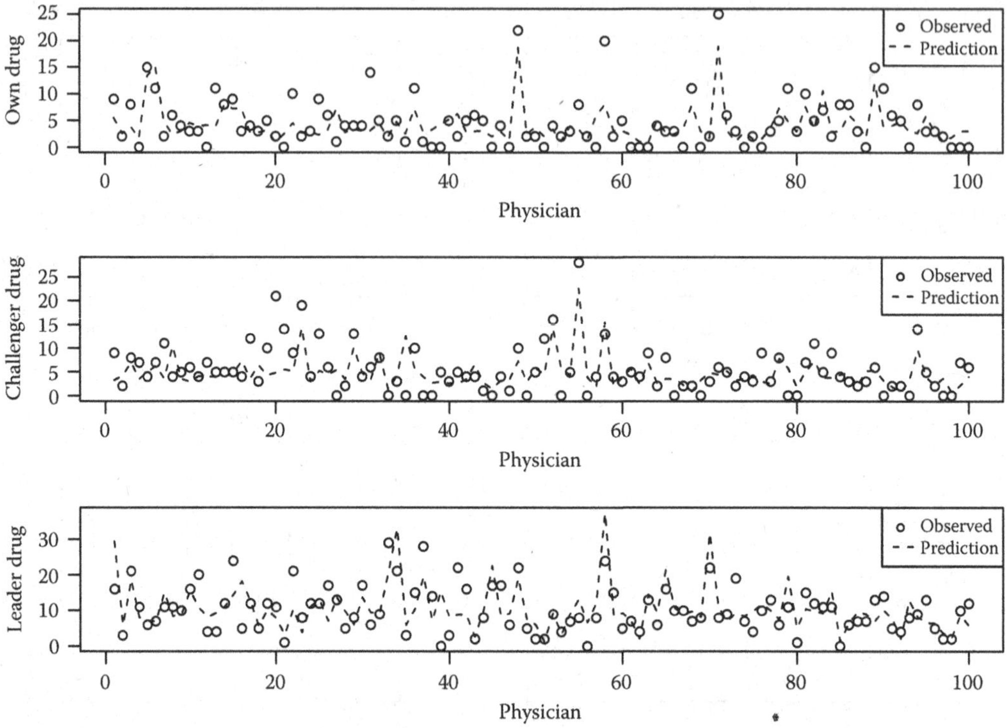

FIGURE 20.3
One-month-ahead predictions for all physicians.

possible to construct other models for multivariate time series of counts. For instance, Ma et al. (2008) describe a multivariate Poisson-lognormal framework for modeling multivariate count time series with an application to transportation safety, while Li et al. (1999) describe a framework for fitting a multivariate ZIP model.

One consideration in fitting multivariate Poisson models is the computation time for calculating the likelihood function, which can increase rapidly with increasing component dimension m. In such cases, it is useful to explore approximations for carrying out Bayesian inference, such as the integrated nested Laplace approximation (INLA), or the variational Bayes approach. The INLA approach has been discussed in Rue and Martino (2007) and Rue et al. (2009); also see Gamerman et al. (2015; Chapter 8 in this volume) Serhiyenho et al. (2015) described an application to traffic safety using R-INLA. The variational Bayes approach has been discussed in several papers including McGrory and Titterington (2007), Ormerod and Wand (2010), and Wand et al. (2011), among others. A study of these approaches for modeling multivariate time series of counts and a comparison with the MCMC approach will be useful.

20.A Appendix

Details on the complete conditional distributions for the HMDM and MDFM models are shown below.

20.A.1 Complete Conditional Distributions for the HMDM Model

The complete conditional density of β_{it} for $i = 1, \ldots, N$ and $t = 1, \ldots, T$ is

$$f(\beta_{it}|\gamma_t, \eta, V_i, W, Y) \propto MP_m(y_{it}|\lambda_{it}) \exp\left\{-\frac{1}{2}(\beta_{it} - \gamma_t)' V_i^{-1}(\beta_{it} - \gamma_t)\right\}.$$

The complete conditional density of γ_t for $t = 1, \ldots, T$ is

$$\begin{aligned} f(\gamma_t|\beta_{it}, \eta, V_i, W, Y) \propto & \prod_{i=1}^{N} \exp\left\{-\frac{1}{2}(\beta_{it} - \gamma_t)' V_i^{-1}(\beta_{it} - \gamma_t)\right\} \\ & \times \exp\left\{-\frac{1}{2}(\gamma_t - G\gamma_{t-1})' W^{-1}(\gamma_t - G\gamma_{t-1})\right\} \\ & \times \exp\left\{-\frac{1}{2}(\gamma_{t+1} - G\gamma_t)' W^{-1}(\gamma_{t+1} - G\gamma_t)\right\}. \end{aligned}$$

The complete conditional density of η is

$$f(\eta|\beta_{it}, \gamma_t, V_i, W, Y) \propto \prod_{t=1}^{T}\prod_{i=1}^{N} MP_m(y_{it}|\lambda_{it}) \exp\left\{-\frac{1}{2}(\eta - \mu_\eta)' \Sigma_\eta^{-1}(\eta - \mu_\eta)\right\}.$$

The complete conditional density of V_i for $i = 1, \ldots, N$ is

$$\begin{aligned} f(V_i|\beta_{it}, \gamma_t, \eta, W, Y) \propto & \prod_{t=1}^{T} |V_i|^{-1/2} \exp\left\{-\frac{1}{2}(\beta_{it} - \gamma_t)' V_i^{-1}(\beta_{it} - \gamma_t)\right\} \\ & \times |V_i|^{-n_v/2} \exp\left\{-\frac{1}{2} tr(V_i^{-1} S_v)\right\}. \end{aligned}$$

The complete conditional density of W is

$$\begin{aligned} f(W|\beta_{it}, \gamma_t, \eta, V_i, Y) \propto & \prod_{t=1}^{T} |W|^{-1/2} \exp\left\{-\frac{1}{2}(\gamma_t - G\gamma_{t-1})' W^{-1}(\gamma_t - G\gamma_{t-1})\right\} \\ & \times |W|^{-1/2} \exp\left\{-\frac{1}{2}(\gamma_0 - m_0)' C_0^{-1}(\gamma_0 - m_0)\right\} \\ & \times |W|^{-n_w/2} \exp\left\{-\frac{1}{2} tr(W^{-1} S_w)\right\}. \end{aligned}$$

20.A.2 Complete Conditional Distributions and Sampling Algorithms for the MDFM Model

For $t = 2, \ldots, T$ and $h = 1, \ldots, H$,

$$f(\boldsymbol{\beta}_{0,t,h}) \propto \prod_{i=1}^{N} \pi_h^{z_{i,t,h}} [p(\mathbf{y}_{i,t}|\boldsymbol{\lambda}_{i,t,h})]^{z_{i,t,h}} \exp\left\{-\frac{1}{2}(\boldsymbol{\beta}_{0,t,h} - \mathbf{G}\boldsymbol{\beta}_{0,t-1,h}-)'\mathbf{W}_h^{-1}(\boldsymbol{\beta}_{0,t,h} - \mathbf{G}\boldsymbol{\beta}_{0,t-1,h})\right\}$$

$$\times \exp\left\{-\frac{1}{2}(\boldsymbol{\beta}_{0,t+1,h} - \mathbf{G}\boldsymbol{\beta}_{0,t,h})'\mathbf{W}_h^{-1}(\boldsymbol{\beta}_{0,t+1,h} - \mathbf{G}\boldsymbol{\beta}_{0,t,h})\right\}.$$

For $i = 1, \ldots, N$ and $h = 1, \ldots H$,

$$f\left(\boldsymbol{\beta}_{1,i,h}\right) \propto \prod_{t=1}^{T} \pi_h^{z_{i,t,h}} \left[p\left(\mathbf{y}_{i,t}|\boldsymbol{\lambda}_{i,t,h}\right)\right]^{z_{i,t,h}} \exp\left\{-\frac{1}{2}\left(\boldsymbol{\beta}_{1,i,h} - \mathbf{a}_{1,i,h}\right)' \mathbf{R}_{1,i,h}^{-1} \left(\boldsymbol{\beta}_{1,i,h} - \mathbf{a}_{1,i,h}\right\}, \text{ and}$$

$$f\left(\boldsymbol{\beta}_{2,i,h}\right) \propto \prod_{t=1}^{T} \pi_h^{z_{i,t,h}} \left[p\left(\mathbf{y}_{i,t}|\boldsymbol{\lambda}_{i,t,h}\right)\right]^{z_{i,t,h}} \exp\left\{-\frac{1}{2}\left(\boldsymbol{\beta}_{2,i,h} - \mathbf{a}_{2,i,h}\right)' \mathbf{R}_{2,i,h}^{-1} \left(\boldsymbol{\beta}_{2,i,h} - \mathbf{a}_{2,i,h}\right)\right\}.$$

For $h = 1, \ldots, H$,

$$f(\mathbf{W}_h) \propto \prod_{t=2}^{T} \exp\left\{-\frac{1}{2}(\boldsymbol{\beta}_{0,t,h} - \mathbf{G}\boldsymbol{\beta}_{0,t-1,h})'\mathbf{W}_h^{-1}(\boldsymbol{\beta}_{0,t,h} - \mathbf{G}\boldsymbol{\beta}_{0,t-1,h})\right\}$$

$$\times \exp\left\{-\frac{1}{2}(\alpha - \mu_\alpha)'\sigma_\alpha^{-1}(\alpha - \mu_\alpha)\right\} |\mathbf{W}_h|^{-n_h/2} \exp\left\{-\frac{1}{2}tr(\mathbf{W}_h^{-1}\mathbf{S}_h)\right\}.$$

The complete conditional density of $\boldsymbol{\pi}$ is

$$f(\boldsymbol{\pi}) \propto \prod_{h=1}^{H} \pi_h^{d_h - 1 + \sum_{i=1}^{N} \sum_{t=1}^{T} z_{i,t,h}},$$

which reduces to $f(\pi_1) \propto \pi_1^{d_1 - 1 + \sum_{i=1}^{N} \sum_{t=1}^{T} z_{i,t,1}} (1 - \pi_1)^{d_2 - 1}$ when $H = 2$.

The sampling algorithms are shown in the following. At the lth Gibbs iteration, (a) we generate $\mathbf{z}^{(l)} \sim f(\mathbf{z}|\mathbf{y}, \boldsymbol{\Theta}^{(l)})$ and (b) we then generate $\boldsymbol{\Theta}^{(l+1)} \sim \pi(\boldsymbol{\Theta}|\mathbf{y}, \mathbf{z})$). Specifically, in step (a), for $i = 1, \ldots, N$ and $t = 1, \ldots, T$, we simulate $\mathbf{z}_{i,t}$ from a multinomial distribution with probabilities

$$p_j = \pi_j^{(l)} [p(\mathbf{y}_{i,t}|\boldsymbol{\lambda}_{i,t,h})]^{z_{i,t,h}} / \sum_{h=1}^{H} \pi_h^{(l)} [p(\mathbf{y}_{i,t}|\boldsymbol{\lambda}_{i,t,h})]^{z_{i,t,h}}$$

for $j = 1, \ldots, H$. When $H = 2$, we simulate $z_{i,t,1}$ from Bernoulli distribution with weight

$$p_z = \frac{\pi_1^{(l)} [p(\mathbf{y}_{i,t}|\boldsymbol{\lambda}_{i,t,h})]^{z_{i,t,h}}}{\pi_1^{(l)} [p(\mathbf{y}_{i,t}|\boldsymbol{\lambda}_{i,t,h})]^{z_{i,t,h}} + \pi_2^{(l)} [p(\mathbf{y}_{i,t}|\boldsymbol{\lambda}_{i,t,h})]^{z_{i,t,h}}}.$$

In step (b), the complete conditional densities of W_h for $h = 1, \ldots, H$ and the mixture proportion π have closed forms; we directly sample $\mathbf{W}_h$ from $IW(n_h^*, \mathbf{S}_h^*)$, where

$$n_h^* = n_h + T$$

$$\mathbf{S}_h^* = \mathbf{S}_h + \sum_{t=2}^{T} \left\{ \left(\boldsymbol{\beta}_{0,t,h} - \mathbf{G}\boldsymbol{\beta}_{0,t-1,h}\right)\left(\boldsymbol{\beta}_{0,t,h} - \mathbf{G}\boldsymbol{\beta}_{0,t-1,h}\right)' \right\}.$$

We sample π from a Dirichlet$\left(d_1 + \sum_{i=1}^{N}\sum_{t=1}^{T} z_{i,t,1}, \ldots, d_H + \sum_{i=1}^{N}\sum_{t=1}^{T} z_{i,t,h}\right)$ distribution when $H > 2$ and from a Beta $\left(d_1 + \sum_{i=1}^{N}\sum_{t=1}^{T} z_{i,t,1}, d_2\right)$ distribution when $H = 2$. The draws are made using a random walk Metropolis–Hastings algorithm.

References

Ahearne, M., Jelinek, R., and Jones, E. (2007). Examining the effect of salesperson service behavior in a competitive context. *Journal of Academy of Marketing Science*, 35(4):603–616.

Aktekin, T., Soyer, R., and Xu, F. (2014). Assessment of mortgage default risk via Bayesian state space models. *Annals of Applied Statistics*, 7(3):1450–1473.

Albert, J. H. and Chib, S. (1993). Bayesian analysis of binary and polychotomous response data. *Journal of American Statistical Association*, 88:669–679.

Blattberg, R. C., Kim, B.-D., and Neslin, S. A. (2008). *Database Marketing: Analyzing and Managing Customers*. Springer Science+ Business Media LLC: New York.

Carlin, B. P., Polson, N., and Stoffer, D. S. (1992). Bayes estimates for the linear model (with discussion). *Journal of the Royal Statistical Society, Series B*, 34:1–41.

Carter, C. and Kohn, R. (1994). On Gibbs sampling for state space models. *Biometrika*, 81:541–553.

Chen, M.-H., Shao, A., and Ibrahim, J. (2000). *Monte Carlo Methods*. Springer-Verlag: New York.

Davis, R. A., Dunsmuir, W. T. M., and Streett, S. B. (2003). Observation-driven models for Poisson counts. *Biometrika*, 90(4):777–790.

Dempster, A. P., Laird, N. M., and Rubin, D. B. (1977). Maximum likelihood from incomplete data via the EM algorithm. *Journal of the Royal Statistical Society, Series B*, 39:1–38.

Diebolt, J. and Robert, C. P. (1994). Estimation of finite mixture distributions through Bayesian sampling. *Journal of the Royal Statistical Society, Series B*, 56:363–375.

Fahrmeir, L. (1992). Posterior mode estimation by extended Kalman filtering for multivariate dynamic generalized linear models. *Journal of the American Statistical Association*, 87:501–509.

Fahrmeir, L. and Kaufmann, H. (1991). On Kalman filtering, posterior mode estimation and Fisher scoring in dynamic exponential family regression. *Metrika*, 38:37–60.

Fruhwirth-Schnatter, S. (1994). Data augmentation and dynamic linear models. *Journal of Time Series Analysis*, 24:295–320.

Gamerman, D. (1998). Markov chain Monte Carlo for dynamic generalized linear models. *Biometrika*, 85:215–227.

Gamerman, D., Abanto-Valle, C. A., Silva, R. S., and Martins, T. G. (2015). Dynamic Bayesian models for discrete-valued time series. In R. A. Davis, S. H. Holan, R. Lund and N. Ravishanker, eds., *Handbook of Discrete-Valued Time Series*, pp. 165–186. Chapman & Hall, Boca Raton, FL.

Gamerman, D. and Migon, H. (1993). Dynamic hierarchical models. *Journal of the Royal Statistical Society Series B*, 3:629–642.

Harvey, A. C. and Fernandes, C. (1989). Time series models for count or qualitative observations. *Journal of Business and Economic Statistics*, 7(4):407–417.

Hu, S. (2012). Dynamic modeling of discrete-valued time series with applications. PhD thesis, University of Connecticut: Storrs, CT.
Johnson, N., Kotz, S., and Balakrishnan, N. (1997). *Discrete Multivariate Distributions*. Wiley: New York.
Kalman, R. E. (1960). A new approach to linear filtering and prediction problems. *Transactions ASME Journal of Basic Engineering*, 82(Series D):35–45.
Kalman, R. E. and Bucy, R. S. (1961). New results in linear filtering and prediction theory. *Transactions ASME-Journal of Basic Engineering*, 83(Series D):95–108.
Karlis, D. and Meligkotsidou, L. (2005). Multivariate Poisson regression with covariance structure. *Statistics and Computing*, 15:255–265.
Karlis, D. and Meligkotsidou, L. (2007). Finite mixtures of multivariate Poisson regression with application. *Journal of Statistical Planning and Inference*, 137:1942–1960.
Kedem, B. and Fokianos, K. (2002). *Regression Models for Time Series Analysis*. Wiley: Hoboken, NJ.
Lambert, D. (1992). Zero-inflated Poisson regression, with an application to defects in manufacturing. *Technometrics*, 34(1):1–13.
Lancaster, T. (2000). The incidental parameter problem since 1948. *Journal of Econometrics*, 95: 391–413.
Landim, F. and Gamerman, D. (2000). Dynamic hierarchical models: An extension to matrix-variate observations. *Computational Statistics and Data Analysis*, 35:11–42.
Li, C., Lu, J., Park, J., Kim, K., Brinkley, P., and Peterson, J. (1999). Multivariate zero-inflated Poisson models and their applications. *Technometrics*, 41:29–38.
Lindley, D. and Smith, A. (1972). Bayes estimates for the linear model (with discussion). *Journal of the Royal Statistical Society, Series B*, 34:1–41.
Ma, J., Kockelman, K. M., and Damien, P. (2008). A multivariate Poisson-lognormal regression model for prediction of crash counts by severity, using Bayesian methods. *Accident Analysis and Prevention*, 40:964–975.
Mahamunulu, D. (1967). A note on regression in the multivariate Poisson distribution. *Journal of the American Statistical Association*, 62:251–258.
McCullagh, P. and Nelder, J. (1989). *Generalized Linear Models*, 2nd edn. Chapman & Hall: London, U.K.
McGrory, C. A. and Titterington, D. M. (2007). Variational approximations in Bayesian model selection for finite mixture distributions. *Computational Statistics and Data Analysis*, 51:5352–5367.
Mizik, N. and Jacobson, R. (2004). Are physicians easy marks? Quantifying the effects of detailing and sampling on new prescriptions. *Management Science*, 50(12):1704–1715.
Montoya, R., Netzer, O., and Jedidi, K. (2010). Dynamic allocation of pharmaceutical detailing and sampling for long-term profitability. *Marketing Science*, 29(5):909–924.
Netzer, O., Lattin, J. M., and Srinivasan, V. (2008). A hidden Markov model of customer relationship dynamics. *Marketing Science*, 27(2):185–204.
Ormerod, J. T. and Wand, M. P. (2010). Explaining variational approximations. *The American Statistician*, 64(2):140–153.
Ravishanker, N., Serhiyenko, V., and Willig, M. (2014). Hierarchical dynamic models for multivariate time series of counts. *Statistics and Its Interface*, 7(4):559–570.
Rue, H. and Martino, S. (2007). Approximate Bayesian inference for hierarchical Gaussian Markov random fields models. *Journal of Statistical Planning and Inference*, 137:3177–3192.
Rue, H., Martino, S., and Chopin, N. (2009). Approximate Bayesian inference for latent Gaussian models by using integrated nested Laplace approximations. *Journal of the Royal Statistical Society: Series B*, 71:319–392.
Serhiyenko, V., Mamun, S. A., Ivan, J. N. and Ravishanker, N. (2015). Fast Bayesian inference for modeling multivariate crash counts on Connecticut limited access highways. *Technical Report, Department of Statistics*, University of Connecticut: Storrs, CT.
Tsiamyrtzis, P. and Karlis, D. (2004). Strategies for efficient computation of multivariate Poisson probabilities. *Communications in Statistics - Simulation and Computation*, 33:271–292.

Venkatesan, R., Reinartz, W., and Ravishanker, N. (2014). ZIP models for CLV based customer management using attitudinal and behavioral data. *Technical Report, Department of Statistics,* University of Connecticut: Storrs, CT.

Wand, M. P., Ormerod, J. T., Padoan, S. A., and Fruhrwirth, R. (2011). Mean field variational Bayes for elaborate distributions. *Bayesian Analysis*, 6(4):847–900.

Weinberg, J., Brown, L. D., and Stroud, J. R. (2007). Bayesian forecasting of an inhomogeneous Poisson process with applications to call center data. *Journal of the American Statistical Association*, 102:1185–1198.

West, M. (1989). *Bayesian Forecasting and Dynamic Models*. Springer-Verlag: New York.

Wikle, C. K. and Anderson, C. J. (2003). Climatological analysis of tornado report counts using a hierarchical Bayesian spatio-temporal model. *Journal of Geophysical Research-Atmospheres*, 108(D24), 9005, doi:10.1029/2002JD002806

Zeger, S. L. (1988). A regression model for time series of counts. *Biometrika*, 75(4):621–629.

Zucchini, W. and MacDonald, I. L. (2010). *Hidden Markov Models for Time Series: An Introduction Using R*, 2nd edn. Chapman & Hall: London, U.K.

21

Long Memory Discrete-Valued Time Series

Robert Lund, Scott H. Holan, and James Livsey

CONTENTS

21.1 Introduction

This chapter reviews modeling and inference issues for discrete-valued time series with long memory, with an emphasis on count series. En route, several recent areas of research and possible extensions are described, including Bayesian methods and estimation issues.

A covariance stationary time series $\{X_t\}$ with finite second moments is said to have long memory (also called long-range dependence) when

$$\sum_{h=0}^{\infty} |\mathrm{Cov}(X_t, X_{t+h})| = \infty.$$

Other definitions of long memory are possible (Guégan 2005). Long memory time series models and applications are ubiquitous in the modern sciences (Granger and Joyeux 1980, Hosking 1981, Geweke and Porter-Hudak 1983, Robinson 2003, Beran 1994, Palma 2007). However, time series having both long memory and a discrete (count) marginal distribution have been more difficult to devise/quantify; literature on the topic is scarce (Quoreshi 2014 is an exception). Here, we overview models for long memory count series, discussing what types of methods will and will not produce long memory features.

One appealing approach for modeling non-Gaussian long-range dependence is proposed by Palma and Zevallos (2010). Here, the authors devise a long memory model where the distribution of current observation is specified conditionally upon past observations. Often, such series are stationary (see MacDonald and Zucchini [2015; Chapter 12 in this volume]). On Bayesian fronts, Brockwell (2007) constructs a model having a general non-Gaussian distribution (including Poisson) conditional on a long memory latent Gaussian process. While Brockwell (2007) does not pursue probabilistic properties of his model, a Markov chain Monte Carlo (MCMC) sampling algorithm is devised for efficient estimation. Also worth mentioning are the parameter-based (process-based) approach of Creal et al. (2013) and the estimating equation approach of Thavaneswaran and Ravishanker (2015; Chapter 7 in this volume).

This chapter proceeds as follows. Section 21.2 shows why some classical approaches will not generate discrete-valued long memory time series. Methods capable of generating long memory count time series are presented in Section 21.3. The special case of a binary long memory series is discussed in Section 21.4. Bayesian methods are pursued in Section 21.5, where some open research questions are suggested. Section 21.6 provides concluding discussion.

21.2 Inadequacies of Classical Approaches

Integer autoregressive moving-average (INARMA) models (Steutel and van Harn 1979, McKenzie 1985, 1986, 1988, Al-Osh and Alzaid 1988) and discrete ARMA (DARMA) methods (Jacobs and Lewis 1978a, 1978b) cannot produce long memory series. A simple first-order integer autoregression, for example, obeys the recursion

$$X_t = p \circ X_{t-1} + Z_t, \quad t = 0, \pm 1, \pm 2, \ldots, \tag{21.1}$$

where $p \in (0,1)$ is a parameter, $\circ$ is the thinning operator, and $\{Z_t\}_{t=-\infty}^{\infty}$ is an independent and identically distributed (IID) sequence supported on the nonnegative integers. Clarifying, $p \circ M$ is a binomial random variable with M trials and success probability p. The thinning in (21.1) serves to keep the series integer valued. While the solution of (21.1) is stationary, its lag h autocovariance is proportional to p^h and is hence absolutely summable over all lags. Higher-order autoregressions have the same autocovariance summability properties; that is, one cannot construct long memory count models with INARMA methods.

Similarly, DARMA methods also cannot produce long memory sequences. A first-order discrete autoregression with marginal distribution π obeys the recursion

$$X_t = A_t X_{t-1} + (1 - A_t) Z_t, \quad t = 1, 2, \ldots,$$

where $\{Z_t\}$ is IID with distribution π and $\{A_t\}$ is an IID sequence of Bernoulli trials with $P[A_t = 1] \equiv p$. The recursion commences with a draw from the specified marginal distribution π: $X_0 \stackrel{\mathcal{D}}{=} \pi$. While any marginal distribution can be achieved, the lag h autocovariance is again proportional to p^h, which is absolutely summable in lag. Moving to higher-order models does not alter this absolute summability.

ARMA methods will not produce long memory series, even in noncount settings. The classical ARMA(p,q) difference equation is

$$X_t - \phi_1 X_{t-1} - \cdots - \phi_p X_{t-p} = Z_t + \theta_1 Z_{t-1} + \cdots + \theta_1 Z_{t-q}, \quad t = 0, \pm 1, \ldots, \tag{21.2}$$

where $\{Z_t\}_{t=-\infty}^{\infty}$ is white noise (uncorrelated in time) with variance σ_Z^2. Here, $\phi_1, \ldots, \phi_p$ are autoregressive coefficients and $\theta_1, \ldots, \theta_q$ are moving-average coefficients. We make the usual assumption that the autoregressive (AR) polynomial

$$\phi(z) = 1 - \phi_1 z - \cdots - \phi_p z^p \tag{21.3}$$

and the moving-average (MA) polynomial

$$\theta(z) = 1 + \theta_1 z + \cdots + \theta_q z^q \tag{21.4}$$

have no common roots (this is needed for solutions of the difference equation to be unique in mean square). When the AR polynomial has no roots on the complex unit circle $\{z : |z| = 1\}$, solutions to (21.2) can be expressed in the form

$$X_t = \sum_{k=-\infty}^{\infty} \psi_k Z_{t-k}, \quad t = 0, \pm 1, \ldots, \tag{21.5}$$

where the weights are absolutely summable (i.e., $\sum_{k=-\infty}^{\infty} |\psi_k| < \infty$). From (21.5), we have

$$\text{Cov}(X_t, X_{t+h}) = \sigma_Z^2 \sum_{k=-\infty}^{\infty} \psi_k \psi_{k+h}, \quad h = 0, \pm 1, \pm 2, \ldots. \tag{21.6}$$

It now follows that

$$\sum_{h=-\infty}^{\infty} |\text{Cov}(X_t, X_{t+h})| \leq \sigma_Z^2 \left(\sum_{k=-\infty}^{\infty} |\psi_k| \right)^2 < \infty.$$

The point is that stationary long memory series cannot be produced by ARMA methods. Should one permit the AR polynomial (21.3) to have a unit root, then solutions to (21.2) will not be stationary (this is the result of Problem 4.28 in Brockwell and Davis 1991).

21.3 Valid Long Memory Count Approaches

One model for a long memory count series fractionally differences and thins as in Quoreshi (2014). Specifically, if $\{\psi_k\}_{k=0}^{\infty}$ is a sequence of real numbers with $\psi_k \in [0, 1]$ for each $k \geq 0$ and $\{Z_t\}_{t=-\infty}^{\infty}$ is an IID sequence of nonnegative-valued counts, then

$$X_t = \sum_{k=0}^{\infty} \psi_k \circ Z_{t-k}, \quad t = 0, \pm 1, \ldots, \tag{21.7}$$

defines a stationary sequence of counts when some conditions are imposed on $\{\psi_k\}_{k=0}^{\infty}$. If $\sum_{k=0}^{\infty} |\psi_k| < \infty$ is required, then $\{X_t\}$ will have short memory. However, if this absolute summability is relaxed—but the weights ψ_k still converge to zero slowly enough to make probabilistic sense of the summation in (21.7) —then the resulting process could have long memory. The covariance structure of $\{X_t\}$ in (21.7) is identical to that in (21.6).

Quoreshi (2014) employed this strategy with the weights $\psi_0 = 1$ and

$$\psi_k = \frac{\Gamma(k+d)}{\Gamma(k+1)\Gamma(d)}, \quad k = 1, 2, \ldots,$$

where $d \in (0, 1/2)$ is a fractional differencing parameter arising in the power series expansion of $(1-B)^{-d}$ (see Hosking 1981). Here, B is the backshift operator, defined by $B^k X_t = X_{t-k}$ for $k \geq 0$. This setup was generalized by Quoreshi (2014) to extract integer-valued solutions to difference equations of form

$$\phi(B)X_t = \theta(B)(1-B)^{-d}Z_t, \quad t = 0, \pm 1, \ldots.$$

Here, the AR and MA polynomials in (21.3) and (21.4) have "probabilistic coefficients": $\phi_\ell \in [0,1]$ for $\ell = 1, \ldots, p$ and $\theta_\ell \in [0,1]$ for $\ell = 1, \ldots, q$. This restriction makes the analysis unwieldy in comparison to classical ARMA methods. For example, in an AR(1) setting, $\phi_1 \in [0,1]$ and it follows that the model cannot have any negative autocorrelations by (21.6).

A completely different approach involves renewal sequences (Cui and Lund 2009, Lund and Livsey [2015; Chapter 5 in this volume]). In this paradigm, there is a random lifetime L supported in $\{1, 2, \ldots\}$ with mean μ and IID copies of L, which we denote by $\{L_i\}_{i=1}^{\infty}$. There is an initial delay lifetime L_0 that may not have the same distributions as the L_i for $i \geq 1$. Define a random walk $\{S_n\}_{n=0}^{\infty}$ via

$$S_n = L_0 + L_1 + \cdots + L_n, \quad n = 0, 1, 2, \ldots,$$

and say that a renewal occurs at time t if $S_m = t$ for some $m \geq 0$. This is the classical discrete-time renewal sequence popularized in Smith (1958) and Feller (1968). If a renewal occurs at time t, set $R_t = 1$; otherwise, set $R_t = 0$. To make $\{R_t\}$ a stationary Bernoulli sequence, a special distribution is needed for L_0. Specifically, L_0 is posited to have the first tail distribution derived from L:

$$P(L_0 = k) = \frac{P(L > k)}{\mu}, \quad k = 0, 1, \ldots. \tag{21.8}$$

Notice that L_0 can be zero (in which case, the process is called nondelayed).

For notation, let $u_t = P(R_t = 1 | R_0 = 1)$ be the time t renewal probability in a nondelayed setup. These are calculated recursively from the lifetime L's probabilities via

$$u_t = P[L = t] + \sum_{\ell=1}^{t-1} P[L = \ell]u_{t-\ell}, \quad t = 1, 2, \ldots, \tag{21.9}$$

where we use the convention $u_0 = 1$. The elementary renewal theorem states that $u_t \longrightarrow E[L]^{-1}$ when L has a finite mean and an aperiodic support set (henceforth assumed). When L_0 has the distribution in (21.8), $E[R_t] \equiv \mu^{-1}$ and the covariance function of $\{R_t\}$ is $\text{Cov}(R_t, R_{t+h}) = \mu^{-1}\left(u_h - \mu^{-1}\right)$ for $h \geq 0$ (Lund and Livsey [2015; Chapter 5 in this volume]).

Renewal theory can be used to extract some process properties. Heathcote (1967) gives the generating function expansion

$$\sum_{h=0}^{\infty} z^h \left(u_h - \mu^{-1}\right) = \frac{1-\psi_{L_0}(z)}{1-\psi_{L_1}(z)}, \quad |z| \leq 1, \tag{21.10}$$

where $\psi_{L_1}(z) = E[z^{L_1}]$ and $\psi_{L_0}(z) = E[z^{L_0}]$. Letting $z \uparrow 1$ and using that $E[L_0]$ is finite if and only if $E[L_1^2]$ is finite, we see that $\sum_{h=0}^{\infty} |u_h - \mu^{-1}| < \infty$ if and only if $E[L_1^2]$ is finite. In the case where $E[L_1^2] < \infty$,

$$\sum_{h=0}^{\infty} \left(u_h - \mu^{-1}\right) = \frac{E[L_1^2] - E[L_1]}{2E[L_1]}.$$

Our standing assumption is that $E[L_1] = \mu < \infty$. Since the lag h autocovariance is proportional to $u_h - \mu^{-1}$, $\{R_t\}_{t=0}^{\infty}$ will have long memory if and only if $E[L_1^2] = \infty$. This, perhaps, is our major result.

Lifetimes with a finite mean but infinite second moment are plentiful. One such lifetime involves the Pareto distribution

$$P(L = k) = \frac{c(\alpha)}{k^{\alpha}}, \quad k = 1, 2, \ldots, \tag{21.11}$$

where $c(\alpha) = 1/\sum_{k=1}^{\infty} k^{-\alpha}$ is a normalizing constant and $\alpha > 2$ is a parameter. A Pareto lifetime L with $\alpha \in (2, 3]$ has $E[L] < \infty$ and $E[L^2] = \infty$.

The tactics mentioned above construct a stationary binary sequence with long memory. A sample path of such a series is shown in Figure 21.1 along with its sample autocorrelations. Lifetimes here were generated using (21.11) with $\alpha = 2.5$.

To obtain other marginal distributions, we superimpose as in Lund and Livsey (2015; Chapter 5 in this volume). Let $\{R_{t,i}\}$ be IID long memory binary renewal sequences for $i \geq 1$. If $M \geq 1$ is fixed and

$$X_t = \sum_{i=1}^{M} R_{t,i}, \quad t = 1, 2, \ldots,$$

then X_t has a binomial marginal distribution with M trials and success probability μ^{-1}. Process autocovariances are

$$\text{Cov}(X_t, X_{t+h}) = \frac{M}{\mu}\left(u_h - \frac{1}{\mu}\right), \quad h = 0, 1, \ldots,$$

and will have long memory when $E[L^2] = \infty$.

Other marginal distributions with long memory can be constructed. For Poisson marginals, suppose that $\{M_t\}_{t=0}^{\infty}$ is a sequence of IID Poisson random variables with mean λ and set

$$X_t = \sum_{i=1}^{M_t} R_{t,i}, \quad t = 1, 2, \ldots. \tag{21.12}$$

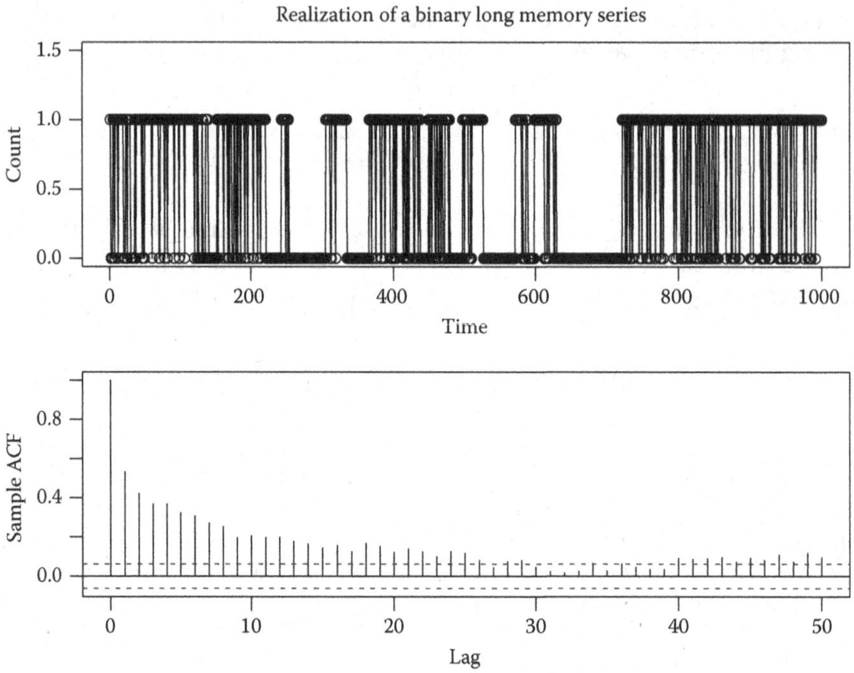

FIGURE 21.1
A realization of length 1000 of a long memory stationary time series with Bernoulli marginal distributions. Sample autocorrelations are also shown with pointwise 95% confidence bounds for white noise.

Then $\{X_t\}_{t=0}^{\infty}$ is strictly stationary with marginal Poisson distributions with mean λ/μ and

$$\mathrm{Cov}(X_t, X_{t+h}) = \frac{C(\lambda)}{\mu}\left(u_h - \mu^{-1}\right), \quad h = 1, 2, \ldots.$$

Lund and Livsey (2015; Chapter 5 in this volume) show that $C(\lambda) = \lambda[1 - e^{-2\lambda}\{I_0(2\lambda) + I_1(2\lambda)\}]$, where

$$I_j(\lambda) = \sum_{n=0}^{\infty} \frac{(\lambda/2)^{2n+j}}{n!(n+j)!}, \quad j = 0, 1,$$

is a modified Bessel function. Again, $\{X_t\}_{t=0}^{\infty}$ will have long memory when $E[L^2] = \infty$.

Figure 21.2 shows a realization of (21.12) of length 1000 along with sample autocorrelations of the generated series. The sequence $\{M_t\}_{t=1}^{n}$ was generated using $\lambda = 5$; the Pareto lifetime in (21.11) was utilized with $\alpha = 2.1$. Since $E[L^2] = \infty$, this model has long memory.

As discussed in Lund and Livsey (2015; Chapter 5 in this volume), generalities are possible; there, it is shown how to construct geometric marginal distributions (and hence also negative binomial marginals) with renewal methods.

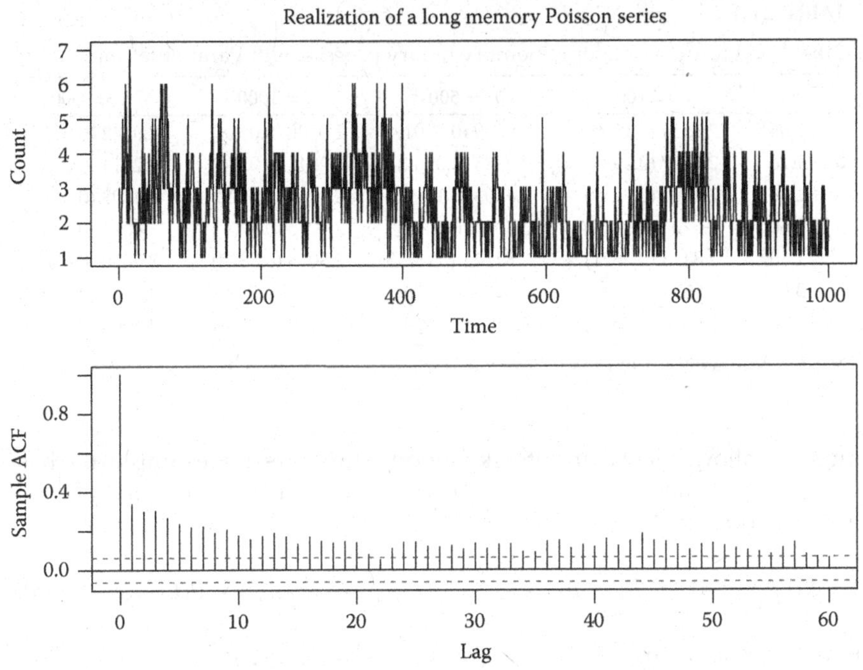

FIGURE 21.2
A realization of length 1000 of a long memory stationary time series with Poisson marginal distributions. Sample autocorrelations are also shown with pointwise 95% confidence bounds for white noise.

21.4 Binary Series

The case of a binary series (binomial with $M = 1$) is worth additional discussion. Here, $X_t = R_{t,1}$ and the covariance function has form

$$\text{Cov}(X_t, X_{t+h}) = \frac{1}{\mu}\left(u_h - \frac{1}{\mu}\right), \quad h = 0, 1, \ldots.$$

For long memory, L needs to have a finite mean but an infinite second moment.

Suppose that $X_1, \ldots, X_n$ constitutes data sampled from a long memory binary process. In this case, likelihood estimators can be constructed. To see this, suppose that $x_i \in \{0, 1\}$ for $i = 1, 2, \ldots, n$ are fixed. Let $\tau_1 = \inf\{t \geq 1 : x_t = 1\}$ be the first x_t that is unity, and inductively define the kth occurrence of unity as

$$\tau_k = \inf\{t > \tau_{k-1} : x_t = 1\}, \quad k = 1, 2, \ldots, N(n).$$

Here, $N(n)$ denotes the number of unit x_ts in the first n indices. For notation, let $\eta_k = \tau_k - \tau_{k-1}$. By the construction of the renewal process, we have

$$P\left(\cap_{t=1}^{n} X_t = x_t\right) = P(L_0 = \tau_1, L_1 = \eta_2, \ldots, L_{N(n)-1} = \eta_{N(n)}, L_{N(n)} > n - \tau_{N(n)}). \quad (21.13)$$

TABLE 21.1

Likelihood results for a long memory binary process with Pareto lifetimes

	$n = 100$	$n = 500$	$n = 1000$	$n = 5000$
$\alpha = 2.05$	2.1554 (0.1509)	2.0809 (0.0701)	2.0708 (0.0548)	2.0542 (0.0299)
$\alpha = 2.20$	2.2627 (0.1939)	2.2085 (0.0962)	2.2048 (0.0710)	2.2014 (0.0331)
$\alpha = 2.40$	2.4503 (0.2259)	2.4103 (0.1016)	2.4069 (0.0722)	2.4003 (0.0326)
$\alpha = 2.60$	2.6400 (0.2445)	2.6085 (0.1081)	2.6075 (0.0767)	2.6017 (0.0341)
$\alpha = 2.80$	2.8433 (0.2671)	2.8114 (0.1167)	2.8053 (0.0822)	2.8014 (0.0367)
$\alpha = 3.00$	3.0534 (0.2919)	3.0155 (0.1270)	3.0017 (0.0892)	3.0016 (0.0398)
$\alpha = 3.50$	3.5827 (0.3733)	3.5156 (0.1599)	3.5110 (0.1126)	3.5006 (0.0501)

Note: Cases with $\alpha \in (2, 3]$ have long memory.

This relationship allows us to compute likelihood estimators. For example, if L has a Pareto distribution with parameter α, then the probability in (21.13) is a function of α and can be used as a likelihood $L(\alpha)$:

$$L(\alpha) = P(L_0 = \tau_1) \times P(L_1 = \eta_2) \times \cdots \times P(L_{N(n)-1} = \eta_{N(n)}) \times P(L_{N(n)} > n - \tau_{N(n)}).$$

We recommend maximizing the log-likelihood $\ell(\alpha) = \log(L(\alpha))$:

$$\ell(\alpha) = \log P(L_0 = \tau_1) + \sum_{i=1}^{N(n)-1} \log P(L_i = \eta_i) + \log P(L_{N(n)} > n - \tau_{N(n)}). \tag{21.14}$$

Table 21.1 summarizes likelihood estimation results when L has the Pareto lifetime in (21.11). Series of various lengths n were generated and the Pareto parameter α was estimated by maximizing the log-likelihood in (21.14). Each table entry reports the average estimator over 1000 simulations; sample standard deviations are listed in parentheses for intuition (there is no theoretical guarantee that these error estimates are finite). While there appears to be some overestimation (bias) in the estimators, biases decay with increasing series length. Estimation precision seems to decrease with increasing α. Overall, the procedure seems to work well. Extensions of these likelihood calculations to settings where $M \geq 2$ constitute an open area of research and appear difficult.

21.5 Bayesian Long Memory Models

Bayesian methods to model general long memory series are an emerging area of research. Recent contributions include Pai and Ravishanker (1996, 1998), Ravishanker and Ray (1997, 2002), Ko and Vannucci (2006), Holan et al. (2009), and Holan and McElroy (2012). The literature on Bayesian methods for modeling count-valued long memory time series is scarce, with Brockwell (2007) being an exception. As such, we briefly outline a conditionally specified (hierarchical) Bayesian approach and suggest possible avenues for further research. In this section, we do not make an attempt to identify the marginal distribution of the constructed series.

The setup here is similar to Brockwell (2007) and MacDonald and Zucchini (2015; Chapter 12 in this volume) and proceeds via a conditional specification. For simplicity, we restrict attention to a conditional Poisson model where the logarithm of a latent intensity parameter is modeled as a Gaussian autoregressive fractionally integrated moving-average (ARFIMA) (p, d, q) process. It is assumed that the data are conditionally independent given the underlying latent Poisson intensity parameter. Specifically, we posit that the conditional distribution of X_t given λ_t is

$$X_t|\lambda_t, \overset{\text{ind}}{\sim} \text{Poisson}(\lambda_t), \quad t = 1, 2, \ldots. \tag{21.15}$$

Let $\lambda_t^* = \log(\lambda_t)$. We model $\{\lambda_t^*\}_{t=1}^{\infty}$ with a zero-mean Gaussian ARFIMA(p, d, q) process satisfying

$$\phi(B)(1 - B)^d \lambda_t^* = \theta(B)\epsilon_t, \quad t = 1, 2, \ldots,$$

where $(1 - B)^d = 1 - Bd - B^2 d(d-1)/2! - \cdots$ is the general binomial expansion, $p, q \in \mathbb{Z}^+$, $d \in (-1/2, 1/2)$, $\{\epsilon_t\}$ is zero-mean white noise, and the AR and MA polynomials are as in (21.3) and (21.4). With $\boldsymbol{\lambda}_n^* = (\lambda_1^*, \ldots, \lambda_n^*)'$ and $\boldsymbol{\Psi} = (\boldsymbol{\Phi}, \boldsymbol{\Theta}, d, \sigma^2)$, where $\boldsymbol{\Phi} = (\phi_1, \ldots, \phi_p)$ and $\boldsymbol{\Theta} = (\theta_1, \ldots, \theta_q)$, the Gaussian ARFIMA supposition implies that

$$\boldsymbol{\lambda}_n^*|\boldsymbol{\Psi} \sim N(\mathbf{0}, \boldsymbol{\Gamma}_n), \tag{21.16}$$

where $\boldsymbol{\Gamma}_n$ is the autocovariance matrix of $\boldsymbol{\lambda}_n^*$. As in Brockwell (2007), it is straightforward to specify a nonzero mean in (21.16); that is, deterministic regressors could be added to (21.16) in a straightforward manner.

With fixed values of the ARFIMA parameters in $\boldsymbol{\Psi}$, $\{\lambda_t^*\}_{t=1}^{\infty}$ is a strictly stationary Gaussian series. It follows that $\{\lambda_t\}_{t=1}^{\infty}$ and $\{X_t\}_{t=1}^{\infty}$ are also strictly stationary. However, the marginal distribution of X_t is unclear. Some computations provide the form

$$P(X_t = k) = \int_0^{\infty} \frac{e^{-\lambda}\lambda^k}{k!} \frac{\exp\{-\frac{1}{2}\frac{\ln(\lambda)^2}{\gamma^*(0)}\}}{\lambda\sqrt{2\pi\gamma^*(0)}}\, d\lambda, \quad k = 0, 1, \ldots,$$

where $\gamma^*(0) = \text{Var}(\lambda_t^*)$. This is a difficult integral to explicitly evaluate, although numerical approximations can be made—see Asmussen et al. (2014) for the latest. The covariance $\text{Cov}(X_t, X_{t+h})$ also appears intractable. It seems logical that $\{X_t\}_{t=1}^{\infty}$ will also have long memory, but this has not been formally verified.

In a Bayesian setting, the time series parameters $\boldsymbol{\Psi}$ are typically treated as random. For example, the distributions of $\boldsymbol{\Phi}$ and $\boldsymbol{\Theta}$ could be taken as uniform over their respective AR and MA stationarity and invertibility regions, d could be uniform over $(-1/2, 1/2)$, and σ^2 would have a distribution supported on $(0, \infty)$. One could take these components to be independent, although formulations allowing dependence between these components are also possible.

In practice, it is convenient to work with an autoregressive setup (i.e., $q = 0$) for $\boldsymbol{\lambda}^*$. Even with this simplifying assumption, several open research questions arise. For estimation, it would be useful to derive efficient MCMC sampling algorithms. One such algorithm is provided by Brockwell (2007). Also, for large n, it might be advantageous to consider approximate Bayesian inference by a Whittle likelihood in lieu of an exact Gaussian likelihood (see Palma 2007, McElroy and Holan 2012). Further computational efficiencies might be obtained from using preconditioned conjugate gradient methods (Chen et al. 2006).

Long memory need not be driven via ARFIMA(p, d, q) structures. For example, $\{\lambda_t^*\}_{t=1}^{\infty}$ could follow a fractionally differenced exponential model (FEXP(q)) as in Holan et al. (2009). Such models could be extended to permit seasonal long memory specifications, including GARMA (Gegenbauer ARMA) (Woodward et al. 1998) and GEXP (Gegenbauer exponential models) (Holan and McElroy 2012, McElroy and Holan 2012). This said, seasonal long memory cases pose additional computational challenges (McElroy and Holan 2012).

21.6 Conclusion

Long memory models for discrete-valued time series are a promising area of research—there is little current guidance on how to extend Gaussian long memory methods. Here, several recent advances were proposed and future research was suggested. Specifically, we pointed out that most classical discrete-valued approaches will not produce long memory series. This motivated methods that will produce long memory count series. These include Quoreshi (2014)'s approach of thinning with a fractional weighting scheme and the renewal theory approach in Cui and Lund (2009) and Lund and Livsey (2015; Chapter 5 in this volume). Binary long memory series were examined in greater detail. Here, maximum likelihood parameter estimates were obtained for a renewal model. In the binomial case of $M \geq 2$ (see Section 21.3), likelihood estimation constitutes an open area of research. Bayesian approaches to discrete-valued long memory time series were also discussed. While we focused on conditional Poisson data with a latent Gaussian intensity parameter, other discrete-valued distributions seem plausible. In the Bayesian context, several avenues for future research were suggested.

In summary, although long memory time series has become a popular topic, little has been previously done for discrete-valued time series. Here, we have detailed the current state of the topic and described several areas for future research. Of particular interest is the exploration of the utility of these models to real-world applications.

Acknowledgments

Robert Lund's research was partially supported by NSF Award DMS 1407480. Scott Holan's research was partially supported by the U.S. National Science Foundation (NSF) and the U.S. Census Bureau under NSF grant SES-1132031, funded through the NSF-Census Research Network (NCRN) program.

References

Al-Osh, M. and Alzaid, A.A. (1988). Integer-valued moving averages (INMA), *Statistical Papers*, **29**, 281–300.

Asmussen, S., Jensen, J.L., and Rojas-Nandayapa, L. (2014). *Methodology and Computing in Applied Probability*. doi: 10.1007/s11009-014-9430-7.

Beran, J. (1994). Statistical methods for data with long-range dependence, *Statistical Science*, **7**, 404–416.

Brockwell, A.E. (2007). Likelihood-based analysis of a class of generalized long-memory time series models. *Journal of Time Series Analysis*, **28**, 386–407.

Brockwell, P.J. and Davis, R.A. (1991), *Time Series: Theory and Methods*, 2nd edn., Springer-Verlag, New York.

Chen, W.W., Hurvich, C.M., and Lu, Y. (2006). On the correlation matrix of the discrete Fourier transform and the fast solution of large Toeplitz systems for long memory time series, *Journal of the American Statistical Association*, **101**, 812–822.

Creal, D., Koopman, S.J., and Lucas, A. (2013). Generalized autoregressive score models with applications, *Journal of Applied Econometrics*, **28**, 777–795.

Cui, Y. and Lund, R.B. (2009). A new look at time series of counts, *Biometrika*, **96**, 781–792.

Feller, W. (1968). *An Introduction to Probability Theory and its Applications*, 3rd edn., John Wiley & Sons, New York.

Geweke, J. and Porter-Hudak, S. (1983). The estimation and application of long memory time series models, *Journal of Time Series Analysis*, **4**, 221–238.

Granger, C.W.J. and Joyeux, R. (1980). An introduction to long-memory time series models and fractional differencing, *Journal of Time Series Analysis*, **1**, 15–29.

Guégan, D. (2005). How can we define the concept of long memory? An econometric survey, *Econometric Reviews*, **24**, 113–149.

Heathcote, C.R. (1967). Complete exponential convergence and some related topics, *Journal of Applied Probability*, **4**, 217–256.

Holan, S., McElroy, T., and Chakraborty, S. (2009). A Bayesian approach to estimating the long memory parameter, *Bayesian Analysis*, **4**, 159–190.

Holan, S.H. and McElroy, T.S. (2012). Bayesian seasonal adjustment of long memory time series. In: *Economic Time Series: Modeling and Seasonality*, Bell W.R., Holan S.H., McElroy T.S. (eds). Chapman & Hall/CRC Press: Boca Raton, FL.

Hosking, J.R.M. (1981). Fractional differencing, *Biometrika*, **68**, 165–176.

Jacobs, P.A. and Lewis, P.A.W. (1978a). Discrete time series generated by mixtures I: Correlational and runs properties, *Journal of the Royal Statistical Society, Series B*, **40**, 94–105.

Jacobs, P.A. and Lewis, P.A.W. (1978b). Discrete time series generated by mixtures II: Asymptotic properties, *Journal of the Royal Statistical Society, Series B*, **40**, 222–228.

Ko, K. and Vannucci, M. (2006). Bayesian wavelet analysis of autoregressive fractionally integrated moving-average processes, *Journal of Statistical Planning and Inference*, **136**, 3415–3434.

Lund, R. and Livsey, J. (2015). Renewal-based count time series. In R. A. Davis, S. H. Holan, R. Lund and N. Ravishanker, eds., *Handbook of Discrete-Valued Time Series*, pp. 101–120. Chapman & Hall, Boca Raton, FL.

MacDonald, I.L. and Zucchini, W. (2015). Hidden Markov models for discrete-valued time series. In R. A. Davis, S. H. Holan, R. Lund and N. Ravishanker, eds., *Handbook of Discrete-Valued Time Series*, pp. 267–286. Chapman & Hall, Boca Raton, FL.

McElroy, T.S. and Holan, S.H. (2012). On the computation of autocovariances for generalized Gegenbauer processes, *Statistica Sinica*, **22**, 1661–1687.

McKenzie, E. (1985). Some simple models for discrete variate time series, *Water Resources Bulletin*, **21**, 645–650.

McKenzie, E. (1986). Autoregressive-moving average processes with negative-binomial and geometric marginal distributions, *Advances in Applied Probability*, **18**, 679–705.

McKenzie, E. (1988). Some ARMA models for dependent sequences of Poisson counts, *Advances in Applied Probability*, **20**, 822–835.

Pai, J.S. and Ravishanker, N. (1996). Bayesian modeling of ARFIMA processes by Markov chain Monte Carlo methods, *Journal of Forecasting*, **15**, 63–82.

Pai, J.S. and Ravishanker, N. (1998). Bayesian analysis of autoregressive fractionally integrated moving-average processes, *Journal of Time Series Analysis*, **19**, 99–112.

Palma, W. (2007). *Long-Memory Time Series: Theory and Methods*, John Wiley & Sons: Hoboken, NJ.

Palma, W. and Zevallos, M. (2011). Fitting non-Gaussian persistent data, *Applied Stochastic Models in Business and Industry*, **27**, 23–36.

Quoreshi, A.M.M.S. (2014). A long-memory integer-valued time series model, INARFIMA, for financial application, *Quantitative Finance*, **14**, 2225–2235.

Ravishanker, N. and Ray, B.K. (1997). Bayesian analysis of vector ARFIMA processes, *Australian and New Zealand Journal of Statistics*, **39**, 295–311.

Ravishanker, N. and Ray, B.K. (2002). Bayesian prediction for vector ARFIMA processes, *International Journal of Forecasting*, **18**, 207–214.

Robinson, P. (2003). Time series with long memory: Advanced texts in econometrics. Oxford University Press, Oxford, England.

Smith, W.L. (1958). Renewal theory and its ramifications, *Journal of the Royal Statistical Society, Series B*, **20**, 243–302.

Steutel, F.W. and Van Harn, K. (1979). Discrete analogues of self-decomposability and stability, *Annals of Probability*, **7**, 893–899.

Thavaneswaran, A. and Ravishanker, N. (2015). Estimating equation approaches for integer-valued time series models. In R. A. Davis, S. H. Holan, R. Lund and N. Ravishanker, eds., *Handbook of Discrete-Valued Time Series*, pp. 145–164. Chapman & Hall, Boca Raton, FL.

Woodward, W.A., Cheng, Q.C., and Gray, H.L. (1998), A *k*-factor GARMA long-memory model, *Journal of Time Series Analysis*, **19**, 485–504.

Index

H

I

K

L

Printed in the United States
by Baker & Taylor Publisher Services